Radiola

The Golden Age of RCA

1919-1929

Radiola

The Golden Age of RCA

~ 1919-1929 ~

Eric P. Wenaas

SONORAN PUBLISHING, LLC
Chandler, Arizona

To my loving Mother and Father,
always close in mind and heart

Sonoran Publishing, LLC, Chandler, Arizona 85226

First printing April 2007

Printed in Canada

Library of Congress Cataloging-in-Publication Data

Wenaas, Eric P., 1942-
Radiola : the golden age of RCA, 1919-1929 / Eric P. Wenaas.
p.cm.
Includes bibliographical references and index. ISBN-13: 978-1-886606-21-0
1. Radio Corporation of America—History. 2. Electronic industries—United States—History.
3. Radio--Receivers and reception—History. 4. Radio broadcasting—United States—History.
I. Title.

HD9696.A3U643 2006
338.7'621384097309042–dc22
2006047119

Table of Contents

Make your home life richer with the magic of the Radiola

A world of entertainment is at your command when you have a Radiola in your home.

The magic of this wonderful instrument brings to you through the year an infinite variety of broadcast programs from the great transmitting stations in New York, Chicago and other cities across the country.

Radio is the most modern thing in this modern world. To be without it is to shut one's door to one of the greatest gifts of twentieth century science.

Purchase of a Radiola is an investment in happiness—an investment that yields rich returns, not only in entertainment of the highest quality, but in knowledge of what the world is thinking and doing from day to day.

And it pays to buy a genuine Radiola—product of the three great companies associated in its manufacture, RCA, General Electric and Westinghouse. Radio instruments carrying the RCA mark and guarantee embody all the world's knowledge of radio design.

RCA RADIOLA 60—Beautifully designed table model of new Radiola Super-Heterodyne with A. C. electric operation. Single control with illuminated dial. Ideal for congested broadcasting areas—highly sensitive for places remote from broadcasting stations. Two-toned walnut veneered cabinet. An instrument of wonderful musical range and tone fidelity. $147 (less Radiotrons)

RCA LOUDSPEAKER 103—De Luxe model of the famous 100A. The new design and tapestry covering make it highly decorative. $37.50

Buy with confidence where you see this sign

RCA RADIOLA

MADE BY THE MAKERS OF THE RADIOTRON

RADIO CORPORATION OF AMERICA • NEW YORK • CHICAGO • ATLANTA • DALLAS • SAN FRANCISCO

Preface

Radiola was the trade name used by the Radio Corporation of America for the broadcast receivers and selected accessories it marketed from the beginning of the broadcast era in the early 1920's until well after the end of the vacuum tube radio era in the latter half of the twentieth century. Radiola was used almost exclusively for receivers marketed by RCA from mid-1922 until the early 1930's shortly after RCA merged with the Victor Talking Machine Co. Thereafter, the Radiola brand name was used less often, displaced by the RCA Victor name and the Nipper icon. Radiola was one of the most successful and widely recognized brand names ever in the United States. As an aside, it was used for radios manufactured by companies in other countries as well, most notably Canada, Australia, England, France, Sweden, and Austria.

The ten-year period from 1919 to 1929 was surely the Golden Age of RCA in radio when Radiola receivers were a dominant force in the broadcast radio market, and changed the course of RCA from its original mission of providing point-to-point communication services to a mission focused on broadcast radio in general and radio receivers and other apparatus in particular. It was also a period when Radiolas scored many firsts for receivers mass-produced in large quantities for the American public—the first superheterodyne circuit, the first to use light-socket power, the first to use an electrodynamic cone speaker, the first to use tubes designed for AC current, the first to incorporate an automatic volume control (AVC), and the first to use the newly-developed screen grid vacuum tube. To be sure, other manufacturers would claim to be first to incorporate some of these features—for example, Experimenters Information Service with their "Model L" kit incorporating a superheterodyne circuit, and the Dynamotive Radio Corporation with their "Dynergy" model incorporating an AC circuit—but very few of these sets were actually manufactured in quantity, and the companies who made them did not stand the test of time. It was RCA who popularized these features in America, selling hundreds of thousands of mass-produced sets at popular prices in the twenties with many features that are still found on modern AM radios today.

From the viewpoint of both the Radiola historian and collector, this decade is surely the most interesting, as well. The Radiolas of this period tend to be more desirable in appearance, and many are certainly more rare than the ones from later periods. Much to the delight of the historian and collector alike, documents and early prototype receivers continue to surface from time to time, shedding new light on the history of this period. There is still much to be discovered, so the diligent historian can expect to be rewarded by his or her research.

The objective of this book is to document the early history of RCA with an emphasis on Radiola broadcast receivers, accessories, and advertising material in the historical context of the formative period of RCA as the broadcast era unfolded. While there has been more written about RCA and Radiolas of the 1920's than any other manufacturer of radio receivers of that era, most of this literature is fragmented, appearing in journals, magazines, and documents which are available only in the archives of selected museums and/or the libraries of private collectors and historians. The book *Radio Manufacturers of the 1920's* by Alan Douglas is unquestionably the best and most complete book addressing RCA receivers of this era, but even this three-volume book has only one chapter devoted to RCA, and all in black and white. There is a substantial amount of material that was not included in his book, either due to space limitations or because the material was not available to the author at the time of its publication in 1991.

Every attempt has been made to locate and cite original documents of the era to ensure accuracy and minimize dependence on secondary sources. In this regard, the author has extensively researched the *Radioana Collection* at the Archives Center, National Museum of American History, Behring Center of the Smithsonian Institution—the *Radioana Collection* being a repository of historical documents collected by RCA historian George H. Clark from Marconi, RCA, Wireless Specialty Apparatus, General Electric and other early radio companies. A number of documents have been discovered in the *Radioana Collection* which shed new light on the early history of the broadcast era, and reveal errors and popular misconceptions that continue to be propagated. The author has also obtained documents and photographs from other museums, most notably the Schenectady Museum & Suits-Bueche Planetarium, the AWA Museum in Bloomfield, New York, and the American Museum of Radio in Bellingham, Washington. In addition, the author has also researched many contemporaneous documents, newspapers, and radio magazines such as *Modern Electrics*, *Wireless Age*, *Radio*, *Radio News*, *Popular Radio*, *Radio Broadcast*, *The New York Times*, and *The Town Crier*, (RCA's publication for their dealers) to name just a few. The author has also accessed the more than 4000 pages of testimony documented by the Federal Trade Commission in their hearings into practices of the RCA radio group, principally RCA, GE, Westinghouse, Western Electric, and Wireless Specially Apparatus. Direct testimony from principals in the early radio industry provided rare insights into the motivation for many of the decisions that were made by RCA, Westinghouse, GE, and Western Electric. Finally, the author acquired and/or observed first hand almost all of the radio apparatus marketed by RCA in the 1920's—as well as a number of prototype receivers.

The first chapter of this book covers the history of the Marconi Wireless Telegraph Company of America, a history which is crucial to set the stage for the creation of RCA and the RCA receiver technology. American Marconi, in fact, mserged with RCA when GE acquired the stock of American Marconi in 1919. The second chapter covers the early history of RCA and the receivers of that time, a time when RCA focused on transoceanic and marine communications. Much of the early broadcast receiver technology was based on the receivers designed for these applications. The third chapter addresses the dawn of the broadcast era with a focus on the receivers designed by Westinghouse before Westinghouse entered into cross-licensing and marketing agreements with RCA. Also addressed are sets manufactured by Wireless Specialty Apparatus and sold by RCA as "special high grade receiving sets," all of which were in fact designed for point-to-point communication before the beginning of the broadcast era.

The next seven chapters focus on the Radiola receivers and accessories RCA offered for sale during the years 1922 through mid-1930. One chapter is devoted to the Radiolas offered in each of the seven RCA radio seasons that occurred during this period. One season, the 1925-27 season, spanned a two year period—1926 being the only year of the decade in which RCA did not introduce a new receiver. The receivers and accessories offered in each season are grouped according to the way in which RCA pictured them in their advertisements and catalogs for that season. In many cases, the receivers and accessories that sold well were held over from a previous year, so that the offerings in any season comprised a mixture of older and newer models. The new-season models are clearly distinguished from the holdover models in these tables, and each model is introduced and discussed in the chapter corresponding to the year in which it was first released.

The main focus of these seven chapters is on the receivers and accessories that were mass produced for and marketed by RCA. These models, which are listed in a table appearing near the beginning of each chapter, are found in official RCA catalogs and distributor price lists of the period, with three notable exceptions—the Radiola Special, the Radiola 24 and the Radiola 50. While these three sets were mass produced for and distributed by RCA, they never appeared in any RCA catalog, RCA distributor catalog or RCA distributor price list known to this author. The three did appear, however, in one or more advertisements in major newspapers and/or national magazines. Other models such as prototypes and model variants, while not specifically listed in the tables, are described in the text and pictured in photographs as deemed appropriate.

The last chapter addresses the Radiolas used in phonographs manufactured by Brunswick-Balke-Collender and the Victor Talking Machine Company from 1924—when RCA first supplied apparatus to Brunswick—to 1929 when Victor merged with RCA and RCA formed the RCA Victor Company. The Radiolas supplied by RCA to these companies were electrically similar to the ones they marketed to the public, but were generally repackaged in one way or another—for example, rearranged on square panels designed to fit into specific models produced by each company. Radiolas used in the earliest sets shared the acoustic horn with the acoustic record pickup mechanism of the phonograph, while Radiolas in later sets used electrical power amplifiers and cone speakers supplied by RCA, which were shared by the phonograph records using a new magnetic pickup.

The main portion of the book is followed by an *Afterwords* and nine appendices. The *Afterwords* describes the direction that RCA took in the years immediately following the decade of the 1920's when RCA became independent of General Electric and Westinghouse. Appendices A through F contain data in summary form on apparatus marketed by RCA: A) RCA Receiver Characteristics, B) Vacuum Tubes Used in Radiola Receivers, C) Release and End Dates for Radiola Receivers, D) Serial Number Data, E) Relative Availability of Radiola Apparatus, and F) RCA Model Designation Schema.

Appendix G focuses on the artists commissioned by RCA to create elaborate and attractive ads appearing in national magazines that were so effective in selling RCA products. While the early ads were relatively simple black and white sketches, they quickly evolved into more elaborate full-color works by artists well known in graphic art circles including D. Hendrickson, H. Botkin, W. Rienecke, and E. Shinn, to name a few. The artist's signatures often appeared in the margin of the ads or at the edges of the drawings.

Appendix H provides information on Radiolas that were manufactured for distribution by companies other that RCA. These Radiolas include models manufactured for sale in the U.S. before RCA registered and enforced its trade name domestically, and those manufactured outside of the U.S., many of which were sanctioned by RCA. Appendix I summarizes the receivers manufactured by Graybar which used Radiola hardware supplied by RCA, but which were not sold under the Radiola name.

Acknowledgments

I would like thank all those who made this book possible. Many thanks go to the Smithsonian Institution and in particular the Archives Center, National Museum of American History, Behring Center, Smithsonian Institution which holds the George H. Clark *Radioana Collection*. It would not have been possible to write this book without access to this material, which provided many valuable insights into the early history of radio at RCA and many of the images that appear in the book. I would like to thank the staff, and in particular Ms. Kay Peterson, who was most helpful in reproducing images and helping with approvals. Thanks also go to other museums and the staff that provided images and valuable information which include the AWA Museum in Bloomfield, NY, the Schenectady Museum & Suits-Bueche Planetarium in Schenectady, NY, and the American Museum of Radio and Electricity in Bellingham, WA.

I would like to thank the many friends who gave me encouragement and advice. In particular I would to thank those who spent significant time reviewing the manuscript and making valuable comments and corrections—Alan Douglas, John Wolkonowicz, John Jenkins, Jeff Lendaro and Floyd Engels. I would also like to thank my wife Karen, not only for helping review the manuscript, but also putting up with the entire project for a number of years. Their efforts are much appreciated, and as a result the manuscript is much improved.

I would like to thank all those who provided images of apparatus, instruction manuals, RCA sales brochures and advertisements, and other materials. Specific credits appear at the end of the caption accompanying each image. There were a few individuals who contributed significantly to the book by providing relevant documents and/or images: Sonny Clutter, Alan Douglas, Joe Fomanek, Stephen Brown, Howard Stone and John Terrey. I am particularly indebted to Sonny Clutter who assisted by post processing a number of images. I appreciate their efforts and thank them once again.

I would like to recognize Tom Rodriguez, who was responsible for the page layout and cover design. I thank him for a job well done, and also for the patience to process the many edits that were made to arrive at the final form of manuscript.

Finally, I would like to thank editor and publisher George Fathauer of Sonoran Publishing, LLC who spent many hours, days, and months over a two-year period editing the manuscript and making many valuable suggestions, comments and corrections. I appreciate his confidence in this project from the outset, his attention to detail, and his encouragement.

Chapter 1

The Stage is Set

1899-1919

Gugliemo Marconi, often characterized as the "Father of Radio," was the visionary responsible for incorporating the Marconi Wireless Telegraph Company of America to construct and operate wireless telegraphy systems in the Americas. (Guglielmo Marconi, Marconi Company, Limited)

$10,000,000 MARCONI COMPANY

Authorized to Operate Lines for Using Wireless Telegraphy Invention Over Land and Sea.

TRENTON, N. J., Nov. 22.—Articles of incorporation were filed here to-day with the Secretary of State of the Marconi Wireless Telegraph Company of America, with an authorized capital of $10,000,000. One-half of the capital is preferred stock, with 8 per cent. non-cumulative dividend.

The company is incorporated to purchase and acquire various inventions and discoveries of letters patent connected with the business of communication by means of wireless telegraphy, and especially to use the Marconi patent. The company is also empowered to construct and operate lines for using this invention over land and sea.

The incorporators are Guglielmo Marconi, London, England; Isaac L. Rice and August Belmont, New York; Clement A. Griscom, Philadelphia, and Robert Goodbody, Paterson, N. J. This company, it is said, has no connection with the Telephone, Telegraph, and Cable Company of America incorporated a short time ago.

The New York Times, Nov. 23, 1899, p.1.

The Radio Corporation of America (RCA) was incorporated on October 17, 1919 shortly after General Electric (GE) reached an agreement with Marconi's Wireless Telegraph Co., Ltd. to purchase its holdings in the Marconi Wireless Telegraph Company of America, hereinafter referred to as American Marconi, or in many cases simply as Marconi. The shareholders of American Marconi voted for a merger with the newly formed RCA as a result of pressure brought to bear on management by the U.S. government. At the time of the forced merger, American Marconi was the dominant wireless communication company in America with virtually no competition. The reason why American Marconi was forced to merge with GE/RCA is a story unto itself about corporate arrogance and miscalculation. It is also a story that provides the necessary background for understanding the state of radio communication at the time RCA was formed, and the direction RCA would take in the formative years before broadcasting dominated the RCA agenda and Radiola became a household name.

The Marconi Monopoly Agenda

From the very beginning of American Marconi on Nov. 22, 1899,[1] the corporate objective was to establish a world-wide monopoly in wireless communications to and from America, its possessions and dependencies, including Cuba, Puerto Rico, Hawaii, the Philippines, Alaska, the Aleutians, and all ships with an American registry. The timing for such a strategy could not have been worse. The Sherman Antitrust Act had been passed by the U.S. Congress a scant nine years before in 1890 with only one dissenting vote, and the executive branch had begun to prosecute the most flagrant of these monopolies under this act. Further, public sentiment was fully behind the prosecution of these monopolies.

Apparently American Marconi was not deterred. It was no secret that Marconi intended to establish a world-wide wireless monopoly, and indeed, Marconi proceeded in this endeavor by instituting policies that could best be described as predatory. Three of the most egregious policies were exclusive leasing arrangements designed to stifle competition, aggressive legal action to deter the manufacture of competing hardware, and extensive lobbying efforts to forestall government regulations detrimental to Marconi monopoly ambitions. American Marconi's policies were largely successful—at least in the short run—but in the end, these policies caused the government so much grief that they turned on American Marconi and its management, and took drastic measures to assure that Marconi would not achieve a monopoly in the Americas. How these three policies led to the demise of the Company is addressed in the remainder of this chapter.

Exclusive Leasing Arrangements

Marconi instituted a policy to provide end-to-end communication services on a lease basis rather than selling communication equipment so that

Marconi could maintain control of the wireless network. To stifle competition, American Marconi refused to exchange messages with users of wireless equipment manufactured by other companies, thereby raising the ante for any company wanting to compete with Marconi. A competitor would not only have to provide the potential user with communication equipment, but would also have to provide connectivity with ground stations having the required geographical coverage. American Marconi had a great advantage in this regard because British Marconi had already established a large world-wide communication network available exclusively to the American Marconi affiliate.

The strategy was clearly successful because with a minimum of litigation and lobbying, American Marconi had become one of the two dominant wireless communication companies in America by 1910—United Wireless Telegraph Co. being the other. However, discontent over this policy surfaced from the very beginning. Marconi initially impressed the U.S. Navy in demonstrations that began on Oct. 26, 1899, and as a result, the Navy requested a quote for the purchase of 20 communication sets of the same type used in the demonstrations.[2] American Marconi, confident with its patent position, stuck to its strategy by refusing to sell the equipment outright, and countered with a long-term lease arrangement asking the Navy for an initial payment of $20,000, followed by annual royalty payments of $10,000 per year for the life of the key patents covering the equipment—which at that time amounted to more that a decade! Not only was this offer expensive, but it would have tied the Navy into a long-term leasing arrangement with equipment that would become obsolete long before the lease term ended.

As an example of how fast the technology evolved, consider the detector—the heart of every wireless receiver. In 1899, American Marconi offered wireless receivers to the Navy employing coherer detector technology, which was both insensitive and temperamental. By 1902, Marconi had developed the more sensitive magnetic detector known affectionately as the "maggie," which was highly reliable—unless the operator forgot to rewind the spring driving the endless band of stranded soft-iron wire; by 1904, it was the standard detector for Marconi installations in America. The Fleming valve detector—more sensitive than the magnetic detector—was used in the Marconi Valve Receiver, which became regular equipment in most of the important Marconi land stations and many of the trans-Atlantic steamers by the beginning of 1912.[3] Circa 1914, American Marconi began using a triode detector in commercial work similar to the audion invented by Lee De Forest in 1906—even though it infringed on De Forest audion patents.[4] De Forest caught Marconi using the triode in commercial work in the course of litigation initiated in late 1914 by Marconi against De Forest and his De Forest Radio Telephone and Telegraph Co. alleging vacuum tube infringement (addressed later in this chapter), whereupon Marconi quickly confessed to the infringement.[5] Marconi introduced receivers with carborundum and cerusite detectors circa 1913, which were more sensitive and less expensive than the magnetic detector—although they were subject to changes in sensitivity from shock and vibration. Marconi chose carborundum and cerusite because, while they were somewhat less sensitive to radio signals than galena crystals, they were also significantly less sensitive than galena to shock and vibration environments found aboard ships. Somewhat later, carborundum and cerusite detectors were incorporated into the ubiquitous Marconi 106 and 106B receivers, which became standard marine receivers for American Marconi's commercial work after 1915.

With such rapid changes in technology, it is difficult to imagine what would have happened had the Navy accepted a long-term lease arrangement with the early coherer technology. It is almost certain that Marconi would not have gratuitously replaced it with the improved technology from time to time at no

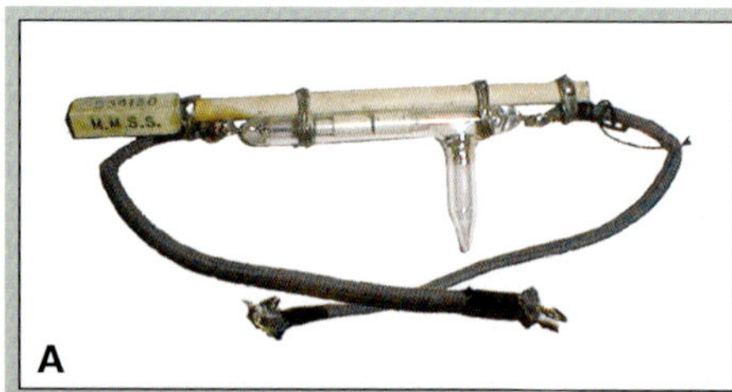

A. Coherer introduced by Marconi circa 1899 (John Jenkins, Sparkmuseum)

B. Magnetic detector introduced by Marconi circa 1904 (John Jenkins, Sparkmuseum)

C. Fleming valve detector introduced by Marconi circa 1912 (John Jenkins, Sparkmuseum)

D. Crystal detector introduced by Marconi circa 1913

Detectors are an example of a wireless technology evolving so fast in the early years that the Marconi equipment would have become obsolete long before the end of the lease period Marconi was offering to the Navy.

The Marconi Coherer Receiver was first used by Marconi as early as 1899. (George H. Clark Collection, Smithsonian Institution)

The Marconi Magnetic Detector replaced the Coherer Receiver circa 1904. (John Jenkins, Sparkmuseum)

additional cost, or without other changes in lease terms unfavorable to the Navy. The Navy clearly understood the implications of the proposed terms, which they considered to be both expensive and dictatorial. They not only declined the offer, but there is no record that they even attempted to negotiate more favorable terms.

Instead, the Navy purchased equipment outright for test and evaluation in 1902 from four foreign manufacturers (Slaby-Arco and Braun-Siemens-Halske of Germany and Ducretet and Rochefort of France), and two American companies (De Forest Wireless Telegraph Co. and Consolidated Wireless Co.).[6] As a result of these tests, the Navy initially purchased a large number of sets from Slaby-Arco, and several American manufacturers. The Navy did not purchase any equipment from Marconi in this time frame—even for further test and evaluation—because they could not come to terms for an outright purchase of the equipment had the tests been successful.

The Navy's actions eventually forced American Marconi to reconsider its position in 1906 when, according to Howeth, Marconi agreed to sell equipment outright to the Navy.[7] It would be more accurate to say that Marconi decided to bid on government contracts for wireless equipment in 1906—not to sell

The Marconi Valve Tuner was the first Fleming valve receiver to be widely deployed by Marconi in both land and ship stations circa 1912—including the Titanic. (George H. Clark Collection, Smithsonian Institution)

outright the equipment leased in commercial activities.[8] According to a statement by Marconi VP Edward J. Nally in a *The New York Times* article appearing in 1918,[9] it was not until then that Marconi changed its policy regarding the sale of its commercial equipment:

> *"Edward J. Nally, Vice President and General Manager of the Marconi Wireless Telegraph Company of America, on his return from Washington, where he went for a conference with the Navy Department, has made known a change in the policy of the Marconi Company under which it will in [the] future sell as well as lease wireless apparatus. The first result of this change has been the sale to the government of the apparatus on some 300 ships now under Government control and all of the small coastal stations which the Marconi Company has maintained in the past, but which it no longer needs under its new policy."*

While Marconi may have decided to bid on government contracts in 1906, the damage had already been done. A number of other manufacturers already had their noses under the tent, and within a few years they—and not Marconi—provided the majority of equipment to the Navy. The Marconi bids were not well received—perhaps because of the high prices or perhaps because of past Marconi policies and practices. Not until 1912 did Marconi begin to provide sets to the Navy under contract, and even then, far fewer than its competitors. Historian George H. Clark prepared a table reproduced here as Table 1-1 which shows the number of sets in operation at Navy stations in four different years.[10] Clark's table, which indicates that Marconi did not sell equipment to the government until 1912, is supported by the American Marconi Annual Shareholder's Report for the period ending 31 January 1913—the first Annual Report to chronicle wireless sales by American Marconi to the government: *"Many sets have been manufactured and delivered to the several Government Departments, for both ship and shore work."*

One can only wonder what would have happened if Marconi had decided to become a partner with the Navy at the outset by developing and selling equipment to meet Navy needs under mutually favorable terms. Marconi

The Marconi 103 Crystal Receiver introduced circa 1913 was one of the first Marconi receivers to use crystal detectors in commercial work. (Howard Stone collection)

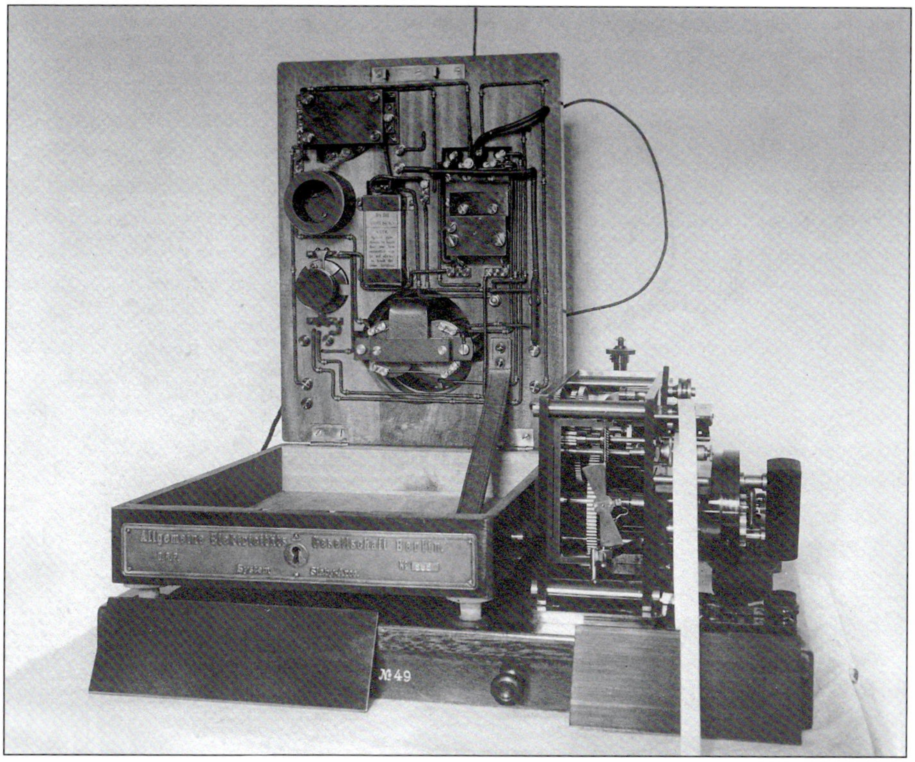

The Navy purchased many Slaby-Arco receivers in the early years including this Coherer Receiver dating from 1903 featuring an ink recorder. (George H. Clark Collection, Smithsonian Institution)

could have defrayed the cost of equipment development under government contracts, and used the resulting technology to great advantage in the commercial markets. Instead, the cost of the wireless stations and experimental work had to be supported by lease revenues from a relatively small customer base. As a result, American Marconi did not make a profit until 1914.[11] It is all but certain that American Marconi became profitable at that time only as a result of government contracts for wireless apparatus produced at the Aldene, New Jersey plant—a plant which had been acquired in 1912 for the purpose of manufacturing equipment for government as well as commercial accounts.[12]

Table 1-1. Sets in operation at Navy Wireless Station through 1912, the year they began to use Marconi sets

Company	1906	1907	1908	1912
Slaby-Arco and Telefunken	60	56	62	90
Shoemaker	18	23	23	25
De Forest	11	8	8	8
Fessenden	5	6	7	23
Stone	5	9	10	10
Massie	3	8	8	10
United Wireless	-	-	-	8
Pierce	-	3	3	3
Marconi	-	-	-	7
Total	**102**	**113**	**121**	**184**

Note: Table prepared by George Clark

Intellectual Property and Patent Litigation

A second part of the Marconi monopoly strategy was to create an extensive portfolio of patents and licensing agreements, and to defend this portfolio from potential infringers by means of litigation. This strategy took on an added dimension in 1910 when Godfrey C. Isaacs, the new managing director of British Marconi, initiated a policy of aggressive litigation against potential patent infringement by competitors. This strategy was highly successful over the next decade, a time during which both British and American Marconi had a number of major patent litigation victories without any notable defeats. However, the patent disputes ultimately resulted in a manufacturing gridlock because the features needed to meet the ever increasing design requirements were in most cases protected by patents held by competing manufacturers. While American Marconi was not the only litigant—and therefore not solely responsible for the resulting gridlock—it was the major litigant (most often as the plaintiff), and the decisions rendered in their favor had a highly visible and largely negative impact on the Navy and the U.S. government.

Of all the lawsuits in which American Marconi was involved, there were three in particular with decisions that shaped the early history of wireless in general and that of American Marconi in particular: Marconi v. United Wireless (1912), Marconi v. De Forest (1916) and Marconi v. E. J. Simon (1918). The outcomes of these suits were considered to be great victories for Marconi at the time, but in the end it became apparent that the victories were largely pyrrhic.

Marconi v. United Wireless Telegraph Co. Litigation

By 1910, American Marconi and the United Wireless Telegraph Co. were the only two sizable wireless communication companies in America. American Marconi controlled the transatlantic shipping communications with only a handful of shore stations and perhaps several hundred ships under contract. United Wireless controlled the majority of the coastal shipping communications, claiming 262 merchant ships under contract and 102 land stations in a brochure dated March 1910. A map from this brochure shows the locations of shore stations either owned or used by United Wireless. Some of the stations listed in the brochure were undoubtedly business offices, because a Navy department publication dated September 1, 1909, only six months before, listed United Wireless as owner of only 76 shore stations world wide.[13] While United Wireless had more than ten times the number of Marconi stations, many of the United Wireless stations were nothing more than a shack sitting at the base of a stripped tree used to support the antenna, such as the station in Roseburg, Oregon. In sharp contrast, the Marconi stations such as

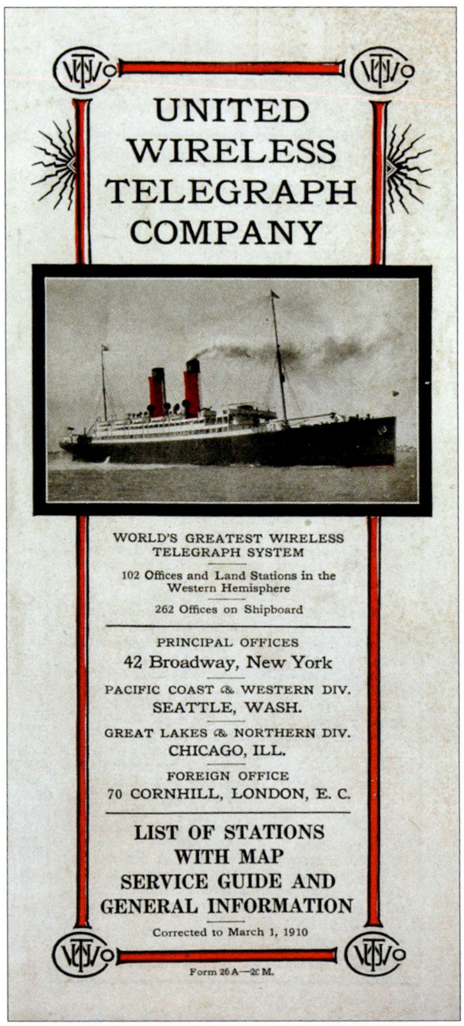

A United Wireless brochure dated 1910 claims the most extensive wireless zone in the world with 102 land stations and 262 ships under contract.

Map of United Wireless stations as of March 1, 1910

the one at Wellfleet were far more sophisticated with much longer ranges.

United Wireless designed and manufactured its own communication equipment, and by 1910, the United Wireless Type D tuner had become the standard for both their ship and shore stations. United Wireless had brazenly infringed on the British Marconi tuning patent in the design of their equipment, using exactly the same technology covered by the famous Marconi 7777 patent, which detailed a method to tune a transmitter and receiver to a common frequency to allow communication without interference from other stations operating on different frequencies.[14] The technology covered by this Marconi patent was a cornerstone of their patent portfolio, and was embodied in the well-known Marconi multiple tuner, a standard for Marconi ship and shore stations at that time.

Emboldened by a recent victory in litigation with British Radio Telegraph and Telephone Company in an English court for infringement of the 7777 tuning patent, American Marconi filed suit against United Wireless in 1911 for a similar infringement. United Wireless had no defense, and rather than contest the allegations, the Company consented to a decree in favor of Marconi in March of 1912.[15] United Wireless, already weakened by mismanagement and the criminal prosecution of key executives for stock fraud, subsequently filed for bankruptcy. British Marconi purchased United Wireless assets on behalf of American Marconi for a payment of $700,000, and gratuitously turned them over to American Marconi. According to an article in the April 12, 1912 issue of *Modern Electrics*, Marconi absorbed about 70 land stations and 500 shipboard stations as a result of the lawsuit.[16] As an aside, United

Wireless had achieved its premier position as a provider of wireless service in 1912 by offering services below cost, or in some cases at no cost, and making up for losses by selling more stock—a classic pyramid scheme!

Two other events occurred circa 1912 which assured Marconi a virtual monopoly on wireless communications in the United States. First, the corrupt Continental Telephone and Telegraph Co. failed sometime in late 1912 or early 1913 as a result of the arrest of the officers in late 1910 and their subsequent prosecution for stock fraud.[17] Continental had been formed in October 1909, and according to an article in *Electrical Review and Western Electrician*, it become active circa May of 1910 by consolidating four wireless entities including Massie Wireless Telegraph Company, Clark Wireless Telephone and Telegraph Company, Pacific Wireless Telegraph Company, and Collins Wireless Telegraph Company.[18] This article—obviously prepared by Continental Wireless for self-serving purposes—was apparently the basis of several historical accounts about Massie Wireless becoming part of Continental Wireless.[19] However, according to an article appearing in the November 22, 1910 issue of *The New York Times*, it is clear that Walter Massie did not believe that the Massie Wireless Telegraph Company was ever a part of Continental Wireless:

> *"The Massie Wireless Telegraph Company, of which Walter W. Massie of this city [Providence. R. I.] is the head, was never controlled by the Continental Wireless Company. Mr. Massie said that he accepted a Directorship in the Continental Company, but at the first meeting he attended, on August 12 [1910], when he learned the true state of the corporation's affairs, he resigned. He said he never invested any money in the company."* [20]

Several additional articles appearing *The New York Times* clearly indicate that Massie was operating as an independent company as late as January 1912.[21]

Thorn Mayes contends that Marconi purchased the West Coast assets of Massie for $5000 before 1912, and that the remaining East Coast assets were absorbed by Marconi sometime in 1912.[22] While the details of this account cannot be confirmed, of the seventeen stations on ships operating out of San Francisco as of June 30, 1911 with systems listed as Massie[23], ten were listed as under Marconi control in *Radio Stations of the United States*, Edition of July 1, 1914.[24] Of the half dozen or so land stations known to be under Massie control prior to 1912, two of these stations located on the east coast at Point Judith and Cape May were listed under Marconi control in the November 12, 1912 edition of the American Marconigraph and also in *Radio Stations of the United States*, Edition of July 1, 1914.[25] Scattered accounts by George Clark in the *Radioana Collection* also indicate that Massie assets were sold to Marconi in 1912, supporting Mayes account. However, of the 17 ship stations operating out of New York with systems listed as Massie as of June 20, 1911, 14 were listed as under control of National Electric Signaling Company (NESCO) in the *Radio Stations of the United States*, Edition of July 1, 1914—and none with Marconi. Whether or not these Massie assets were first distributed to Marconi in 1912, it is clear that most of the Massie stations on ships out of New York came under control of NESCO by 1914. It is equally clear that the Massie Wireless assets were in fact distributed—most likely by the end of 1912—and so with the demise of Continental and Massie Wireless, four additional competitors of Marconi disappeared from the scene.

The second event was a judgment against the National Electric Signaling Company resulting from a lawsuit by co-founder Reginald A. Fessenden against the remaining principals for breach of contract when they terminated his employment the year before. The judgment of $406,175 forced the company into receivership on August 5, 1912,[26] and while it was able to operate for a number of years in receivership, it did not have the financial wherewithal to be a major competitor. NESCO was reorganized in 1917 as the International

United stations were more numerous than those of Marconi but far less sophisticated—often consisting of nothing more than a shack near a tree which was used to support the antenna.

While the Marconi stations were fewer in number than those of United Wireless, they were generally much more sophisticated with higher power and greater range.

Radio Telegraph Company, and the wireless assets of this company were eventually acquired by Westinghouse in the early twenties.

Thus did American Marconi became the dominant wireless company in America with little competition—except possibly from the United Fruit Company. In 1912, the United Fruit Company operated radios on its 20 or so ships, and also operated a number of supporting ground stations. While United Fruit did not provide wireless services to other companies, it acquired the Wireless Specialty Apparatus Company in 1912, which did compete with Marconi in the manufacture and sale of wireless equipment. United Fruit created Tropical Radio as an entity to operate their ship-to-shore stations in 1913, but there is little or no evidence that Tropical Radio entered into direct competition with Marconi for customers—other than with United Fruit, of course.[27]

These events of 1912 were positive in one respect. The demise of the corrupt United Wireless and Continental Wireless Companies restored integrity to the U.S. radio industry, which had been plagued by corruption and stock manipulation schemes for many years. Marconi, the largest surviving entity, was not involved in the many stock schemes of its competitors, and had a reputation for integrity and honesty. On the downside, American Marconi had essentially achieved a monopoly in wireless communications, which likely constituted a violation of American anti-trust laws, and certainly was resented by a number of independent ship owners—some of whom sued Marconi over unreasonable rates and lease terms.[28]

To illustrate the degree to which Marconi dominated wireless communications in the period from 1913 until the American entry into WWI, RCA historian George Clark assembled a table showing the owners and operators of ship and shore stations based on data from the *Radio Stations of the World*, the edition of July 1, 1916 (see Table 1-2).[29] Clark concluded that

Marconi owned and operated 416/691 or 60 % of all commercial stations in the U.S.—but of the companies that provided services to non-owners, Marconi owned and operated a much larger percentage—416/485 or 86 % of the stations. Stations categorized by Clark in this table as "other" were for the most part owner-operated stations by owners such as American-Hawaiian Steamship Co., Standard Oil of New Jersey, etc.

Table 1-2. The number of wireless stations in operation by company in 1916, a time when American Marconi was dominant

Company	Ship Stations	Shore Stations	Total
Marconi W. T. Co. of America	362	54	416
Tropical Radio Co. (United Fruit)	20	2	22
Atlantic Communications Co.	3	2	5
Federal Telegraph Co.	5	7	12
National Electric Signaling Co.	20	3	23
E. J. Simon	1	–	1
Kilbourne and Clark	1	–	1
Mutual Telephone Co.	–	5	5
Other ships, miscellaneous	161	–	–
Other shore stations, 45 miscellaneous	2	0	6
Total	**573**	**118**	**691**

Note: Data from *Radio Stations from the United States*, July, 1916, arranged by George Clark

The fledgling communication industry in the U.S. was now firmly in the hands of American Marconi, a company with foreigners holding a majority of shares—an issue that would become of great concern for national security as the clouds of war gathered over Europe. It was only a few short years later that England declared war on Germany at the onset of WWI. One of the first acts of the war on the part of the British was to sever a German-owned transatlantic cable at the Azores on August 5, 1914—the dominant means of communication between Germany and the U.S.[30] Because the U.S. was still neutral at that time,

The United Wireless Type D tuner, the standard for United ship and shore installations, used a tuning technology that directly infringed on the famous Marconi 7777 patent. (John Jenkins, Sparkmuseum)

The Marconi Multiple Tuner was the heart of the Marconi system using a method covered by the 7777 patent to tune a transmitter and receiver to a common frequency to allow communication without interference from other stations operating on different frequencies. (John Jenkins, Sparkmuseum)

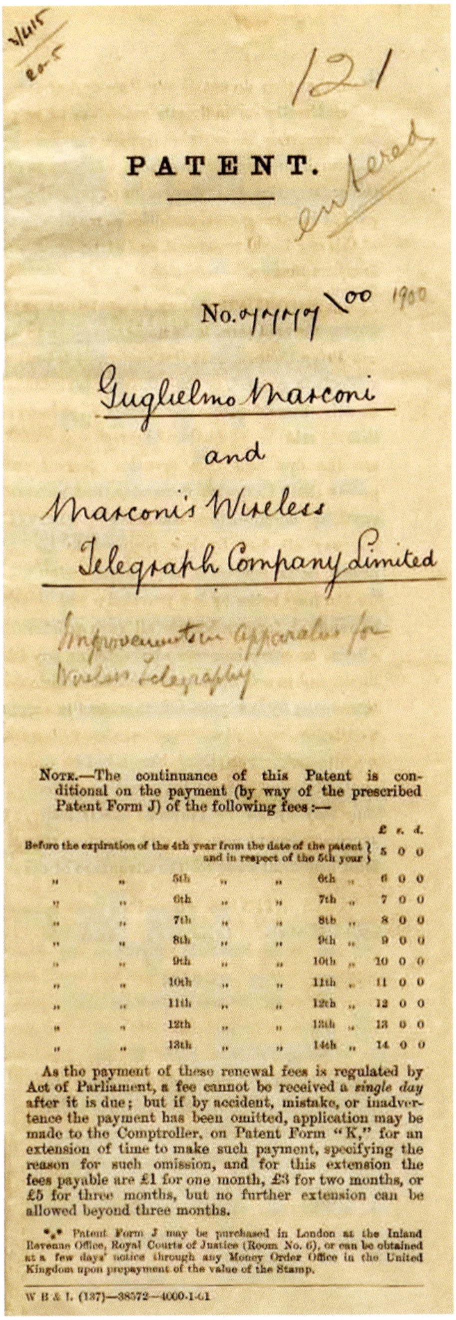

121

PATENT.

entered

No. 7777 '00 1900

Guglielmo Marconi

and

Marconi's Wireless Telegraph Company Limited

Improvements in apparatus for Wireless Telegraphy

NOTE.—The continuance of this Patent is conditional on the payment (by way of the prescribed Patent Form J) of the following fees:—

		£	s.	d.
Before the expiration of the 4th year from the date of the patent	and in respect of the 5th year	5	0	0
" " 5th "	" 6th "	6	0	0
" " 6th "	" 7th "	7	0	0
" " 7th "	" 8th "	8	0	0
" " 8th "	" 9th "	9	0	0
" " 9th "	" 10th "	10	0	0
" " 10th "	" 11th "	11	0	0
" " 11th "	" 12th "	12	0	0
" " 12th "	" 13th "	13	0	0
" " 13th "	" 14th "	14	0	0

As the payment of these renewal fees is regulated by Act of Parliament, a fee cannot be received a *single day* after it is due; but if by accident, mistake, or inadvertence the payment has been omitted, application may be made to the Comptroller, on Patent Form "K," for an extension of time to make such payment, specifying the reason for such omission, and for this extension the fees payable are £1 for one month, £3 for two months, or £5 for three months, but no further extension can be allowed beyond three months.

*** Patent Form J may be purchased in London at the Inland Revenue Office, Royal Courts of Justice (Room No. 6), or can be obtained at a few days' notice through any Money Order Office in the United Kingdom upon prepayment of the value of the Stamp.

W B & L (137)—38572—4000-1-01

American Marconi was transformed into the largest communication company in America in 1912 as a result of successful litigation against United Wireless based on the famous "four-sevens" patent, which was actually a British Marconi patent filed in England. (Bodleian Library, University of Oxford)

the U.S. government and the public at large considered this to be a most provocative act. The dangers of relying on communication links controlled by a foreign entity had just entered the national consciousness.

Marconi v. De Forest Litigation

In November 1914, American Marconi filed a suit against De Forest and his De Forest Radio Telephone & Telegraph Co. claiming the De Forest three-element audion detector infringed on the Fleming two-element valve patent, and in response, De Forest counterclaimed that the Fleming valve infringed his audion patents. It was difficult to predict the outcome of this litigation, because on the one hand the three-element audion was truly a revolutionary new discovery, but on the other hand the third element was useless without the other two elements, which were covered by the Fleming patent. In essence, it was possible that the court could find that the De Forest patent for his audion was valid, but at the same time it could find that De Forest could not manufacture it without infringing on the Fleming patent. De Forest must have realized this, because in his counter claim De Forest cited his patents on detection using electrodes in an open flame, a patent that predated the Fleming valve patent. In this way, De Forest hoped to either invalidate the Fleming patent, or at the very least, have the court determine that the Fleming valve infringed on his earlier flame patents.

The litigation continued for an extended period of time with a large volume of expert testimony supporting both sides. However, the experts testifying on behalf of Marconi were most ingenious in demonstrating the equivalency between the two-electrode valve and the three-electrode audion, both in theory and in practice. Roy Weagant, an engineer for Marconi, developed a clever demonstration board using the two simple receiver circuits, one shown in the third figure of the De Forest patent No. 841386 and the other shown in the first figure of the Fleming patent No. 803684.[31] The board had switches that allowed one to quickly switch between the receiver circuit with Fleming valve and the receiver circuit with the De Forest audion so that the performance as determined by the strength of the audio signal in a pair of headphones could be directly compared for the court.

The Marconi engineers must have selected the circuit in this particular De Forest patent because they realized that in this patent De Forest described

Marconi engineer Roy Weagant developed a clever demonstration board containing two receiver circuits appearing in the Fleming and De Forest audion patents, and was able to convince Judge Mayer that the performance of the two tubes when used as detectors was equivalent.

an audion in which he placed the grid on the outside of the tube where it is least effective, and also because this particular De Forest audion patent made claims as a detector of oscillations rather than as an amplifier of radio signals. As one might expect, Weagant was able to demonstrate that the circuit with the Fleming value was as sensitive a detector —and generally more so—than the De Forest circuit using an audion with the grid on the outside of the tube. Incredibly, the De Forest side did not object to the use of this particular audion configuration in the demonstration—possibly because De Forest had made the mistake of testifying that all his audions were essentially the same, and that they all worked on the same principle (a principle De Forest claimed was different than that of the Fleming valve).

Thus did Weagant and the Marconi side convince the court that the Fleming valve and audion were essentially equivalent in the detection of oscillation waves. American Marconi won a stunning victory in 1916, when District Judge J. M. Mayer ruled that the audion infringed the Fleming patent and dismissed De Forest's counter claim that the Fleming valve infringed upon the audion patents.[32] The judge enjoined both companies from manufacturing the triode without mutual permission, while Marconi was free to manufacture the Fleming valve. The injunction against De Forest was suspended pending appeal, but the Company failed to meet certain conditions of the suspension, and so the injunction became effective. The De Forest appeal was rejected by the Circuit Court of Appeals in May of 1917, so the injunction became permanent.[33]

In early 1917, Marconi requested that Judge Mayer extend the decree to cover the audion, not only when used as a detector, but also when used as an amplifier and an oscillator. Judge Mayer modified the decree to cover the audion when used as an amplifier for radio frequencies in radio receivers (as opposed to audio frequencies used in telephone circuits), but ruled that the question of audion infringement when used as an oscillator would require a new trial. The Marconi Company began proceedings on this issue, but the U.S. entered WWI before the trial began, and the proceedings were suspended for the duration of the war. The proceedings were reinstituted after the war, and on July 7, 1919 Judge Mayer ruled that the audion when used as an oscillator

In order to demonstrate equivalent performance between the Fleming valve and De Forest audion when used as a detector, Marconi engineer Roy Weagant cleverly selected a version of the audion for the demonstration with the grid on the outside of the tube as depicted in Figure 3 of De Forest's patent 841,386. (George H. Clark Collection, Smithsonian Institution)

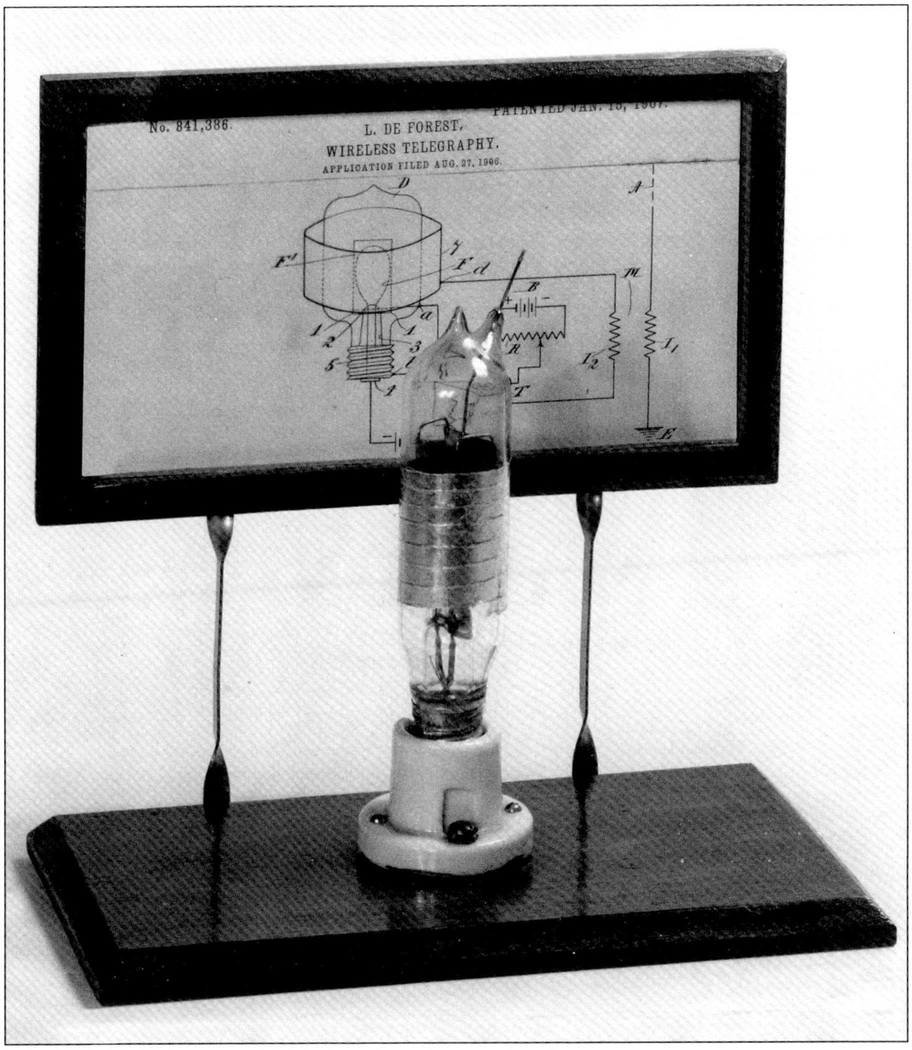

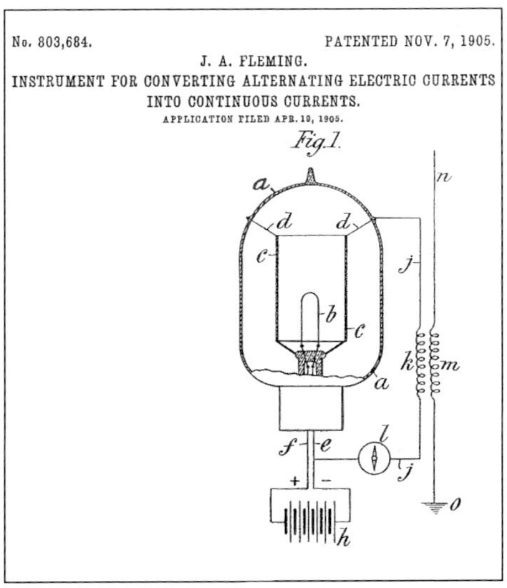

During the lawsuit filed by Marconi claiming the De Forest audion infringed on the Fleming diode, Marconi engineer Roy Weagant used the circuit shown here from the Fleming patent 803,684 to demonstrate equivalence of diode and audion performance.

also infringed on the Fleming detector patent.[34]

The patent situation for the triode was further complicated during the war when other companies working for the military made significant improvements on the triode, which they subsequently patented. By 1919 the patent situation was so confused that it would require a realignment and consolidation of interests before tube development could resume in earnest. Both U.S. government and Navy officials were acutely aware of the patent problem—not only for the triode but also for many other radio circuits and devices—and the need to take drastic measures to resolve it.

The Marconi-De Forest litigation over the audion had profound ramifications, not only because the triode was so crucial to the further development of radio, but also because as a result of the litigation, neither party could manufacture the tube without permission of the other. In fact, neither company did give permission for the manufacture of the triode for several years until after WWI when an agreement was reached in June of 1919 that allowed Moorhead Laboratories of San Francisco—founded by Otis B. Moorhead circa 1915—to manufacture tubes restricted to amateur and experimental use, and made Marconi the sole sales and distributing agent.[35] Moorhead was selected by both Marconi and De Forest because of Moorhead's facilities, experience, and proprietary manufacturing techniques developed during the war while manufacturing vacuum tubes for the government.

Marconi advertisements for these triodes began to appear almost immediately after the agreement was reached as evidenced by the

As a result of agreements between De Forest and Marconi, the Marconi V.T. triode finally became available circa July 1919, but only for amateur and experimental work.

The Marconi V. T. tube on the left was manufactured by Moorhead under licenses from Marconi and De Forest between mid-1919 and mid-1920, while the Electron Relay tube on the right was manufactured by Moorhead after July 30, 1920 but without the required license, which Marconi had terminated immediately following the merger with RCA.

accompanying ad for Marconi V.T. tubes appearing in July of 1919.[36] This particular ad identified both De Forest and Fleming patent numbers, and asserted that its use was restricted to amateurs, laboratories, schools, and experimenters. Ads appeared for two types of Moorhead tubes, Class I (also known as Type B) with a soft vacuum designed for use as a detector, and Class II (also known as Type A) with a harder vacuum designed for use as an amplifier. There were a number of different markings on both the bulbs and bases of these Moorhead tubes made for distribution by Marconi during the period from June 1919 through July 30, 1920—the effective date of cancellation of the agreement by Marconi at the behest of RCA.[37] The A tubes were manufactured with both clear and gold bulbs—most were cylindrical but a small number were spherical. Most bulbs carried the marking *Moorhead Laboratories, San Francisco*. The B tubes were all clear, had the same markings as the A tube, and like the A tube most were cylindrical.

Moorhead continued to make tubes even after Marconi cancelled the license, the most common being an "Electron Relay" tube with a bulb marked ER followed by a serial number. None of these tubes carried the Moorhead name, but many continued to carry the Marconi and De Forest logos as well as references to the Fleming and De Forest patents on the base—even though Moorhead no longer had a license to manufacture these tubes. These tubes were marketed by two distributors specifically set up by the Moorhead Company, the Atlantic Radio Supplies Co. and the Pacific Radio Supplies Co.

Marconi v. E. J. Simon Litigation

In May 1910, the Navy took the position that only those patents that had been established as valid by the highest courts of the land would be considered as adjudicated, and that the liability for supplying the Navy with equipment covered by patents must be borne by the contractor. However, in June of 1910, Congress passed legislation that granted the owner of a patent the right to sue the government for reasonable compensation whenever a device covered by a patent is used by the government without a license. Undaunted, the Navy continued to purchase equipment without cognizance of patents, and the suppliers believed themselves to be immune from prosecution for supplying the government with equipment that infringed on patents.

In early 1915, the Navy circulated a request for proposals for a number of wireless sets, and E. J. Simon responded to the proposal with a design that used technology covered by Marconi patents. As a result, American Marconi sought an injunction against Simon to stop them from supplying equipment to the Navy that contained the four-circuit tuning covered by the 7777 patent. Simon's defense was based on the 5 June 1910 Act and their assertion of the "right of eminent domain" invested with the government. On November 18, 1915, Judge Charles M. Hough presiding in the U.S. District Court for the Southern District of New York denied the motion for an injunction, and ruled that Marconi must sue the government to recover damages.[38] The ruling was appealed to the U. S. District Court of Appeals for the Second Circuit, where the decision of the lower court was upheld. Based on this ruling, the manufacturers of wireless equipment felt confident that they could supply the government with any type of radio equipment without concern for litigation or liability arising from potential patent infringements. The Navy was happy because they were able to acquire equipment to meet their design specifications from any manufacturer of their choice without concerns or impediments arising from patent infringements.

American Marconi appealed this ruling all the way to the U. S. Supreme Court, where it was reversed in a stunning victory on March 4, 1918.[39] Suddenly, the manufacturers found themselves potentially liable for all the infringements that had taken place, and there had been many since the lower court ruling several years before. The manufacturers threatened to stop supplying the government with wireless equipment unless the government could find some way to indemnify them from the potential liability arising from patent infringements. The Navy was more than a little concerned about this turn of events, and a few weeks later on March 29, 1918 the acting Secretary of the Navy, Franklin D. Roosevelt, began to notify the wireless manufacturers and other government contractors such as the Sperry Gyroscope Company by letter that the Navy would accept certain liabilities arising out of patent infringements.[40]

While the Marconi interests were delighted, the elation was short lived because on July 1, 1918—just three months after the first indemnification letter went out—Congress passed a new law that reinstated the basic intent of the 1910 act absolving the manufacturers of patent infringement liability when dealing with the government. The manufacturers' only recourse for patent infringements was now through the United States Court of Claims, and

26817-88:8
S/N/DAR

NAVY DEPARTMENT

WASHINGTON

May 2, 1918.

Gentlemen:-

The Department has considered your letter of April 24, 1918, advising it that in view of the recent decision of the United States Supreme Court in the case of William Cramp and Sons Ship and Engine Building Company, petitioner, vs. International Curtis Marine Turbine Company, et al., (393) and Marconi Wireless Telegraph Company of America, petitioner, vs. Emil J. Simon, (168), it would be disastrous to you financially to proceed with the manufacture and delivery of search lights, gyroscopic compasses, gyroscopic apparatus for torpedoes, gunfire control instruments and the like called for by your existing contracts with the Department or other contracts that it might award you unless the Department affords you protection against litigation and recovery of damages by the owners of patents alleged to control such apparatus or essential features thereof.

On consideration of said decisions of the Supreme Court and the law applicable to the matters affected thereby and the obligations placed on you by your existing contracts, the Department, having in view the Government's interests and the exceptional public necessity that the apparatus called for by your contracts be procured as soon as possible, hereby directs and requires you to proceed uninterruptedly and expeditiously with the fulfillment of said contracts, using and embodying in such apparatus any patented inventions necessarily required for that purpose. In order that you may be protected against litigation for any purpose by reason of alleged infringement of patents, the Department hereby guarantees to hold and save you harmless from and against any and all suits of which you shall give it immediate notice that may be brought against you for or on account of any alleged infringement of patents in the manufacture and production in accordance with the provisions and requirements of said contracts of search lights, gyroscopic compasses, gyroscopic apparatus for torpedoes, gunfire control instruments and the like, and your contracts are each and severally hereby modified accordingly, it being understood however that such modification does not release you from any obligation in said contracts or any of them that you shall hold the Government harmless against claims on account of the use of patented inventions.

Very respectfully,

Franklin D. Roosevelt

Acting Secretary

Sperry Gyroscope Company,
Manhattan Bridge Plaza,
Brooklyn, New York.

Table 1-3. Awards recommended by the Interdepartmental Radio Board in 1921 for patent infringements on wireless equipment acquired by the government circa WWI

Patent Holder	Recommended Award ($)
Marconi Wireless Telegraph Co. of America	$1,253,389
International Radio Telegraph Co.*	711,452
American Telephone and Telegraph	615,334
E. H. Armstrong	89,624
H. M. Horton	75,000
E. J. Simon	30,273
Lowenstein Radio Company	22,893
Wireless Specialty Apparatus Co.	22,450
Dublier Condenser Co.	18,194
John Firth	15,561
National Electric Supply Co.	8,875
Sperry Gyroscope Co.	5,000
Louis Cohen	1,271
American Mechanical Improvement Co.	384
Total	**$2,869,700.00**

**Successor of the National Electric Signaling Co.*

Marconi's competitors were free to supply the Navy with infringing technology, secure in the knowledge that Marconi could not sue them. Marconi could not have been happy about this turn of events.

After the war, seven companies submitted claims totaling $14,860,000 pursuant to the July 1918 legislation, while 17 others demanded monetary awards without specifying amounts.[41] The Interdepartmental Radio Board, which was assigned the responsibility for evaluating claims, recommended awards totaling $2,896,700 distributed as indicated in Table 1-3, with about two-thirds to be paid by the Navy Department and one-third by the War Department. These recommendations are interesting in several respects. First, the government finally recognized the validity of the many wireless patents and their obligation to reimburse the patent holders for infringements. Second, in recommending that American Marconi be awarded almost half the total amount owed, they implicitly recognized that Marconi held the largest number of the most important wireless patents. Third, the total amount of $2.9 million in damages represented about 7.2% of the total cost of radio apparatus purchased during the time period addressed by the Board, a respectable license fee, which was essentially pure profit for all the companies. Finally, while $2.9 million may not seem like a particularly large award today, considering the annual salary of the average worker in 1920 was about $1200, it would have supported over 2400 workers for a year!

Lobbying Against Government Control

A third element of the Marconi strategy was to lobby government representatives and elected officials to oppose legislation or executive orders that might restrict its business objectives. While lobbying is a time-honored tradition used by many corporations to achieve their goals, it usually takes place well out of the public eye, and with a modicum of tact. American Marconi raised the stakes by airing grievances in public, more often than not using a very confrontational style. The public forum of choice was initially the New York daily newspapers, and later *The Wireless Age* magazine, a magazine owned by American Marconi, staffed by Marconi executives, and generally used as a propaganda organ for both American and British Marconi.

An example of Marconi's confrontational style is captured in a letter from American Marconi to Commander Francis M. Barber, a Navy wireless representative, in response to an interview with Barber appearing in the June 18, 1902 edition of the *New York Herald*.[42] Barber had commented on the difficulties that the Navy was having in coming to terms with the Marconi Company for acquisition of their wireless equipment for test and evaluation, and also made comments marginalizing the Marconi patent position. The close of the letter from American Marconi reads:

> *"We trust, therefore, that in justice to this Company you will, in case you have been misrepresented, deny the accuracy of the interview, demand that the statement you made, if any, be correctly published in the New York Herald and its Paris edition, and protest against being made, as you have undoubtedly been made, the medium of interested attacks upon the Marconi system."*

As a result of the Supreme Court decision in favor of Marconi, acting Secretary of the Navy, Franklin D. Roosevelt began to notify the wireless manufacturers and other government contractors by letter that the Navy would accept certain liabilities arising out of patent infringements. (George H. Clark Collection, Smithsonian Institution)

Marconi executives presumably chose the public media to articulate the Company position with the intent of swaying public opinion, thereby increasing pressure on government officials to act favorably on wireless legislation and in other dealings with the company. However, it is not clear why Marconi executives used such a confrontational style to achieve their goal. This confrontational strategy appears to have achieved little except to create resentment, antagonism, and mistrust at the highest levels of government, which almost certainly contributed to decisions made by government officials in 1919 that were highly unfavorable to American Marconi.

The first comprehensive set of radio laws and regulations were gathered together and published in a single hardback volume by the Department of Commerce.

Opposition to Regulation

In 1904, the Navy began a concerted effort to bring control of wireless communication in the U.S. under the Department of the Navy.[43] In a report for the President, the Navy General Board recommended that all government stations be placed under control of the Navy, and that legislation be passed to regulate all commercial stations. The President approved these recommendations, and in July 1904 the Navy assumed control of government stations pursuant to executive order. The Marconi interests, fearing government regulation of private stations, immediately launched a campaign to defeat any legislation that might be proposed to regulate wireless communication. While their lobbying efforts were initially successful, it began a decade-long struggle between the private wireless interests led by Marconi on the one hand and government wireless interests led by the Navy on the other.

The next major threat for government regulation arose out of an international meeting in Berlin convened in 1906 to resolve a number of issues relating to world-wide communication, most notably the need for inter-communication among all stations. The resulting treaty known as the Berlin Convention of 1906 was adopted by all participants except the U.S. The Marconi interest vehemently opposed this treaty–primarily because of the requirement for intercommunication, which would have severely hampered the Marconi strategy for a world wide monopoly. The Senate's failure to ratify this treaty in 1908 was a high point for the Marconi interests, and demonstrated their ability to lobby effectively at the highest levels. However, the failure to ratify this treaty simply intensified efforts on the part of the Navy and other departments to regulate wireless communication.

The first legislation passed to regulate wireless communication was Public Law 262, commonly known as the Radio Ship Act of 1910, but since it did not require intercommunication and was innocuous to private interests in most respects, Marconi did not even bother to oppose this legislation. The sinking of the Titanic on April 15, 1912 prompted the quick passage of Public Law 238 on July 23, which amended the Radio Act of 1910. The provisions of this amendment focused on safety, such as the one requiring two skilled operators on board, one of which had to be on duty at all times while the vessel was being navigated.

Even before the sinking of the Titanic, the U.S. government had been pressing for more comprehensive regulation. The upcoming London International Convention of 1912 for radio regulations provided additional impetus to the Navy and U.S. legislators to seek ratification of the Berlin Convention of 1906 prior to this meeting. Despite vigorous opposition from the Marconi interests, the Berlin Convention was finally ratified by the U.S., and Public Law 264 was approved August 13, 1912, finally making intercommunication compulsory. While this was a major setback for the Marconi monopoly strategy, Marconi had all but achieved its goal of creating a monopoly by acquiring the assets of United Wireless a few months prior to passage of this legislation.

The last piece of legislation of this era to regulate wireless–or radio, a term the Navy had decided by then to use in favor of wireless–was the London International Convention of 1912. This Convention, which superceded the Berlin Convention of 1906, was ratified by the Senate on January 22, 1913 and took effect on July 18, 1913. This legislation added some relatively minor provisions to Public Law 264, which was based on the Berlin Convention, but did not invalidate any of its provisions. All of the existing radio laws and regulations in effect in 1914 were then gathered together and published in a single hardback volume.[44]

Opposition to Censorship

With the outbreak of WWI in Europe, President Woodrow Wilson issued a proclamation on August 5, 1914 imposing censorship on all wireless stations to enforce the U.S. policy of strict neutrality. The proclamation specifically delegated enforcement of this order to the Secretary of the Navy, directing him to take whatever actions he deemed necessary. American Marconi, alone among all wireless companies, not only refused to obey the executive order, but openly challenged the legality of the President's executive order by publishing exchanges with the Secretary of the Navy in the pages of *The Wireless Age* magazine. The September 1914 issue chronicled written communications that took place between Marconi's President John W. Griggs and the Secretary of the Navy in the month of August 1914.[45] On August 12, Griggs sent a telegram to Secretary Josephus Daniels insisting that the Navy had no right to censor Marconi transmission, and requested Daniels to provide the legal authority for doing so. The next day Secretary Daniels responded by inviting Griggs or his representative to Washington to discuss the executive order of the President dated August 5. Griggs declined the invitation in a long letter to Secretary Daniels dated August 19, insisting that Marconi was following the letter of the law, and that the Navy was exceeding its authority under the law by imposing censorship.

On September 2, the Siasconset Station relayed a communication from the British cruiser Suffolk to a private individual in New York requesting supplies and newspapers.[46] The Navy intercepted the message, and believing it to violate neutrality, asked for an explanation from American Marconi. They also threatened to close the station, pending a suitable explanation. Griggs responded on September 9, 1914 with a long diatribe, first claiming that the message was in fact neutral, and further taunted Daniels by stating once again that, in any event, he did not have the authority to close the station. American Marconi then sought an injunction from the courts to prevent the government from interfering with Marconi operations. Secretary Daniels promptly ordered the station to be closed effective at noon of September 25, 1914 without waiting for a court hearing, and notified Marconi of his order by letter. American Marconi responded on September 25th with a brazen letter of its own containing the following language:[47]

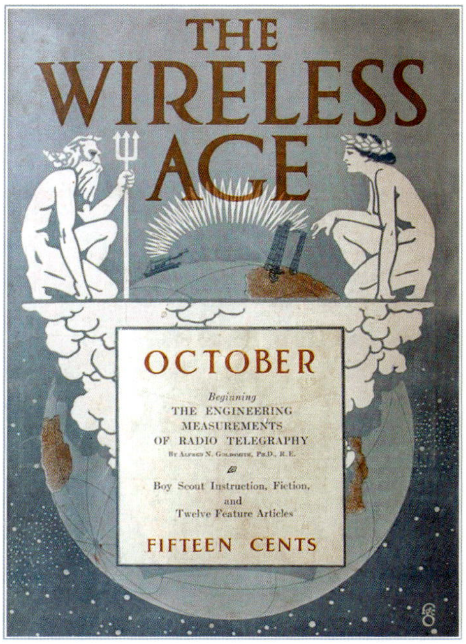

From the very first issue of *The Wireless Age* published in October of 1913, Marconi used their magazine to shape public opinion and pressure government officials to support Marconi policies.

> *"I acknowledge receipt of your letter of instructions relative to the cessation of all radio communications at Siasconset, Mass., and would ask if you are prepared to carry out you order by force.*
>
> *'MARCONI WIRELESS TELEGRAPH COMPANY OF AMERICA'*
>
> *I request instructions. E. B. Nixon"*

The stage was set for an historic–if not physical–confrontation between Marconi and its employees on the one hand and the Navy on the other. Fortunately for all concerned, American Marconi reconsidered its position because without any further communication from the Navy, E. B. Nixon sent another letter to the Navy about an hour later containing the following language:

> *"The Marconi Company withdraws the letter previously sent and the station was closed at 1:00 P.M.*
>
> *E. B. Nixon"*

The court subsequently refused to consider Marconi's injunction request, citing lack of jurisdiction over the executive orders of the President. Marconi did not appeal, and the station remained closed for a number of months thereafter. It was allowed to reopen January 16, 1915, but only after the Navy issued new rules for the operation of the station, and the Marconi Company issued a statement that they were acceptable.[48]

The censorship incident was an incredible miscalculation on the part of American Marconi. They had everything to gain and nothing to lose by bowing

The government closed the Marconi station at Siasconset for four months during the period after the onset of WWI, but before the U.S. entered the war, because Marconi refused to allow their messages to be censored.

to government censorship at the direction of the President, as all the other wireless companies, in fact, did. American Marconi not only directly defied and challenged both the Navy and the President of the United States, but they did it in such an overt and haughty fashion. Of all people, President Griggs, a former U.S. Attorney General, should have known of the broad powers granted to the President—particularly in times of national peril—and should have been able to predict the outcome of this altercation well in advance.

Opposition to Government Monopoly

By the end of 1912, American Marconi had effectively achieved its goal of creating a wireless communication monopoly in America, and in response, the government concluded that it was not in the best interests of America to have a wireless communication monopoly under foreign control. In November of 1916, the government proposed new legislation that would substantially increase control over that authorized by Public Law 264 passed in 1912. Over the strenuous objections of the Marconi interests, the enabling legislation was introduced in the House by Congressman Joshua W. Alexander. Secretary of the Navy Daniels stated that the intent of the Alexander Bill, as the proposed legislation became known, was to eliminate commercial interests from ship-to-shore communications initially, and eventually from all wireless communications. In essence, the Bill proposed that a government monopoly be formed to take over wireless communication for both government and commercial purposes. Once again, the Marconi interests led the opposition during the congressional hearings on the Alexander Bill held in early 1917, and once again Marconi used the pages of *The Wireless Age* magazine to appeal to the public to oppose this legislation.[49]

For some time, American Marconi had been trying to project an image that it was an American company with U.S. citizens as owners. For example, in the April 1916 issue of *The Wireless Age*, American Marconi asserted that 21,664 shareholders out of a total of 23,027 were residents of the United States.[50] In reality, personal notes of Marconi Managing Director Edward J. Nally made prior to the creation of RCA indicated that 65% of the outstanding shares of American Marconi was held by foreign shareholders.[51] Obviously, there was a significant difference between a majority of shareholders being U.S. residents and majority of shares being held by U.S. citizens. American Marconi was never able to provide an affidavit requested by the government to affirm that the majority of American Marconi shares were in fact held by U.S. citizens.

Before the hearings on the Alexander Bill were concluded, the U.S. entered WWI, and the government seized control of all U.S. wireless stations useful for naval and commercial communications by Executive Order on April 7, 1917, and closed all other commercial and amateur stations.[52] The Navy assumed control of shore stations, and the U.S. Shipping Board and the Railroad Administration assumed control of ships, including their wireless stations. The hearings were suspended, and the government had achieved a virtual monopoly without any enabling legislation—at least for the duration of the war.

During the war, the Navy made further moves to strengthen its position in the quest for a permanent government monopoly after the war.[53] First, the Navy negotiated to purchase the patents and shore stations of the Federal Telegraph Company, concluding the purchase of patents, three high-power stations, and two coastal stations for the sum of for $1.5M on May 15, 1918. Next, the Navy urged the Shipping Board to purchase the leased ship stations, and shortly thereafter the Shipping Board and the Railroad Administration both requested the Navy to purchase the ships stations under their respective control against their account. American Marconi refused to sell the leased ship stations without also selling the associated coastal stations—which would have been a financial burden without the corresponding ship stations with which to communicate—and the Navy was most happy to accommodate American Marconi on this score. The following transaction for a total of $1.45M was concluded November 30, 1918, a mere 19 days after the armistice:

Shipping Board: For radio installations on 267 vessels	$519,000
Railroad Administration: For radio installations on 63 vessels	$141,200
Navy: For 45 shore stations, including Ketchikan, Juneau-Astoria Alaskan circuit, the South Wellfleet (obsolete) high-power station, and the leased short radio stations	$798,500

It is surprising that Marconi would offer to sell all of its shipboard stations and most of its land stations, leaving it with only four high-power stations. The rationale can be found in the announcement of the transaction by Edward Nally appearing in the January 1919 issue of *The Wireless Age*, an excerpt from which follows:[54]

> *"This transfer does not in any way affect the status of the company's large stations for transoceanic communication, but it is believed that it will have met every need of the Government for control of domestic wireless and that there will now be no occasion for the Government seeking secure control or ownership of the large high power stations which were built with the principal object of rendering to the public international transoceanic wireless service in competition with cables, which is by far the most important part of the Marconi Company's activities."*

Clearly, American Marconi hoped this sale would forestall any action on the part of the Navy to revive the quest for a government monopoly, and they were willing to sacrifice the coastal business segment in an attempt to protect the transoceanic communication segment, one they believed would be the most lucrative. Unfortunately for Marconi, just as this transaction was being concluded, Secretary Daniels successfully revived hearings on the Alexander Bill, which if passed, would allow the Navy to perpetuate the wireless communication monopoly it had enjoyed during the war.

Again the Marconi interests led a vigorous opposition to the bill, this time aided by the midterm elections, which delivered control of the House and Senate to the Republicans. With the Republicans in firm control of the Congress, there was virtually no chance of enacting any legislation that would put wireless communication under government control. In fact, during the hearings, several congressmen severely rebuked Navy representatives for purchasing the Marconi and Federal Telegraph stations without authorization or approval from the Congress. Ironically, the purchase of these stations had hurt, rather than helped, the Navy's cause. In January of 1919 the Bill was unanimously tabled by committee vote, and never again considered.

The Final Curtain

Secretary Daniels never fully gave up on his dream of creating a government monopoly under Navy control, but he was a pragmatist, and so he also proposed the concept of a government-sanctioned monopoly under the control of a private American-owned company. However, before he was able to promote this concept, American Marconi was about to preempt this option by obtaining from General Electric sole rights to the Alexanderson alternator, the key technology at that time needed for transoceanic communication. When the news of this proposed agreement reached the Navy, it triggered a series of events that brought down the final curtain on American Marconi.

Certain Navy officials, with the apparent knowledge and consent of President Wilson, approached GE and persuaded them on patriotic grounds to decline an offer from Marconi to purchase the alternators. The Navy proposed sanctioning an American company to create a monopoly for wireless communication, and suggested that GE take the lead in creating such a company. The Navy agreed to develop and sponsor the necessary legislation to charter the monopoly, and indeed such a draft was completed by May 30, 1919.

When Vice President Nally of Marconi was informed of the GE decision to withhold the Alexanderson alternators, he arranged a meeting with the Navy for clarification. At that meeting Nally was informed that the Navy was encouraging GE to form an American wireless company that would directly compete with Marconi, and he was encouraged to join GE in establishing such a company. Both American and British Marconi, understanding their options were very limited, entered into negotiations with GE to sell their interest in American Marconi. It is notable that GE acted to form a new company without any enabling legislation to charter a monopoly. Perhaps GE thought such legislation was not necessary, perhaps they decided to seize the moment, fearing that passing any such legislation would be a lengthy and uncertain process, or perhaps they feared that the Congress would agree to a government-owned monopoly that Secretary Daniels continued to seek.

During the negotiation period, it became clear that both the Navy and U.S. government officials had concerns about American Marconi that went beyond foreign ownership, the pretext used by these officials to promote an American-owned company. A reference to this concern is contained in a letter from Owen Young, Vice President of General Electric to Marconi Vice President Nally dated October 23, 1919:[55]

> *"I must speak very frankly in order that you may fully understand the background for my point of view. Since Mr. Davis announced to the Navy Department that you were to be the president and operating head [of] the new company [RCA], there has been very grave opposition in Washington, in*

different quarters, to our whole enterprise, on the ground that that indicated merely that the Marconi Company was going to take over the General Electric interests rather than the General Electric interests, through the new radio company, taking over the Marconi Company. This criticism has been made not only in important places in the Executive Department, but it has also permeated in influential places in Congress..."

A reading of Navy correspondence and testimony in this era reveals the Navy was annoyed with American Marconi for their opposition to government regulation, and their aggressive patent stance, as well. The Navy's annoyance with Marconi even surfaced in a congratulatory letter from Navy Captain D. W. Todd to Nally dated February 14, 1920 after he became president of RCA:[56]

General Electric was about to grant American Marconi sole rights to the Alexanderson alternator (shown here with inventor Ernst Alexanderson), the key technology for trans-oceanic communication at the time, when the Navy intervened and convinced General Electric to create a wireless communication company as an alternative market for the alternator. (George H. Clark Collection, Smithsonian Institution)

"From the time when the Marconi Company opposed any radio regulation, the necessity for which was clearly apparent in 1911, it has been difficult for many of us who have had to do with radio matters in the navy to deal sympathetically with that company."

These concerns were ultimately laid to rest by assurances to the Washington establishment that GE would maintain control of the new company through a Board of Directors dominated by GE, in which GE officers were to hold half the board seats headed by GE Vice President Owen Young as Chairman of the Board. The proposed by-laws of the new company also contained a number of features designed to further please the government and ensure continued American control and ownership. Only U.S. citizens could be elected Directors and Officers. A government representative would be allowed to attend meetings and participate in the management of the company. Not more than 20% of the stock could be held and voted by foreigners without restriction, and these shares were marked on their face "Foreign Share Certificates." The remaining shares carried a clause on the face of the certificates restricting ownership to loyal U.S. citizens.

Nally and GE attorney Albert G. Davis of GE sailed for England in June of 1919 to negotiate the purchase of American Marconi shares held by British Marconi. On September 5, 1919, after several months of intensive negotiations and diplomatic discussions, British Marconi agreed to sell the 364,826 shares they held in American Marconi to GE for millions of dollars in cash and other concessions. The sale of this stock was contingent upon shareholder approval of the merger of American Marconi into the new radio corporation.

The Radio Corporation of America was incorporated in the state of Delaware on October 17, 1919, and on October 22, American Marconi entered into a preliminary agreement with GE for such a merger. The preliminary agreement referenced two additional agreements, a General Electric-RCA Main Agreement and a Marconi-RCA Main Agreement, which contained the terms and conditions of the proposed merger. The preliminary agreement required Marconi to recommend and to urge upon its stockholders approval and consent for the execution by the Marconi Company of Marconi-RCA Main Agreement to effect the merger. A circular dated October 22 over the signature of John Griggs, President of American Marconi, containing the full proposal was sent to shareholders. Shareholders were urged to approve the proposal and return their proxies prior to the special shareholder meeting scheduled for November 20 for the purpose of voting on the proposal. The full text of the circular was published in the November 1919 issue of *The Wireless Age* magazine.[57]

American Marconi shareholders voted overwhelmingly for the proposed exchange on November 20, and both the Marconi-Radio Main Agreement and the General Electric-RCA Main Agreement dated November 20, 1919 were signed, almost 20 years to the day after American Marconi was formed. Shortly thereafter, shareholders in American Marconi were able to exchange their shares for the preferred stock of the Radio Corporation for an equal amount, par for par, and in addition received one share of the common stock

SHARES $5.00 EACH
CERTIFICATE FOR LESS THAN 100 SHARES
045790
25
MARCONI WIRELESS TELEGRAPH COMPANY
OF AMERICA
INCORPORATED UNDER THE LAWS OF THE STATE OF NEW JERSEY
This is to Certify, that GEORGE RANDALL JOHNSON
MAR 25 1920
TWENTY - FIVE
CAPITAL STOCK
TREASURER.
VICE-PRESIDENT.
AMERICAN BANK NOTE COMPANY, NEW YORK.

S74878
100 SHARES
COMMON STOCK
00000
100
RADIO CORPORATION OF AMERICA
INCORPORATED UNDER THE LAWS OF THE STATE OF DELAWARE.
This is to Certify that
is the owner of ONE HUNDRED fully paid and non-assessable shares
WITHOUT NOMINAL OR PAR VALUE OF THE COMMON STOCK
of Radio Corporation of America
COMMON
THE NEW YORK TRUST COMPANY.
THE CORPORATION TRUST COMPANY.
CORPORATION OF AMERICA DELAWARE 1919
SPECIMEN
SECRETARY.
PRESIDENT.
AMERICAN BANK NOTE COMPANY.

Shares of the Marconi Wireless Company of America were exchanged for shares in RCA, and the outstanding Marconi stock certificates were cancelled in early 1920.

of the new company for each share held in American Marconi. Marconi stock certificates were cancelled and retired, but they still circulate today among collectors. RCA stock certificates bore the following restrictions even into the nineteen eighties:

> *"The voting rights represented by this certificate shall be transferable only to loyal citizens of the United States and/or corporations formed under the laws of the United States or for and under the laws of one of the states of the United States free from foreign control or domination and not in any foreign interest."*

American Marconi continued to operate for a number of months after the agreements were signed to tidy up loose ends, mostly in the form of claims pending against the U.S. government. At a special meeting on April 6, 1920, the shareholders of the Marconi Wireless Telegraph of America voted to dissolve the company, and on August 2, 1920, a certificate of dissolution was filed in New Jersey, thereby officially bringing down the final curtain on one of the most colorful phases in the development of wireless communication in the Americas.

Epilog

In two short decades, American Marconi and the Navy had come to dominate radio in the United States, but neither could give a good account of their activities during this period. American Marconi for its part refused to cooperate with the Navy, attempted to achieve a monopoly enforced by the courts through litigation, indulged in predatory practices to stifle competition, and opposed even minimal government regulation that was in the best interests of everyone but perhaps Marconi. The Navy for its part ignored patents and intellectual property rights specifically sanctioned by laws passed by Congress, attempted to hijack the radio industry by endeavoring to create a government monopoly to the exclusion of private industry—notwithstanding the fact that the Navy had never discovered an important radio principle or filed a relevant patent, misappropriated funds to purchase radio stations from private enterprise under duress and without congressional approval, further abused its power by interfering with private business dealings and ultimately forced American Marconi to merge with RCA—almost certainly violating a number of federal laws in the process.

All stock certificates issued to shareholders in the new Radio Corporation of America bore the highly unusual legend restricting voting rights only to loyal acitizens of the United States and/or corporations formed under the laws of the United States free from foreign control or domination.

The cynic would say that they both deserved each other and their respective fates in the history of radio. American Marconi was forced to merge with RCA and ceased to exist as a separate entity. The Navy was rebuked by congress for attempting to create a government monopoly and for misappropriating funds by purchasing radio stations during the war without congressional consent, was criticized for poor management of the numerous radio stations they operated during the war, was forced to return numerous radio stations to private industry after the War, incurred millions of dollars in damages for patent infringements, and never again had significant influence in the development or regulation of radio and other communication technologies thereafter.

Endnotes and References for Chapter 1

1 The Marconi Wireless Telegraph Company of America was registered as a subsidiary of British Marconi on this date but it was not until April 1, 1902 that it was incorporated in the State of New Jersey as a public company.

2 An excellent summary of the early problems between American Marconi and the Navy with contemporaneous references can be found in the following reference: Captain Linwood S. Howeth, USN (Retired), *History of Communications-Electronics in the United States Navy*, (United States, Government Printing Office, Washington, 1963) Chapters III and IV.

3 "The Marconi Valve Receiver," *Modern Electrics*, February 1912, Vol.4, No. 11, pp. 779-780.

4 De Forest spelled his surname "de Forest," but this spelling leads to a number of problems when distinguishing between the man and his companies–his companies were invariably spelled with a capital D. Many publications–including *The New York Times* and De Forest publications–adopted the convention of using a capital D for all forms including the man himself; for example, see the following catalog: *De Forest Radio Equipment*, (The De Forest Radio Telephone & Telegraph Co., 2nd Ed., 1922) p. 4.

5 It was no secret that E. E. Bucher, a Marconi engineer, was performing highly successful experiments at sea with a two-stage receiver employing audion tubes as early as 1913: E. E. Bucher, "How to Conduct a Radio Club," *The Wireless Age*, Vol. I, No. 4, January 1914, pp. 274-280. It was not until after RCA was formed in late 1919 that the vacuum tube patents were sorted out and the triode became the dominate detector of choice.

6 See Chapter IV of Howeth, History of Communication, for a discussion and specific references to the Navy's early purchase of equipment for test and evaluation.

7 Howeth, History of Communication, p. 147.

8 It should be noted that Marconi did prepare a catalog circa 1915 for sales of surplus equipment to private individuals and experimenters, but with the restriction that it could not be used for commercial work.

9 "Marconi Sells Apparatus," *The New York Times*, November 28, 1918, p. 11.

10 *Radioana*, George H. Clark, "Analysis of Companies from whom the U.S. Navy purchased Wireless Sets in 1906, 1907, and 1908," Document SRM 5 811 dated November 2, 1939; Clark used data from *"Wireless Stations of the World"* from 1906, 1907, and 1908 to generate the first three columns, and data from a letter on file at the Bureau of Equipment, National Archives Washington, D. C. for the year 1912. The results of Clark's analysis differ slightly from a similar analysis performed by Howeth (*Howeth, History of Communication*, p. 106) for the year 1906 because Clark adjusted his table for composite sets in stations where the receiver and transmitter were of different types.

11 Howeth reports that American Marconi first made a profit in 1910, but a close reading of the annual report indicates that the profits were for operations only and these profits did not cover other costs of the Corporation such as those associated with Marconi land stations and experimentation: Howeth, *History of Communication*, p. 147; the American Marconi annual report for the year ending December 31, 1914 was the first to show a profit that was carried to the balance sheet.

12 The acquisition of the Aldene, N. J. plant was reported in the American Marconi annual report ending January 31, 1913, which covered activities for eleven months of calendar year 1912.

13 *List of Wireless Telegraph Stations of the World*, Department of the Navy, Bureau of Equipment, Corrected to September 1, 1909, (Government Printing Office, Washington, D.C., 1909).

14 The famous "Four Sevens" Patent awarded to Marconi on April 26, 1900 in England was also valid in the United States as well.

15 "Marconi Absorbs United Wireless; Inventor's Suit for Infringement Comes to an End and Merger Is Announced," *The New York Times*, March 26, 1912, p. 15.

16 *Modern Electrics*, April, 1912, p. 27.

17 The Continental Wireless stock fraud was highly publicized in late 1910: Postal Raids Show Vast Stock Frauds," *The New York Times*, November 22, 1910, pp. 1-2; the principals were convicted of fraud on January 9, 1913: "Wireless Men Guilty," *The New York Times*, January 10, 1913, p. 8.

18 "Wireless Companies Consolidate," *Electrical Review and Western Electrician*, May 21, 1910, p. 1044.

19 For example, Thorton Mayes contends that Massie Wireless became part of Continental Wireless: Thorn L. Mayes, *Wireless Communication in the United States, The Early Development of American Radio Operating Companies*, (The New England Wireless and Steam Museum, Inc., East Greenwich, R. I. 02818, 1989) pp. 72-100.

20 *The New York Times*, November 22, 1910, p. 2.

21 "Wireless Supplants Cable [-] Government Gives Contract to Massie Company for Block Island Service," *The New York Times*, January 13, 1912, p. 1.

22 Thorn L. Mayes, *Wireless Communication in the United States, The Early Development of American Radio Operating Companies*, pp. 79, 100.

23 *Annual Report of the Commissioner of Navigation of Commerce and Labor for the Fiscal Year Ended June 30, 1911*, (Submitted November 18, 1911 from Commissioner Eugene Tyler Chamberlain to Secretary Charles Nagel).

24 *Radio Stations of the United States*, Edition July 1, 1914 (Department of commerce Bureau of Navigation Radio Service, Government Printing Office, 1914).

25 *The American Marconigraph*, Vol. 1, No.1 November, 1912.

26 "Name Receivers for Wireless Company," *The New York Times*, August 6, 1912, p. 12.

27 The author reviewed *Radio Stations of the United States* published by the Department of Commerce, Bureau of Navigation for the years 1914, 1916, and 1919, and could find no ships with stations controlled by Tropical Radio Co. other than those owned by United Fruit.

28 For example: "Shipowners Fight Marconi Co. Rates," *The New York Times*, October 14, 1914, p. 17.

29 *Radio Stations of the World*, (Department of Commerce, Bureau of Navigation, Government Printing Office, Washington, Ed. July 1, 1916).

30 "Our German Cables Cut at the Azores," *The New York Times*, August 6, 1914, p. 6.

31 Testimony of Roy Weagant in Marconi Wireless Telegraph Co. of America v. De Forest Radio Telephone & Telegraph Co. et al., New York, February 29, 1916, 10:30 A. M., pp. 878-880.

32 "Marconi Wireless Telegraph Co. of America v. De Forest Radio Telephone & Telegraph Co. et al.," *The Federal Reporter*, Vol. 236, (West Publishing Co., St. Paul, 1917) 942-955.

33 "Marconi Wireless Telegraph Co. of America v. De Forest Radio Telephone & Telegraph Co. et al.," *The Federal Reporter*, Vol. 243, (West Publishing Co., St. Paul, 1917) 560-567.

34 "Marconi Wireless Telegraph Co. of America v. De Forest Radio Telephone & Telegraph Co. et al.," *The Federal Reporter*, Vol. 261, (West Publishing Co., St. Paul, January-March, 1920) 393-395.

35 "Radio Vacuum-Tube Litigation is Settled." *Radio Amateur News*, Vol. 1, No. 2, August 1919, p. 77.

36 *The Wireless Age*, Vol. 6, No. 9, July 1919, p. 35.

37 Gerald F. J. Tyne, *Saga of the Vacuum Tube*, (Prompt Publications, 1st Edition, 3rd printing, Berkeley Heights, NJ, 1994) pp. 167-180; Bill Condon, "Moorhead and His Tubes," *Tube Collector*, Vol. 5, No. 2, April 2003, pp. 3-26 (an online version of this article with color photographs can be found at http:/www.bill01a.com/articles/Moorhead.htm).

38 "The Marconi Quit Against Simon," *The Wireless Age*, Vol. 3, No. 3, December 1915, p. 209.

39 U. S. Supreme Court, Marconi Wireless Tel. Co. of America v. Simon, 246 U. S. 46 (1918); see for example: http://laws.findlaw.com/us/246/46.html.

40 *Radioana*, Letter from Franklin D. Roosevelt, Acting Secretary of the Navy, to Sperry Gyroscope Company dated May 2, 1918, CWC 120 109, 1918.

41 Howeth, *History of Communications-Electronics in the United States Navy*, p. 376; Howeth cites Files, Bureau of Engineering, National Archives, Washington, D. C.

42 Letter, dated July 1902, Cmdr. F. M. Barber, USN (retired), to Chief of the Bureau of Equipment; letter dated 21 June 1902, from Executive Committee, Marconi Wireless Telegraph Co. of America, to Commander Barber, files, Bureau of Equipment, National Archives, Washington, D.C. See Howeth, *History of Communications-Electronics in the United States Navy*, p. 46.

43 Howeth, *History of Communications-Electronics in the United States Navy*, Chapter VII: The Origins of Regulation.

44 *Radio Laws and Regulations of the United States*, Department of Commerce, Bureau of Navigation; July 27, 1914 Edition, Government Printing Office, Washington, D.C. 1914.

45 "The Censorship of Messages," *The Wireless Age*, Vol. 1, No.12, September 11, 1914, pp. 964-966.

46 "The Censorship Situation," *The Wireless Age*, Vol. 2, No. 1, October, 1914, pp. 8-10.

47 "Marconi Station Seized by Daniels," *The New York Times*, September 26, 1914, p. 4.

48 "Siasconset Station Reopened," *The Wireless Age*, Vol. 2, No. 5, February, 1915, p. 392.

49 "The New Radio Legislation," *The Wireless Age*, January 1917, pp. 226-240; "Government Ownership of Wireless," *The Wireless Age*, February 1917, pp. 300-308; "Dangers of the Proposed New Radio Bill," *The Wireless Age*, March 17, 1917, pp. 374-387; "The Control of Wireless," *The Wireless Age*, June 1917, p. 651.

50 "The American Marconi Company Annual Report," *The Wireless Age*, Vol. 3, No.7, April 1916, p. 453.

51 *Radioana*, Letter from to Edward J. Nally to GE Vice President Albert G. Davis dated April 1st, 1924 transmitting personal notes made before the creation of RCA-including a list of shareholders by country and shares held.

52 "Government Seizes Whole Radio System [-] Navy Takes Over All Wireless Plants It Needs and Closes All Other," *The New York Times*, April 8, 1917, p. 2.

53 *History of the Bureau of Engineering Navy Department During the World War*, under direction of Edwin Denby, Sec. Navy, Navy Department, Office of Naval Records and Library, (Government Printing Office, Washington, D.C. 1922) pp. 110-115.

54 "Marconi to Sell and Lease ship Equipment," *The Wireless Age*, Vol. 6, No. 4, January 1919, p. 8.

55 *Radioana*, Letter from Owen Young, Vice President of General Electric to Vice President Nally dated October 23, 1919, SRM 4 658, 3 pages.

56 *Radioana*, Personal Letters of E. J. Nally loaned to G. H. Clark December 1938, Letter No. 7 from Navy Captain D. W. Todd to RCA President Nally dated February 14, 1920. p. 3.

57 "A New and Powerful Wireless Company," *The Wireless Age*, Vol. 7, No. 1, November 1919, pp. 10-12.

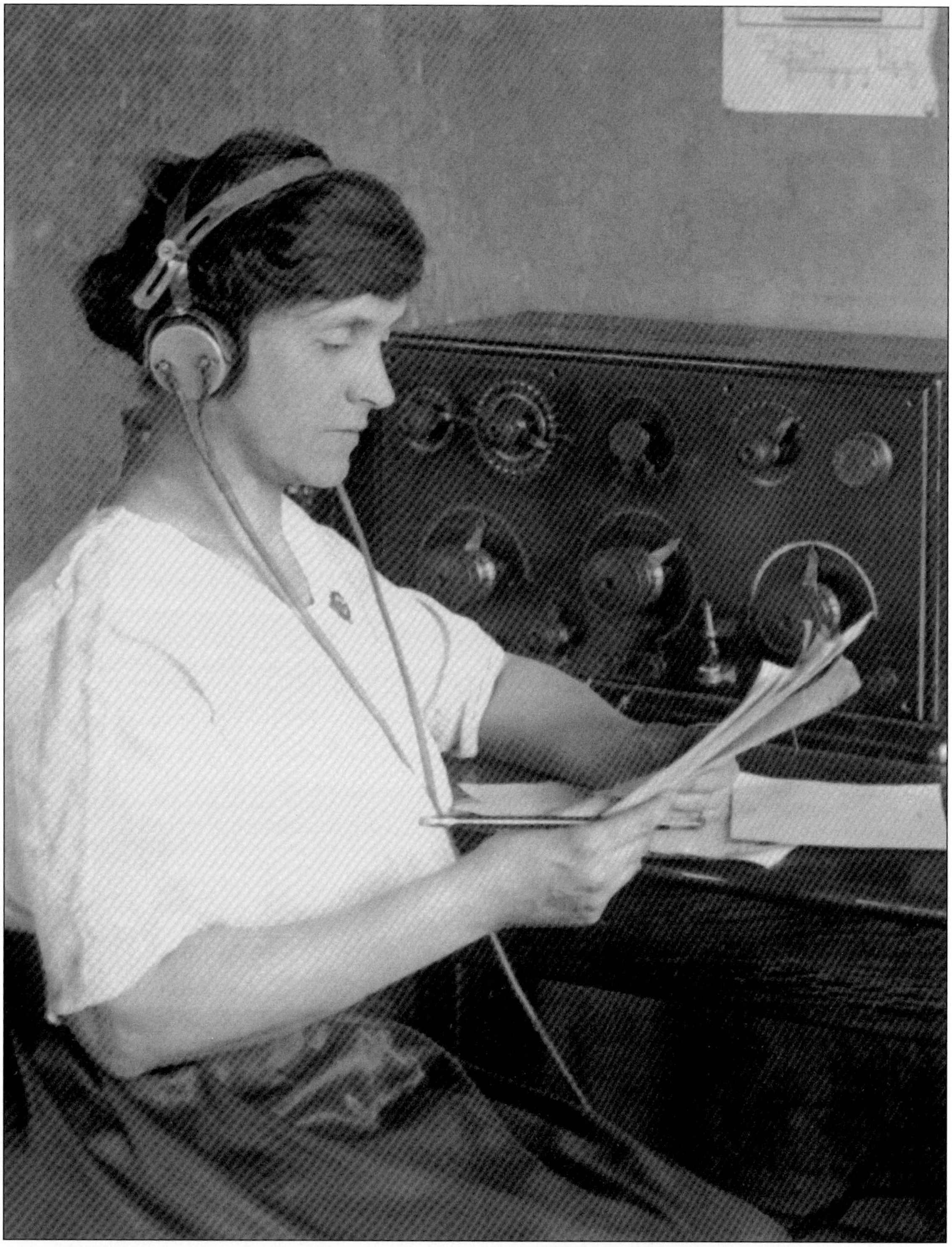

An unidentified young lady operates the earliest version of a Marconi 106 receiver dating from 1915 at a wireless station.

Chapter 2

Wireless Becomes World Wide

1919-1920

RCA was formed by General Electric out of the need to create a market for the Alexanderson alternator in particular, and radio apparatus in general. The agreements between RCA, GE, and American Marconi granted RCA the rights to operate and maintain communications systems, market radio products and services, and develop new radio technologies—but specifically excluded RCA from manufacturing radio-related products. As a result, the manufacturing operations from the American Marconi plant in Aldene, New Jersey were transferred to GE in Schenectady shortly after Marconi was absorbed into RCA. Marconi's senior management, most notably General Manager Edward J. Nally, Secretary and Comptroller Charles J. Ross, Treasurer George S. De Sousa, and General Counsel John W. Griggs, became the core of the new RCA management, with Nally as the new RCA President and the others retaining parallel positions at RCA. The executive offices which Marconi occupied in the Woolworth Building at 233 Broadway in New York became the new executive offices of RCA.

In the first year following its incorporation, RCA focused on developing a world-wide wireless communication network, building on the larger stations owned by American Marconi prior to WWI. The major corporate objectives were to reactivate the Marconi sites confiscated by the Navy during the war,

The American Marconi manufacturing plant at Aldene, New Jersey, had been modernized and expanded to support the war effort a few years earlier. It was closed when operations were transferred to General Electric in Schenectady shortly after the merger. (George H. Clark Collection, Smithsonian Institution)

replace antiquated hardware at these sites, expand the network by developing new circuits to other foreign countries, build a new world-wide communication center on Long Island—aptly named Radio Central, and develop a ship-to-shore communication service. RCA also began the practice of cross-licensing radio technologies with other companies to create a patent pool which would assure access to technologies needed for state-of-the-art radio components, transmitters, receivers, and other related accessories.

The American Marconi headquarters located in the Woolworth Building at 233 Broadway in New York became the new headquarters for RCA.

RCA Trademarks and Monograms

With former Marconi General Manager Edward J. Nally as the new President of RCA, it was business as usual for the Marconi employees who became RCA employees. In fact, it was said by one employee that the only difference he noticed was the company name on the paycheck. Indeed, the first order of business was to develop the trademarks and monograms that would appear on paychecks, correspondence, announcements, and other business documents—but most importantly in advertising and catalogs. President Nally to his credit clearly understood the significance of trademarks and monograms to the commercial success of the company. He took personal interest in developing them, many of which are still instantly recognized throughout the world today.

A number of years after he retired, Nally confided to RCA historian George H. Clark in a hand-written letter that he had worked out all of the trademarks and slogans with company architect R. C. Edwards.[1] Describing himself in the third person, he wrote: *"Glad to see that the first President of RCA is, at least, given the credit for zigzagging the tail of the RCA trademark! As a matter of history, he alone worked out with Mr. Edwards all of the slogans, trademarks, etc."* Changing to the first person, Nally went on to say:

> *"I also worked out and used in Marconi days the slogans, or circles*
> *Shore to Ship*
> *Ship to Shore*
> *Ship to Ship*
> *I designed the first messenger uniform with its trimming of Marconi red."*

Marconi manufacturing operations at Aldene were transferred to the huge General Electric facility in Schenectady, New York in mid-1920.

MARCONI

MARCONI WIRELESS TELEGRAPH CO. OF AMERICA

The globe motif used by RCA in the early years had its origins in the Marconi organization, appearing in magazines, on correspondence, and on other documents published by Marconi. (George H. Clark Collection, Smithsonian Institution)

The monograms and trademarks created during Nally's presidency are shown in Table 2-1, grouped in three categories according to whether they contain the World Wide Wireless globe logo, consist of monograms, or are trademarks for specific products. Within each of these three categories, the entries are listed in order of the date of first use, which was either specified in the registration documents or deduced from advertisements, catalogs, and other documents of the day.

World Wide Wireless Globe Trademark

The first official RCA trademark was announced in the June 1920 edition of *World Wide Wireless*, an RCA house organ for employees that succeeded the *Marconi Service News*. The new trademark consisted of a world globe with an impression of North America on the upper half, and the words WORLD WIDE WIRELESS on the lower half (see first entry in Table 2-1). The globe was

Table 2-1. Logos, monograms and trademarks developed by RCA in the early 1920s, grouped by the World Wide Wireless globe logo, RCA monograms, and specific product trademarks

	Trademark, Monogram or Logo	Registration Number	Filing Date	Date of First Use in Commerce	Proximate Date of Last Use
World Wide Wireless Globe Logos	WORLD WIDE WIRELESS / RADIO CORPORATION OF AMERICA	Unknown, if any	Unknown, if any	June, 1920	Never used on products or in advertising
	WORLD WIDE WIRELESS	Unknown, if any	Unknown, if any	June, 1920	1928
	Radio (WORLD WIDE WIRELESS) Corporation of America	Unknown, if any	Unknown, if any	May, 1921	December, 1923
RCA Monograms	RC	150,947	April 28, 1921	October 25, 1920	August, 1922
	RCA	167,591	September 22, 1922	August 15, 1922	January, 1968
	"Via RCA"	509,804	January 31, 1948	1923*	After 1989
Product Trademarks	Radiotron	150,948	April 28, 1921	October 25, 1920	Well after 1962
	Radiola	152,961	October 29, 1921	September 27, 1921	Well after 1962

* A 1948 registration document specifies 1923 as the first date of use, but "Via RCA" was in use as early as 1921; it is unknown if there was an earlier registration document.

enclosed in a square box with the company name Radio Corporation of America enclosed in a rectangle directly below the square.

Both the globe and the phrase WORLD WIDE WIRELESS had their origins in the Marconi organization predating RCA. The globe motif was used by American Marconi for many years on business documents and memos, on Marconigram forms, and in publications including *The Wireless Age* magazine and the *Marconi Service News* for employees. In later years, the *Marconi Service News* sported a more stylized version with a border that is evocative of a wax imprint. The "world wide wireless" phrase was also regularly used by American Marconi on Marconigram forms and advertisements, and became the title of the news feature appearing in each edition of *The Wireless Age* magazine beginning with the June 1917 edition. It was an obvious step for Nally to combine the phrase and the globe to form the new trademark.

Oddly enough, after the splashy introduction in *World Wide Wireless*, the new trademark never again appeared in any other publication that has been discovered to date. It was, however, prominently displayed on the edifice of certain buildings that accepted RCA radiograms from the public. The globe logo by itself *without* the rectangular enclosures or corporate name (see the second entry in Table 2-1) was used on letterheads and other business documents for a number of years, but it was not widely used in advertising. It also appeared prominently on the front of every RCA annual report from incorporation until 1928, the year it was finally replaced by the RCA monogram—long after world wide wireless communication had been overshadowed by broadcast radio. The globe trademark died hard, most likely because the old guard at RCA had strong ties to their roots in world wide wireless.

Another version of the globe with two zigzag lightning bolts extending from the globe was widely used in advertisements, catalogs, and business documents beginning circa May 1921. This version was accompanied with the RCA name spelled out, and with the globe placed between the words "Radio" and "Corporation" as shown in the third entry of Table 2-1. It was used extensively in advertising until mid-1923, when it disappeared from most ads for radio apparatus. By the end of the year it had been totally replaced by the ubiquitous RCA monogram described below.

The stylized form of the Marconi globe appeared at the top of the employee's *Radio Service News*, the last issue of which shown here was dated December 1919, coinciding with the merger of Marconi into RCA.

The first RCA trademark, announced in the June 1920 edition of *World Wide Wireless*, appeared at message windows and on the façade of buildings used by RCA for sending and receiving radiograms but never again appeared in print. (George H. Clark Collection, Smithsonian Institution)

RCA Monograms

The initials of the Company—with or without periods and spaces—were often used as a shorthand notation from the outset (e.g., RCA, R C A, R.C.A.). However, the first monogram proper was developed by Mr. R. C. Edwards, an RCA architect, sometime in 1920.[2] The monogram consisted of the two letters R and C overlaid and enclosed within a circle as shown in the fourth entry of Table 2-1. According to the trademark application, the monogram was first used in commerce on October 25, 1920, and appeared in RCA's *The Wireless Age* magazine for the first time in the December 1920 issue in an ad for RCA Radiotron tubes.[3] It is interesting to note that the trademark filing date for the RC monogram is the same as that for the Radiotron, and that the registration numbers for the two are in immediate sequence. Clearly, the RC monogram was developed to coincide with the introduction of the Radiotron trademark, although the RC monogram was used in advertising, catalogs, and on many products in addition to tubes sold by RCA until mid-1922.

Sometime in early 1922, it was suggested that the monogram be altered and expanded by adding "A" as a third letter. Mr. Edwards redesigned the monogram accordingly, but when RCA's President Nally saw the proposed design, he ordered that a zigzag lightning flash be added to the final "A."[4] Thus, the RCA monogram took the form shown in the fifth entry of Table 2-1, a form which has persisted and remains highly familiar even to this day. One of the first places this monogram appeared was on an RCA Wholesale Distributor's Discount Schedule dated August 21st, 1922. At the same time, it also began to appear in RCA advertisements as an adjunct to—but not a replacement for—the trademark with the lightening bolt. However, by mid-1923 the lightening bolt trademark began to disappear from advertising, and by the end of the year it had completely disappeared, leaving the RCA monogram within a circle as the principal symbol of RCA. It should also be mentioned that the zigzag line began to appear about the same time on RCA initials without the outer circle. Examples of this form can been seen in the phrase "Via RCA" printed on early RCA Radiogram forms and Radiogram Tariffs documents such as the ones which appear later in this chapter.

The phrase "*Via* RCA" was introduced in July, 1921 by Nally in conjunction with a change in title on RCA message blanks from Marconigram to Radiogram.[5] The initial version used a simple non-stylized font, while later versions were stylized like the one shown in Table 2-1. Both forms are shown in RCA documents pictured later in this chapter. The only filing document found for this trademark is one dated July 31, 1948, long after it was first used in business. The filing document indicates this phrase was first used in business in 1923, but it was in fact used as early as 1921.

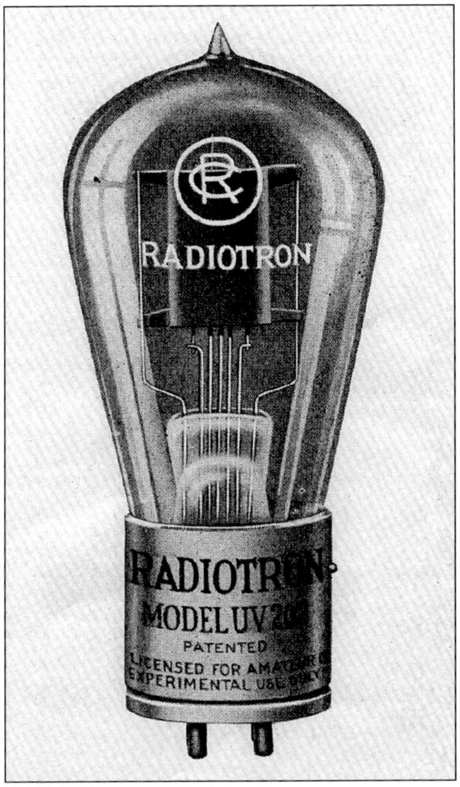

Shortly after April of 1921, RCA began to market its vacuum tubes with the new Radiotron trade name and the new RC monogram. (*Radio Apparatus for Amateur and Experimental Use*, RCA catalog, September 1921)

Product Trademarks

In addition to the globe and monograms, RCA also developed trademarks for selected products, most notably vacuum tubes and broadcast receiver apparatus. The trademark "Radiotron" for use with RCA vacuum tubes was filed on the same date as the RC monogram, April 28, 1921. Thereafter, vacuum tubes were described as Radiotrons, and the RC monogram was stamped on vacuum tubes as well as on a host of other radio-related devices.

The Radiola trademark was filed on October 29, 1921, and while the registration claimed the first usage in commerce was on September 27, 1921, the first commercial RCA apparatus bearing the name did not appear on the market until mid-1922. There were other manufacturers who used the Radiola name in late 1921 and early 1922 (see Appendix H), but use of the Radiola name by others quickly ceased in the United States after the trademark was officially registered on March 7, 1922.

The Beginnings of World Wide Wireless Communication

President Nally outlined his agenda for RCA in the January 1920 edition of *The Wireless Age* magazine:[6] *"The principal aim and purpose of the Radio Corporation will be the establishment and maintenance of transoceanic and long distance overland communication."* The plan to achieve this goal was to be implemented in two phases. The first phase was the reactivation and modernization of the transoceanic stations that RCA had inherited from Marconi in the merger. The second phase was the construction of a new communication center—Radio Central—on Long Island in proximity to New York City, which was to become the focal point for all RCA communications worldwide.

The RCA plan was a daunting one because at the outset RCA owned only five long distance stations, and the government still had possession of these stations. Worse yet, most of the equipment in these stations was obsolete and would have to be replaced. Finally, because most of the experienced Marconi employees had joined the service during the war, RCA had available only a skeleton crew with any experience in operating wireless communication systems.

Marconi Stations Reactivated

The initial circuits that RCA would reactivate were for the most part dictated by the five stations they inherited from Marconi as listed in Table 2-2.[7] Because of the limited range of pre-war stations, it was the general practice to

Table 2-2. The first five transoceanic stations RCA inherited from Marconi along with the corresponding overseas stations that formed each complete circuit

Circuit	RCA Stations in U.S. (Nov. 1919)			Corresponding Overseas Stations		
	Transmitter	Receiver	Separation (mi)	Transmitter	Receiver	Separation (mi)
New York to Marconi–GB	New Brunswick, NJ	Belmar, NJ	62	Carnarvon, Wales	Towyne, Wales	62
New York to Norway	Marion, MA	Chatham, MA	30	Stavanger,* Norway	Naerboe, Norway	30
San Francisco to Hawaii	Bolinas, CA	Marshall, CA	40	Koko Head, HI	Kahuku, HI	60
Hawaii to Japan	Kahuku, HI	Koko Head, HI	60	Funabashi, Japan	Funabashi, Japan	Single Station
New York to Germany	Tuckerton, NJ	Belmar, NJ**	43	Nauen, Germany	Nauen, Germany	

Note: Data taken from Elmer E. Bucher, *Practical Wireless Telegraphy*, (New York, N. Y., Wireless Press, 1921) pp. 288-307

*Station was located at Hinna, a suburb of Stavanger
**Receiver was co-located at Tuckerton prior to WWI

site stations in pairs to form a single circuit. The five U.S. stations RCA inherited from American Marconi were originally designed to operate with the five corresponding overseas stations shown in Table 2-2.

The first four U.S. stations listed in the table were owned by Marconi from the outset, while Tuckerton was originally owned by French and German interests. Tuckerton was turned over to the U.S. government as part of the war reparations agreements after the war, and subsequently sold to Marconi prior to the merger with RCA. To allow simultaneous transmission and reception (i.e., duplex operation) without interference, the general practice at the time was to locate the receivers 30-60 miles from the transmitter. The receiver for the Tuckerton station was originally co-located with the transmitter, but was moved to Belmar, NJ by the Navy during the war.

The radio frequency generators at these stations were obsolete except for the New Brunswick Station, which had been equipped with both a 50 kilowatt and a 200 kilowatt continuous wave Alexanderson alternator during the war at the request of the Navy. The subsequent transmissions from this site had conclusively demonstrated the efficacy and superiority of the Alexanderson technology to the whole world. Commissions and representatives from a number of European countries had visited this site with a view towards procuring alternators for their stations.

RCA began advertising trans-oceanic radio service to Great Britain at 17 cents a word on March 1, 1920, the same day the first circuit was opened. (*The New York Times*, March 1, 1920)

Communications Begin

By presidential order, the huge transoceanic stations were turned over to RCA at the stroke of midnight on February 29, 1920. The first circuits to open on March 1, 1920 were from New York to British Marconi in Great Britain, San Francisco to Hawaii, and Hawaii to Japan. Three messages were sent from the New Brunswick station to Great Britain with much fanfare, the first from Edward J. Nally addressed to Godfrey C. Isaacs, Managing Director of British Marconi. The three messages were reproduced in an article published in the

The huge Tuckerton, New Jersey station with an 850 foot tower was originally owned by French and German interests, but it was confiscated by the government and turned over to Marconi after the war. (George H. Clark Collection, Smithsonian Institution)

March 1920 issue of *The Wireless Age*,[8] where the following observation was made with regard to the first message:

> *"The message marked an epoch in wireless history. Never before has there been established direct communication between the United States and England. Before the war the United States communicated with England by way of Canada. During the war the only wireless communication with England was under government control."*

The circuit to Norway through the Marion Station was officially opened on May 17, and circuits to Germany and France through the Tuckerton Station were officially opened on August 1 and December 14 respectively. While the circuit from San Francisco was officially opened on March 1, the station at Bolinas did not actually commence operations until October. Prior to the October opening of Bolinas, message traffic for RCA on this circuit was handled by the Naval Communications Service.

In addition to reactivating communication circuits, RCA also immediately began the arduous task of replacing obsolete equipment in facilities both in the U.S. and abroad. By the end of 1921, GE had manufactured 21 of the 200kW Alexanderson alternators,[9] and by early 1922 had installed two in each of the seven stations shown in Table 2-3, along with new or modified antennas, grounding systems, couplers and receivers. According to the RCA Annual Report for 1922, the rehabilitation of all high-powered and marine stations was completed in 1922.

Radio Central Extends RCA Communication Capacity

At the first Annual Shareholder Meeting held on May 4, 1920, Edward J. Nally announced to the RCA shareholders that the Board of Directors of RCA had approved the purchase of a ten-square-mile tract of land at Rocky Point on Long Island approximately 70 miles from New York City for purposes of constructing a large centralized group of stations consisting of six units—five devoted to communication with Buenos Aires, France, Germany, Scandinavia, Italy and Poland, and a sixth to be held in reserve for emergencies or bad weather.[11] The huge complex was to become the world's largest radio station, known as Radio Central. Twelve antennae were proposed for the six transmission units, each antenna consisting of six 400-foot towers extending radially outward for 1.5 miles from a central power house at the hub. Each tower was to hold up 16 wires suspended from a 150-foot cross bar at the top, and each antenna was to be driven by a 200kW 18 KHz Alexanderson alternator located in the central power house.

Table 2-3. The seven stations in which two Alexanderson alternators had been installed by early 1922

Station	Wavelength (m)	Frequency (kHz)	Call Letters
New Brunswick, NJ	13,600	22.06	WII
Marion, MA	11,620	25.81	WSO
Tuckerton, NJ	16,900	17.75	WGG
Kahuku. HI	16,300	18.40	KIE
Bolinas, CA	13,300	22.56	KET
Carnarvon, Wales	14,100	21.27	MUU
St. James, New York*	16,465	18.22	WQK

Note: Data from *Commercial and Government Radio Stations of the United States* and *The Wireless Age*[10]

*Radio Central was listed as St. James, N. Y. in the government list of radio stations.

RCA purchased a ten-square-mile tract of land on long Island approximately 70 miles from New York City for the purpose of construction Radio Central, the world's largest communication station with twelve antennae extending radially outward from a central power house containing the Alexanderson alternators. The diagram below shows how each of the twelve planned radial spokes of the antennae at Radio Central was assigned to communicate with stations at the different geographical locations shown. (George H. Clark Collection, Smithsonian Institution)

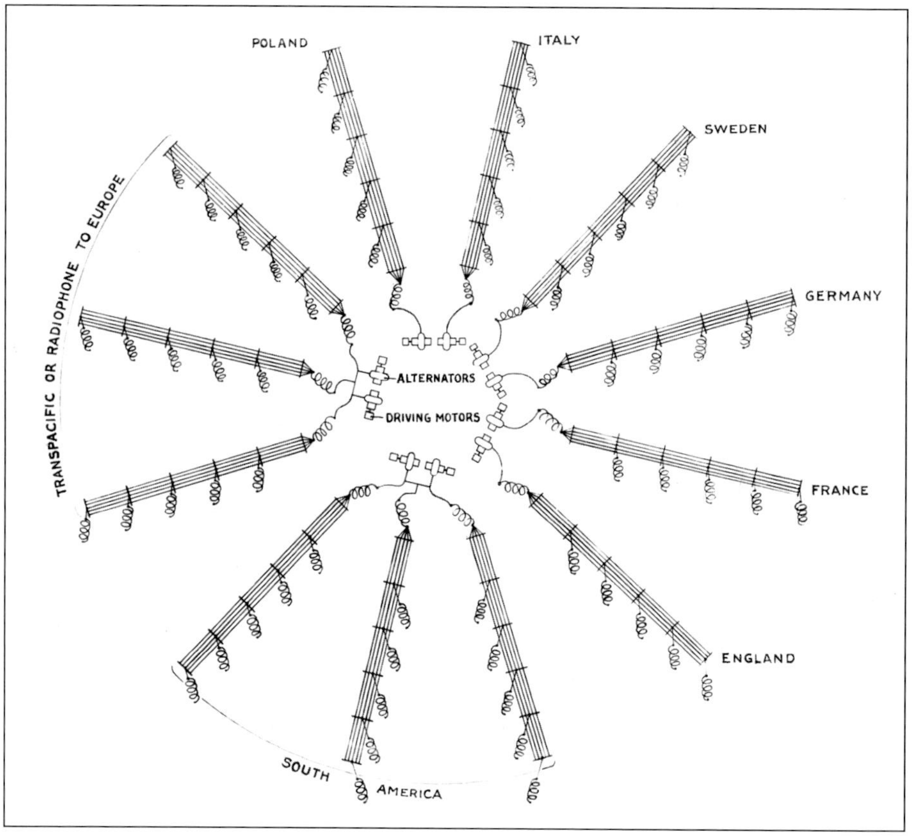

Six 400-foot towers in a radial spoke were spaced over a distance of 1.5 miles at Radio Central, standing as silent sentinels for transoceanic communications. (George H. Clark Collection, Smithsonian Institution)

Commercial communication transmissions were to be initiated at the Central Office of RCA located at 64 Broad Street in downtown New York City by using automated tape transmitters as well as ordinary hand keys. The messages would be sent over land lines from New York City to Radio Central where they would automatically operate relays to apply the power that would transmit the dots and dashes associated with Morse code used at that time for communications.

Simultaneous reception from multiple stations was to be accomplished with a single antenna located at Riverhead, L. I. sixteen miles away from the transmitter at Rocky Point, L.I. The antenna was to be nine miles (14 km) long, supported on ordinary thirty-foot telephone poles. The length of the antenna

Each barrage receiver at Riverhead, L. I. had three stages of RF amplification, a two-element synchronous detector, and two stages of audio amplification; the resulting audio output was transmitted over landlines to 34 Broad Street in New York City. (George H. Clark Collection, Smithsonian Institution)

was approximately one wavelength long in order to provide directionality (i.e., gain), a feature which was used to minimize interference from both transmissions at nearby Radio Central as well as static which plagued earlier transoceanic communications. These signals were to be automatically transferred over land-lines to the Central Office, with final delivery effected by messenger service from the Central Office or its three branch offices, two located on Broadway and one on Fifth Avenue.

The first phase of the project, consisting of two of the twelve antennae, was officially completed on November 5, 1921, the date Radio Central was opened with much fanfare by a message from the then President Warren G. Harding to the world.[12] The twelve huge towers forming the first two antennae were truly impressive, and RCA often referred to them as "silent sentinels." These two antennae at Radio Central, along with the existing RCA transoceanic stations, were fully capable of satisfying communication requirements at the time. Construction of the remaining antennae were planned to coincide with the construction of additional transoceanic stations in foreign countries to meet the expected growth in international radio traffic.

Riverhead began operations with only four receiving sets, although provisions were made for up to nine receivers for nine different stations. The four receivers were adequate not only for the two transmitters at Radio Central, but also for the other RCA transatlantic stations. The receivers were of the "barrage" type developed during the war to reduce interference from unwanted transmitters. Each receiver had three stages of RF amplification, a two-element synchronous detector, and two stages of audio amplification. The resulting audio output was transmitted over landlines to 34 Broad Street, where it was received either via telephone and decoded by individual operators, or by tape machine for higher rates with multiple operators decoding the tape.

Westinghouse Ambitions Thwarted

Westinghouse was a major competitor of GE in most electrical disciplines including wireless communication even before WWI. During the war, both had lucrative wireless contracts with the government, and both were looking for post-war markets for their respective wireless products developed during the war. GE had developed the world-wide-wireless strategy as a result of the merger with Marconi and the creation of RCA as a surrogate to market their Alexanderson alternators, but Westinghouse had not yet formulated a strategy to sell receivers, their area of expertise developed on government contracts during the war.

As Westinghouse executives cast about for ideas, they found their options to be extremely limited. In the end, they decided their best strategy was to enter into direct competition with GE as an "also ran" in the transoceanic wireless race, but they did not have access to all the necessary technologies to build a competitive communication system. First, they tried to purchase rights to the patent portfolio of International Radio Telegraph Company, successor to the respected National Electric Signaling Company (NESCO), which held patents on the rotary spark gap and Armstrong superheterodyne receiver, among many others. When they were rebuffed, Westinghouse changed their strategy by borrowing a page from GE, and offered financial backing and a cross-licensing arrangement patterned on the GE-RCA model to create an organization to market transoceanic communication systems in direct competition with RCA.

The two companies formed a third company on May 22, 1920 in which Westinghouse gained controlling interest in return for a $2.5M investment. In order to retain name recognition and good will of the existing company, the new company was named "The International Radio Telegraph Company", which differed from the existing company only by the word "The," a legal finesse that would have delighted the most discriminating attorney. Thereafter, the initial "The" was used not only in corporate documents but also on their receiving equipment (see figure below). Following the pattern set by the GE-RCA arrangement, Westinghouse assumed responsibility for the manufacture of wireless equipment while International Radio assumed responsibility for marketing and wireless operations.

It was unlikely that this pact caused much alarm in the executive offices of GE or RCA, because while Westinghouse was a worthy competitor for GE, International Radio Telegraph was no match for RCA. International Radio had experience with coastal shipping communications and owned a handful of coastal stations in the northeast, but they did not have the experience, the transmitter technology, the hardware, or the overseas business connections needed to effect transoceanic communications. In stark contrast, RCA had an experienced staff from American Marconi, five operational transoceanic

The name change from International Radio Telegraph Co. to The International Radio Telegraph Co. occurring when Westinghouse gained control of the radio assets of the former company is reflected on the names engraved at the top of two otherwise identical heterodyne receivers—one made by the original company before the acquisition and one made by the successor company afterwards. (John Jenkins, Sparkmuseum)

Table 2-4. The 37 transoceanic communication circuits opened by RCA in the 1920's

Circuit	Opening Date
Hawaii	Mar. 1, 1920
Japan	Mar. 1, 1920
Hawaii-Japan	Mar. 1, 1920
Great Britain	Mar. 1, 1920
Norway	May 17, 1920
Germany	Aug. 1, 1920
France	Dec. 14, 1920
Italy	Aug. 10, 1920
Poland	Oct. 4, 1923
Argentina	Jan. 25, 1924
Sweden	Dec. 1, 1924
Dutch East Indies (Java)	Jul. 16, 1925
Brazil	May 3, 1926
French Indo-China	Sep. 15, 1926
Holland	Nov. 1, 1926
D. West In. (St. Martin)	Jun. 21, 1927
Philippines	Jun. 27, 1927
Dutch Guiana	Aug. 9, 1927
Columbia	Aug. 12, 1927
Venezuela (Maracay)	Aug. 18, 1927
Belgium	Oct. 3, 1927
Puerto Rico	Oct. 10, 1927
Hong Cong via Manila	Oct. 18,1927
Turkey	Dec. 10, 1927
Venezuela (Maracaibo)	Jan. 22, 1928
Shanghai via Manila	Feb 21, 1928
Canada (Montreal)	Mar. 10, 1928
Portugal	Apr. 2, 1928
Australia via Montreal	Jun. 15, 1928
D. West In. (Curacao)	Aug. 4, 1928
Liberia	Sept. 1, 1928
Cuba	Dec. 4, 1928
Fiji	May 1, 1929
Spain	Aug. 1, 1929
Puerto Rico	Aug. 12, 1929
Costa Rica	Sept. 3, 1929
Syria	Sept. 3, 1929

Note: Data from the RCA Annual Reports for the years 1924 through 1929

stations, the Alexanderson alternator—which was believed to be the most efficient long-range transmission technology at the time—and most importantly, exclusive agreements with virtually all the major countries abroad with interest in transoceanic communication with the U.S.

Apparently neither Westinghouse nor International Radio was aware of the traffic agreements that RCA President Nally had laboriously negotiated over the previous six months until Samuel M. Kintner, president of International Radio, traveled to Europe in early summer of 1920 in an attempt to line up partners for the new enterprise.[13] He was sorely disappointed to find he had been preempted by RCA, who had already arranged exclusive agreements for transoceanic message traffic. International Radio was forced to look for niche domestic markets, despite the international flavor implied by the corporate name. It is surprising that a corporation such as Westinghouse would commit the $2.5M that it did to an enterprise with a business plan that focused on transoceanic communications without first determining if there was a viable market share available to the enterprise.

RCA succeeded in thwarting Westinghouse ambitions to compete head-to-head with RCA in the point-to-point communications, and by the end of the decade RCA dominated this market. RCA opened approximately 37 transoceanic circuits in the 1920's, and innumerable marine stations both on land and aboard ships, as well. No other American company was a match for RCA, with its capital resources, patent position, technical expertise, and the will to dominate the world-wide communication business. RCA was obviously proud of this accomplishment because they listed the opening dates of these circuits at the front of their Annual Report each year from 1924 through 1929 (see Table 2-4).

Messengers and the Messages

The wireless transmission and reception facilities were obviously crucial to the transoceanic communication business, but equally so was the retail distribution system for the messages at either end of the circuit. In this regard, RCA was at a great disadvantage in comparison to its cable company competitors, who over many years had opened offices in cities and towns across the U.S., and in fact all over the world. The two major cable companies in the U.S.—Western Union and the Postal Telegraph-Cable Company—employed many thousands of uniformed messenger boys to pick up and deliver messages.

RCA's strategy was to locate its message collection and distribution offices in a few key cities in the United States and leverage off the existing infrastructure of the cable companies for distribution to other cities in the U.S. and abroad. RCA was able to make an agreement with Western Union to distribute messages between the U.S. and the Pacific region because Western Union did not have a competing cable in the Pacific. One could go to any Western Union office to place a radiogram to the Pacific via RCA. However, Western Union would not agree to a similar arrangement for messages to Europe and Africa, because they had competing cable service to this region. Thus, for several years RCA could receive messages from Europe via wireless and distribute them in the U.S. over Western Union or Postal Telegraph-Cable Company cable lines just like any other cable customer, but neither cable company would accept messages at their local offices destined for transmission to Europe on RCA wireless links.

Finally, in September of 1922, RCA was able to negotiate an arrangement with Postal Telegraph to handle RCA messages destined for Europe.[14] After September of 1922, RCA advertised widely that messages to the Pacific could be placed at Western Union offices, and messages to Europe and Africa could be placed at Postal Telegraph offices. It is not clear why it took so long to negotiate such an arrangement with Postal Telegraph because they did not have competing transoceanic cable service.

Soon after RCA negotiated arrangements with the Postal Telegraph-Cable Company to send and receive RCA radiograms, RCA placed numerous large ads in newspapers and magazines such as this one from the *Saturday Evening Post* to advertise the new arrangement. (*Saturday Evening Post*, November 18, 1922)

THE SATURDAY EVENING POST
75
"Via RCA"
RADIO CORPORATION OF AMERICA
233 BROADWAY NEW YORK
"Via RCA"
How to send your Radiograms
Write your message for foreign countries or ships at sea on a Red Radiogram blank if available. If not, use the Postal Telegram blank.
After the address insert the route "Via R C A." No charge will be made for this insertion and it will insure your message going forward as a Radiogram at radio rates.
Use the same registered addresses and code books you now use. The regulations covering Radiograms and Cablegrams are the same.
Ring the Postal call-box for a messenger; telephone your Radiogram or, if more convenient, file it direct at the nearest R C A or Postal Office.
Radiogram tariff books and pads of Red Radiogram blanks may be obtained at any Postal Telegraph or at any R C A Office. Ask for them.
NOTE: In New York, San Francisco and Washington, the Radio Corporation of America will continue as heretofore to maintain its own collection and distribution system.
—Arc de Triomphe
—Near East
Paris 'round the Corner
HOW far away seemed Paris in the old days! Paris with its Boul' Miche, the heart of Montmartre, where the sun puts out the café lights in the early Paris morning; with its Luxembourg, where the Three Musketeers tested their blades against the Cardinal's men; with its streets where Marie Antoinette took her last tragic ride, and with its Arc de Triomphe standing nobly under the brilliant morning light, the symbol of Napoleon's dreams in the height of his splendor. Not only the Paris that our soldiers came to know, but London, Berlin, Christiania—all of Europe, Africa, India and the Near East—are brought nearer to the people of the United States through the arrangement recently effected by the Radio Corporation of America, whose sending and receiving stations are the most powerful in the world, with the Postal Telegraph-Cable Company, giving the American public everywhere land line access to World-Wide Wireless.
WHEN President Harding sent a Radiogram of good-will from Radio Central at Rocky Point, Long Island — newest and most powerful radio station in the world, constructed and operated by the Radio Corporation of America — twenty-one foreign countries received and recorded it instantaneously.
That Radiogram made radio history. It signalized America's complete participation in the field of international communication — a field where the national interests of the United States hitherto had been subordinate to the interests of foreign countries. It meant that no foreign nation ever again could interfere with America's free intercommunication with the rest of the world.
Radio Corporation of America
WORLD WIDE WIRELESS
233 Broadway, New York
World-Wide Wireless

Actually, it is curious that either cable company would negotiate any type of arrangement for the convenience of RCA, given that RCA was in direct competition with their core business. However, competition between cable and wireless communication companies was restrained for both economic and technological reasons. First, the rapid increase in long-distance communication traffic was sufficient to support both RCA and the transoceanic cable companies. Second, the overseas cables were operating at or near capacity at that time, and could not have easily handled the additional transoceanic business that RCA was able to generate with it lower rates. Third, the cable companies generated additional revenues by distributing RCA messages on the less-congested domestic cables outside the regions served directly by RCA.

RCA customers could arrange for message pickup by a uniformed messenger, who in turn delivered them to the main RCA office in selected cities such as New York. (George H. Clark Collection, Smithsonian Institution)

The Messengers

Taking a page from the cable companies, RCA hired and trained a cadre of uniformed messenger boys who were dressed in brown uniforms with red trim. However, unlike their cable competitors who provided a similar service in hundreds of cities, RCA initiated messenger service in New York City and later extended it to only a few other large cities including Washington, D.C. and San Francisco. Advertisements of the day in cities served by messengers instructed users to either file messages at the nearest RCA office or call for message pickup by a uniformed messenger, while potential customers in other cities were advised to file messages at Western Union offices, and later at Postal Telegraph offices as well.

The Message Rates

RCA's pricing strategy was to set the rates at a level of 30-40% less than cable rates.[15] RCA was able to do this because it cost approximately \$2M to build a long-wave station versus \$4M to lay a transoceanic cable. Virtually all published rate sheets listed not only the cost per word but also the savings, and while not explicitly stated, the savings were in comparison to cable rates. One RCA advertisement pointed out that *"a business house that sends only six twenty word messages a day to Germany saves over \$1800 a year by using Radio."*

RCA hired cadres of messenger boys dressed in brown uniforms with red trim, first in New York City and later in Washington, D.C and San Francisco. (George H. Clark Collection, Smithsonian Institution)

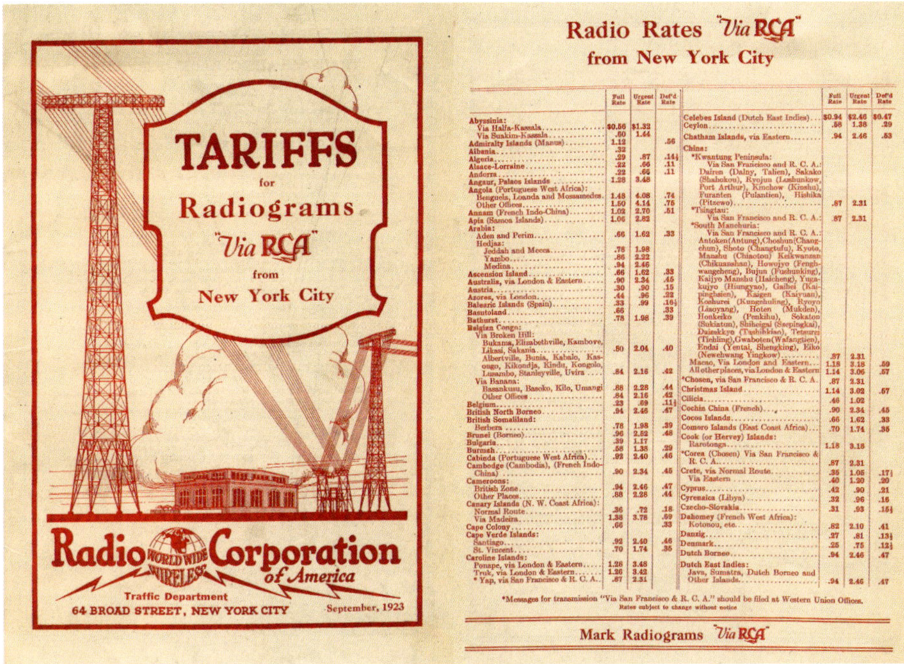

RCA rate sheets of the day appeared somewhat like international airline schedules of today, not only with long lists of rates for numerous far-off and exotic places, but also with different rates for different classes of service. (George H. Clark Collection, Smithsonian Institution)

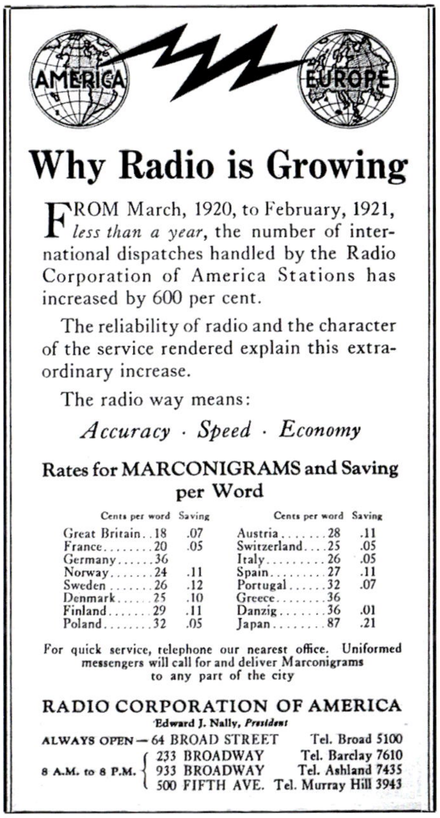

RCA advertisements for radiograms emphasized cost per word as compared implicitly to rates charged by cable companies.

Table 2-5. Annual RCA revenues generated from transoceanic communications for years 1920 through 1928

Year	Revenue from Transoceanic Communications
1920	$0.90M
1921	$2.14M
1922	$2.91M
1923	$3.19M
1924	$3.36M
1925	$3.41M
1926	$3.60M
1927	$3.93M
1928	$4.59M

Note: Data from RCA Annual Reports from 1920 through 1928

The rate sheets of the time appeared somewhat like international airline schedules of today—one far-off and exotic destination after another with a smorgasbord of rates akin to economy, coach and first class. By 1923, there were basically four rates: the urgent rate, the full or ordinary rate, the deferred rate, and the "radioletter" ratc. Messages were generally sent in that order with the radioletter having the lowest priority. The urgent rate was 2-3 times the full rate, and the deferred rate was about half the full rate. The radioletter for most destinations outside of the receiving city were forwarded to the final destination by mail.

The RCA business model for transoceanic communication was compelling, at least on paper. At a transmission speed of 100 words per minute,[16] RCA could transmit a maximum of 72,000 words per day over each link—assuming the link was down for as many as 12 hours a day for maintenance and experiments! At a charge of $0.25 per word, each link could theoretically generate $6.57M in revenue annually, amounting to a whopping $33M for all five links. However, actual revenues from transoceanic communications during the twenties were much less than that, as shown in Table 2-5. The reason that RCA never generated revenues anywhere near the theoretical potential was the result of a capacity that far exceeded demand.

While demand never approached capacity, still RCA's revenues increased rapidly in thc first few years after the war. By 1922, RCA captured 20-30% of the total transatlantic traffic and 40-50% of the transpacific traffic, with the cable companies carrying the rest.[17] The cable companies also experienced growth as a result of the explosive demand for transoceanic communication immediately following the war, although RCA received a disproportionate share of this new business due to both their lower rates and congestion on overseas cables.

The Messages

In American Marconi days, the messages were transmitted and received with the aid of message forms such as the first Marconigram pictured here. Outgoing messages would be written on these forms by customers or company clerks, and then passed to the radio operator from transmission. Incoming messages would be written or typed on these forms by the radio operator and sent on to the recipient.

The forms used by RCA to transmit and receive messages are interesting from an historical perspective because they provide direct evidence of the obvious, such as the evolution of RCA trademarks and monograms, as well as indirect evidence of the less obvious—corporate political intrigue at the highest level. Correspondence between Edward Nally and RCA historian George H. Clark clearly indicates that Nally took a personal interest in these forms, and was responsible for the content and the many changes thereto over his three year tenure as president of RCA.

Nally clearly used an American Marconi message blank as a basis for the design of the first RCA message blank, Form No. 1, believed to be the first form of this type used by RCA. The most notable feature of the early RCA form is that it retained the American Marconi signature title "MARCONIGRAM"—despite the fact that RCA was no longer affiliated with Marconi, and that the

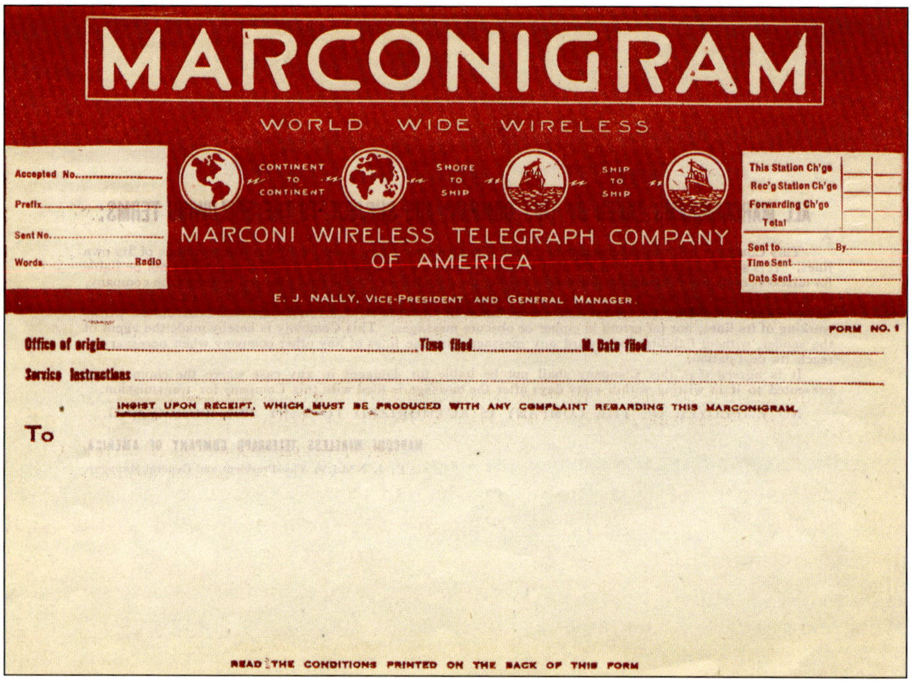

MARCONIGRAM

WORLD WIDE WIRELESS

Accepted No.
Prefix
Sent No.
Words ... Radio

CONTINENT TO CONTINENT — SHORE TO SHIP — SHIP TO SHIP

This Station Ch'ge
Rec'g Station Ch'ge
Forwarding Ch'ge
Total
Sent to ... By ...
Time Sent
Date Sent

MARCONI WIRELESS TELEGRAPH COMPANY OF AMERICA

E. J. NALLY, VICE-PRESIDENT AND GENERAL MANAGER.

FORM NO. 1

Office of origin ... Time filed ... M. Date filed

Service Instructions

INSIST UPON RECEIPT, WHICH MUST BE PRODUCED WITH ANY COMPLAINT REGARDING THIS MARCONIGRAM.

To

READ THE CONDITIONS PRINTED ON THE BACK OF THIS FORM

Marconigram forms developed by General Manager Nally at American Marconi formed the basis for future RCA forms.

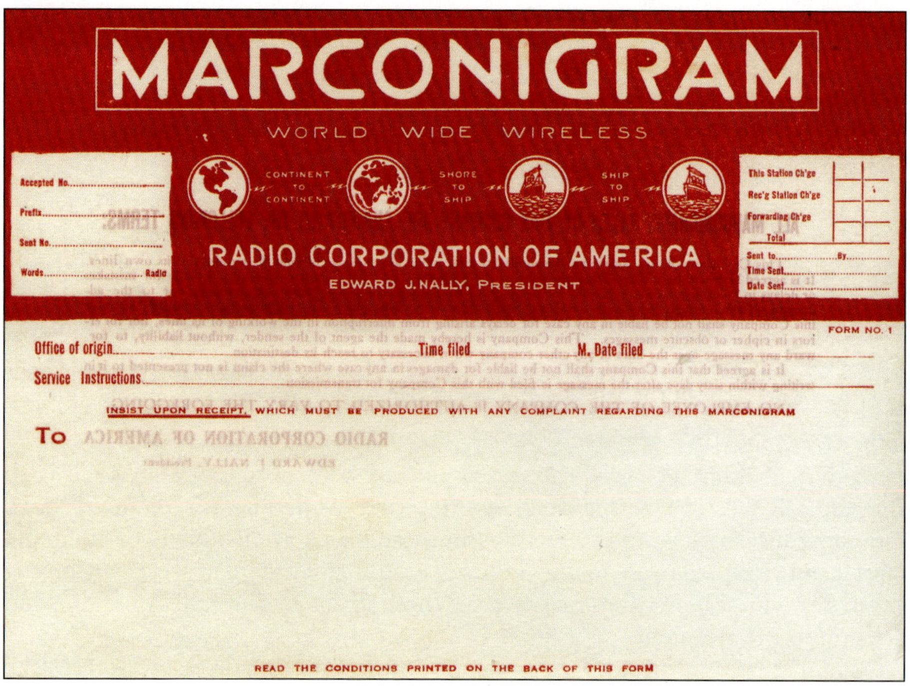

MARCONIGRAM

WORLD WIDE WIRELESS

Accepted No.
Prefix
Sent No.
Words ... Radio

CONTINENT TO CONTINENT — SHORE TO SHIP — SHIP TO SHIP

This Station Ch'ge
Rec'g Station Ch'ge
Forwarding Ch'ge
Total
Sent to ... By ...
Time Sent
Date Sent

RADIO CORPORATION OF AMERICA

EDWARD J. NALLY, PRESIDENT

FORM NO. 1

Office of origin ... Time filed ... M. Date filed

Service Instructions

INSIST UPON RECEIPT, WHICH MUST BE PRODUCED WITH ANY COMPLAINT REGARDING THIS MARCONIGRAM

To

READ THE CONDITIONS PRINTED ON THE BACK OF THIS FORM

The first RCA message blank was almost identical to the last message blank used by American Marconi, the only significant difference being the name of the company.

Marconigram title was still used by British Marconi on their forms. The RCA form also retained the Marconi red color as well as the familiar circles bracketing the slogans designed by Nally during his tenure at American Marconi. At first glance, the new RCA forms looked just like the old Marconi forms, an appearance that Nally obviously intended, and one which appeared to be a great source of pride to him.

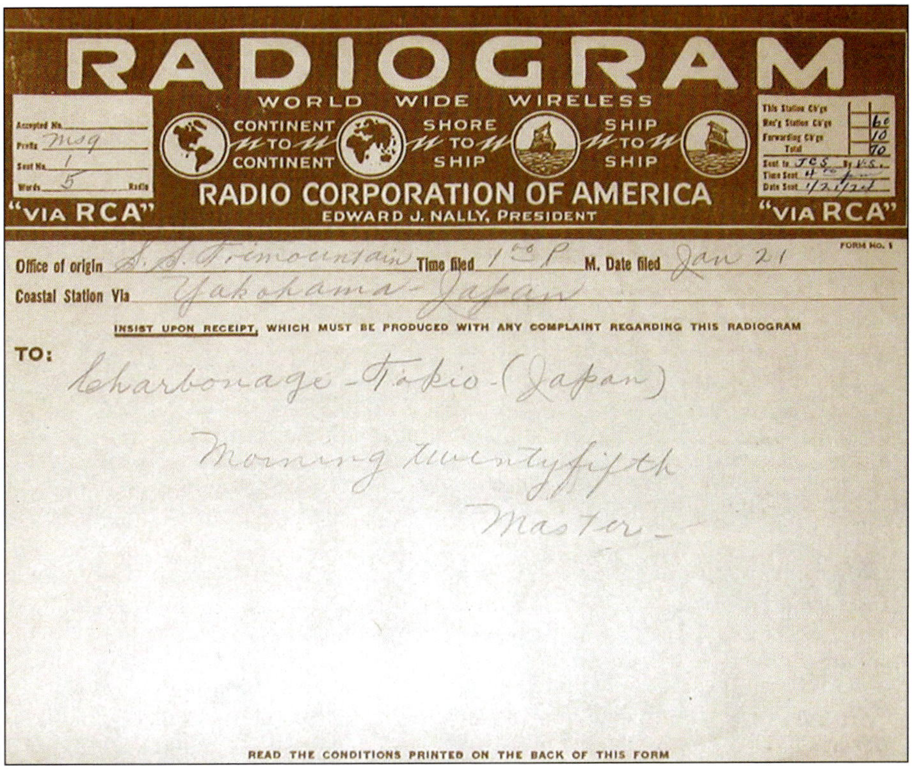

Nally was forced to change the title on RCA message blanks from Marconigram to Radiogram and the color from Marconi red to light brown; the phrase "Via RCA" was also added to assure messages were routed through the RCA system. (George H. Clark Collection, Smithsonian Institution)

The color of the message blank returned to Marconi red, the lightening bolts were added to letters in the "Via RCA," and the president's name was removed from the front.

FAST DIRECT

RCA

CLASS OF SERVICE DESIRED	
FOREIGN	DOMESTIC
FULL RATE	FULL RATE
CDE	DAY LETTER
URGENT	SERIAL SERVICE
DEFERRED	NIGHT MESSAGE
RADIO LETTER	NIGHT LETTER

NO.	CASH OR CHG.
NUMBER OF WORDS	
TIME FILED	

RADIOGRAM

R.C.A. COMMUNICATIONS, INC.

A RADIO CORPORATION OF AMERICA SERVICE

TO ALL THE WORLD — BETWEEN IMPORTANT U. S. CITIES — TO SHIPS AT SEA

Send the following Radiogram "Via RCA" subject to terms on back hereof, which are hereby agreed to

109 Congress Street Boston Phone: Liberty 8864

FULL-RATE MESSAGE UNLESS MARKED OTHERWISE

Sender's Name and Address

(Not to be transmitted)

Beginning in 1929, message blanks bore the corporate title "R.C.A. Communications, Inc." to reflect the creation of a communication services subsidiary in early 1929, quietly signaling that point-to-point communications had become secondary to broadcast radio.

Powerful sources at the highest levels within RCA took issue with this form and ultimately forced Nally to change its title and color. In a letter to RCA historian George H. Clark in later years, Nally reveals:[18]

> *"The 'Marconi' Red was my suggestion. It was changed to a dark color by Navy bureaucrats, and the same meddlesome force insisted upon dropping the name Marconigram, and I had to order the destruction of numbers of forms, sent & recd."*

The "meddlesome force" in Nally's letter almost surely referred to Admiral W. H. G. Bullard, who regularly represented the government at meetings of the Board of Directors throughout the twenties and into the thirties. Clearly, the enmity that existed between Navy officials and Nally during his tenure at American Marconi did not end with the creation of RCA, which is not surprising because these problems went far beyond the issue of foreign control or influence—the key issue used by the Navy as a subterfuge in a failed attempt to create a government monopoly in wireless.

In July 1921, Nally announced the change from the prominent "MARCONIGRAM" heading to "RADIOGRAM," and the form was printed in a brown hue. It is likely that the brown hue was forced on Nally not only to create a contrast to the red of the Marconigrams, but also to tie in with the brown uniforms of the RCA messenger boys. The phrase "Via RCA" was added—perhaps to further distinguish it from a Marconigram, but more likely to assure that the radiogram, a more generic term than Marconigram, was transmitted by RCA rather than by another competing radio or cable service.

Soon thereafter, the form was once again printed in a bright red, which then became a signature color for RCA. Nally ordered the addition of zigzag flashes to the letters "V" and "A" in "Via RCA" on the new forms, thus beginning the evolution of the now-ubiquitous RCA monogram enclosed in a circle with a flash attached to the A. The president's name and title disappeared from the front of the form, and instead was moved to the rear. Forms printed after 1922 bore the name of James G. Harbord, who became the new president of RCA on Jan 1, 1923 following Nally's resignation.

Further changes in these forms were unremarkable from an historical viewpoint until 1929, when R.C.A. Communications, Inc. was created as a subsidiary to handle transoceanic communications.[20] For many decades thereafter, the subsidiary name "R.C.A. Communications, Inc." appeared prominently in the heading on transoceanic communication forms, replacing Radio Corporation of America. The relegation of the transoceanic communications business to subsidiary status reflected the fact that broadcast radio had come to dominate the RCA agenda, and that transoceanic communications produced a small fraction of the total RCA revenues. By 1928 the annual revenues from transoceanic communications were only $4.6M, while total RCA revenues from all sources were $101.8M.[21]

Marine Communications

As mentioned in the previous chapter, American Marconi was coerced to sell its coastal stations to the Navy, a transaction which took place shortly before the merger with RCA. Undaunted by the lack of coastal stations at the time of the Marconi merger, RCA initiated ship-to-shore service in 1920 utilizing the existing high-power stations. RCA was able to obtain contracts for a large number of ship stations in the first few years following the war for two reasons. First, there was a large demand for radio from the merchant marine owners as ships transitioned from military to private control after the war. This demand was accelerated by owners' dissatisfaction with radio service and equipment provided by the Navy during and immediately following the war. Second, RCA achieved a virtual monopoly on the distribution of vacuum tubes for radio communications as a result of cross-licensing agreements signed among RCA, AT&T, Western Electric, and GE on July 1, 1920. RCA then refused to sell tubes to companies in competition with them in the field of radio communications for the merchant marine, giving RCA a huge edge in this business. This policy rekindled hostility between Navy officials and top RCA managers, the same managers who, in fact, pursued similar restrictive policies at American Marconi. It was ironic that the Navy urged RCA, GE and AT&T to enter into an arrangement whereby vacuum tube patents could be shared so that tubes could finally be produced and supplied to the merchant marine without fear of infringement litigation, and as a result of the agreement, the tubes were then denied to much of the merchant marine.

RCA changed this policy as a result of pressure applied by the Navy to both RCA and GE. The change in policy was announced in the August 1922 issue of *The Wireless Age* magazine,[22] gratuitously characterizing the change as a response to "the call of humanitarianism." However, the restrictions on the use of the vacuum tubes were still sufficiently onerous that any competitor agreeing to the licensing terms was at a severe disadvantage. First, the tubes were offered at $10 each for commercial use versus the $5-6 retail price for Radiotron receiver tubes sold to amateurs and experimentalists at the time. Second, while licensing agreements allowed the use of RCA tubes on ships, they did not allow for their use on shore equipment–precisely where they were needed the most.[23]

The immediate success of marine communications in 1920 led RCA to establish a separate Marine Department in 1921. The scope of the RCA radiomarine services in the twenties–as reflected by the number of ships under contract, shore stations in operation, and revenues from marine services–is shown in Table 2-6.

Table 2-6. RCA radiomarine activity from 1920 through 1929 showing a marked increase in revenues as a result of the merger with the Independent Wireless Telegraph Company on January 1, 1928

Year	No. Ships	No. Stations	Revenues
1920	>700*	2	~$553,000
1921	~705**	6	$553,298
1922	705	6	$630,084
1923	759	5	$738,140
1924	700	5	$742,345
1925	729	6	$735,179
1926	822	8	$828,167
1927	844	10	$990,954
1928	1360	14	$1,536,544
1929	1395	15	not reported

Note: Data from RCA annual Reports from 1920 through 1929

* Deduced from 1921 statement that 1921 results represented a decrease from 1920

** Deduced from 1922 statement that 705 ships represented a very slight increase over 1921

In 1921, RCA acquired the four coastal stations and a number of ship stations of The International Radio Telegraph Company.[24] The purchase of these stations largely offset a significant reduction in existing RCA ship stations resulting from a depression in the shipping industry that year. Business remained relatively constant in the intervening years until December 30, 1927 when RCA purchased the assets of the Independent Wireless

Telegraph Company.[25] As a result, RCA increased the number of ships under contract from 844 in 1927 to 1360 ships by the end of 1928, and the number of shore stations increased from 10 to 14.

Despite this success, marine operations produced only a small fraction of the total corporate revenue over the decade—less than 1% in 1927, the year marine operations were relegated to RCA Radiomarine, Inc., a wholly-owned subsidiary created for that purpose.[26] While the marine business may not have been significant in terms of revenue, it was important to RCA in terms of public relations, influence and prestige. It also served as a springboard to other markets, most notably broadcast radio. In this regard, the first receivers RCA offered for sale to the public for radio broadcast reception were receivers designed and used for marine service.

Marine Receivers

Marine receivers manufactured by Marconi, GE, and Wireless Specialty in the late teens and early twenties are of particular interest to the subject of this book not only because a number of them were sold by RCA for marine and broadcast use, but also because some of the early broadcast receiver designs were based on marine receivers. The marine receivers manufactured by Marconi and GE for use by RCA—but never marketed to the public for broadcast radio—are addressed here. The early broadcast receivers designed by GE were clearly adaptations of one or more of these marine receivers.

None of the broadcast receivers designed by either Westinghouse or Wireless Specialty for RCA was based on marine designs, but certain Wireless Specialty marine receivers including the IP-500 and IP-501 were, in fact, sold by RCA for use in broadcast reception. These receivers and associated accessories are addressed in next chapter covering the earliest receivers marketed by RCA for broadcast reception.

Marconi Type 106 Receiver and Variants

In the eight year period from 1912 to 1919, American Marconi designed 22 receivers numbered consecutively from 101 to 122, although not all models were put into production. Most, if not all, the sets with numbers above 112 were designed for government applications. For example, the 113 was adopted by the Navy and became the CM-294 with variants A, B, and C. Marconi manufactured the CM-294 up to the time of the RCA merger, and GE continued to manufacture this set for sale by RCA to the Navy for several years after the merger.

The 101, the 103 and the 106 receivers were the most successful models in the commercial market, and therefore are the best known today. The Marconi 106 and its variants B, C, and D were produced in much larger numbers than any other receiver produced by Marconi, and perhaps by any other manufacturer at the time. The Navy's SE-143 receiver may have been more plentiful, but it was made by several manufacturers.

Production of the 106 started in early 1915 at the Marconi Aldene plant in New Jersey, and continued there until production was transferred to GE Schenectady in mid-1920. GE continued production of the 106B—the variant in production by Marconi at the time of the merger with RCA—and subsequently designed and manufactured the 106C receiver and 106D model upgrades that were sold by RCA for marine use in succeeding years.

Marconi Type 106 Receivers

The Marconi 106 receiver is basically a crystal set of the loose coupler type in which tuning was accomplished by sliding one large coil within a second coil, supplemented by adjusting two variable capacitors. It operated on wavelengths of 200 to 3500 meters—a range which incidentally covered the band of 200 to 600 meters used in early broadcast radio. This range was too short for use in transoceanic communications, however, where wave lengths ranged from 14,000 to 23,000 meters.

The earliest version of the Marconi 106 can be identified by the nickel-plated brass buzzer cover which is held in place by two screws clearly visible on the face of the housing. (AWA Museum)

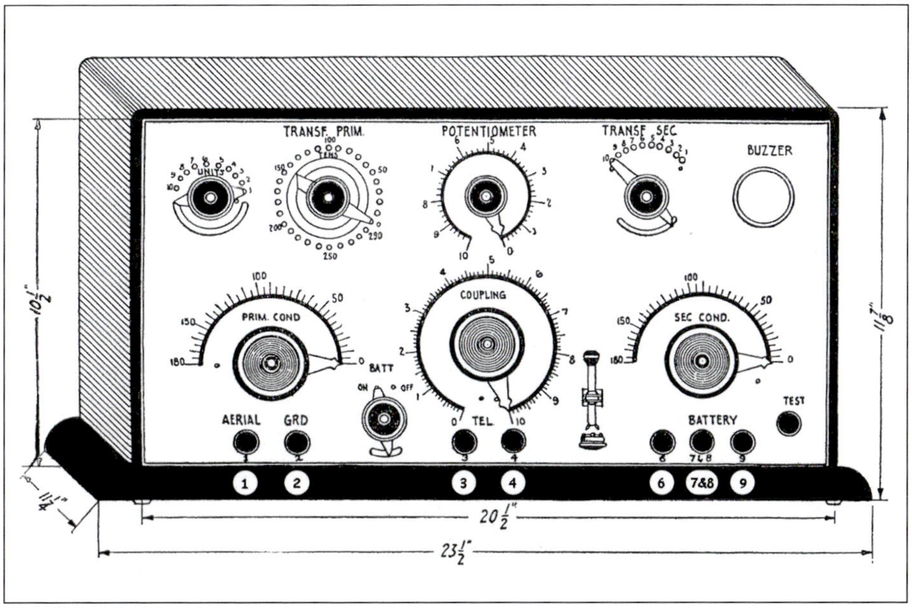

The numbering system for the binding posts on the early sets is quite confusing with the crystal detector cup assigned #5 and the single common battery terminal assigned both #7 & #8. (modified figure from *Electrical Experimenter* Vol. V, No. 2, June 1917, p. 112)

The 106 receiver was supplied with two types of detectors—cerusite crystal and carborundum. Cerusite crystal was perhaps more sensitive than carborundum, but carborundum was more permanent in its adjustment, and was not disturbed as much by vibration, ship motion, static, and loud signals. The detector material appropriate to the existing conditions was selected and inserted in the single crystal holder mounted vertically on the front panel.

The receiver also had a buzzer circuit to assist in selecting the most sensitive position of the cat's whisker on the detector surface. A button on the front surface of the panel at the lower right closed a circuit with an external battery, causing the buzzer located at the upper right of the panel to generate RF energy, which was coupled into the RF portion of the receiver circuit and rectified by the detector. The resulting tone could be heard in the earphones, thus allowing the operator to adjust the detector for maximum sensitivity even without the presence of a radio signal. This feature was particularly important for ship board applications where radio signals were often few and far between.

The Marconi 106 sets manufactured during WWI used etched dial plates in lieu of panel engraving to characterize controls on the front panel. (Jim Kreuzer collection)

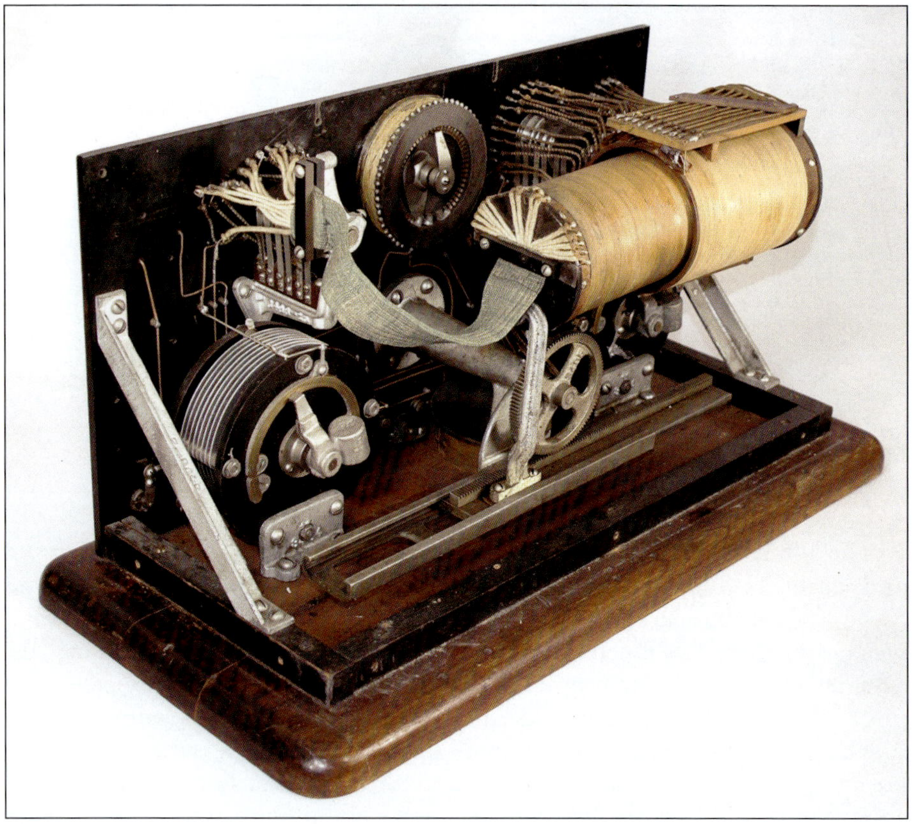

The Marconi 106 was of the loose-coupler type in which tuning was accomplished by sliding one large coil within a second coil, supplemented by adjusting two variable capacitors.

The earliest version of the 106 receiver can be easily identified by the nickel-plated brass buzzer cover which is held in place with two screws clearly visible on the face of the housing. This set has seven binding posts along the lower portion of the front panel-four posts positioned in two pairs to the left of the crystal detector and three posts equally spaced to the right as shown in the

All of the Marconi 106 receivers had a Marconi nameplate affixed to the top of the cabinet similar to the one pictured here, while a smaller number of sets dating from 1915-1916 to November 1917 had a second plate referencing a license with the National Electric Signaling Company as shown here.

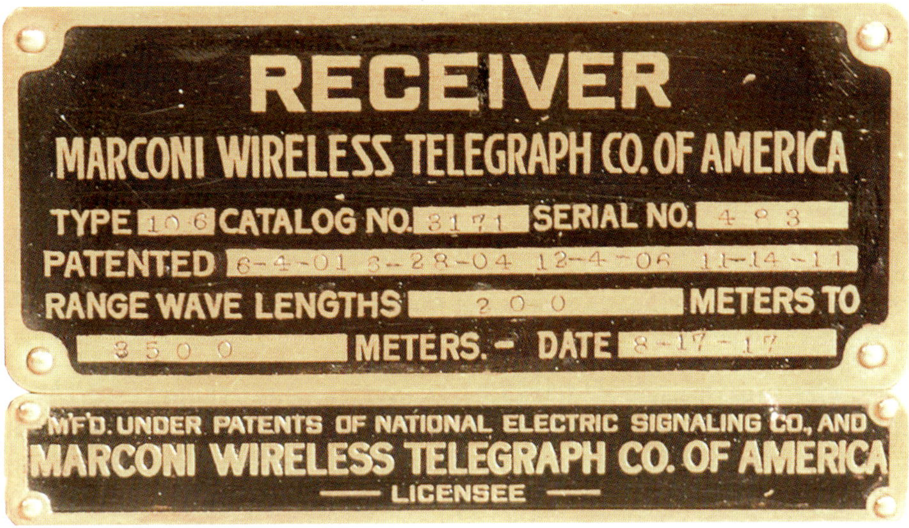

figure. The first four posts were assigned the numbers 1 through 4, but the bottom of the detector—not the next binding post—was assigned the number 5. The three posts to the right of the detector were assigned numbers 6, 7 & 8, and 9 respectively. The middle post was assigned the two numbers 7 & 8 because it was the common battery connection for both the buzzer circuit and the crystal detector bias circuit.

The receiver was soon modified in two relatively minor ways, both of which are easily recognizable with a quick glance at the front panel. First, an additional binding post was added to the right side of the detector for the buzzer battery so that the operator could reverse bias the carborundum detector without disturbing the buzzer battery connections (see page 50). The four binding posts on the right side of the detector were then numbered 6 through 9 respectively.[27] Second, the nickel-plated buzzer cover with screw fasteners was replaced by a black cover that could be removed with a simple twist, thus allowing rapid access to the buzzer for adjustments. Drawings and documents dated June and August 1915 reflect these new design features.[28]

The next major change to the front panel occurred as a result of WWI. Before WWI, the labels for the various dials and connectors on the face of the 106 were engraved in the panel, as they had been on predecessor sets including the 101 and 103. However, during the war the engraving equipment was used on war contracts with the government, so etched metal plates and dials were substituted to designate the various panel control functions.[29] An example of a 106 made during war with the etched dial plates is shown on page 50. Note the four binding posts on the right and the black buzzer cover referenced in the previous paragraph. The metal plates came in two versions—one with a brass appearance, and one with a black and white appearance—both of which are shown together in the accompanying figures to the left.

A drawing bearing a number 3901-A dated August 5, 1918 shows a revision to the 106 in which the wire from one end of the detector potentiometer is moved to a center tap added to the potentiometer. This change—which is not obvious by inspection from the outside—would have allowed the operator to reverse bias the detector by adjusting the potentiometer.

There was at least one more obvious modification of the 106 in which another binding post was added to the front panel just to the left of the crystal detector, bringing the total number of binding posts to nine (grouped from left to right as a set of two, a set three, and a set four). This binding post was connected to the base of the crystal detector (which had already been labeled as terminal #5). It is likely that this terminal, the new terminal #5, was used in conjunction with terminal #4 to connect a more sensitive vacuum tube detector

Two types of etched plates were used in lieu of engraved panels on the controls of the Marconi 106 receivers during WWI—one with a brass appearance (above) and one in black and white (below).

in place of the crystal detector. An example of this late 106 set converted to a 106D is pictured on page 56.

All of the Marconi 106 sets described above had a name tag attached to the top of the cabinet with the following information: type number, catalog number, serial number, patent dates, operating range of wavelengths, and date of manufacture. At some point after Marconi settled a law suit with the National Electric Signaling Company (NESCO) in late 1914 by purchasing a license for NESCO patents, a second nameplate was attached to 106 sets referencing this license (see page 51). It is not known exactly when the second nameplate was added, but a number of early sets manufactured in 1915 have been observed without this license plate, while later sets with dates of 1916 and 1917 have been observed with this license plate. The nameplate referencing the NESCO license appeared on 106 receivers until November of 1917—the month the assets of NESCO were transferred to the International Radio Telegraph Company and NESCO ceased to exist. Indeed, a set with the date 8-17-17 was found with the plate referencing the NESCO license while a set with the date 11-7-17 was found which did not have the plate.

The author was unable to locate any Company records with the exact number of Marconi 106 sets manufactured. However, the highest serial number on any nameplate observed by this author is 751 with a date of 8-19-19—predating the merger of Marconi with RCA by only a few months. Since American Marconi was manufacturing 106B sets by that date, it safe to assume that the total number of 106 sets manufactured by Marconi was close to 751.

Marconi Type 106A Receivers

According to Robert S. Palmer, a Marconi engineer who worked on a number of receivers—including the 106—from 1915 to 1920, there were one or more versions of the 106 which were referred to as a 106A:[30]

> *"The original 106 had engraved panel[s], and then during [the]1917 war all the engraving machines were busy on war orders so during that period, the etched metal plates were used—some called these the "A" series, but later I saw a blue print of a model modified in other ways that was marked 106A."*

At the end of the war, Marconi brought out the Quenched Multiple Spark (or QMS), a small ½ kW transmitter made specifically for yachts and small ships; it was packaged in a varnished hardwood cabinet with square corners and a base flush with the front panel to match the companion 106B receiver. (John Jenkins, Sparkmuseum)

Despite Palmer's assertion about the existence of a 106A, none of the many sets seen by this author have a 106A designation on either a name tag or on the panel. It appears that all the sets with the etched dials have nameplates with the Marconi 106 designation. Furthermore, the set with what appears to be the final change to the 106—the addition of the ninth actual binding post—has a nameplate with a date of 8-19-19 and the designation "Type 106." The date of 8-19-19 is significant because it is believed to be one of the last 106 receivers made by Marconi before RCA absorbed Marconi in late 1919, and so it is unlikely that there was ever a model 106A.

Marconi/GE Type 106B Receivers Type

At the end of the war, Marconi brought out a small ½ kW transmitter named the Quenched Multiple Spark (or QMS) transmitter in a varnished hardwood cabinet made specifically for yachts and small ships. The 106B receiver was

The Type 106B is easily distinguished from the 106 by the straight lines of the wood cabinet, the addition of a second crystal detector on the front panel, and the placement of the binding posts on the rear.

The interior of the Type 106B has the same general appearance as the 106 with the notable exception of the nine binding posts on the rear.

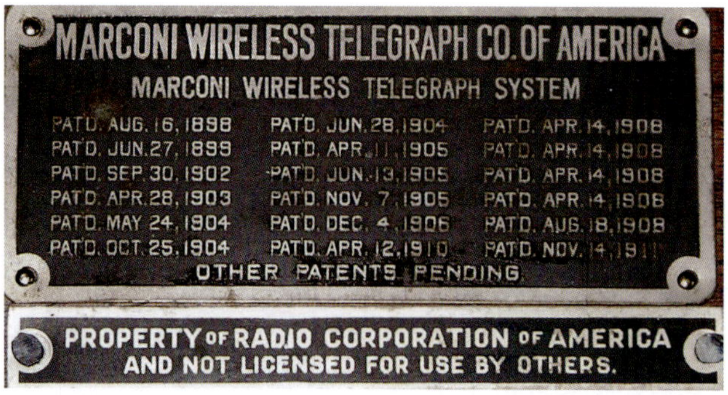

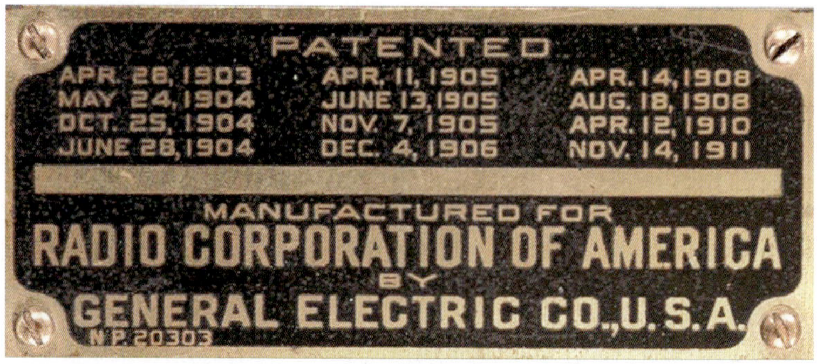

The 106B sets manufactured before the RCA-Marconi merger have nameplates affixed to the top of the cabinet referencing the Marconi Wireless Telegraph Co. of America and a second nameplate added after the merger referencing Radio Corporation of America, while sets manufactured after the merger have a single nameplate referencing Radio Corporation of America and General Electric Co., U.S.A. without any reference to Marconi. (Collections of Eric Wenaas and Howard Stone)

designed to go with this transmitter, and was offered in a matching varnished hardwood case with square corners and a base that was flush with the front panel—in sharp contrast to the 106 receivers, all of which had a case with rounded corners and a base that extended approximately an inch out from the receiver cabinet on all sides. The binding posts were moved from the front panel to a strip on the rear, and front panel was once again engraved, as it was in the early versions of the 106. Unlike the 106, the the designation of this set (Type 106B) was clearly engraved in the center of the face panel.

The more substantive changes to the front panel were the addition of a second vertical crystal holder so both the carborundum and crystal detectors were readily available to the operator, and a repositioning of the binding posts to the rear in a strip configuration. A 106B can readily be identified in a photograph of the interior from the rear by the characteristic nine binding posts readily visible without the cover. One pair of these binding posts was used to connect to an external vacuum tube detector with its greater inherent sensitivity and stability.

Production of the 106B began in early 1919 at the Marconi Aldene plant, and continued there until mid-1920 when all the production equipment was transferred to GE in Schenectady, New York. Production of the 106B continued there until early 1922 when the 106C designed by GE went into production.

Two distinctly different types of labels and nameplates have been observed on the 106B sets depending on where and when the sets were manufactured. Sets manufactured by Marconi before the merger with GE in November 1919 have the name Marconi Wireless Telegraph Co. of America, Type 106B, and the serial number engraved in the upper middle portion of the panel. In addition, there is a nameplate attached to the top listing a number of patents but with no reference to any model or type number. Some, but not all, of these earlier sets have a second nameplate attached to the top stating "Property of Radio Corporation of America and not licensed for use by others," which was obviously added by RCA later.

The 106B sets manufactured after the Marconi-RCA merger have the RCA and GE names and the Type 106B designation engraved on the front panel without any reference to the Marconi company. In addition, there is a nameplate affixed to the top labeled "Manufactured for Radio Corporation of America by General Electric Co., U.S.A." but without any reference to the model type.

The 106B sets manufactured before the RCA-Marconi merger have front panels engraved with Marconi Wireless Telegraph Co. of America, while 106B sets manufactured after the merger have panels engraved with Radio Corporation of America and General Electric Co., U.S.A. without any reference to Marconi. (Collections of Eric Wenaas and Howard Stone)

GE Type 106C Receiver

In early 1922, RCA introduced the regenerative 106C model with a modifier or tickler system mounted in a small box which was placed on the top of the receiver.[31] The modifier consisted of a variometer with the fixed side in series with the grid of the secondary coil circuit, and the variable rotor side connected in series with the place circuit. A 3-pole, double-throw (3PDT) switch was added to the front panel to switch operation from the crystal/carborundum detectors to the external vacuum detector and amplifier circuits. Three additional binding posts were added to the rear strip to accommodate

The Type 106C is easily distinguished from the 106B by the double-throw switch which is positioned towards the right center of the front panel.

The interior of the Type 106C is similar to the 106B with the notable exception of twelve binding posts on the rear as compared to nine on the 106B.

connections to these external circuits. The switch on the front and the three additional binding posts for a total of twelve in the rear are the most notable physical differences between the 106B and 106C receivers.

The 106C was initially used in conjunction with the AD-1527 detector unit and one or two stages of the AA-1528 audio amplifier unit. However, neither of these units was designed specifically for the 106C. The AD-1527 and AA-1528 were actually part of a complete receiving set designated the ER-1620 introduced in 1922 as a follow on to the 106 series (see General Electric ER-1620 Receiving Equipment below). Later, the 106C was used with the AA-1400, a unit introduced in mid-1922 which contained both the detector and amplifier functions in a single box.[32] According to Palmer, the AA-1400 was made in two versions, a military version with a BC designation in red-colored metal boxes, and a commercial version with pointed knobs in dark green metal boxes.[33]

The markings on the 106C receivers are similar to those found on the later 106B sets. The front panel is engraved with the RCA and GE company names and the model Type 106C. The nameplate attached to the top with the inscription "Manufactured for Radio Corporation of America by General Electric Co., U.S.A." is the same nameplate used on the 106B.

GE Type 106D Receiver

Towards the end of 1922, GE developed a modification kit to convert the 106 and variants to a regenerative receiver. The old secondary circuit, with its track and gear arrangement, was replaced by an entirely new secondary circuit with a coupling coil hinged to the primary coil. The secondary coil was mounted vertically (to the left as viewed from the rear) in a fixed position with

The Type 106D, a modified version of the 106, 106B or 106C is easily recognized by a new circular nameplate on a concentric shaft labeled "Tickler" and "Coupler," with "Type 106D" appearing prominently at the top of the nameplate—such as the Type 106 converted to a Type 106D shown here. (John Jenkins, Sparkmuseum)

The Type 106D used an entirely new secondary circuit with a coupling coil hinged to the primary coil which replaced the old track and gear arrangement. (John Jenkins, Sparkmuseum)

its tickler coil on top. Both the tickler and coupling coils are operated with concentric knobs mounted on a new concentric shaft, which replaced the single shaft that had been used to vary the coupling by means of rods and ball joints. A new circular nameplate was centered on the concentric shaft with the labels "TICKLER" and "COUPLER" identifying the functions of the respective knobs. The words "TYPE 106D" appeared prominently at the top of the nameplate. The exterior modifications are evident on the 106D shown in an accompanying figure, a set that was originally a very late 106. Clearly visible are the grouping of the three binding posts immediately to the left of the detector and the brass-toned version of the etched metal plated referred to in the pervious discussion on the Marconi 106.

Another noticeable change on the front panel was the addition of a phone jack in the lower right corner for operator earphones. The phone jack was placed in the buzzer button hole on the 106, while a new hole was drilled in the 106B and 106C panels. This phone jack matched the phone jacks on the panels of the AD-1527, AA-1528 and AA-1400 units, with which the 106D was also used. The new phone jack allowed the operator to use the earphones with either the vacuum tube sets or crystal detector without having to use binding posts.

Three new pairs of binding post appear at the top of most—but not all—panels of the 106D receivers. These pairs of binding posts were connected to the primary, secondary and tickler circuits, and were used in conjunction with external load coils. The shorting wires would be removed from the three pairs of binding posts, and the loading coils would be connected to increase the operating wavelength range of the receiver. Presumably these sets were used on transoceanic liners that communicated with transoceanic stations such as RCA's WQL at Radio Central on Long Island which operated at a wavelength of 16,465 meters. The sets without provision for loading coils were used in locations where communication at long wavelengths was not required, such as aboard ships operating along the coast or in the Great Lakes.

The 106D conversions were available at the factory, but most conversions were performed in the field with factory-provided conversion kits at repair depots, and in some cases aboard ships. Some of the 106D conversions can be found on 106C receivers, which is curious because the 106C was already a regenerative receiver. Perhaps the 106D had better performance characteristics than the 106C, or perhaps the new 106D receivers produced at the factory used a combination of existing parts and panels from the 106C receiver in conjunction with parts from the new 106D conversion kits.

The nameplates appearing on the top of the 106D receivers were generally unchanged from the original nameplates that were affixed to the 106, 106B, or 106C receivers before the conversion. Since the 106D model was engraved on a dial affixed to the front panel, and since all the sets converted to a 106D already had a nameplate, there was no need to add a different nameplate to the top. However, it is possible that a nameplate with the RCA name was added to an occasional 106D when converted from a 106 or 106B set whose original nameplate had no reference to RCA, but this author has not yet seen one.

The 106D, the last of the 106 models, was succeeded by newer sets such as the ER-1620 introduced by RCA in 1921. However, the venerable 106D receiver remained in operation until WWII, at which time it was banned because of the ease with which regenerative receivers could be detected by enemy submarines.

General Electric ER-1620 Receiving Equipment

The ER-1620 receiving set was the first commercial receiver completely designed by GE for sale by RCA.[34] Intended primarily for marine communications in the 250 to 3000 meter band, this receiver went into production in 1921 as a follow on to the Marconi 106 series. The complete receiving set consisted of the following units: an AR-1529 tuned coupler, an AD-1527 vacuum tube detector, two AA-1528 single-stage vacuum tube amplifiers, an AD-692 crystal detector,

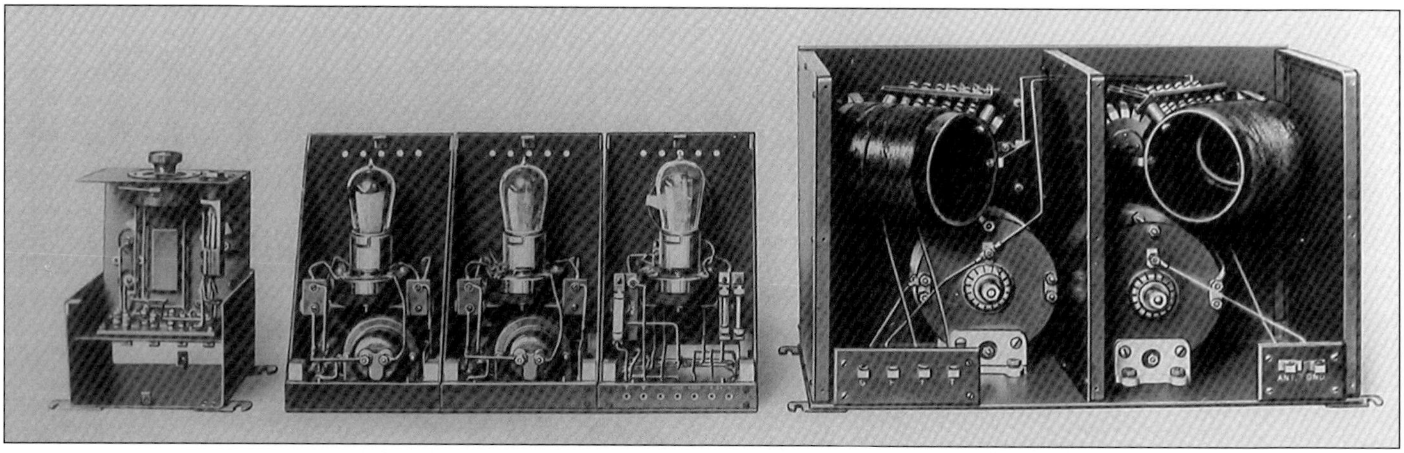

The ER-1620 consisted of six separate units including a tuned coupler, a vacuum tube detector, two single-stage tone amplifiers, a backup crystal detector, and a headset. (George H. Clark Collection, Smithsonian Institution)

and two Western Electric type CW-834 headsets. This set is particularly relevant to the subject of Radiola receivers, in that a number of the early Radiola broadcast receivers made by GE for RCA can be traced directly to this set.

The most striking feature of this set is the metal cabinet–known as an EZCASE–used to enclose and isolate each of the functional circuits. The metal cases not only provided electromagnetic shielding, but also provided easy access to the interior, and a rugged housing that could easily be bolted down in harsh environments such as those found on board ships. Similar to the 106B

The AR-1529 tuned coupler was the heart of the ER-1620 receiver, the first complete receiver designed by GE for marine use.

The AR-1529 tuned coupler consists of a primary and secondary coil shielded from each other by a metal plated dividing the cabinet in two.

The nameplate for each unit carried the model number of the respective unit, but oddly, there was no reference to the complete ER-1620 radio receiver on the nameplate or anywhere else on the receiver. (Courtesy of A. J. Link)

and 106C, terminal strips on the rear of the receiver provided connectivity between units, as well as connections for batteries, antenna, and ground. However, unlike previous receivers, the separate metal detector and amplifier units were connected together with short metallic bus wires that extended between the sides of adjacent cabinets.

The receiver was of the regenerative type with an option for either a vacuum tube or crystal detector, and one or two stages of audio amplification. The nameplate on the AA-1528 audio amplifier identified the set as a "Tone Amplifier." The set was designed for the more sensitive UV-200 detector tube and two UV-201 amplifier tubes, although the UV-201 could also be used as the detector tube. The degree of amplification and indeed the type of detector could be selected by plugging the headphones into the appropriate jack on the crystal detector unit, the vacuum tube unit, or either of the amplifier units.

The AD-692 crystal detector was a backup to the vacuum tube detector, and could be used with the AR-1529 coupler to form a complete receiver.

General Electric Duplex Radio Telephone Receiver Model 2RF/D/21F/3-8/3RDA/A

General Electric developed a prototype duplex (simultaneous two-way communication) radio telephony system for RCA in 1922 to satisfy the desire by steamship passengers to converse with persons on shore.[35] Tests with this system were made on board the S.S. America on two transatlantic trips in which they communicated with a shore station at Deal Beach, N.J. operated by

A duplex radio system with a prototype receiver, the first superheterodyne set designed and manufactured by GE, was developed to satisfy the desire by steamship passengers to converse with persons on shore. (George H. Clark Collection, Smithsonian Institution)

The duplex receiver was designed such that each functional unit was placed in its own shielded compartment to reduce interference. (George H. Clark Collection, Smithsonian Institution)

The S.S. Leviathan shown here leaving the New York Harbor was the first ocean liner to offer regular ship-to-shore radio telephone service to its passengers and crew beginning in late 1929. Note the long-wire radio antenna suspended between the fore and aft masts.

the Western Electric Company. Test participants included RCA, GE, AT&T, and Western Electric.[36] The shore station was connected by a land line to a telephone switchboard in New York City where connections could be made to any part of the then-existing telephone system. Logs from these tests showed that approximately 125 people used this system. While the tests were deemed highly successful at the time, it would be another seven years before the S.S. Leviathan became the first ocean liner to offer regular radio telephone service to its passengers and crew in late 1929.[37]

This set is significant because it was the first superheterodyne receiver manufactured by GE—the forerunner for commercial superheterodyne receivers to be manufactured for RCA. The receiver had seven separate units mounted on the rear of a metal panel that opened by swinging outward. Metallic dividers were placed in the cabinet to shield each of the seven functional units from each other, the seven functional units being the antenna circuit, RF Amplifier, H.F. Oscillator, H.F. Detector, IF Oscillator, IF circuit, and Detector Audio Amplifier. A total of eight UV-201 tubes were used in the various circuits.

The Short Wave Revolution

At the beginning of the 1920's, the short wave portion of the radio spectrum below 1000 meters was believed to be virtually worthless by both the government regulators and commercial radio operators alike, and so naturally the lower portion of this band was assigned to the amateur radio operators. It wasn't long before amateurs were reporting long distance reception with short waves, although such communication was not generally reliable.

Experimentation by the larger organizations including RCA, Westinghouse, and British Marconi began in earnest in the mid-1920's, spurred on by results from the amateurs. With the advent of the vacuum tube, it was possible to generate relatively high-power radio waves at the shorter wavelengths, much shorter than ever would have been possible with the Alexanderson alternator technology. By 1926, the technical problems had been solved, and it became possible to communicate reliably at short wavelengths with a fraction of the power and antenna size required for long-wave communication. Not only that, but also the speed was much higher than for long wave, equaling that of the best cable systems of the day.

RCA began to install short wave equipment at Radio Central, and plans for the remaining long antenna systems for long wavelength communication were unceremoniously dropped. Rates for radio communication dropped dramatically, and cable companies were no longer able to compete. Many cable companies went out of business, while others consolidated, retaining only the most profitable routes. The short wave revolution was a fait accompli—short-wave radio had become the technology of choice for long distance point-to-point communication. However, by this time broadcast radio had gripped the nation and RCA as well—so much so that point-to-point communication services contributed only about 5% to total RCA revenues by the end of the 1920's. Broadcast radio had taken central stage at RCA, and it was broadcasting that would dominate the RCA agenda for decades to come.

Endnotes and References for Chapter 2

1 *Radioana*, Letter from E. J. Nally to George H. Clark regarding trademarks dated Nov. 4, 1944.

2 George H. Clark, "Monograph on Monograms," *Relay*, November 1944, pp. 21-23.

3 *The Wireless Age*, Vol. 8, No. 3, December 1920, p. 43.

4 George H. Clark, "Monograph on Monograms," *Relay*, p. 22.

5 *Radioana*, "Activities of the Radio Corporation of America, Period — April, May, June 1921" RCA memo dated July 1, 1922.

6 "Radio Corporation of America Plans Commercial Wireless," *The Wireless Age*, Vol. 7, No. 4, January 1920, p. 7.

7 "A New and Powerful Wireless Company," *The Wireless Age*, Vol. 7, No. 2, November, 1919, p. 10.

8 "Commercial Radio Across Atlantic and Pacific Opens," *The Wireless Age*, Vol. 7, No. 6, March 1920, p. 10.

9 *General Electric Review*, Vol. XXV, No. 1, January 1922, p. 29.

10 *Commercial and Government Radio Stations of the United States, Department of Commerce, Bureau of Navigation*, (Washington D.C Government Printing office, June 30,1922 Edition); "America's Busiest Radio Station," *The Wireless Age*, Vol. 9, No. 12, September 1922, p. 1955.

11 "Radio Corporation Holds First Meeting," *The Wireless Age*, Vol. 7, No. 9, June 1920, p. 7.

12 "The Opening of Radio Central," *The Wireless Age*, Vol. 9, No. 3, December 1921, p. 21.

13 "Wide Expansion of Trans-Ocean Wireless Assured," *The Wireless Age*, Vol. 8, No. 1 October, 1920, pp. 7-8.

14 "Filing Radiograms with the Postal," *The Wireless Age*, Vol. 10, No. 1, October 1922, p. 55.

15 "Thousand Word-a-Minute World-Wide Radio Station," *The Wireless Age*, Vol. 7, No. 10, July 1920, p. 7.

16 Ibid.

17 RCA Annual Report, 1922.

18 *Radioana*, Letter from E. J. Nally to George H. Clark regarding trademarks dated Nov. 4, 1944.

19 *Radioana*, "Activities of the Radio Corporation of America, Period - April, May, June 1921" memo dated July 1, 1922.

20 RCA Annual Report, 1929.

21 RCA Annual Report, 1928.

22 "R.C.A. Makes Tubes Available for Ship Work," *The Wireless Age*, Vol. 9, No.11, August 1922, p. 53.

23 The shipboard transmitters had much lower power than the shore stations, so the most sensitive detectors were most needed at the shore stations.

24 "Radio Corporation of America Buys International Company," *The Wireless Age*, Vol. 9, No. 1, October 1921, p. 15.

25 RCA Annual Report, 1927.

26 According to the RCA Annual Report for 1927, gross income from Marine Service for 1927 was $990,954 versus total gross corporate income of $65,082,074.

27 Marconi Wireless Telegraph Company of America blueprint No. 3910-A dated June 2, 1915.

28 Ibid; Marconi Wireless Telegraph Company of America blueprint No. 3901-A dated August 15, 1918.

29 Robert S. Palmer, "Marconi 106 Tuner," private collection of A. J. August, undated.

30 Robert S. Palmer, private handwritten letter addressed to an unidentified individual "OM," private collection of A. J. August, undated.

31 Robert S. Palmer, "Marconi 106 Tuner," private collection of A. J. August.

32 George E. Sterling, *The Radio Manual*, (New York, D. Van Nostrand Co., 1928) pp. 451-459.

33 Robert S. Palmer, "Marconi 106 Tuner," private collection of A. J. August.

34 "Some Developments in the Electrical Industry During 1921," *General Electric Review*, Vol. XXV, No. 1, January 1922, pp. 29-32.

35 "Duplex Radio Telephone Transmitter with Special Reference to Tests on the S. S. America," *General Electric Review*, Vol. XXV, No. 8, August 1922, pp. 477-483.

36 "Radio Connects Ship and Shore Telephone," *The Wireless Age*, Vol. 10, No. 3, December 1922, pp. 41-43.

37 "Voices on Shore Find Ships at Sea," *The New York Times*, December 22, 1929, p. XX11.

August, 1921 — 25 Cents

The WIRELESS AGE

Volume 8 — Number 11

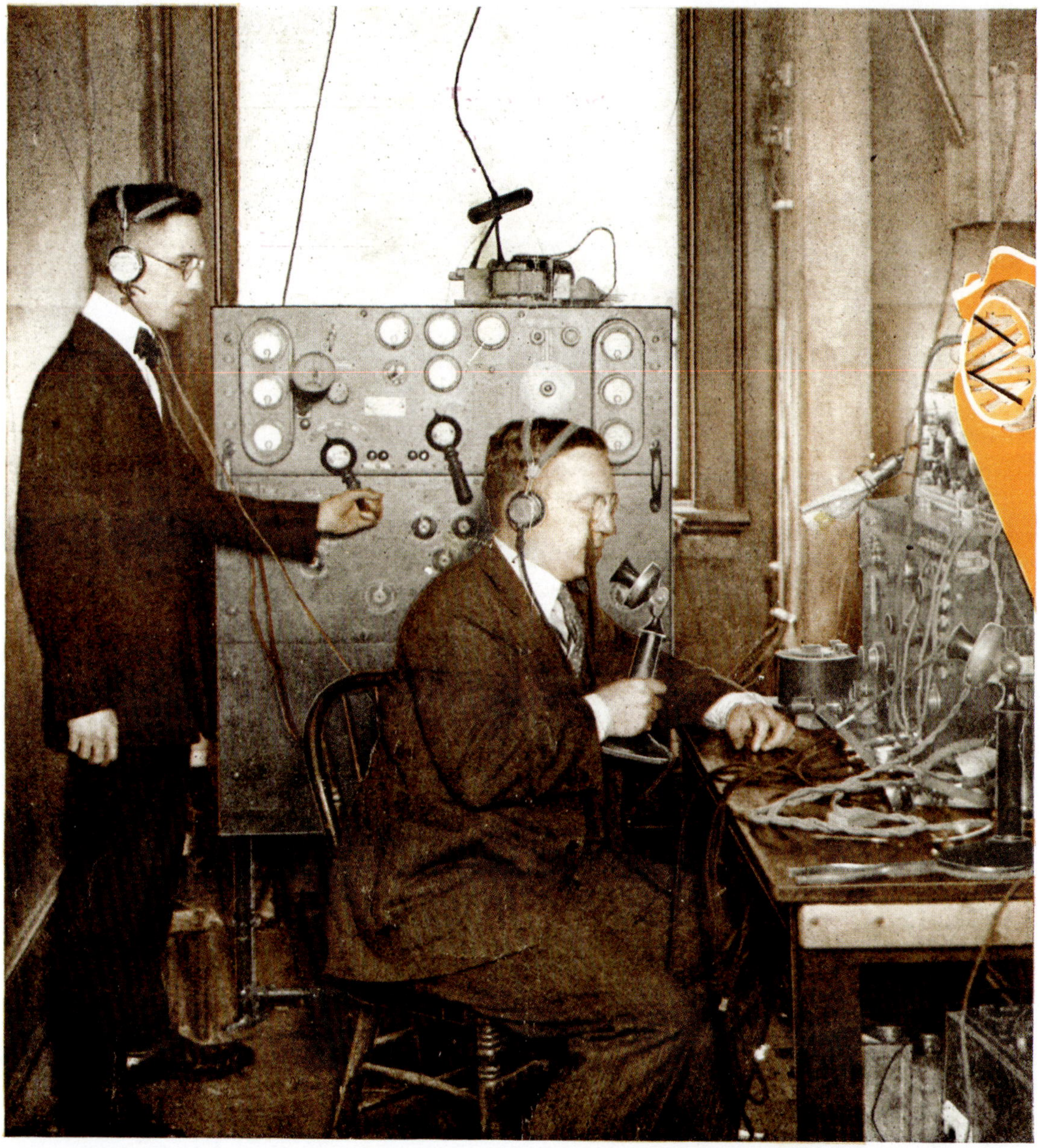

The radiophone transmitter that broke all records, being tested on the day before the fight by the President of the National Amateur Wireless Association

300,000 Persons Hear Voice Describe Fight

How the Radiophone Served Greatest Audience With Aid of Amateurs

And Many Exclusive Features in This Issue

The broadcast of the Dempsey-Carpentier fight on July 2, 1921 from the temporary transmitter site of fledgling Westinghouse station WJZ at the Lackawanna Railway terminal in Hoboken, NJ was one of the most celebrated experimental broadcasts in U.S. radio history.

Chapter 3

The Broadcast Era Begins

1920-1921

By the end of 1920, RCA had consolidated its position as the dominant force in wireless transoceanic and marine communication in America. RCA's major rival, Westinghouse, was unsuccessful in its attempt to "muscle in" on RCA's wireless business, and so it turned to other potential opportunities in the radio field, most notably opportunities involving wireless telephony.[1] Westinghouse explored a number of potential applications involving point-to-point wireless telephony, but never could identify one with a business model that would provide an acceptable return on investment.

In the fall of 1920, Westinghouse management suddenly became interested in broadcast radio, and while the concept of broadcast radio was not new, no individual or company up to that time had the key elements required to make broadcast radio a success: vision, resources, and the willingness to risk those resources to achieve the vision. H. P. Davis, a Vice President of Westinghouse, was the visionary who persuaded Westinghouse management to pursue broadcast radio by committing the resources required not only to develop and manufacture the necessary transmitters and receivers, but also to create multiple broadcast radio stations and develop program content for those stations. Westinghouse began transmitting from its first broadcast station, assigned the call letters KDKA, on November 2, 1920, the date historically used to mark the beginning of the broadcast era.

Westinghouse correctly anticipated a demand for broadcast receivers, pursuing their development in early 1921 despite the fact that they could not manufacture and market them without infringing on patents held by RCA—most notably those covering the triode vacuum tube. RCA and General Electric were in the same predicament because they too could not manufacture and market state-of-the-art receivers without infringing on the Armstrong regeneration patent owned by Westinghouse, but unlike Westinghouse, GE did not elect to develop broadcast radio receivers for fear of patent infringement litigation.

Both organizations saw the wisdom of joining forces, so it was not long before Westinghouse signed marketing and cross-licensing agreements with RCA and GE patterned after the RCA/GE agreements, whereby RCA became the sole marketing organization for radio equipment manufactured by both GE and Westinghouse, but now in a ratio of 60% to 40% respectively. However, because Westinghouse was willing to risk capital to develop receivers without the necessary patents, they were in a position to offer receivers for RCA to market at the time the agreements were signed in June of 1921, while GE would not have their first broadcast receiver ready for market—a crystal set at that—until the first part of 1922. As a result, Westinghouse dominated the manufacturing of broadcast receivers marketed by RCA during the first two years of the broadcast era, and it would not be until the 1923-4 season that GE was able to catch up.

Westinghouse Casts About for a Role in Wireless

Westinghouse first became interested in wireless during WWI as a result of contracts from the U.S. Signal Corps to design and manufacture 75 small portable transmitters and 150 portable receivers.[2] Frank Conrad of Westinghouse and Donald G. Little, a Bureau of Standards employee representing Signal Corps interests, worked out the design for a transmitter designated as the SCR-69. The companion SCR-70 receiver was designed largely by Conrad himself at a workshop located in his garage at home. The system was tested using two stations licensed for wartime use, one located at Westinghouse and one located in Conrad's garage about five miles from the plant. According to Westinghouse Vice President H. P. Davis, the two stations were assigned the call letters 2WE and 2WM,[3] but Frank Conrad in an interview with RCA historian George Clark recounted that his call letters were 3WE.[4]

After the war, the radio contracts at Westinghouse all but disappeared, and both Conrad and Little, who became a Westinghouse employee immediately following the war, were reassigned to other duties. However, their interest in radio telephony, which was stimulated by the war efforts, lingered. With Westinghouse's blessings, Conrad continued to experiment with transmitter and receiver designs in his garage shop at home where he was often joined by Little. In late 1919 and early 1920, Conrad set about to design a receiver for Westinghouse so the company could communicate between facilities in Pittsburgh, Newark and Cleveland by code, with radiophone service as a sideline.[5] He completed his design by mid-1920 as evidenced by two patent applications with filing dates of June 3 and July 7, 1920.[6] This receiver, later designated the RC, would be the first receiver sold to the public specifically for broadcast radio reception, although broadcast radio was not yet on the minds of the Westinghouse management at the time of its development.

The SCR-70 receiver was largely designed by Frank Conrad working out of his garage workshop during WWI. (John Jenkins, Sparkmuseum)

Westinghouse Aligns Itself with International Radio

Westinghouse first tried to compete directly with RCA in the trans-oceanic and marine markets by acquiring The International Radio Telegraph Co., a company already entrenched in marine communications with access to the technology suitable for trans-oceanic communications. However, Westinghouse and International Radio were quickly out maneuvered by RCA in the trans-oceanic markets, so Westinghouse spent much of 1920 casting about for other roles, focusing mainly on domestic applications. An announcement appearing in the *Electrical Review* a few months after the International Radio acquisition outlined the ambitious Westinghouse strategy:[7]

> *"The Westinghouse Electric & Manufacturing Co. has acquired a controlling interest in the International Radio Telegraph Co., and in the future will manufacture the radio equipment heretofore produced by the latter organization....The Radio company operates and maintains Wireless stations on ships, and has shore stations at Newport, New London, Brooklyn, Cape May, with others under construction in Maine and Massachusetts."*
>
> *"The Westinghouse Company will be interested in both the manufacturing and the operating branches of wireless telegraphy and telephony. A separate factory has been equipped at East Springfield, Mass., for the manufacture of all types of modern radio apparatus, from the largest transmitting and receiving sets required by transoceanic stations to the small sets used by the amateur. Apparatus will not only be supplied for the land, sea, ship, airplane, military, and naval services, but special attention will be paid to the development of new uses, such as for railroads, power companies, mines, lumber camps, ranches and farms, and for inter-works communications of large industries. A staff of research and design engineers has been built up and important fundamental work is underway."*

While this strategy sounded impressive, in fact, none of these potential opportunities ever came to fruition. Don Little, Frank Conrad's assistant, gave an oral account of their pursuit of two such opportunities in an interview conducted in the mid-1950s:[8]

> *"Westinghouse thought there might be a use for radio telephone in dispatching and controlling tugboats around a harbor like New York. Accordingly, arrangements were made with the New Haven Railroad, which operated a fleet of tugs around New York, to try out the idea."*
>
> *"One set was installed on shore at the terminal of the old Westchester Electric Railroad at 125th Street and the Harlem River. The other set was installed on a New Haven Railroad tug. Our experiments were not particularly successful. We communicated with the tug at a distance of possibly a mile, but that was about the limit. The antennas were small and low and the space on a tug for an antenna for wavelengths around 500 meters was just not suitable. The tug was too small and antenna was too small to be an efficient radiator of those frequencies."*
>
> *"The same sets were used again a little later. One of them was installed at the Bush Terminal of the International Radio Telegraph Company in Brooklyn and the other put on one of the Fall River Line steamers that went nightly to Newport and Boston. The experiments there were more successful, probably because larger and higher antennas were possible. We were able to communicate at night between the ship and the Bush Terminal to a distance of roughly 100 miles, which was pretty good for those days."*

Presumably the receivers described in these experiments were the ones Conrad had designed in late 1919 and early 1920 for point-to-point communication. Unfortunately, neither of these experiments led to a viable commercial application.

A third potential commercial opportunity was described by Vice President Davis in a speech entitled "The Early History of Broadcasting in the United States," the text of which was published in 1928.[9] A portion of that text describes the opportunity:

> *"In seeking a revenue returning service, the thought occurred to broadcast a news service regularly from our ship-to-shore stations to the ships. This thought was followed up, but nothing was accomplished because of the negative reaction obtained from those organizations which we desired to supply with this news service. However, the thought of accomplishing something which would realize the service referred to still persisted in our minds."*

Apparently this opportunity did not even reach the experimental/ demonstration phase. In fact, Westinghouse could not develop a meaningful business model for any of the opportunities they identified during the first three quarters of 1920.

Westinghouse Stumbles into Broadcast Radio

Westinghouse spent the better part of a year searching earnestly for a role in wireless, but to no avail. Westinghouse stumbled quite serendipitously into broadcast radio towards the end of 1920 as a direct result of Frank Conrad's experimentation and transmissions from his garage. This story has been recounted numerous times, with each account differing from the next, and all being at odds in one detail or another with an oral account given by Frank Conrad and W. W. Rogers to RCA historian George Clark circa 1940 found in Clark's *Radioana Collection* at the Smithsonian.[10] The most startling

Broadcast radio was born in Frank Conrad's garage where his amateur station 8XK was located, and where he first began broadcasting his evening concerts using a phonograph and borrowed records. Later in life, Frank Conrad was often referred to as the Father of Broadcasting or the Father of Radio as a tribute to his role in creating the broadcast radio industry. (George H. Clark Collection, Smithsonian Institution)

revelation in Conrad's account is that he began his "radio concerts" much earlier than is generally believed—well before he could have been reassigned his special 8XK license after the war. The account that follows incorporates some of Conrad's recollections using selected citations from the George Clark interview document.

Frank Conrad Transmits from His Garage Station

Frank Conrad began radio transmissions from his garage under the call sign 8XK after receiving a Special Land Station license from the Department of Commerce in the summer of 1916.[11] Conrad's transmissions under 8XK continued until the onset of WWI when all amateur stations including 8XK were closed by the government. However, the Navy granted Conrad and Westinghouse special authorization to test radios under development for the military, and so transmissions between Conrad's garage and the Westinghouse plant in Pittsburgh continued throughout the war. After the war, Conrad continued the transmissions from his garage on his special wartime license until the ban on amateurs was lifted and his special license was reissued. However, according to George Clark's account of the oral interview, Conrad began to use a phonograph to generate an audio signal to test transmission and receiving equipment well before the amateur ban on *both reception and transmission* were lifted:

> *"During the War, the Navy gave him permission to carry on with his experimental work, for they wanted him to solve some problems, such as an airplane transmitter, a Navy receiver, and the like. His call was then 3WE* (sic) [the footnote reads:*assigned by U.S. Navy]. The '3' meant that the station was in the third Naval District. The 'WE' may have stood for 'Westinghouse', but that is not necessarily so, as far as Dr. Conrad knows today.*[12]
>
> *This experimentation gave him permission to use a phonograph for sending out voice signals for testing. People within radio distance had bootleg receivers, and soon they began picking up Conrad's transmission. They then began to telephone him and say 'We have some friends in this evening. They do not believe us when we tell them we can pick up music out of the air. Won't you please play some selections for them?'*
>
> *This was a hit and miss proposition, some nights a lot called up, other nights no one did. Conrad, always willing to oblige, granted their requests for a time, but he soon got tired of this, so he set a sort of schedule, playing selections on Wednesday and Saturday evenings, telling his friends to 'have their visitors in' on those nights. Wednesday he sent 8 to 10 p.m.; Saturdays 8 a.m. [probably should be p.m.] to 5 a.m.** [a handwritten footnote reads: **5 a.m. Sunday]. Still people kept calling him on the telephone, and a few, who had bootleg transmitters, even called him by radio. In other words, he was popular."*

It is well known that the amateur ban on transmission was lifted on October 1, 1919, but it is less well known that there was also a ban on reception, and that this ban was lifted earlier on April 15, 1919.[13] Therefore, if Conrad's "scheduled concerts" were received by bootleg receivers as he recounted, Conrad must have initiated these scheduled concerts before April 15, 1919, and not on various dates between October 1, 1919 and April 1920 as cited by most authors. More to the point, if the scheduled concerts occurred before April 19, 1919, they must have been initiated using Conrad's special wartime license and not the amateur license 8XK—as asserted by virtually all historians—because the 8XK call sign could not have been reassigned to Conrad before July, 1919 at the earliest, the month that the first special license was issued after the war.[14]

NAVY DEPARTMENT
Naval Communication Service
Office of the Director.
Washington, April 14, 1919.

Editor Electrical Experimenter
Sir:

The Acting Secretary of the Navy authorizes the announcement that effective April 15, 1919, all restrictions are removed on the use of radio receiving stations other than those used for the reception of commercial radio traffic. This applies to amateur stations, technical and experimental stations at schools and colleges, receiving stations maintained by jewelers or others desirous of receiving time signals, receiving stations maintained by manufacturers of radio apparatus, etc.

The restrictions on transmitting stations of all types are still in effect, as are the restrictions on stations operated regularly for the reception of commercial radio traffic. Both of the above classes of stations will be permitted to resume operation as soon as the President proclaims that a state of peace exists.

Attention is invited to the fact that all licenses for transmitting stations have expired, and that it will be necessary, when peace is declared, for the owners of these stations to apply to the Department of Commerce for new licenses.

Very respectfully,
(Signed) E. B. Woodworth,
Commander, U. S. Navy, Assistant Director Naval Communications.

The ban on radio reception was lifted on April 15, 1919, long before the ban on transmission was lifted on October 1, 1919. (*Electrical Experimenter*, June 1919, p. 131)

George Clark's account of the oral interview continues by describing what happened when the ban on reception was lifted on April 15, 1919:

> *"He kept boosting his power as his regular work went on, right up to the end of the official ban on radio, 1919. Up to then not much had been said about his sending, and about the illegal receiving, for people did not want to advertise that they were using their receivers in defiance of the Government's ban, but as soon as the ban was lifted, Conrad got a veritable deluge of letters asking him to 'send.'*
>
> *Again he obliged. The phonograph supplied most of the 'talent' and in addition he had his friends sing, give talk, etc. He even read them news from the newspaper, thus becoming the first news commentator. Gypsy Smith, the evangelist, was one guest star at the Conrad station; also the High School band—his children were attending school at the time, and acted as agents to procure the band for Daddy. So things went up to November, 1919. By this time the power had been increased to 200 watts, but the programs were still given only twice a week."*

This account is in stark contrast to that of other historians, in particular the account of noted radio historian Gleason Archer who states that Conrad's 8XK station was relicensed in April 1920.[15] While Archer does not cite a source for this date, it is an obvious reference to the Department of Commerce's Radio Service Bulletin #37 dated 5/1/1920 (covering April 1920) in which Frank Conrad and station 8XK first appeared after the war. However, this date is far too late as evidenced by the amateur radio magazine *QST* which regularly chronicled the reception of his 8XK transmissions each month beginning with the January 1920 issue in the "Calls Heard" feature. His signals were regularly received throughout the northeastern U.S., the Midwest, and occasionally as far away as California. Obviously, the 8XK entry in the May 1, 1920 Radio Service Bulletin indicating the license was granted in April of 1920 was in error, and indeed historian Thomas White reported in his on-line book *United States Early Radio History* that the May 1, 1920 issue made a number of errors by re-listing as new additions a number of stations that had appeared as additions in earlier bulletins.[16]

The "official version" of Conrad's broadcasts published by Westinghouse in various documents over the years seems to place his first phonograph concerts circa October 1919:[17]

> *"He [Conrad] built a small radio receiver to hear time signals from the Navy transmitter at Arlington, Va. A year later he built a transmitter over his garage in Wilkinsburg. The station was licensed 8XK and was first listed in Government records August 1, 1916.*
>
> *The United States declared war on Germany on April 6, 1917. The following day all radio amateurs were placed under a security ban. Dr. Conrad's facilities were used under special authorization, to test military radio equipment being built by Westinghouse at East Pittsburgh for both the U. S. and British Governments.*
>
> *Although the Armistice was signed November 11, 1918, it wasn't until October 1 of the following year that the security ban on radio amateurs was lifted.*
>
> *By October 17, 1919 Conrad staged another innovation for radio by playing phonographs over the air to the delight of hams."*

The Westinghouse version implied that Conrad's "innovation for radio by playing phonographs over the air" began only several weeks after the amateur ban on transmission was lifted on October 1, 1919, but it is much more probable that Conrad actually used phonograph records to test his radio designs throughout much of the war. In light of Conrad's recollections in his interview, the most logical explanation of the Westinghouse account can be found in words attributed to Conrad in the Clark interview: *"Up to then not much had been said about his sending, and about the illegal receiving, for people did not want to advertise that they were using their receivers in defiance of the Government's ban..."* Clearly, Westinghouse did not want to admit or advertise in print that Conrad had been using his special wartime license to broadcast concerts to the delight of hams who were not supposed to be receiving radio signals. The Westinghouse version cited above is worded very carefully, and does not actually conflict with Conrad's account. It simply says that *"By October 17, 1919 Conrad staged another innovation...,"* which does not preclude his innovation from actually occurring much earlier–before the April 15, 1919 ban on reception was lifted as Conrad clearly states in his interview. It would seem that the apparent Westinghouse obfuscation coupled with the clear error in the May 1, 1920 Radio Service Bulletin has muddled the time lines of Conrad's concerts–an important part of the early history of KDKA.

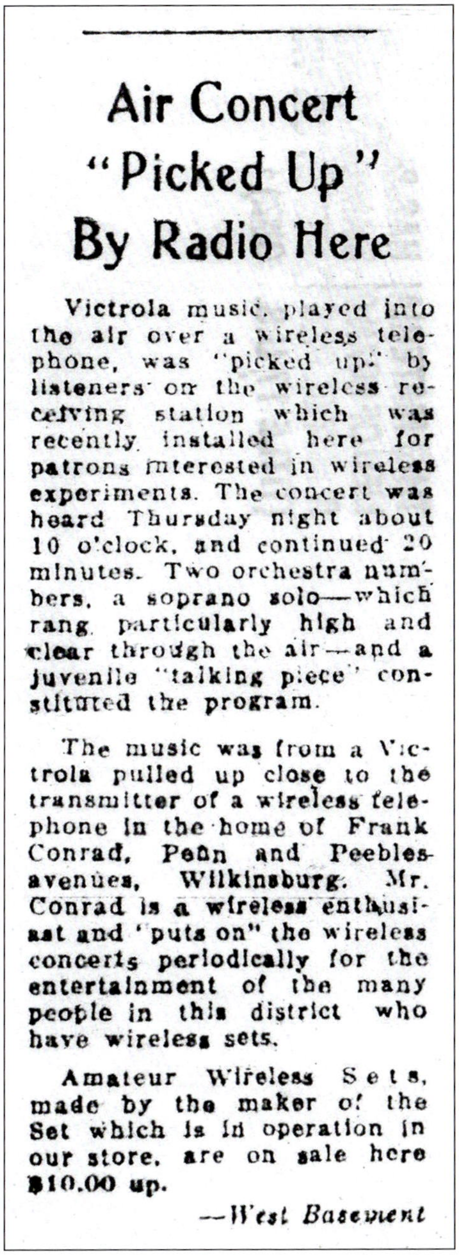

Air Concert "Picked Up" By Radio Here

Victrola music, played into the air over a wireless telephone, was "picked up" by listeners on the wireless receiving station which was recently installed here for patrons interested in wireless experiments. The concert was heard Thursday night about 10 o'clock, and continued 20 minutes. Two orchestra numbers, a soprano solo—which rang particularly high and clear through the air—and a juvenile "talking piece" constituted the program.

The music was from a Victrola pulled up close to the transmitter of a wireless telephone in the home of Frank Conrad, Penn and Peebles avenues, Wilkinsburg. Mr. Conrad is a wireless enthusiast and "puts on" the wireless concerts periodically for the entertainment of the many people in this district who have wireless sets.

Amateur Wireless Sets, made by the maker of the Set which is in operation in our store, are on sale here $10.00 up.

—*West Basement*

The Horne Department Store ad entitled "Air Concert 'Picked Up' by Radio Here," a very small and unassuming three-paragraph ad located at the bottom center of a full page ad for house wares and clothes, is credited for triggering Westinghouse's interest in broadcast radio. (*Pittsburgh Sun*, September 29, 1920)

Serendipity Intervenes on Behalf of Westinghouse

Regardless of when Conrad began his phonograph concerts, it was clear that by the fall of 1920 Conrad and his concerts were immensely popular among amateurs.[18] In the midst of all these concerts, an event happened in September of 1920 that would set the course of broadcast history. The Horne Department store in downtown Pittsburgh arranged to purchase and resell wireless receivers to the public specifically to hear Conrad's concerts. A small ad entitled "Air Concert 'Picked Up' by Radio Here" was placed in the *Pittsburgh Sun* on September 29, 1920 touting amateur wireless sets "on sale here $10 and up." For that price the radios must have been either crystal sets or surplus equipment, and contrary to some accounts, they were not provided or manufactured by Westinghouse.

In fact, Westinghouse management was just as surprised as they were delighted by the ad. Conrad had once before suggested to Vice President Davis in mid-1920 that Westinghouse should consider using radio to broadcast programs to the public to publicize Westinghouse, but Davis did not act on his

The Horne Department Store in Pittsburgh placed an advertisement for radio receivers intended specifically to pick up Conrad's concerts.

suggestion at that time. This time, however, when Davis heard of the advertisement, his reaction was swift and decisive. In Davis' own words:[19]

> *"An advertisement of a local department store in a Pittsburgh paper, calling attention to a stock of radio receivers which could be used to receive programs sent out by Dr. Conrad, caused the thought to come to me that the efforts that were then being made to develop radio telephony as a confidential means of communication were wrong, and that instead its field was really one of wide publicity, in fact, the only means of instantaneous collective communication ever devised. Right in our grasp, therefore, we had that service which we had been thinking about and endeavoring to formulate."*

Davis quickly succeeded in convincing Westinghouse management that broadcast radio was the key to their future in wireless. Radio broadcasts would stimulate sales of radio receivers among the public, and Westinghouse was in the perfect position to provide those receivers, thanks in no small part to Conrad's receiver design work. Davis quickly set about to inaugurate regular broadcast service, seizing upon the Harding-Cox election on November 2 to debut the new broadcast station, calculating that a large audience would be interested in hearing the returns over the air. With only a month or so between the decision to proceed based on the September 29 newspaper ad and the November 2 Election Day, Westinghouse set about to build a broadcast station with the broadcast room and transmitter co-located on top of the Westinghouse building in East Pittsburgh. Conrad and his staff toiled to assemble a 100 watt transmitter patterned after the one in his garage workshop.

On October 16, just two weeks before the broadcast, someone remembered to apply for a commercial license. Shortly thereafter, Westinghouse received permission by telephone to broadcast under the temporary license 8ZZ in case the actual license was not received in time. The license bearing the letters KDKA was issued on October 27, but it was not received at Westinghouse in time for the election-day broadcast.[20]

The schedule was so tight that Westinghouse did not have sufficient time to properly test the transmitter before Election Day arrived. As a result, Conrad stayed at his garage on the evening of the 2nd manning his 8XK station, prepared to back up the fledgling Westinghouse station in case of problems. As fate would have it, the new station performed flawlessly, and the event was a complete success. Election returns were funneled through the *Pittsburgh Post*

The first broadcast from KDKA on November 2, 1920 was captured in this well-known photograph that has been used in historical documents and memorabilia celebrating the inauguration of broadcast radio. (George H. Clark Collection, Smithsonian Institution)

Frank Conrad's receiver—later designated the RA/DA and RC after certain modifications—appears on the far shelf in the picture capturing the first broadcast by KDKA on November 2, 1920. (George H. Clark Collection, Smithsonian Institution)

and relayed to the station via telephone. Radio listeners found out long before the rest of the nation that Harding handily beat Cox.

According to Donald Little, a Mr. Rosenberg, was the so-called announcer,[21] but according to the Conrad/Rogers interview: *"In working the election Night, Conrad was [the] announcer. He used calls 8Z [i.e., 8ZZ] and KDKA more or less interchangeably during that night. Conrad was at his garage; the studio was in the offices of the Pittsburgh Post. The line from the studio went to Conrad's house and then on to Pittsburgh, and if anything had happened to the 'big' set at KDKA Conrad was ready to shift instantly to his own transmitter in his garage."*

This broadcast was memorialized in a now-famous photograph which has been used many times in magazine and newspaper articles, books, and on coins, cards, and other memorabilia to commemorate the inauguration of KDKA,[22] widely billed as the world's first broadcast radio station. In the photograph, one can clearly see Conrad's receiver on the bench in the center of the photograph. The receiver is a prototype of what later became the RA tuner and the DA detector and amplifier, which when placed in a single cabinet, became the RC receiver.

This receiver was referred to—but not photographed—in the September 1920 *QST* feature article on 8XK.[23] The schematic diagram of Conrad's receiver published in that article is virtually identical with the RC receiver diagram first published by Westinghouse, with several minor changes which were most likely made to accommodate the needs of the consumer. The most significant changes were in the B+ power supply and the telephone jack configuration. Conrad's receiver used a 60 volt B+ supply which was applied to all three plates, whereas the production model used two 22½ volt batteries to apply 45 volts to the amplifier tube plates and 22½ volts to the detector plate.

Conrad's receiver had only one telephone jack, while the production model had three filament control jacks which allowed the operator to select the number of stages of audio amplification desired (zero, one, or two) by inserting the phone plug into the appropriate phone jack. These jacks simplified the control panel by eliminating the need for separate filament switches for detector and amplifier tubes. With this system, inserting the phone plug not only connected the headset to the appropriate circuit, but also provided power to the appropriate filaments all in one operation, thus saving both battery and tube life.

KDKA Ushers in the Broadcast Era

While the Westinghouse broadcast was declared a complete success, it was not the first broadcast, nor did it receive the same amount of attention as previous broadcasts, most notably De Forest's November 1916 election night broadcast and subsequent concerts and news reports over 2XG,[24] the Bureau of Standards broadcasts and concerts throughout 1919 on station WWV,[25] and Madam Nellie Melba's June 15, 1920 "world" concert over the Marconi station at Chelmsford, England.[26] The Westinghouse inaugural broadcast by comparison received scant notice in the press, amounting to a four-line article appearing in the November 6, 1920 issue of the *Electrical Review*.[27] So, what was different about this broadcast, and why did it usher in the broadcast era where previous and more notable broadcasts did not?

What set Westinghouse and KDKA apart from the rest was 1) Westinghouse's commitment to regularly scheduled broadcasting that continued beyond the inaugural event, 2) the development of interesting program content that fascinated listeners, and 3) the development and marketing of low-cost, easy-to-operate receivers that enabled the general public to listen. Within several days after the inaugural broadcast, Westinghouse was operating every day—albeit on a very limited basis—with the call sign KDKA. Within a few weeks transmitter power was boosted to 500 watts, and by the end of 1921 Westinghouse had opened three additional stations including WJZ in Newark, N.J., WBZ in East Springfield, Mass., and KYW in Chicago, Ill.[28]

Westinghouse also dedicated resources to develop and improve programming. Harold W. Arlin was selected as the first radio announcer for KDKA in early 1921, and Thomas H. Cowan was selected as the first announcer for WJZ later that year.[29] Initially, they both relied on playing phonograph records, reading newspaper headlines and community bulletins, recapping sports scores, giving weather forecasts, explaining Arlington time signals, and so forth. Cowan is credited with developing a general program mix which added orchestral music, humorous speeches, special programs for women and also for children, religious programming including church services, performers and singers including stars from Broadway and vaudeville, and opera.

While a great boom in broadcast radio lay ahead, it did not come immediately after the inaugural broadcast, or even in the first year thereafter, for that matter. No new broadcast stations were licensed in the first six months following the KDKA broadcast, and a total of only eight broadcast stations were licensed in the following six month period ending in November, 1921, three of them owned by Westinghouse.[30] However, on December 1, 1921 the Department of Commerce assigned 360 and 485 meter wavelengths specifically for broadcasting under the category of "Limited Land Stations," and in that month alone, 20 stations were licensed with frequencies reserved for broadcast stations. By the end of March of 1922, there were 138 such stations licensed, and by March 1, of 1923 there were 556 such stations. The year 1922 would usher in the great broadcast radio craze.

Westinghouse Implements a Bold Strategy to Supply Broadcast Receivers

The third element setting Westinghouse apart from others was their plan to provide a line of affordable broadcast receivers to allow even the most technically unsophisticated to "listen in." Westinghouse was able to meet the demand for broadcast receivers immediately following the KDKA inaugural broadcast by commercializing and marketing the RA/DA receiver (later combined to form the RC receiver) originally designed for point-to-point communication—not withstanding an IEEE paper published in December 1922 where Conrad described it as being principally suited for broadcast radio:

> *"...[one] of those types of receiving apparatus which are adapted for reception over a limited range of wave length, and which depend for their operation on such manipulation as can be successfully carried out by persons unfamiliar with the technique of radio apparatus. Their principal field of application is the reception of broadcast radio telephone signals."* [31]

Serendipitously, Conrad's three major design goals for point-to-point communication between Westinghouse facilities were also optimum for broadcast radio reception by a technically unsophisticated public:

- simple adjustments that do not interact with each other,
- selectivity such that the signal strength from another equal or possibly more powerful station separated by 10 KHz should be below audibility, and
- sensitivity limited by static interference, fading, etc., rather than by actual lack of response.

However, Westinghouse must have realized that the RC was too expensive for the average broadcast radio listener in the long run, so in order to meet the demands of a broader range of listeners with varying requirements and means, Westinghouse cleverly set about to design a series of receivers ranging from a simple inexpensive crystal set to a one-tube regenerative receiver with a companion two-stage amplifier, which when put together, was similar in design, if not performance, to the higher-priced RC.

Westinghouse designed and produced four different models in an 18 month period from June 1921 to January 1923, each mounted in a rectangular box with a similar appearance and dimensions. A fifth model incorporating a phonograph to be used with the Aeriola radio line was under development, but it was dropped after the RCA/Westinghouse agreements were signed in June 1921.[32] The four models were the Aeriola Jr. crystal set, the one-tube Aeriola Sr. regenerative receiver, the two-tube AC audio amplifier, and the two-tube Radiola RS regenerative receiver with one stage of audio amplification. These models provided the public with an inexpensive crystal set and three more-sensitive regenerative receiver combinations with one, two and three tubes. Pertinent characteristics of these four models are compared with each other and with characteristics of the RA/DA/RC technology in Table 3-1.

The Aeriola Jr. crystal set at $25 ($32.50 with antenna) was certainly less expensive than the RA tuner using the DB detector at $85.50, and was priced competitively with other crystal sets on the market at the same time such as the De Forest Everyman, which also sold for $25 comparably equipped. The range of this and most other crystal sets for reliable reception was on the order of 10-25 miles, although stations from much greater ranges could be heard at certain times of the night and under favorable atmospheric circumstances. The Aeriola Sr. regenerative receiver was designed for those who wanted reception from greater distances (nominally 100-150 miles), but like the crystal set, this receiver was unable to actuate a loudspeaker. The AC amplifier was introduced for use with the Aeriola Sr. to permit the actuation of loudspeakers and/or to extend the range of reception. The combined price of $143 for the two units compared favorably to the three-tube RC with a price tag of $213. The RS receiver augmented the regenerative tuning circuits of the Aeriola Sr. with one stage of audio amplification, resulting in a set which was described in a Westinghouse brochure as being designed "*primarily for those who feel that the cost of a receiver embodying two stages of amplification is beyond their means, yet wish to hear their local stations on a loud speaker.*"

The receivers introduced in 1920 and 1921 are described in the following paragraphs of this chapter—including the RC and variants, the Aeriola Jr., the

Table 3-1. The line of affordable receivers originally envisioned by Westinghouse

	Model	Apparatus Type	Tubes	Nominal Range[a]	List Price	Date Introduced
RC Technology	RC Receiver (Radiola RC)[c]	Regenerative Receiver	3	500 miles	$125 without tubes	December 1920
RC Technology	RA Receiver (with DB detector)	Crystal Receiver	None	See text	$70	August 1921
RC Technology	RA Tuner	Tuner	None	N/A	$65	December 1920
RC Technology	DA Detector-Amplifier	Detector plus two-stage audio amplifier	3	N/A	$65 without tubes	December 1920
Westinghouse Broadcast Receiver Strategy[b]	Aeriola Jr. (RE) (Radiola Junior)[c]	Crystal Receiver	None	25 miles	$25[d]	July 1921
Westinghouse Broadcast Receiver Strategy[b]	Aeriola Sr. (RF); (Radiola Senior)[c]	Regenerative Receiver	1	100-150 miles	$75[e]	December 1921
Westinghouse Broadcast Receiver Strategy[b]	Phonograph for Aeriola radio line	Development discontinued after RCA/Westinghouse agreement signed in June 1921				
Westinghouse Broadcast Receiver Strategy[b]	Aeriola Sr. Amplifier (AC); (Radiola AC)[c]	Two-Stage Audio Amplifier	2	N/A	$68	August 1922
Westinghouse Broadcast Receiver Strategy[b]	Radiola RS	Regenerative Receiver	2	See text	$85	January 1923

[a] Nominal ranges as suggested in Andre Catalog R-2-A are listed here to indicate relative sensitivity
[b] The apparatus in shaded rows were introduced by RCA after the Westinghouse-RCA agreements
[c] Name changed by RCA in late 1922 to incorporate the new Radiola trade name; no change in design
[d] Price included earphones
[e] The price advertised by Westinghouse included headphones, antenna, and tubes, and batteries

Aeriola Sr. and the Radiola Senior. The AC amplifier and RS receiver listed in the shaded portion of Table 3-1 were introduced in 1922 and 1923 respectively, and therefore are addressed in subsequent chapters according to the year in which they were introduced.

Westinghouse Markets the RA, DA and RC Receiver

The RA was a single-circuit tuner unit, the DA was a detector-amplifier unit with two stages of audio amplification, and the RC was a regenerative receiver with the RA and DA units combined in a single cabinet. It was common practice at the time to house the tuner and detector/amplifier in separate cabinets for several reasons. First, the tuner could be used separately as a less-expensive crystal set in conjunction with an external crystal detector, and second, the separate units by themselves did not constitute a regenerative receiver and therefore did not infringe on the regeneration patient awarded to Edwin H. Armstrong on October 6, 1914. Westinghouse acquired rights to this patent on November 4, 1920, and therefore was able to manufacture the RC as a complete regenerative receiver in a single package without fear of infringement litigation.

According to historian George Clark, the first RC receiver was sold to W. L. Chubb, a Westinghouse employee, for use at a gathering to hear the election returns transmitted on the first KDKA broadcast on November 2, 1920.[33] This receiver was undoubtedly one of a number of engineering models made for the first broadcast because the first production models made in Pittsburgh were dated 11-30-20, clearly after the first KDKA broadcast in early November.

Many of the early RA and DA units manufactured in Pittsburgh were housed in separate cabinets and sold separately as RA tuners and DA detector-amplifiers.

The interior of the RA was essentially unchanged throughout its life cycle except for some very minor rerouting of wires.

Other *Radioana Collection* documents indicate, the early production models were first advertised and sold in the Pittsburgh area—it was said that the Horne Department Store was one of the first to sell this receiver. It is not known exactly when the first production units were sold, but it could not have been earlier than the November date on the first production units, or later than April 1921, the month they first appeared in national advertisements.[34]

Most of the late RA and DA units manufactured in East Springfield were combined into one cabinet and sold as RC receivers.

According to an interview with Westinghouse engineer Donald G. Little on March 27, 1951, about 1500 RA and DA units were manufactured at the Shadyside plant in Pittsburgh and were put on the market about Christmas 1920.[35] From serial number evidence presented later, both individual RA and DA units and RC's were manufactured during the first production run.

A companion DB detector designed to convert the RA tuner into the RA crystal receiver was introduced to the public in the August 1921 issue of *The Wireless Age* for $5. Retail prices for the RA/DA/RC units were first advertised

The tube sockets on the early DA chassis such as the one shown here with Moorhead tubes were positioned such that the key slots all faced the same direction, and the binding post caps on the rear were fiber (left). The center socket on later DA chassis such as the one shown here with WD-12 tubes was rotated—presumably to avoid some type of unwanted interaction, and the fiber binding-post caps were replaced with metal caps (right).

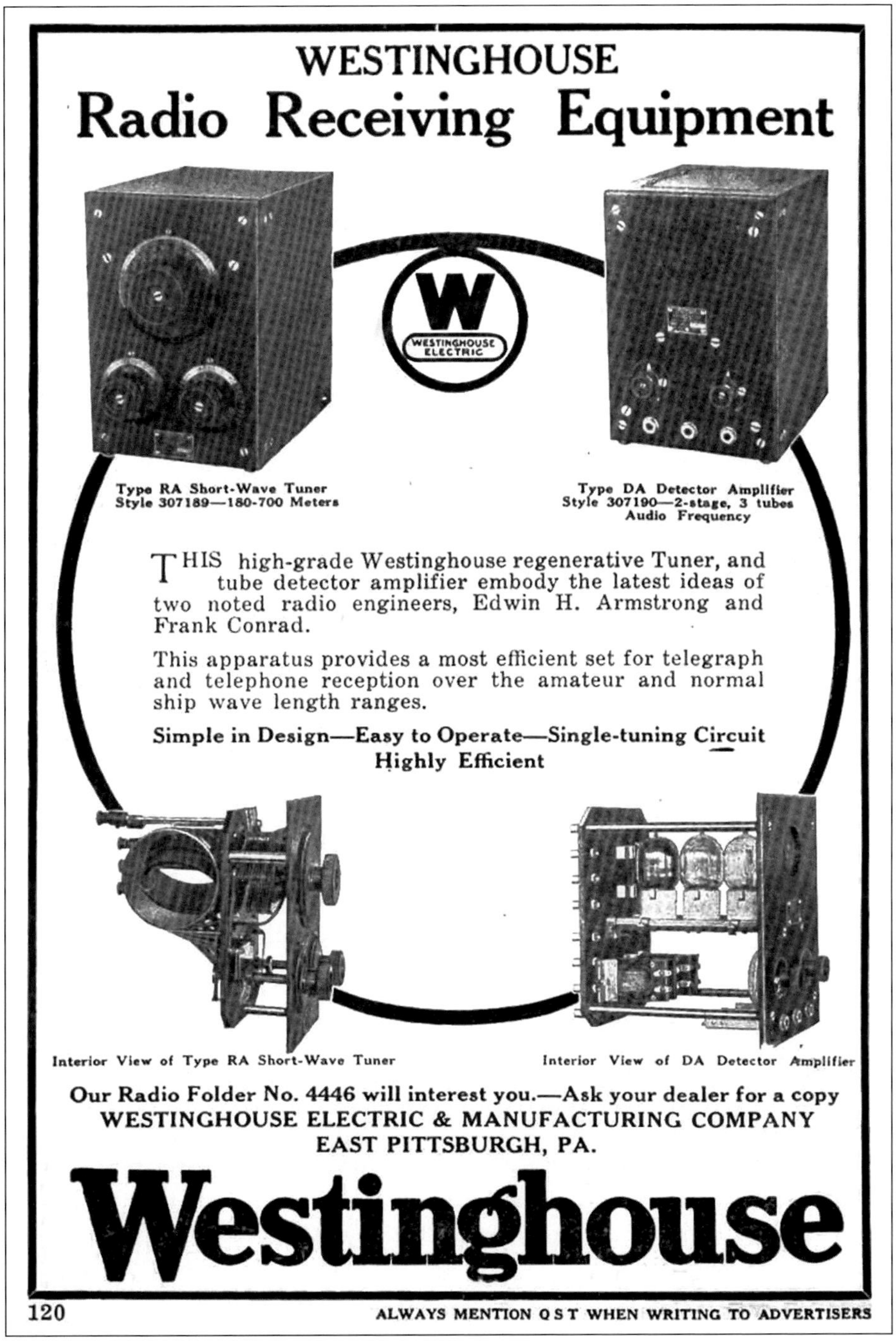

The first national advertisements for the RC technology appearing in April and May of 1921 pictured the RA tuner and DA detector-amplifier without referring to the RC. (*QST*, April, 1921, p. 120)

in the September and October issues of *The Wireless Age*.[36] The RA was advertised at $65, the DA without tubes at $65, the RC without tubes at $125, a complete RA crystal receiver with a DB detector, headphones and antenna at $85.50, and a complete RC receiver with four tubes, a Type CB loading coil, batteries, and headphones with adaptor at $216.

RCA began to advertise radio equipment manufactured by Westinghouse in December 1921,[37] although Westinghouse continued to advertise the RA, DA and RC until circa March 1922. When RCA became the exclusive distributor for Westinghouse products circa January 1922, they tweaked the prices by adding $3 to the price of RA and DA for a total of $68 apiece, by adding $5 to the price of the RC without accessories for a total of $130, and by adding $4.50 to the price of the RA crystal receiver for a total of $90. They maintained the $216 price of the complete RC but did not include a spare vacuum tube—

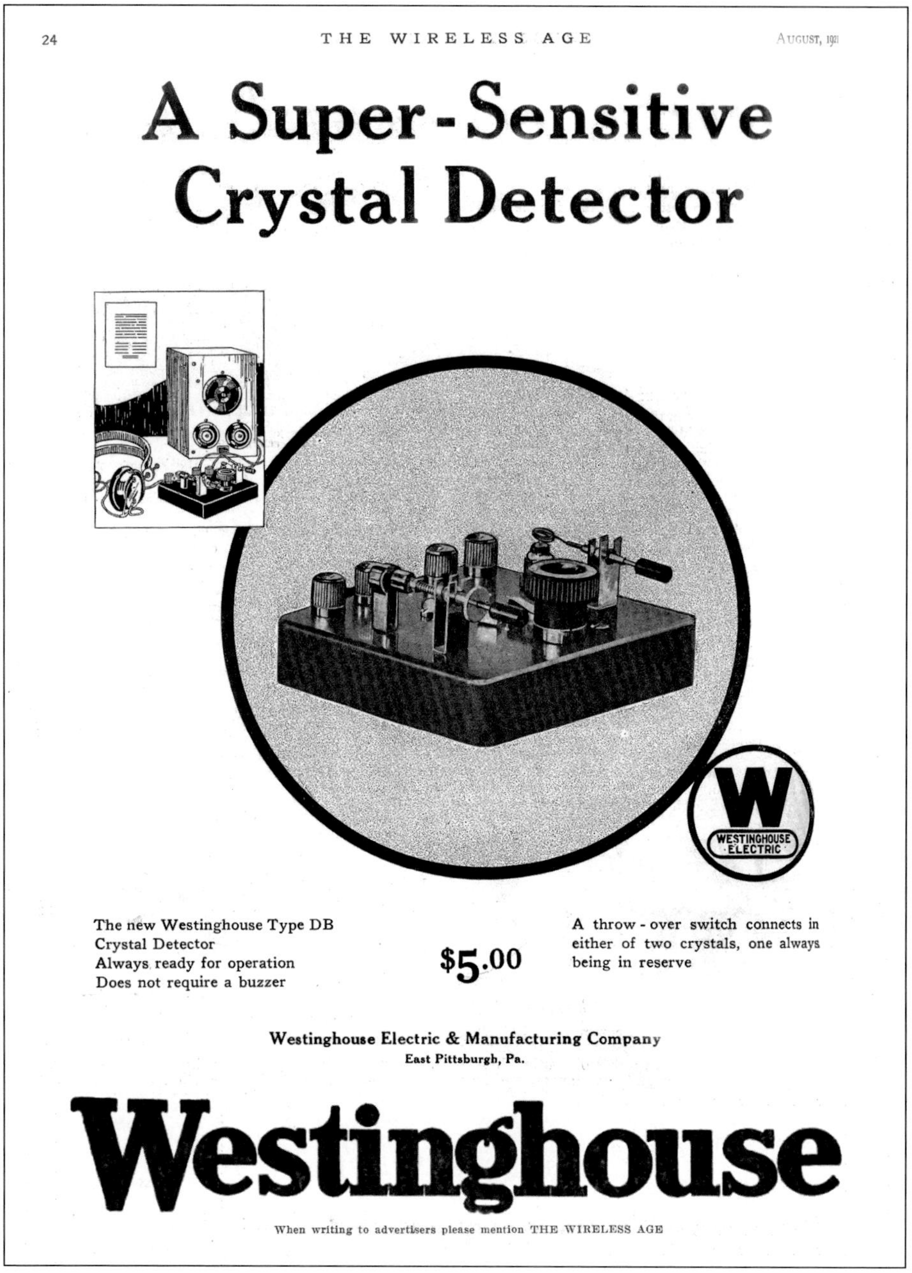

The first advertisements for the "Super-Sensitive" Type DB detector pictured it with the RA tuner, a combination which was later referred to as the RA crystal receiver. (*The Wireless Age*, August, 1921, p. 24)

effectively increasing the price by $6.50. The Westinghouse name rarely appeared in RCA's national advertising, although it was still prominently used in brochures and distributor catalogs for some time thereafter.

The key technical features of the RC receiver were as follows:

- A regenerative vacuum tube to increase sensitivity and selectivity,
- An inductive coupling method for regeneration to achieve constancy of regeneration with varying wavelength,
- A single-knob selection of wavelength by mounting tuning coil and variable capacitor on a single shaft, and
- A "vernier" wavelength adjustment consisting of small variable capacitor in parallel with the main tuning condenser to compensate for small errors in the desired ratio of inductance to capacitance as the receiver is tuned over its range.

The resulting receiver design was excellent and performed well. It could be operated with the detector alone or with one or two stages of amplification by

RCA began the exclusive distribution of Westinghouse receivers circa January 1922 with advertisements in national magazines that rarely mentioned the Westinghouse name. (*The Wireless Age*, January 1922, p. 4)

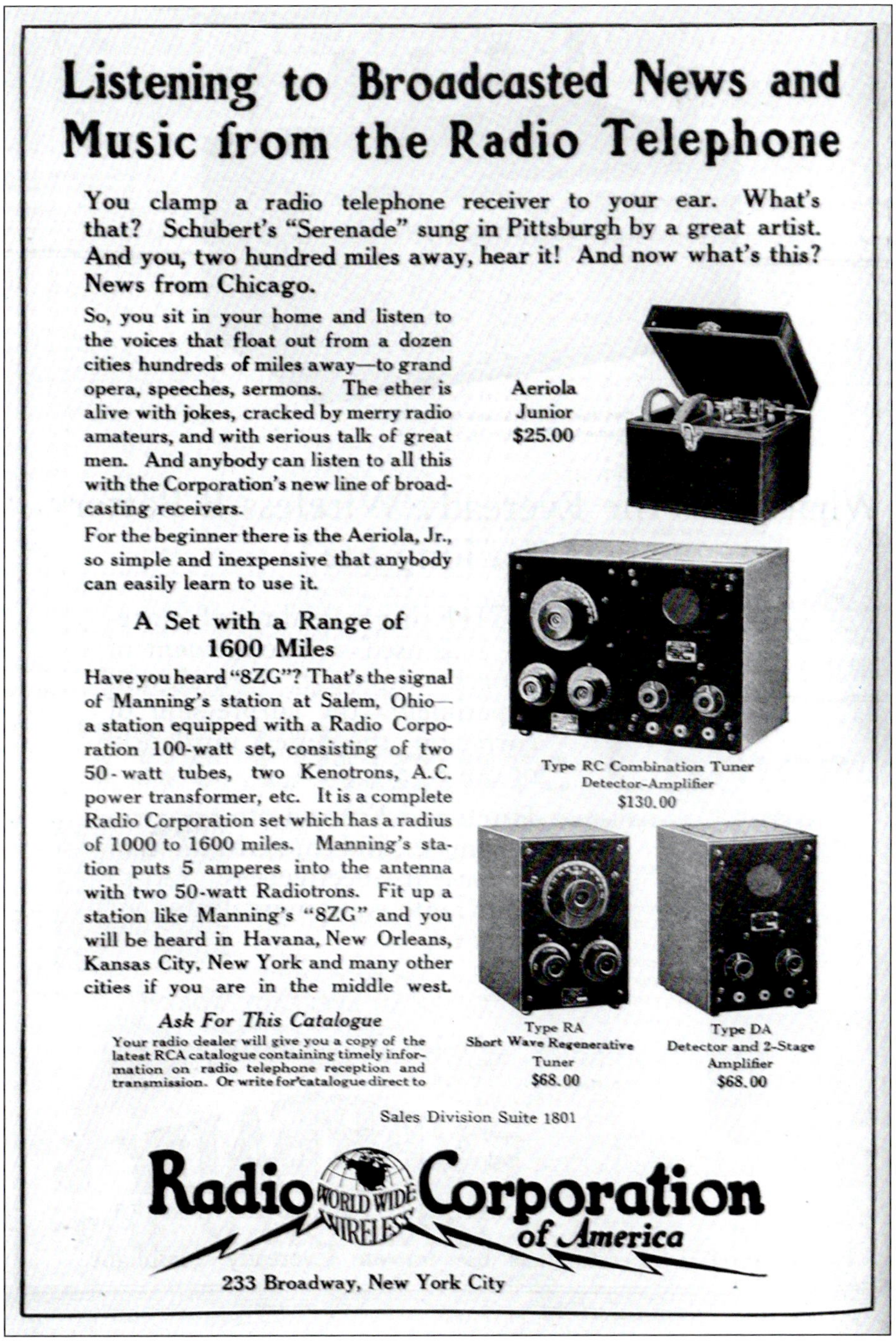

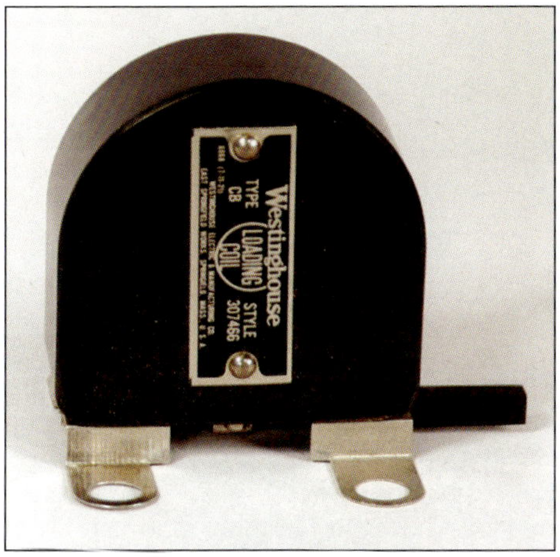

The CB loading coil was offered as an accessory for those who wanted to listen in on time signals, weather forecasts and other transmissions from stations operating with wavelengths between 1600 and 2700 meters.

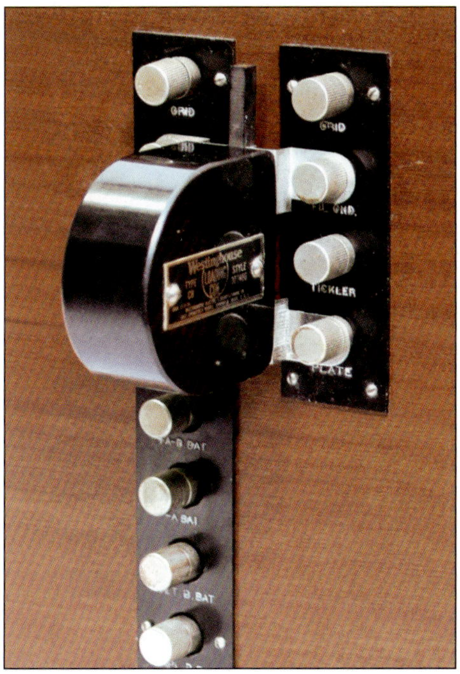

The CB loading coil was positioned on the rear binding posts between the RA and DA units.

selecting the appropriate telephone jack for the headset. It was easy to use, and as sensitive and selective as the more complex designs of the day. It produced less interference than most other regenerative sets of the day because the regeneration circuit did not need to be readjusted after each tuning, a procedure which generally caused the set to radiate a strong signal, invariably producing an annoying howl in neighboring receivers.

While the RA tuner was designed to operate between wavelengths of 180 to 700 meters for the reception of radio telephone (e.g. broadcast radio), amateur and ship stations, a Type CB loading coil could also be purchased for the reception of time signals, weather forecasts, and other types of transmissions from stations operating on wavelengths of 1600 to 2700 meters.[38] The CB coil with Style No. 307466 and a production date of 7-11-21 was placed on the rear of the RA tuner as shown in the accompanying photograph.

RCA changed the name of the RC to the Radiola RC in late 1922, and ads for the Radiola RC began to appear in March, 1923.[39] However, there were no RA, DA, or RC receivers with the Radiola name on the panel because none was manufactured after RCA made the name change. While the earliest instruction sheets had only the Westinghouse name, later revisions had both the Westinghouse and RCA name and RCA logo. The RA/DA/RC sets were on the market until July 1923, when RCA sold off their entire inventory of sets below cost to Gimbel Brothers, who in turn sold them at a deep discount in their New York City store.

RA/DA Production Runs

A large number of RA and DA units were made in the 2½-year period from December 1920 to mid-1923 in six separate production runs, and while the exact number made in each production run—or the total for that matter—is not known, the serial numbers provide a clue. Westinghouse numbered the individual RA and DA units with serial numbers presumably beginning with one, and stamped a "production run" date on each serial number plate. As a result, the total number of sets can be inferred from a sampling of serial numbers for each type of set, and the number in each production run can be inferred from a sampling of serial numbers having the same production run dates.

In a sampling of serial numbers and production dates from approximately 288 RA and DA units (both separate units and combined into RC's), the highest serial number observed to date is close to 143,000. However, this does not mean that there were 143,000 units manufactured because of two complicating factors: 1) there were approximately 28,000 receivers of other types (Aeriola Grand, Radiola Grand, Radiola Senior, and Radiolas DR, AC, RS, AR, and RT) manufactured during the same period with interleaved serial numbers, and 2) in the last production run there were approximately 31,400 RA units without serial numbers paired with DA units having serial numbers to make RC's. It turns out quite serendipitously that these two factors

The early RA instructions showed how the tuner was used in the RA-DA Receiving Set (below left). Instruction sheets for the RA (top right) were also prepared to describe how the RA tuner was used with the DB detector as a complete RA Receiving Set. The early DA instructions (below right) showed how the detector-amplifier was used in the RA-DA Receiving Set.

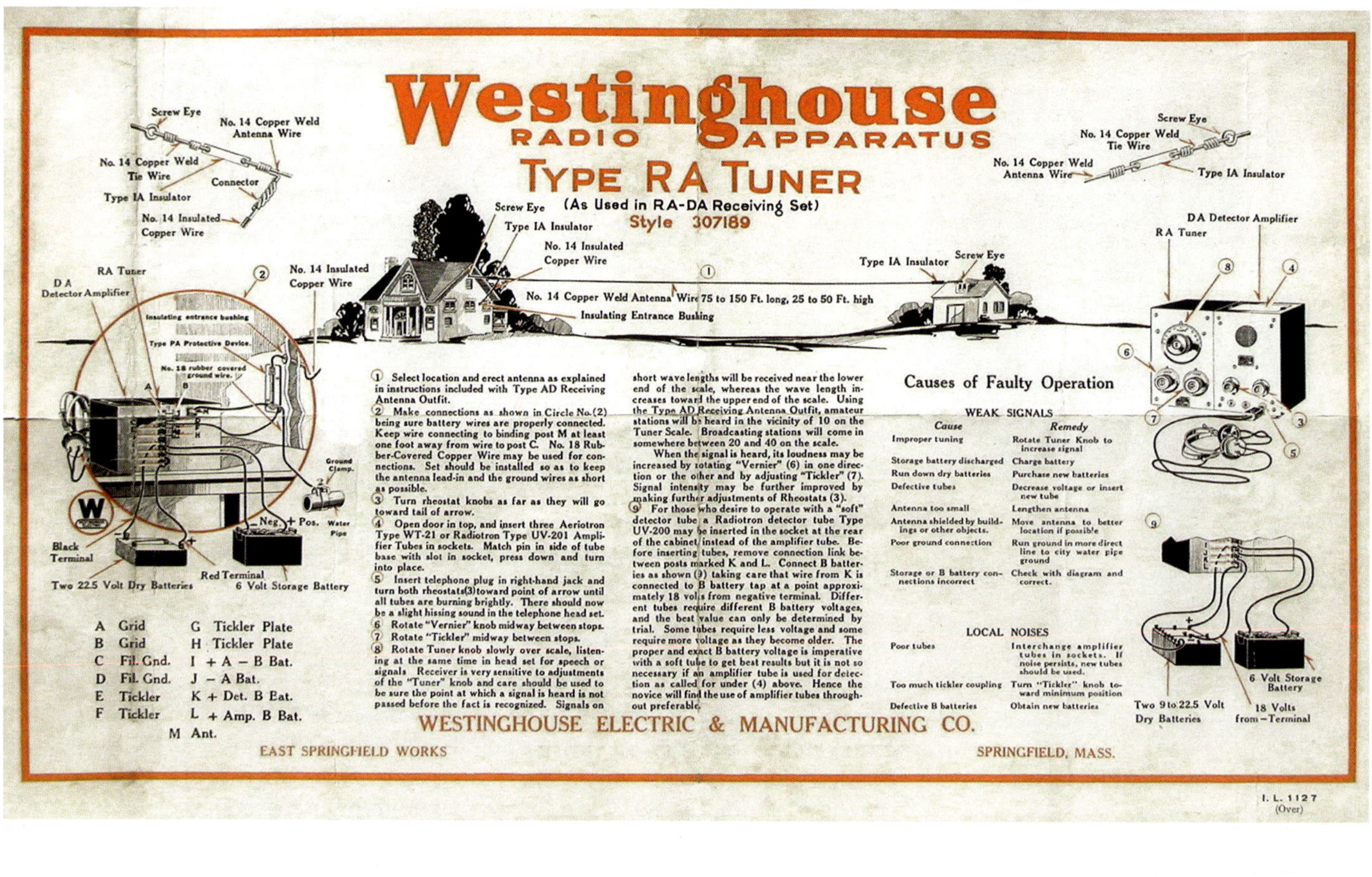

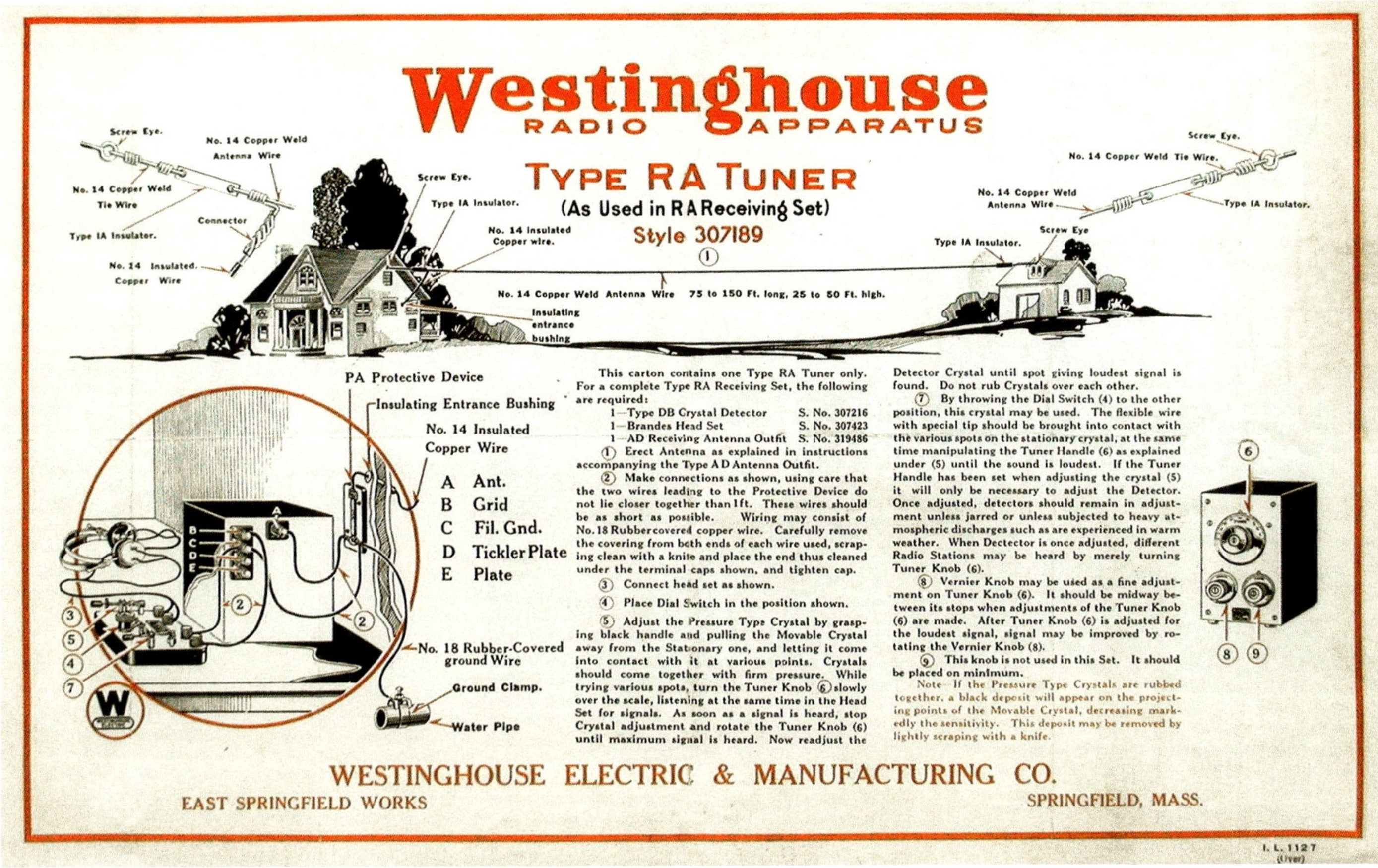

This carton contains one Type RA Tuner only. For a complete Type RA Receiving Set, the following are required:

1—Type DB Crystal Detector S. No. 307216
1—Brandes Head Set S. No. 307423
1—AD Receiving Antenna Outfit S. No. 319486

(1) Erect Antenna as explained in instructions accompanying the Type A D Antenna Outfit.

(2) Make connections as shown, using care that the two wires leading to the Protective Device do not lie closer together than 1 ft. These wires should be as short as possible. Wiring may consist of No. 18 Rubber covered copper wire. Carefully remove the covering from both ends of each wire used, scraping clean with a knife and place the end thus cleaned under the terminal caps shown, and tighten cap.

(3) Connect head set as shown.

(4) Place Dial Switch in the position shown.

(5) Adjust the Pressure Type Crystal by grasping black handle and pulling the Movable Crystal away from the Stationary one, and letting it come into contact with it at various points. Crystals should come together with firm pressure. While trying various spots, turn the Tuner Knob (6) slowly over the scale, listening at the same time in the Head Set for signals. As soon as a signal is heard, stop Crystal adjustment and rotate the Tuner Knob (6) until maximum signal is heard. Now readjust the Detector Crystal until spot giving loudest signal is found. Do not rub Crystals over each other.

(7) By throwing the Dial Switch (4) to the other position, this crystal may be used. The flexible wire with special tip should be brought into contact with the various spots on the stationary crystal, at the same time manipulating the Tuner Handle (6) as explained under (5) until the sound is loudest. If the Tuner Handle has been set when adjusting the crystal (5) it will only be necessary to adjust the Detector. Once adjusted, detectors should remain in adjustment unless jarred or unless subjected to heavy atmospheric discharges such as are experienced in warm weather. When Dectector is once adjusted, different Radio Stations may be heard by merely turning Tuner Knob (6).

(8) Vernier Knob may be used as a fine adjustment on Tuner Knob (6). It should be midway between its stops when adjustments of the Tuner Knob (6) are made. After Tuner Knob (6) is adjusted for the loudest signal, signal may be improved by rotating the Vernier Knob (8).

(9) This knob is not used in this Set. It should be placed on minimum.

Note—If the Pressure Type Crystals are rubbed together, a black deposit will appear on the projecting points of the Movable Crystal, decreasing markedly the sensitivity. This deposit may be removed by lightly scraping with a knife.

WESTINGHOUSE ELECTRIC & MANUFACTURING CO.

EAST SPRINGFIELD WORKS SPRINGFIELD, MASS.

I. L. 1127 (Over)

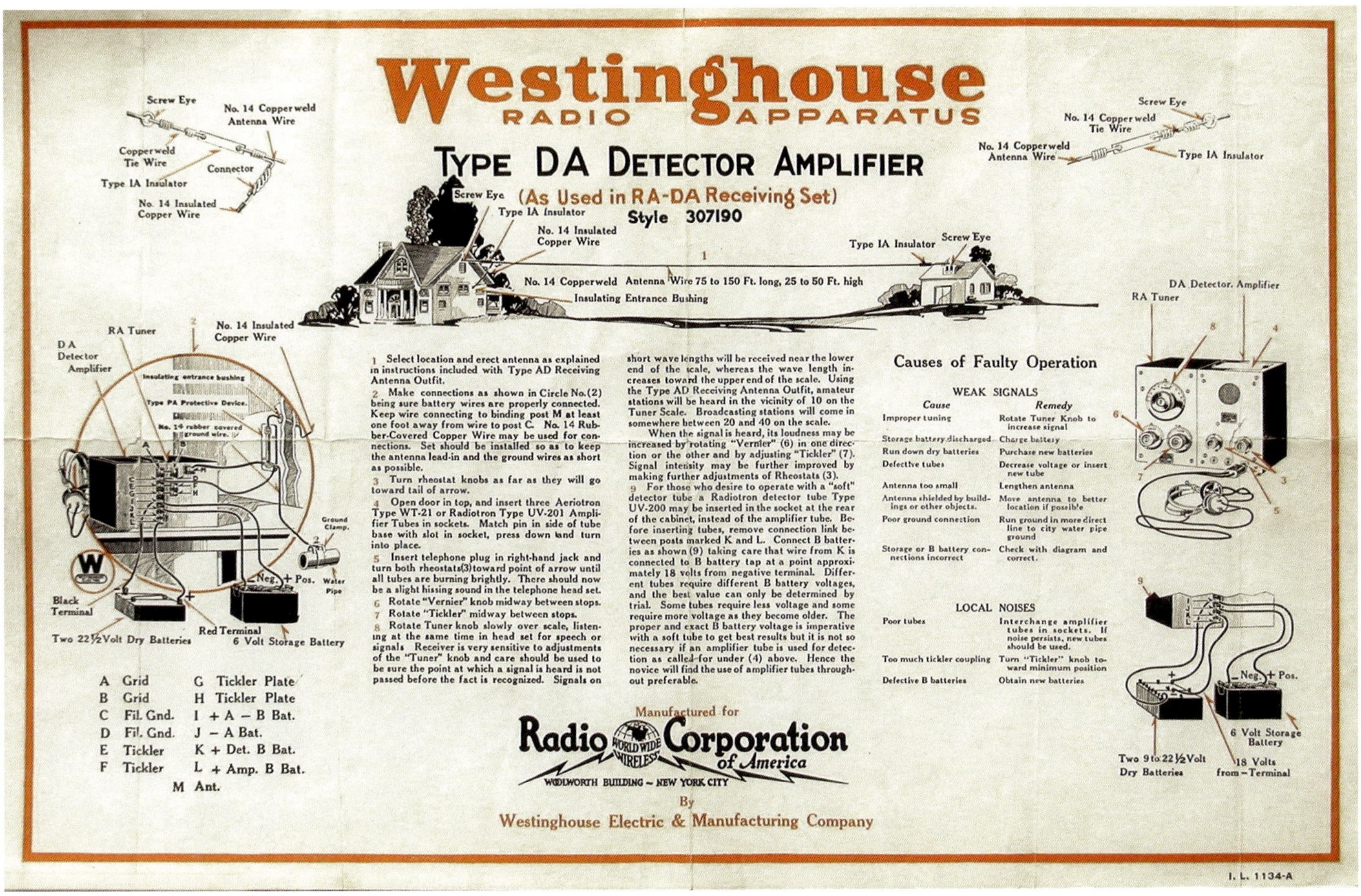

1 Select location and erect antenna as explained in instructions included with Type AD Receiving Antenna Outfit.

2 Make connections as shown in Circle No. (2) being sure battery wires are properly connected. Keep wire connecting to binding post M at least one foot away from wire to post C. No. 14 Rubber-Covered Copper Wire may be used for connections. Set should be installed so as to keep the antenna lead-in and the ground wires as short as possible.

3 Turn rheostat knobs as far as they will go toward tail of arrow.

4 Open door in top, and insert three Aeriotron Type WT-21 or Radiotron Type UV-201 Amplifier Tubes in sockets. Match pin in side of tube base with slot in socket, press down and turn into place.

5 Insert telephone plug in right-hand jack and turn both rheostats(3) toward point of arrow until all tubes are burning brightly. There should now be a slight hissing sound in the telephone head set.

6 Rotate "Vernier" knob midway between stops.

7 Rotate "Tickler" midway between stops.

8 Rotate Tuner knob slowly over scale, listening at the same time in head set for speech or signals. Receiver is very sensitive to adjustments of the "Tuner" knob and care should be used to be sure the point at which a signal is heard is not passed before the fact is recognized. Signals on short wave lengths will be received near the lower end of the scale, whereas the wave length increases toward the upper end of the scale. Using the Type AD Receiving Antenna Outfit, amateur stations will be heard in the vicinity of 10 on the Tuner Scale. Broadcasting stations will come in somewhere between 20 and 40 on the scale.

When the signal is heard, its loudness may be increased by rotating "Vernier" (6) in one direction or the other and by adjusting "Tickler" (7). Signal intensity may be further improved by making further adjustments of Rheostats (3).

9 For those who desire to operate with a "soft" detector tube a Radiotron detector tube Type UV-200 may be inserted in the socket at the rear of the cabinet, instead of the amplifier tube. Before inserting tubes, remove connection link between posts marked K and L. Connect B batteries as shown (9) taking care that wire from K is connected to B battery tap at a point approximately 18 volts from negative terminal. Different tubes require different B battery voltages, and the best value can only be determined by trial. Some tubes require less voltage and some require more voltage as they become older. The proper and exact B battery voltage is imperative with a soft tube to get best results but it is not so necessary if an amplifier tube is used for detection as called for under (4) above. Hence the novice will find the use of amplifier tubes throughout preferable.

Causes of Faulty Operation

WEAK SIGNALS

Cause	*Remedy*
Improper tuning	Rotate Tuner Knob to increase signal
Storage battery discharged	Charge battery
Run down dry batteries	Purchase new batteries
Defective tubes	Decrease voltage or insert new tube
Antenna too small	Lengthen antenna
Antenna shielded by buildings or other objects.	Move antenna to better location if possible
Poor ground connection	Run ground in more direct line to city water pipe ground
Storage or B battery connections incorrect	Check with diagram and correct.

LOCAL NOISES

Cause	*Remedy*
Poor tubes	Interchange amplifier tubes in sockets. If noise persists, new tubes should be used.
Too much tickler coupling	Turn "Tickler" knob toward minimum position
Defective B batteries	Obtain new batteries

Manufactured for

Radio Corporation of America

WOOLWORTH BUILDING – NEW YORK CITY

By

Westinghouse Electric & Manufacturing Company

I. L. 1134-A

The early production line of Westinghouse receivers including the RA units shown here was very labor intensive with little automation. (George H. Clark Collection, Smithsonian Institution)

approximately cancel–for quite unrelated reasons–so the total number of RA and DA units turns out to be about 144,800 (see Appendix D).

The first production runs of RA, DA, and RC units were made in Pittsburgh with name plates dated 11-30-20, while all units thereafter were manufactured in East Springfield, Mass. with name plates bearing five

Westinghouse transferred responsibility for manufacturing broadcast radio apparatus to its East Springfield, Mass. facility.

NEW ENGLAND WESTINGHOUSE FACTORY, EAST SPRINGFIELD, MASS.

Table 3-2. RA and DA production quantities and locations grouped by production dates appearing on name plates

Production Date	Production Location	Unit Type	Estimated Number of Units Produced[a]
11-30-20	Pittsburgh, PA	RA and DA	1700
5-18-21	East Springfield, MA	DA only	1700
8-2-21	East Springfield, MA	RA and DA	4900
9-28-21	East Springfield, MA	RA only	1700
12-14-21	East Springfield, MA	RA and DA	72,000
5-29-22	East Springfield, MA	RA and DA	62,800[b]
		Total RA and DA	144,800

[a]Most of these units were paired and sold as RC receivers
[b]Half are RA units without serial numbers paired with DA's and sold as RC's

different production dates between 5-18-21 and 5-29-22. Approximately 1700 RA and DA units were initially made in Pittsburgh before manufacturing was transferred to East Springfield.

The numbers of RA and DA units manufactured within each of the six production dates are summarized in Table 3-2. The methodology used to

Each of the five different RA serial number plates has a unique design which can be used to readily identify the production run date at a glance. A. 11-30-20 B. 9-28-21 C. 8-2-21 D. 12-14-21 E. 5-29-22

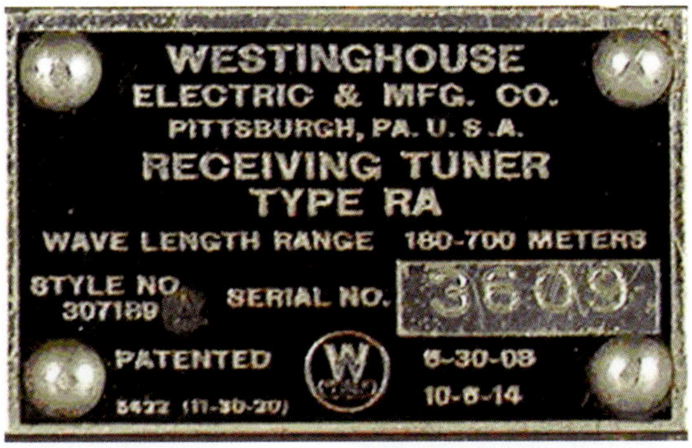

A

B

C

D

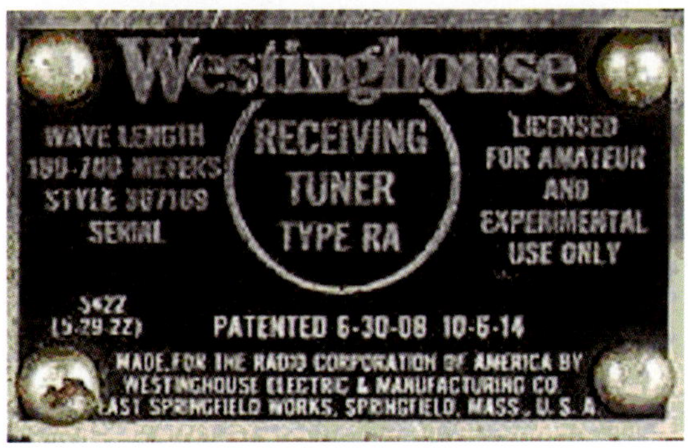

E

A

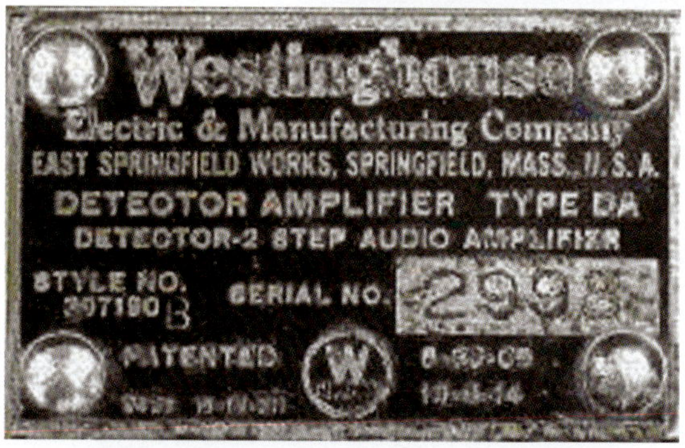

B

C

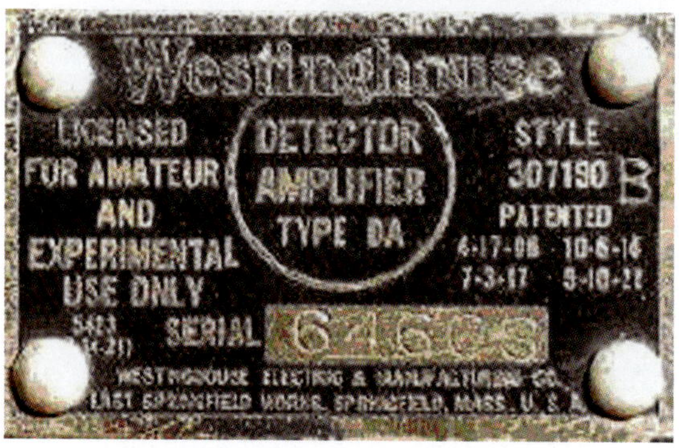

D

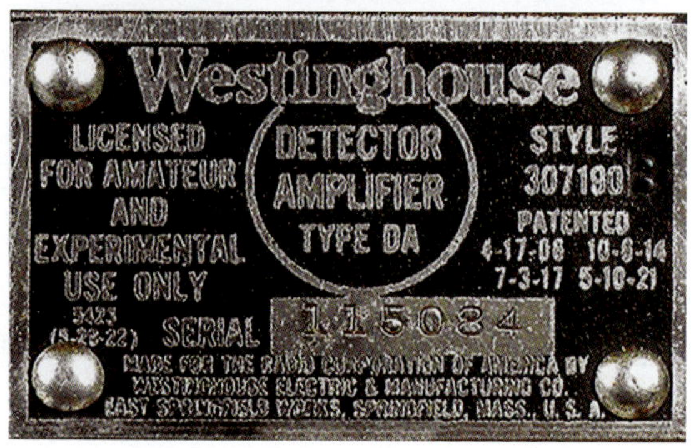

E

Each of the five different DA serial number plates has a unique design which can be used to readily identify the production run date at a glance. A. 11-30-20 B. 5-18-21 C. 8-2-21 D. 12-14-21 E. 5-29-22

arrive at these results is explained in Appendix D. The numbers of units manufactured in the first four production runs are substantially less than in the last two, and as a result these units are more desirable from the collector's viewpoint. The units manufactured in Pittsburgh are the first and among the rarest, and therefore are the most desirable of all. While there is some uncertainty in the numbers manufactured within each of the six groups, the total estimated number of 144,800 units manufactured is believed to be reasonably accurate.

While the total number of individual RA and DA units manufactured can be determined from the observed serial numbers with some accuracy, determining the number of RA and DA units paired together and sold as RC receivers is somewhat more complicated. The upper bound to that number is 72,400, which is arrived at by assuming that all of the 144,800 sets indicated in Table 3-2 are paired together. However, a number of these units were sold individually as RA's and DA's. This fraction can be estimated by taking the ratio of a) the number of all RA and DA units in the serial number data base sold individually to b) the number of all RA and DA units in the data base sold either individually or paired as RC's. Based on the available serial number data base, this fraction is 50% for the first group dated 11-30-20, 5% for the last group dated 5-29-22, and 22% for the remaining units considered as one group. Using this technique, the total number of RC sets manufactured is estimated to

Table 3-3. The total number of RA's and DA's combined into RC's

Group Date	Total Number of RA's and DA's Manufactured	Percentage Sold Individually	Total RA & DA Sold Individually	Total RA & DA Sold as RC's	Total Number of RC's
11-30-20	1700	50%	850	850	425
5-29-22	62,800	5%	3140	59,660	29,830
All others	80,300	22%	17,666	70,664	35,332
Totals	144,800	15%	21,656	123,144	65,587

be 65,587 as indicated in Table 3-3.

This number of RC's is substantially smaller than the 80,000-85,000 estimated by Douglas,[40] who also used serial numbers. Douglas began with an upper bound of 100,000 RC's instead of the 72,400 bound estimated here.[41] His upper bound included a large number of RC's that are now known to be other types of sets with interleaved serial numbers. He then made an educated guess that the upper bound should be reduced by 15,000-20,000 to account for *both* the existence of other models with interleaved serial numbers and the sales of individual RA's and DA's. Based on a more detailed analysis of a larger serial number data base, it appears that his upper bound should have been reduced by about 13,000 or so to account for other receiver types, and an additional 21,600 or so to account for sales of individual RA's and DA's.

DA Design Changes

The DA's in the first production run in Pittsburgh were assigned the Style Number 307190 and dated 11-30-20, while the Style Number for all DA's manufactured in East Springfield was altered slightly by adding a "B." There were numerous design changes made to the DA throughout the life of this unit. Many of the design changes were cosmetic in nature, intended primarily to decrease cost, but some were clearly caused by functional requirements. Other changes were made over time, some coinciding with a new production run, and

Table 3-4. Design features of the DA detector-amplifier evolved in each of the four DA production runs following the original production run dated 11-30-20

DA Design Feature Changes from 11-30-20	5-18-21	8-2-21	12-14-21	5-29-22
Copper shield on cabinet walls changed to ferromagnetic metal	✓	✓	✓	✓
Rear terminals changed from fiber to metal	✓	✓	✓	✓
Labels added to phone jacks	✓	✓	✓	✓
Tube socket design changed and all sockets rotated by 45°	✓	✓	✓	✓
Center tube socket rotated by 180°		✓	✓	✓
Panel mounting brackets changed from solid blocks to L-brackets		✓	✓	✓
Knob diameter decreased; arrowhead pointer design added			✓	✓
Mortised hinges on lid changed to surface mounted			✓	✓

others implemented in the middle of a production run. These design changes for the DA units are summarized in Table 3-4, along with the production run or runs in which each change became prevalent.

In addition to these design changes, there were also changes in the supplier for individual parts such as the telephone jacks and transformers. While the changes in most of these parts are not readily apparent to the eye, a change in the physical appearance of the audio transformer was significant, and differs markedly from the transformer that appears in all of the early ads. These transformers were changed sometime during the second production run, and later changed back to the original. The novice collector might believe the alternate transformer was an after-market replacement for the notoriously fragile factory-installed transformer, but this is not necessarily the case. Other minor variations have been observed such as rounded corners on the lids of the tube access hole on a few sets. The reason for this type of variation on such a small number of sets is unknown.

There were also some minor wiring changes, perhaps the most well known being a change to reduce distortion by inserting a "C" battery in the grid circuit of the amplifier tubes. The RC was originally designed to have 45 volts applied to the plate of the amplifier tubes, but the desire for increased power to drive loudspeakers led to increases of plate voltage first to 68 volts and then to 90 volts. This increase in plate voltage changed the operating point of the amplifier tubes such that the output became distorted. The distortion could

The more obvious changes to the front panel of early DA detector-amplifiers (top left) were the addition of labels to the phone jacks and a change in pointer design on the filament rheostats of later panels (top right). The mortised hinges on earlier DA lids (below) were changed to surface-mounted hinges with both round and square shapes on later lids (center and right units).

be remedied by inserting a "C" battery in the grid circuit of the amplifier tubes to reposition the operating point to a more linear portion of the operating characteristic. The wiring diagram for this change is shown in the RCA Victor Service Notes under the section entitled Radiola RT.[42] Later, the WD-12 tubes were substituted for the UV-200 or UV-201-A tubes, and the distortion then occurred at voltages as low as 45 volts. RCA published articles on how the RC owner could retrofit his or her set with a "C" battery to reduce the distortion,[43] and many sets can be found with wiring changes made by owners to accommodate a "C" battery.

Variations in the thickness of the front panels have been reported,[44] but measurements on a number of panels from each production run do not show any marked variation. Since the front panels were hand sanded and polished, it is certainly possible that some sets have thicker panels than others, but it does not appear to be an intentional design change. Also, a decidedly brown hue has been observed on some of the front panels, but there does not seem to be any pattern with regard to different production runs, and so it is possible that a small number of the panel changed color over time as the result of exposure to some environmental factors such as sunlight.

RA Design Changes

There were a number of design changes made to the RA tuner over the life of this unit as indicated in table 3-5. The RA's in the original production run made in Pittsburgh were assigned Style No. 307189 and dated 11-30-20, while the Style Number for all subsequent RA's manufactured in East Springfield

Table 3-5. Design features of the RA tuner evolved in each of the four RA production runs following the original production run dated 11-30-20

RA Design Feature Changes from 11-30-20	9-28-21	8-2-21	12-14-21	5-29-22
Copper shield on cabinet walls changed to ferromagnetic metal	✓	✓	✓	✓
Rear terminals changed from fiber to metal	✓	✓	✓	✓
Wiring rerouted away from coil; minor change to spiral connectors	✓	✓	✓	✓
Panel mounting brackets changed from solid blocks to L-brackets		✓	✓	✓
Dial skirts changed from Micarta to metal; dial print size decreased			✓	✓
Westinghouse logo added to large center dial				✓

was altered slightly by adding an "A." The first changes to the RA indicated in the table were actually made on sets manufactured in Pittsburgh near the end of the 11-30-20 production run, most likely in preparation for the transfer of production to East Springfield, Mass. The few RA's made in Pittsburgh with these changes have serial numbers in the mid-3600s, and are virtually indistinguishable from those made in the first production run in East Springfield dated 9-28-21—with the notable exception of the details on the name plate.

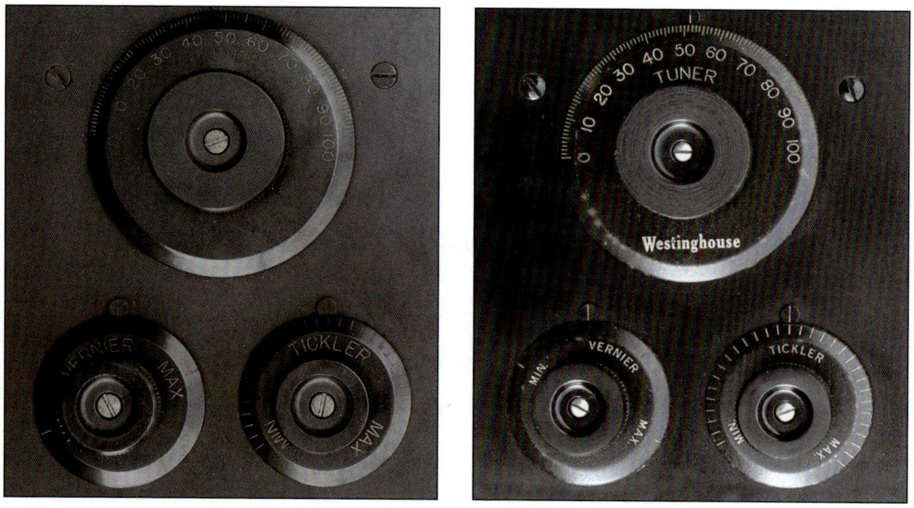

The more obvious changes on the front panel of the early RA tuner (left) were a change in print size on the dials and the addition of the Westinghouse name on the tuning dial of later panels (right).

Vacuum Tubes for the RC

The RC was designed in late 1919 and early 1920 when the only triodes available were the Moorhead tubes licensed under an agreement with Marconi and De Forest. Shortly after RCA was formed, this license was revoked, and the availability of triodes for the RC became of increasing concern to Westinghouse. Westinghouse purchased a large stock of Moorhead tubes for the DA units, and these tubes were sold with the early sets by distributors in 1921. A few months after the Westinghouse-RCA agreements were signed in mid 1921, distributors began selling the UV-200 and UV-201 Radiotron tubes with the DA unit for the detector and amplifier functions respectively. The equivalent Cunningham C-300 and C-301 tubes were also often sold by distributors for the DA unit. Distributors also offered the RC with the UV-201A Radiotron, a tube with a thoriated tungsten filament requiring less filament current which became available in 1922.

In the spring of 1923, RCA began marketing the RC with the new Westinghouse WD-12 tubes, which were identical to the WD-11 tubes except with a Navy standard base used in the RC. These tubes drew significantly less filament current so that the radio could be operated on dry cells rather than storage batteries. The price for a complete set "the New Way" using dry cells was advertised at $142.50, as compared to a price of $207.50 "the Old Way" using storage batteries. This advertising approach was mostly a marketing ploy

Three different tubes were used in the RC over its lifetime including the Marconi-De Forest tubes manufactured by Moorhead (left), the UV 200 and 201 tubes manufactured by GE (center), and the WD-12 tubes (right) manufactured by Westinghouse.

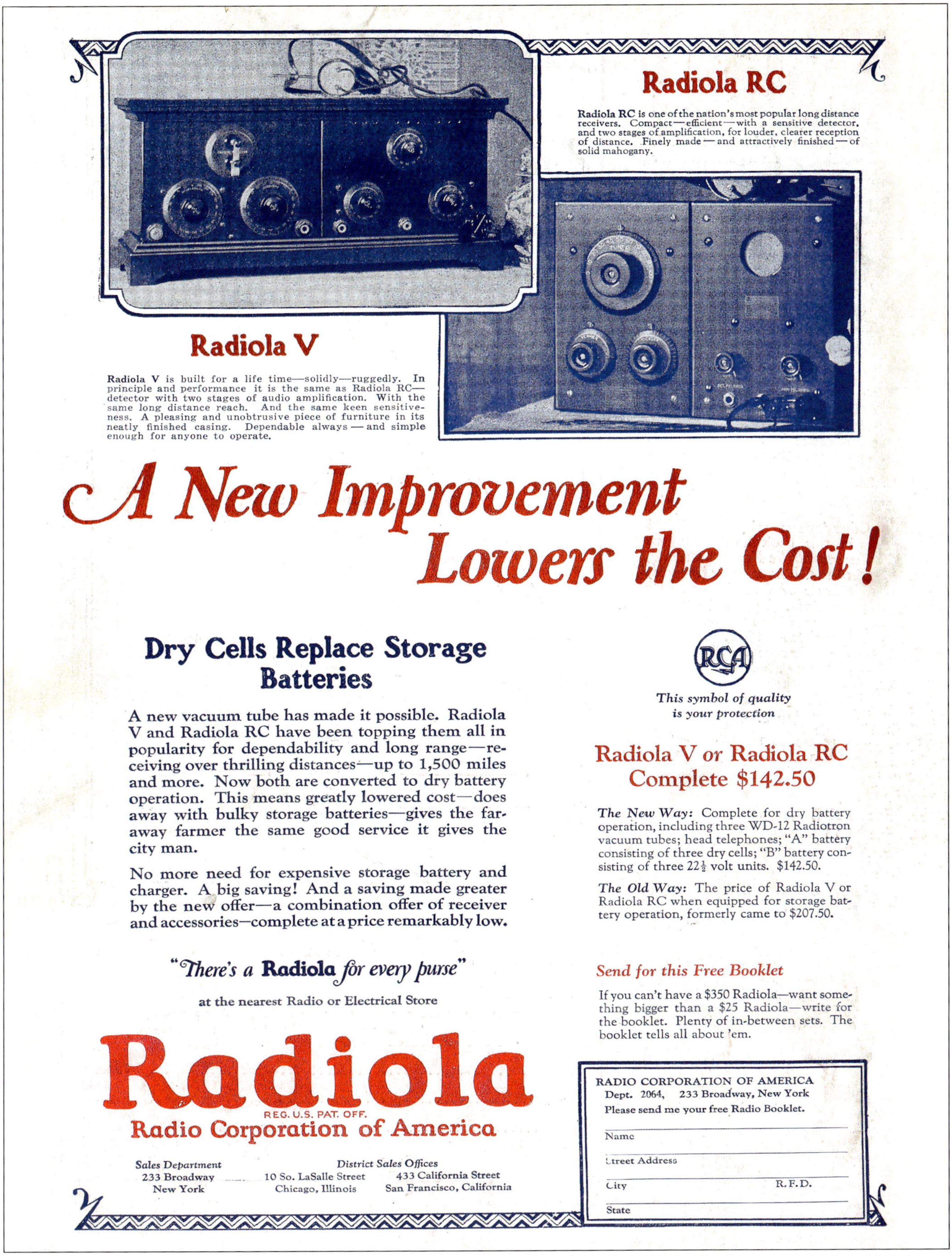

RCA tried to clear out a glut of unsold RC's in early 1923 by substituting the WD-12 dry cell tube for the storage battery UV tubes, and selling the new combination including dry cell batteries at the reduced price of $142.50. (*The Wireless Age*, May 1923, back cover)

by RCA in an attempt to sell a glut of RC's, which resulted from a large order that RCA placed for RC's just before the big slump in radio sales during the later half of 1922.

Over its lifetime, the DA detector-amplifier was offered with three distinctly different types of tubes made by three different manufacturers: the Marconi-De Forest tubes manufactured by Moorhead, the UV series and/or equivalent Cunningham tubes manufactured by GE, and the WD-12 tubes manufactured by Westinghouse. This distinction can be attributed not only to the popularity of the RC—which was actively advertised and sold over a 2½-year period—but also to the rapid change in tube technology and patent positions during that period.

The RC is Discontinued

By 1923, the RC was basically obsolete, and the marketing ploy with WD-12 tubes in early 1923 to clear out the RC sets did not result in substantial sales.

RCA made arrangements with Gimbel Brothers Department Store to clear out 20,000 RC sets in July of 1923 at $59.95, well below RCA's cost. (*The New York Times*, July 23, 1923, p. 7)

RCA distributors such as the Stuyvesant Radio Corporation were forced to meet or beat the Gimbels clearance sale prices. (*The New York Times*, August 4, 1923, p. 5)

Consequently, RCA made arrangement with Gimbel Brothers in New York to sell 20,000 RC's for $30 a piece complete with phones, batteries, and three WD-12s.[45] Gimbels placed an ad in the *The New York Times* on July 23, 1922 offering the 20,000 sets for $59.75 each,[46] and at that price, they sold all 20,000 in less than a week! In order to fill backorders, Gimbels arranged to purchase several thousand RA's and DA's remaining in the RCA inventory, and sold them

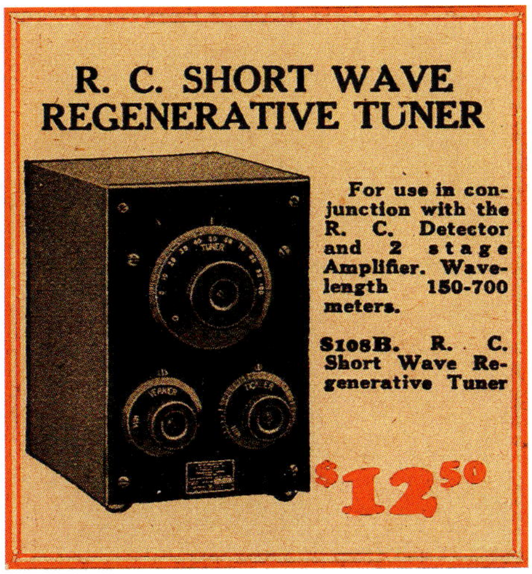

Distributors who were unwilling or unable to sell their stocks of discontinued receivers at clearance sale prices were forced to sell their remaining stock to radio salvage companies for less than twenty cents on the dollar. (Chicago Salvage Stock Store Catalog, Chicago, IL, 1926, p. 42)

together as an *"RC of the RA + DA type slightly different (but equally as good as) those we sold last week."*[47]

While the RC receiver was touted as one of the most popular and successful receivers of the early 1920s, it was not quite as popular or successful as has been previously portrayed. While there were an estimated 65,587 RC receivers manufactured, RCA sold 20,000 complete sets including batteries, three WD-12 tubes, and earphones at clearance for $30, far below RCA's cost of about $70 for the RC receiver alone without accessories. With the accessories, it is estimated that RCA lost well over $1 million on this clearance sale.

However, the carnage did not stop with RCA, because there were a large number of RC's still in the distribution channel at the time of the clearance sale, and these too were also sold at or below the $59.75 price. For example, within the week after the Gimbels ad appeared, Stuyvesant Radio Corporation, an RCA distributor, placed an ad in *The New York Times* offering their inventory of 100 RC sets for $2 less than the Gimbels $59.75 price. The distributors had little choice but to match RCA's close out price at the time or to sell the inventory later to radio salvage companies at even lower prices.[48] These clearance sales were a sore point with distributors because they generally paid significantly more for their inventory than they were able to charge after the clearance prices were advertised, invariably resulting in large losses for the distributors.

According to the 1923 RCA Annual Report there were 116 distributors with 173 distribution centers as well as thousands of retail dealers—many of which must have had a significant inventory of RC's. If each distributor had about 100 RC's in stock and each retail dealer had another 2-3 RC's in stock as well, then there would have been another 10,000-15,000 sets in the distribution channel as of July 1923. Consequently, as many as 30,000-35,000 RC's were actually sold at a loss in 1923, meaning that only about half the total number of RC's manufactured was actually sold because of "popular demand." In fact, since the retail price of the accessories alone far exceeded the $59.75 sale price, a cynic might argue that half of all RC's manufactured were purchased for the accessories and not for the RC at all.

RCA almost certainly lost money on the totality of their RC sales over the 1½-year period they marketed the RC. Between the 20,000 RCA sold at a loss and the 10,000 or so sets Westinghouse sold before RCA's involvement beginning in 1922, RCA sold at most 33,000 RC's at full profit. Since the losses on each of the 20,000 sets sold at clearance was significantly higher than the profit from each sold at full price, it is unlikely that the profits on the 33,000 sets covered the losses on the 20,000 sets. It may well be that RCA broke even or made a slight profit if the sales of RA and DA units are considered.

RC Wannabe

A number of sets have surfaced in the U.S. with an external appearance similar to the Westinghouse RC but with name plates missing and no other identifying marks. The interior components of these sets were markedly different from those used in the RC, and the metric hardware clearly points to a European genesis. One such set pictured here was found with three UV-201-A tubes and two large black cylindrical audio transformers. It is believed to be a Telefunkon D (with an "o"), Type E304 manufactured by Telefunken in Germany for export only—this type of regenerative set was not allowed in Germany at the time.

The same set has been reported in Norway with a name plate on the top stating in English "Made in Germany" and identifying the set as Type E304.[49] The author reported that it had three Telefunken RE58 tubes, but that he did not know the manufacturer. The photograph in the article is exactly the same—both inside and outside—as the set pictured here. The Radiomuseum web site based in Europe has also identified a set with the same physical appearance (except for slightly different knobs) as a Telefunkon D (Type E304).[50] This picture appears to be taken from a German publication which identified the set

The exterior of the Telefunkon D, Model E304 is almost identical to the Westinghouse RC.

The components used in the Telefunkon D, Model E304—most notably the transformers and capacitors—are distinctly different from those used in the Westinghouse RC.

as a Telefunkon D, and listed a wavelength specification of 200-600 meters (versus 180-700 meters for the Westinghouse RC).[51]

Not only was the Telefukon D, E304 made in Germany, it was also made in Sweden by Svenska Aktiebolaget Trådlös Telegrafi (SATT), a Swedish company owned by Swedish AEG-Telefunken (see Appendix H for an ad with a photograph). SATT actually manufactured two models, a transformer-coupled model E304 circa 1923-25, and a resistance-coupled model E305 circa 1927 evocative of the Westinghouse DR. The Swedish Model E304 had a stated operating range of 150 to 4,500 meters, while the E305 had a stated operating range of 150 to 1,750 meters. However, since these sets with Swedish markings were intended only for domestic consumption, it is unlikely the set pictured here was manufactured by SATT.

According to a long-time SATT employee, RCA licensed the RC technology to Telefunken in a program to help German industry recover from WWI. (Evidence of licensing arrangements is given in Appendix H under Sweden.) The licensing arrangement would almost surely have precluded these sets from being imported into the U.S. where they would have competed directly with the RC sets distributed by RCA. Nevertheless, a number of German sets have found their way to the U.S. over the years where they appear frequently at auction and in flea markets.

The DB Crystal Detector for the RA Tuner

Westinghouse introduced a "super-sensitive" DB crystal detector to the public in advertisements appearing circa August 1921, and the combination of the DB detector and RA tuner was subsequently sold as the RA crystal receiver. RCA priced the resulting crystal set at $90, which was substantially less expensive than the regenerative RA-DA combination RCA priced at $216, but was still more expensive than competing crystal sets by other manufacturers, which were typically selling for $20-$40 or less. The DB detector was marketed through the 1923 season and then discontinued.

The DB detector actually consisted of two different types of detectors mounted on a Micarta base with a selector switch to select the appropriate detector, and a telephone condenser located underneath the base and wired in parallel with the output for the earphones. The input terminals were connected to the "GRID and "FIL GND" terminals on the rear of the RA unit. One detector was a galena crystal with a cat's whisker, while the second detector was a Perikon detector.[52] The Perikon, an acronym extracted somewhat torturously from **Per**fect P**ick**ard **Con**tact, was discovered and patented by Greenleaf Whittier Pickard while at Wireless Specialty Apparatus.

The Perikon detector was typically zinc oxide (zincite) mounted in a fixed cup with a copper pyrite such as chalcopyrite on a spring-loaded rod. It was

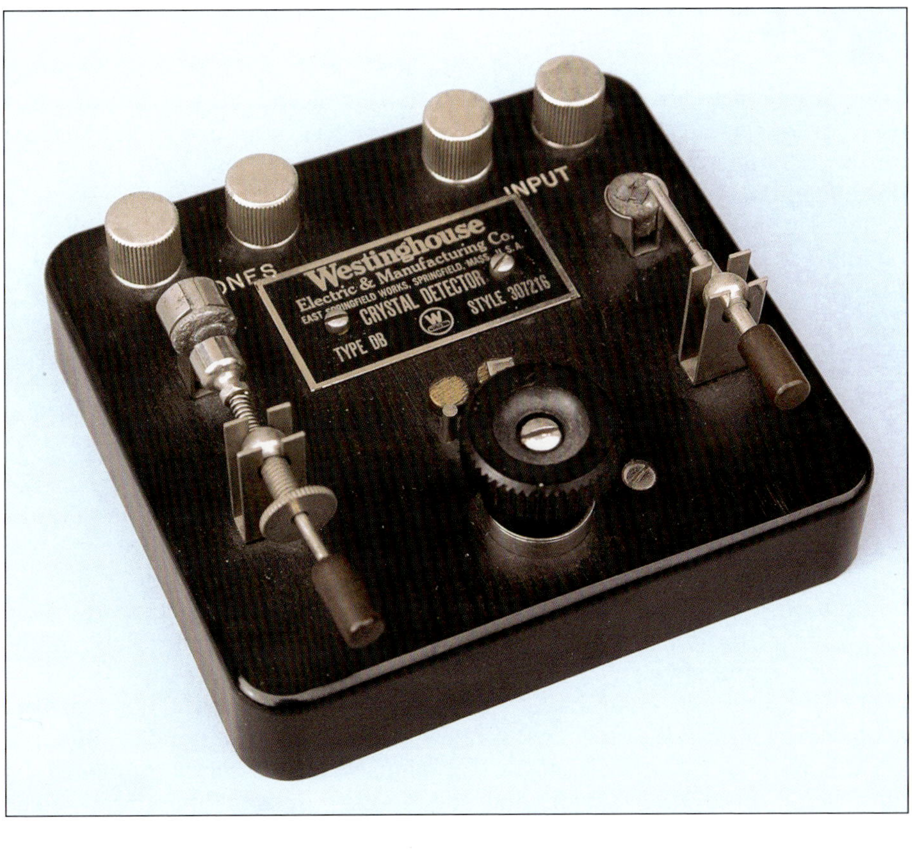

The earliest DB detector did not have a date of manufacture on the name plate.

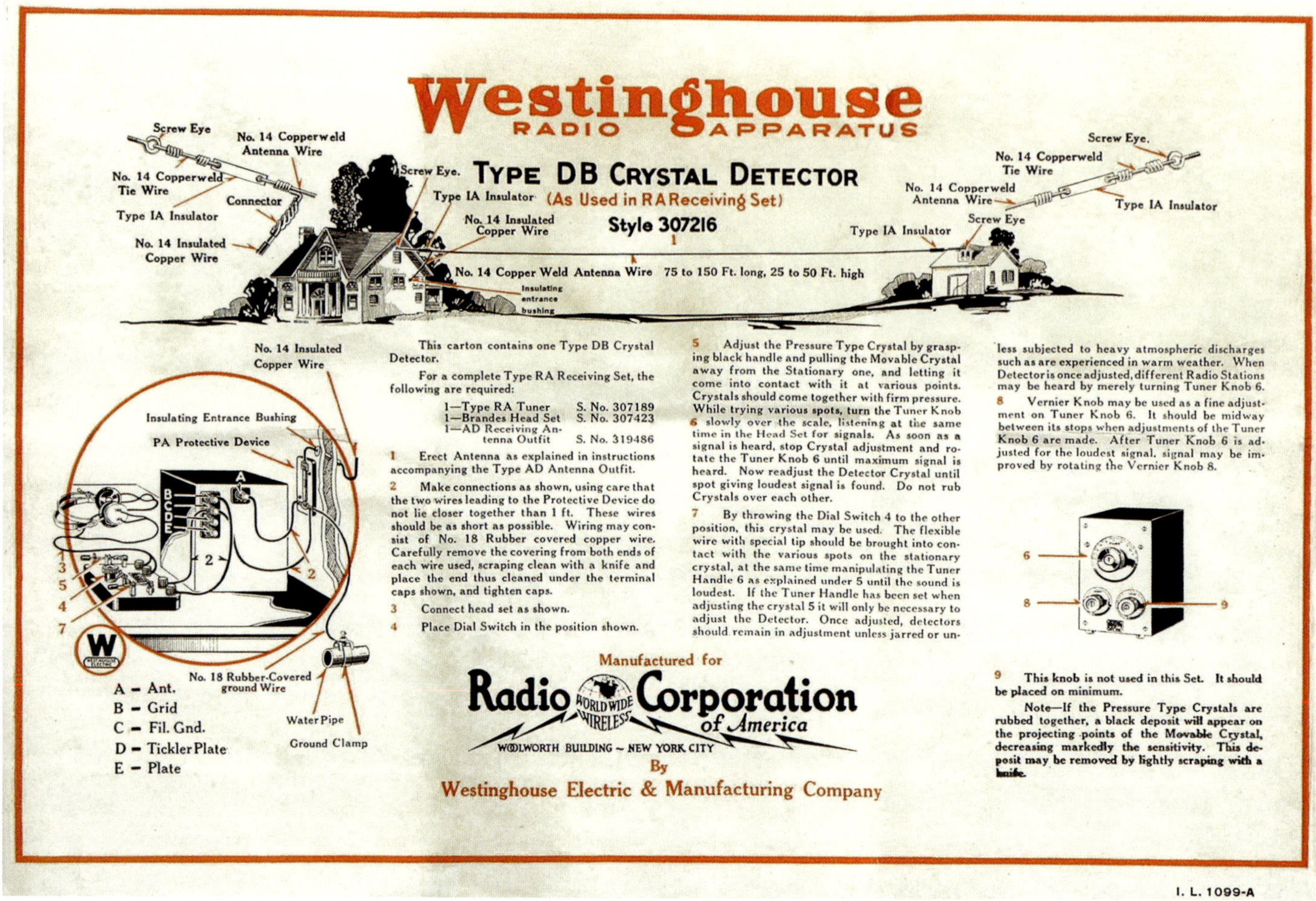

Westinghouse
RADIO APPARATUS

TYPE DB CRYSTAL DETECTOR
(As Used in RA Receiving Set)
Style 307216

This carton contains one Type DB Crystal Detector.

For a complete Type RA Receiving Set, the following are required:

1—Type RA Tuner	S. No. 307189
1—Brandes Head Set	S. No. 307423
1—AD Receiving Antenna Outfit	S. No. 319486

1 Erect Antenna as explained in instructions accompanying the Type AD Antenna Outfit.

2 Make connections as shown, using care that the two wires leading to the Protective Device do not lie closer together than 1 ft. These wires should be as short as possible. Wiring may consist of No. 18 Rubber covered copper wire. Carefully remove the covering from both ends of each wire used, scraping clean with a knife and place the end thus cleaned under the terminal caps shown, and tighten caps.

3 Connect head set as shown.

4 Place Dial Switch in the position shown.

5 Adjust the Pressure Type Crystal by grasping black handle and pulling the Movable Crystal away from the Stationary one, and letting it come into contact with it at various points. Crystals should come together with firm pressure. While trying various spots, turn the Tuner Knob 6 slowly over the scale, listening at the same time in the Head Set for signals. As soon as a signal is heard, stop Crystal adjustment and rotate the Tuner Knob 6 until maximum signal is heard. Now readjust the Detector Crystal until spot giving loudest signal is found. Do not rub Crystals over each other.

7 By throwing the Dial Switch 4 to the other position, this crystal may be used. The flexible wire with special tip should be brought into contact with the various spots on the stationary crystal, at the same time manipulating the Tuner Handle 6 as explained under 5 until the sound is loudest. If the Tuner Handle has been set when adjusting the crystal 5 it will only be necessary to adjust the Detector. Once adjusted, detectors should remain in adjustment unless jarred or unless subjected to heavy atmospheric discharges such as are experienced in warm weather. When Detector is once adjusted, different Radio Stations may be heard by merely turning Tuner Knob 6.

8 Vernier Knob may be used as a fine adjustment on Tuner Knob 6. It should be midway between its stops when adjustments of the Tuner Knob 6 are made. After Tuner Knob 6 is adjusted for the loudest signal, signal may be improved by rotating the Vernier Knob 8.

9 This knob is not used in this Set. It should be placed on minimum.

Note—If the Pressure Type Crystals are rubbed together, a black deposit will appear on the projecting points of the Movable Crystal, decreasing markedly the sensitivity. This deposit may be removed by lightly scraping with a knife.

Manufactured for
Radio Corporation of America
WOOLWORTH BUILDING ~ NEW YORK CITY
By
Westinghouse Electric & Manufacturing Company

I. L. 1099-A

Instructions for the DB detector referenced the RA tuner.

less sensitive to movement and vibration, and therefore provided a more reliable detector function, but it was also less sensitive than the galena crystal detector. Thus, the two were complementary—the Perikon detector was used for shorter-range but more-reliable reception, while the crystal detector was used for longer-range but less-reliable reception. Perikon detectors appeared on virtually all crystal sets marketed by Westinghouse and RCA (including those manufactured by GE for RCA).

There were at least five different versions of the DB detector made in three different production runs, the earliest having no date of manufacture, the next two bearing the date (7-27-21), and the last two bearing the date (12-14-21). There is very little difference among the three production runs except the manner in which the binding posts were labeled and the design of the holes intended to accept mounting screws. All three production runs used bases with engraved labels inlaid with white ink to mark binding post functions, but bases in the second and third production runs—those with dates—were also made with the labels embossed or molded directly into the Micarta. The first version without a production date had no holes for mounting screws, whereas all four dated versions have two holes, one on either side on the base. Two of the bases with the molded labels have a slight indent in the wall directly under the screw holes to provide clearance for the screws used to attach the detector to a table top, whereas the other two bases with holes do not. The reason for these minor variations is unknown.

It is interesting to note that a "vertical-shaft" version of the detector was depicted in an early Westinghouse catalog and brochure, but none is known to exist, and it is unlikely that this version was ever mass produced or marketed.[53] Both this version and the mass-produced "horizontal" shaft

FIG. 6—TYPE DB CRYSTAL DETECTOR

Crystal Detector

Style No.	Description	Dimensions Height	Width	Length
307216	Crystal Detector complete with two crystals	4¼″	5″	5″

(Order by Style Number)

An alternative form of the DB detector with vertical control shafts appears in early Westinghouse sales literature but was apparently never manufactured or marketed. (*Radio Receiving Apparatus for Short Wave Telegraphy and Telephony*, Westinghouse Circular 1641, N.D., p. 4)

DB detector dated 7-27-21 with engraved labels

DB detector dated 7-27-21 with stamped labels

DB detector dated 12-14-21 with engraved labels

DB detector dated 12-14-21 with stamped labels

There were at least four additional versions of the DB detector manufactured, each slightly different from the original and from each other.

version had the same Style No. 307216. The DB detector was on the market until the end of the 1923 season.

The DR Detector-Amplifier Enigma

The DR detector-amplifier is a resistance-coupled version of the transformer-coupled DA. The two are virtually identical with regard to form, fit, and function—from the exterior appearance, it would be difficult to distinguish one from the other were it not for the name plate, which clearly identifies one as a DR and the other as a DA. The most striking visual differences in the interior of the DR are the presence of two large dull-black resistors and two smaller glass resistors in lieu of the two audio transformers, and three WR-21 tubes with European bases in lieu of the Moorhead or GE tubes with Navy standard bases. A few of the DR units have surfaced with the Navy standard base,[54] presumably to accommodate the rebased WR-21 which Westinghouse designated a UV-201—not to be confused with the GE UV-201.

Only a hand full of DR's have surfaced, and most bear serial numbers in the mid-10,500 range, the Style No. 319570, and a production date in 1921.[55] Based on a relatively large gap in RA/DA serial numbers between 10,000 and 11,500, it is estimated that about 800 DR's were manufactured (see Appendix D).

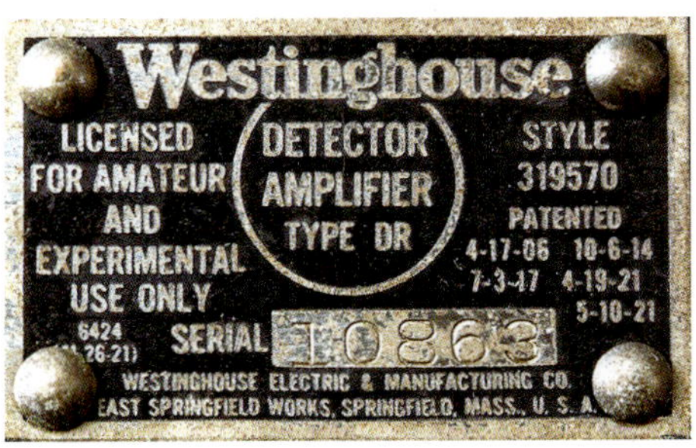

The exterior appearance of the resistance-coupled DR is virtually identical to the transformer-coupled DA, and is distinguishable from the exterior only by the name plate.

The interior of the DR is markedly different from the DA, with the most obvious differences being the large glass resistors in lieu of the audio transformers and the European tube sockets in lieu of the Navy standard sockets with the tall collars.

The DR was never nationally advertised, nor did it ever appear in RCA distributor price sheets, so there is some question about whether it was ever offered for sale to the public. These circumstances alone would not have precluded its sale to the public—it must be remembered that the Radiola Special was never nationally advertised nor did it appear in any RCA price list. The relatively large number of sets (estimated at 800) and the presence of serial numbers would indicate they are production units. Of course, it is also possible that the number of sets manufactured could be far less than the 800-unit estimate, and Westinghouse was known to apply serial numbers to engineering models at that time. The most direct evidence that the DR was manufactured and offered for sale by RCA is the existence of a DR instruction sheet similar in appearance to the DA instruction sheet.

A significant clue to the mystery is the fact that both the DA and the DR appeared on RCA Stock Sheet forms that were pre-printed in 1922 and used well into 1923. These forms were used by RCA for several purposes including inventory control and back-order tracking. The inventory and back-order status for each item were penciled in on these forms on a monthly basis. The fact that both the DA and the DR appear on these pre-printed forms is an indication that RCA had, in fact, made a decision to manufacture and distribute both simultaneously.

It is difficult to understand why RCA would have ever made the decision to market the DR in direct competition with the commercially successful DA—particularly with its identical appearance and function. It might have made sense if there had been performance issues with the DA, but there were no known performance issues with the DA in particular, or with transformer-coupled amplifiers in general. One possibility is the cost savings afforded by a resistance-coupled set, which does not require the comparatively expensive transformers. In this case, the cost tradeoff would have been between two audio transformers versus two large glass resistors and several capacitors. The net

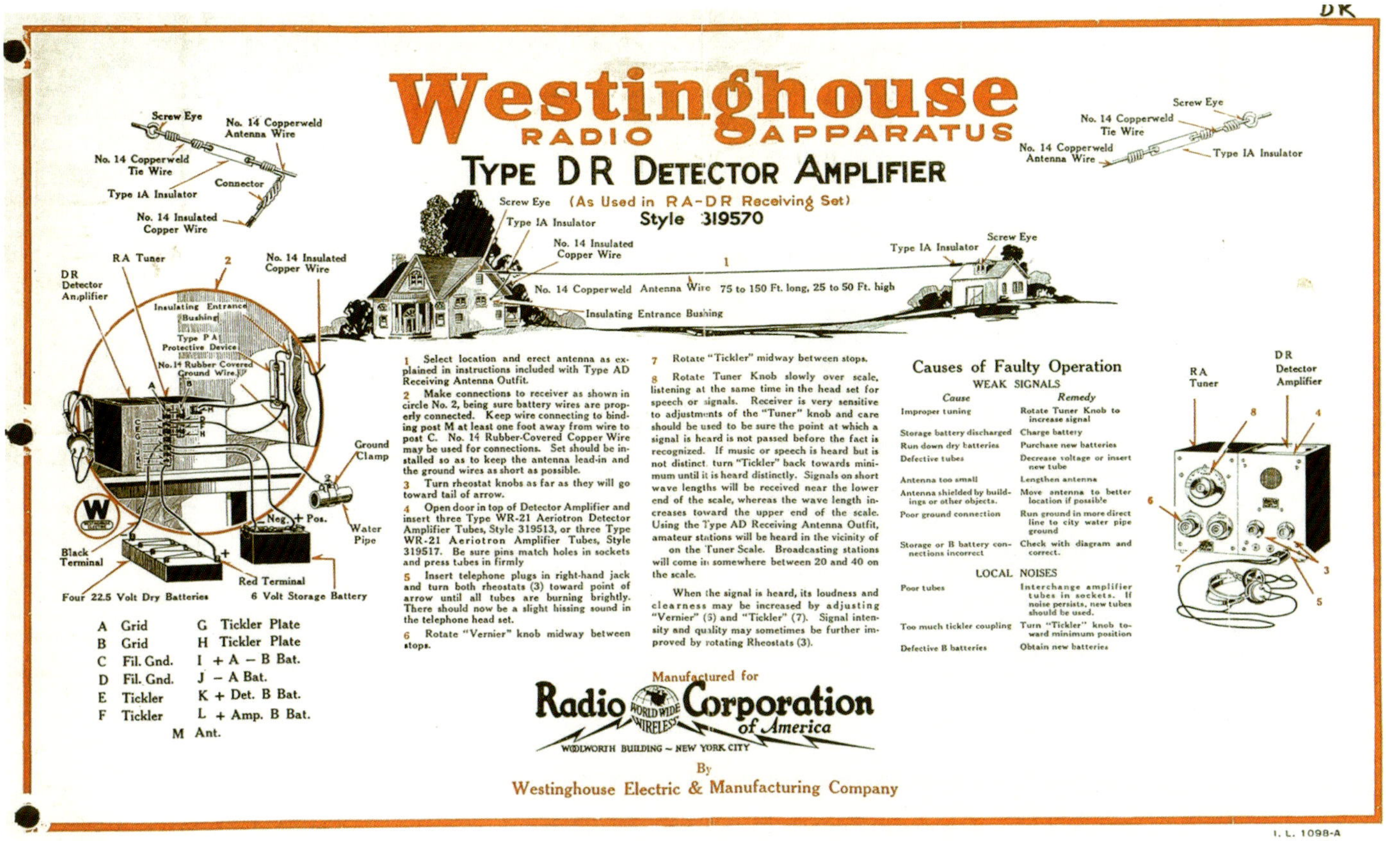

Westinghouse
RADIO APPARATUS
TYPE D R DETECTOR AMPLIFIER
(As Used in RA-DR Receiving Set)
Style 319570

No. 14 Copperweld Antenna Wire 75 to 150 Ft. long, 25 to 50 Ft. high

1 Select location and erect antenna as explained in instructions included with Type AD Receiving Antenna Outfit.

2 Make connections to receiver as shown in circle No. 2, being sure battery wires are properly connected. Keep wire connecting to binding post M at least one foot away from wire to post C. No. 14 Rubber-Covered Copper Wire may be used for connections. Set should be installed so as to keep the antenna lead-in and the ground wires as short as possible.

3 Turn rheostat knobs as far as they will go toward tail of arrow.

4 Open door in top of Detector Amplifier and insert three Type WR-21 Aeriotron Detector Amplifier Tubes, Style 319513, or three Type WR-21 Aeriotron Amplifier Tubes, Style 319517. Be sure pins match holes in sockets and press tubes in firmly

5 Insert telephone plugs in right-hand jack and turn both rheostats (3) toward point of arrow until all tubes are burning brightly. There should now be a slight hissing sound in the telephone head set.

6 Rotate "Vernier" knob midway between stops.

7 Rotate "Tickler" midway between stops.

8 Rotate Tuner Knob slowly over scale, listening at the same time in the head set for speech or signals. Receiver is very sensitive to adjustments of the "Tuner" knob and care should be used to be sure the point at which a signal is heard is not passed before the fact is recognized. If music or speech is heard but is not distinct turn "Tickler" back towards minimum until it is heard distinctly. Signals on short wave lengths will be received near the lower end of the scale, whereas the wave length increases toward the upper end of the scale. Using the Type AD Receiving Antenna Outfit, amateur stations will be heard in the vicinity of on the Tuner Scale. Broadcasting stations will come in somewhere between 20 and 40 on the scale.

When the signal is heard, its loudness and clearness may be increased by adjusting "Vernier" (5) and "Tickler" (7). Signal intensity and quality may sometimes be further improved by rotating Rheostats (3).

Causes of Faulty Operation

WEAK SIGNALS

Cause	Remedy
Improper tuning	Rotate Tuner Knob to increase signal
Storage battery discharged	Charge battery
Run down dry batteries	Purchase new batteries
Defective tubes	Decrease voltage or insert new tube
Antenna too small	Lengthen antenna
Antenna shielded by buildings or other objects.	Move antenna to better location if possible
Poor ground connection	Run ground in more direct line to city water pipe ground
Storage or B battery connections incorrect	Check with diagram and correct.

LOCAL NOISES

Cause	Remedy
Poor tubes	Interchange amplifier tubes in sockets. If noise persists, new tubes should be used.
Too much tickler coupling	Turn "Tickler" knob toward minimum position
Defective B batteries	Obtain new batteries

A Grid — G Tickler Plate
B Grid — H Tickler Plate
C Fil. Gnd. — I + A − B Bat.
D Fil. Gnd. — J − A Bat.
E Tickler — K + Det. B Bat.
F Tickler — L + Amp. B Bat.
M Ant.

Manufactured for
Radio Corporation of America
WOOLWORTH BUILDING ~ NEW YORK CITY
By
Westinghouse Electric & Manufacturing Company

I. L. 1098-A

The existence of a DR instruction sheet with the the Radio Corporation of America name and logo at the bottom, and the Westinghouse identification I.L. 1098-A—with an "A" indicating a second printing—is strong evidence that the DR was in fact marketed and distributed in quantity by RCA.

savings in cost at the wholesale level would have amounted to only a few dollars per unit at most, and this savings would most likely have been shared between both RCA and Westinghouse. Another possibility is that Westinghouse pushed this design to provide a market for its unique WD-21 tubes with a high plate resistance, which was required in an efficient resistance-coupled circuit.

Aeriola Jr. and Aeriola Sr. Receivers

The Aeriola name was selected for the early line of broadcast receivers made by Westinghouse from among a number of suggestions submitted by its employees. It would be difficult to miss the association of this selection with the Aeriola player piano made by the Aeolian Company of New York from 1895 to circa 1905, or the association of the Aeolian Company name with the ancient Greek Aeolian harp, for that matter. It is likely that all of the Westinghouse receivers—in addition to the Junior, the Senior, and the Grand—would have been named Aeriola had the RCA agreements not been signed, but instead RCA selected the Radiola trade name for their entire line of receivers.

In addition to the more familiar Aeriola name, Westinghouse continued to assign a "Type" designation consisting of two letters to their sets. In fact, they continued to assign the two-letter types even to the Radiola sets made for RCA. In the following paragraphs, the two-letter types are enclosed in parenthesis after each model name.

Aeriola Jr. (Type RE)

The Aeriola Jr. was introduced for $25 including headphones in advertisements which first appeared circa July of 1921,[56] just after Westinghouse entered into the cross-licensing agreement with RCA that specified all Westinghouse radio products would be distributed exclusively through RCA. As part of the agreement, Westinghouse continued to sell the Aeriola Jr. though its own distribution channels until early 1922, when RCA began to advertise and market all Westinghouse radio products including the Aeriola Jr.

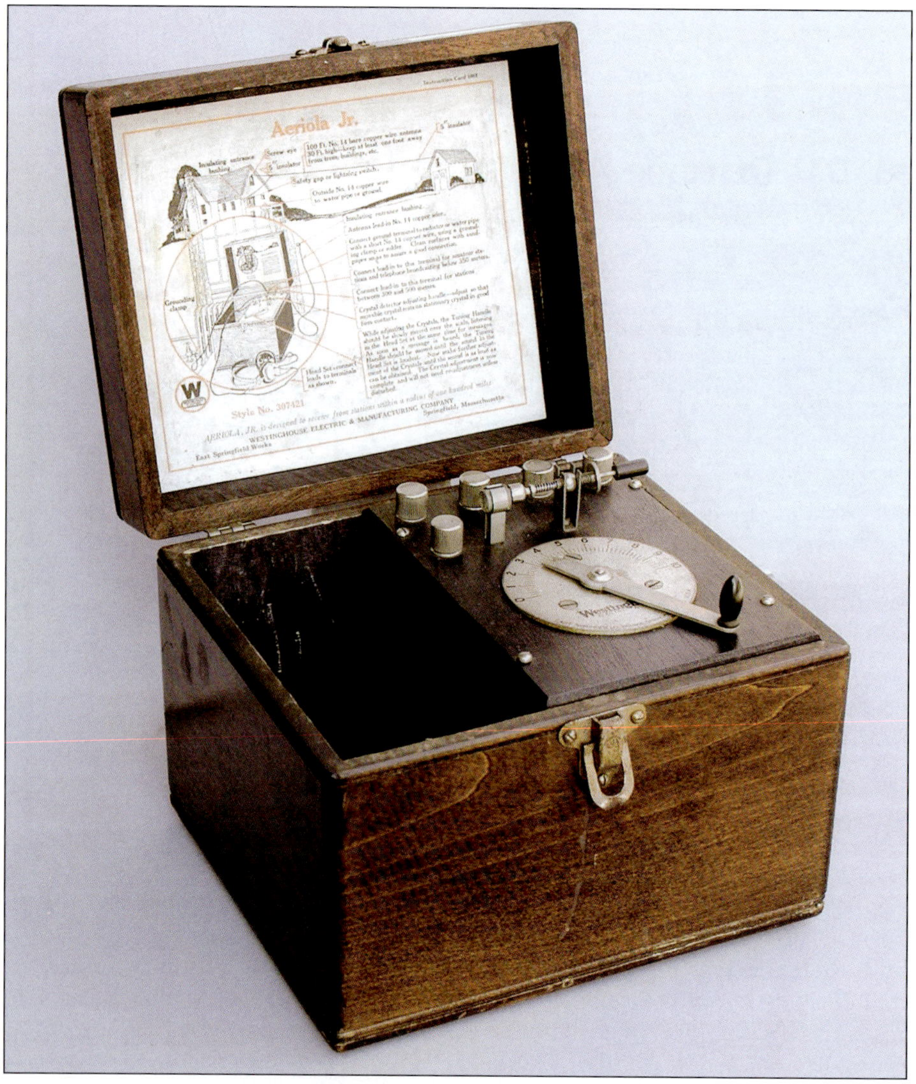

The earliest Aeriola Jr. crystal set can be readily identified by the distinctly oval shape of the tuning handle, while later versions had a cylindrically shaped tuning handle.

The Aeriola Jr. was inductively tuned with a minimum number of parts including a variometer, several capacitors and a Perikon detector.

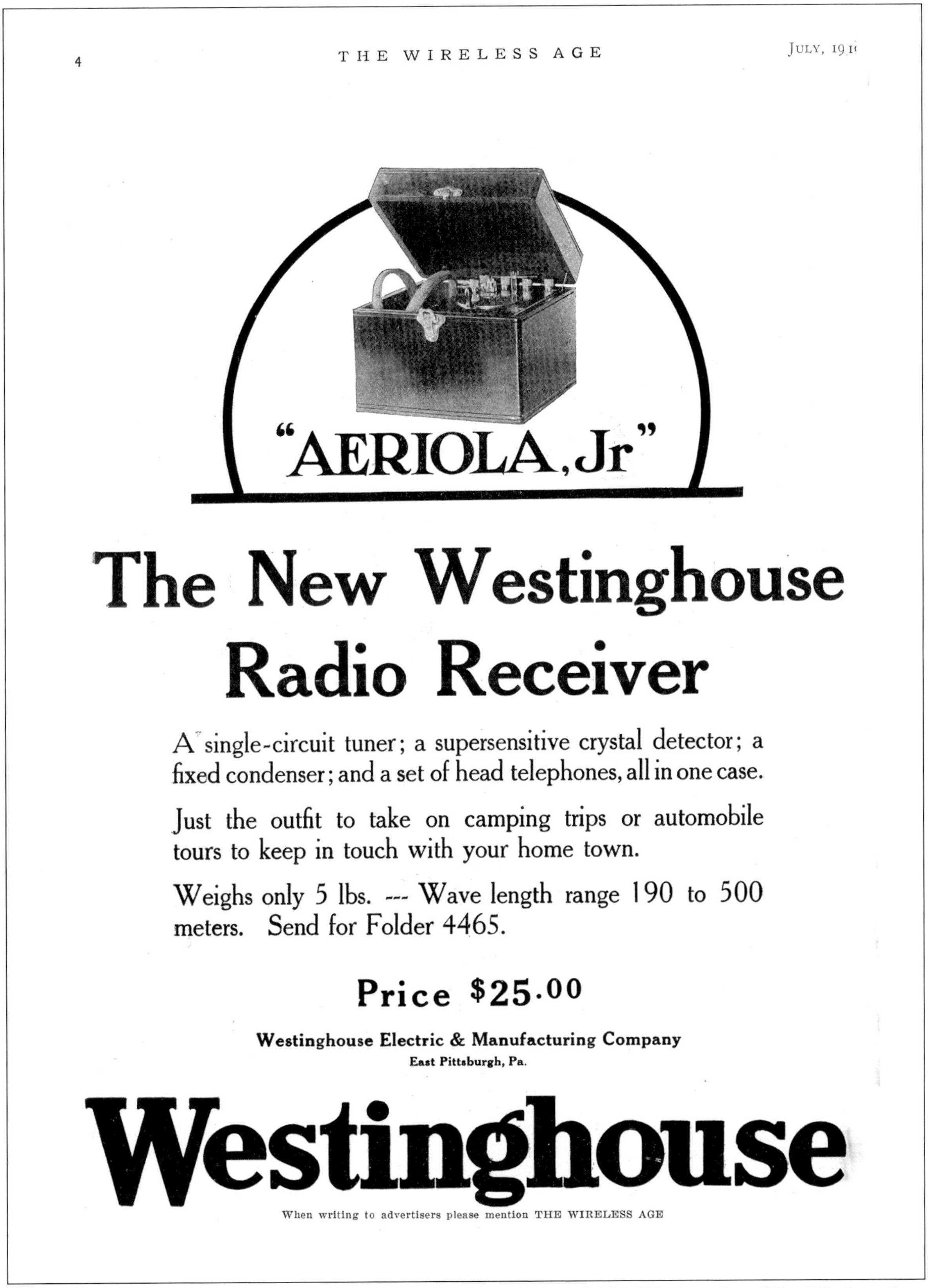

The first ad for the Aeriola Jr. was placed in the July 1921 issue of *The Wireless Age*. (*The Wireless Age*, Vol. 8, No. 10, July 1921, p. 4)

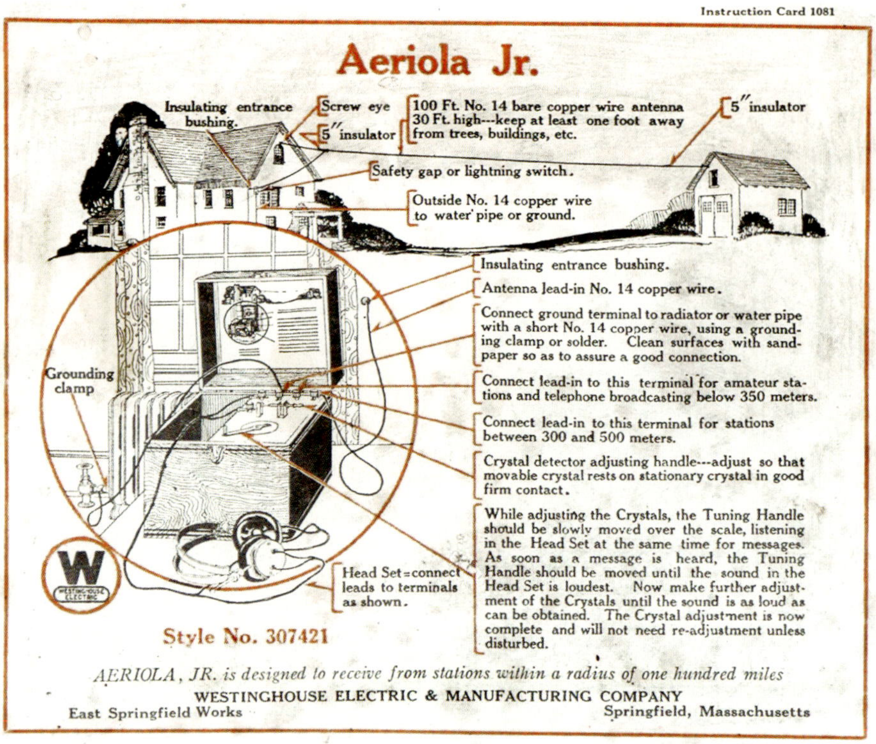

Instruction Card 1081

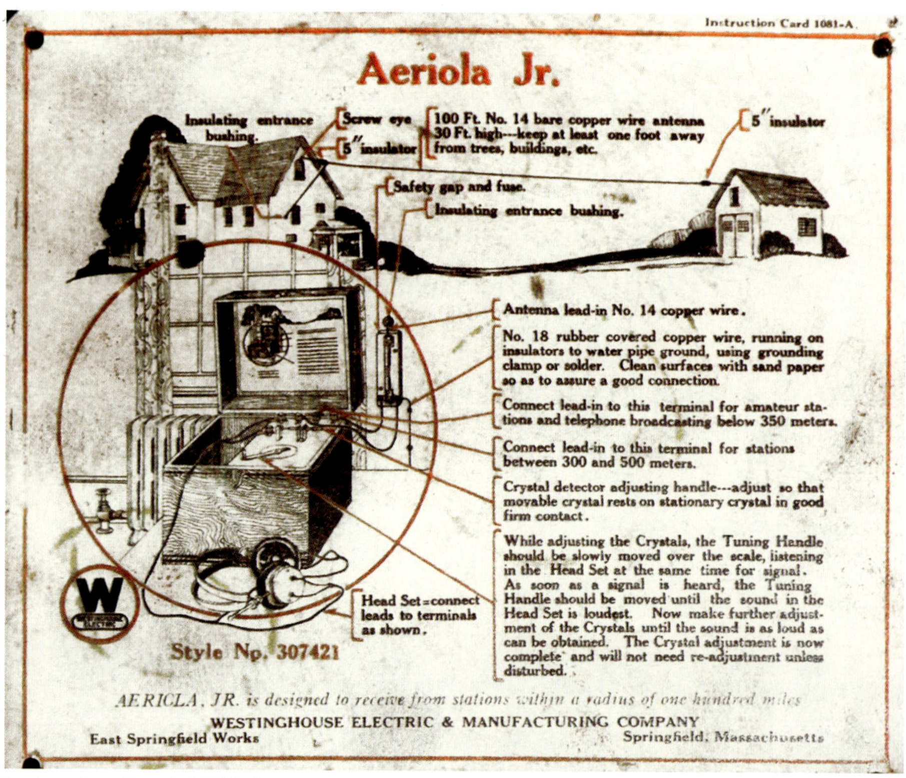

Instruction Card 1081-A

Six different instruction sheets were tacked to the lid of the Aeriola Jr. throughout its lifetime, the fourth with suffix C being virtually indistinguishable from the third with suffix B (see text); the second and third with suffixes A and B were the most common while the first (without a suffix) and the last with suffix E appear to be the rarest.

All Aeriola Jr. sets were given the designation "Style No. 307421," a designation which was retained on all sets even after RCA assumed responsibility for distributing the set. The later Aeriola Jr. receivers marketed by RCA can be distinguished from the earlier sets sold by Westinghouse by the prominent RCA logos on the instruction labels tacked to the inside lid of each set. There was a

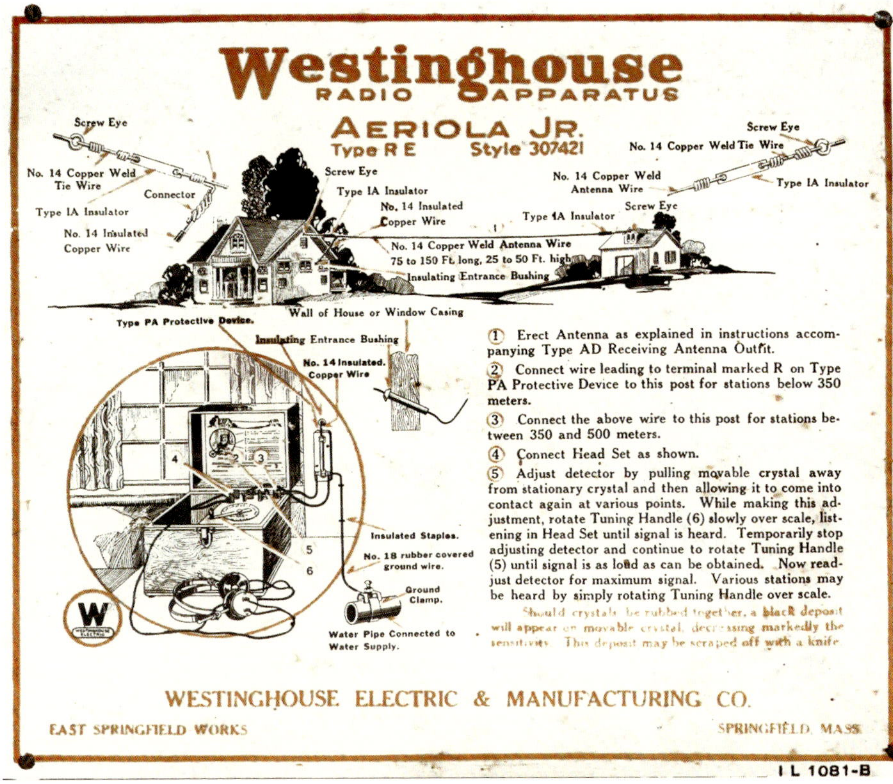

I. L. 1081-B

I. L. 1081-D

total of six different instruction labels, the first with the designation "Instruction Card 1081." The next variation in the instruction card was designated Instruction Card 1081-A, but each subsequent variation thereafter was designated I. L. 1081 followed by a dash and a letter from B to E. All instruction cards up to the letter C have Westinghouse logos, while those with the letters D and E have RCA logos.

I. L. 1081-E

The reason for the change from "Instruction Card" to the initials "I. L." is unknown, but one could speculate that I. L. stands for "Instruction Label."

The six instruction sheets are noticeably different with the exception of I. L. 1081-C, which is virtually identical with I. L. 1081-B. One almost undetectable change was made in the depiction of the instruction sheet which appears in the circle. The three lines connecting the instructions to the figure of the Aeriola Jr. appearing in the circle in the "B" version have been removed in the "C" version. This change was missed when the lines on the actual instructions (outside the circle) were removed when going from the "A" version to the "B" version, and so it was corrected by creating the "C" version. The most common forms of the instruction sheets appear to be the "A" and "B" versions.

Regardless of which label appears in the lid of an Aeriola Jr., it appears that all sets are virtually identical—with one minor exception. The wood handle on the earliest version has a distinctly oval shape (which can be seen in the accompanying photograph), while the tuning handles on all subsequent sets were cylindrical in shape. The crystal set itself was situated in a poplar wood box with a compartment for storing the head sets. According to the 1922 catalog *Radio Enters the Home*, all metal parts were nickel-plated: *"The tuning elements and the crystal detector are mounted on a black panel which forms the top of the receiving set when the cover is lifted. All the metal parts are finished in nickel."* [57] While the catalog specifies that the metal parts are finished in nickel, a spectral analysis of the tuning dial confirms what many collectors have known for some time, namely that the dials are plated with a thin film of silver—too thin to stand up under normal wear and aging. As a result, virtually all of the dials found on Aeriolas today show significant wear or damage. Those that now appear in mint condition were almost certainty replated.

The set was inductively tuned, containing a minimum number of parts including a variometer, two capacitors (one with a center tap), a Perikon detector similar to the one used on the DB, and a set of headphones. According

The Aeriola Jr. was discontinued at the end of 1922 with a clearance sale with a reduced price of $15. (Julius Andre & Sons advertisement, Christmas 1922)

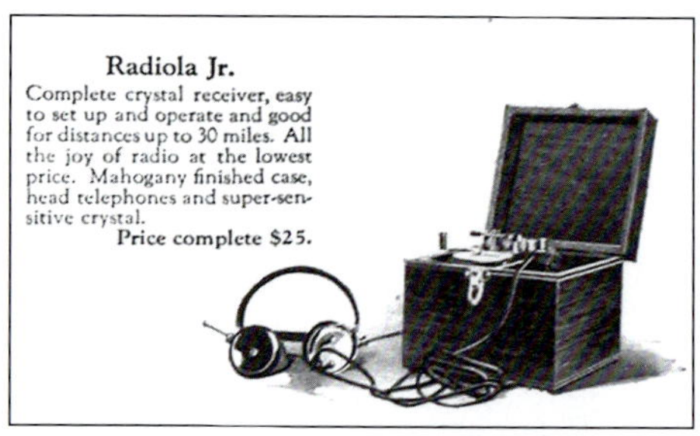

The Radiola Jr. appeared in the *Saturday Evening Post* with the entire 1923 line—even though the decision to drop it from the line had already been made. Note that the image of the Radiola Jr. was reversed in this ad. (*Saturday Evening Post*, February 3, 1923, p. 91)

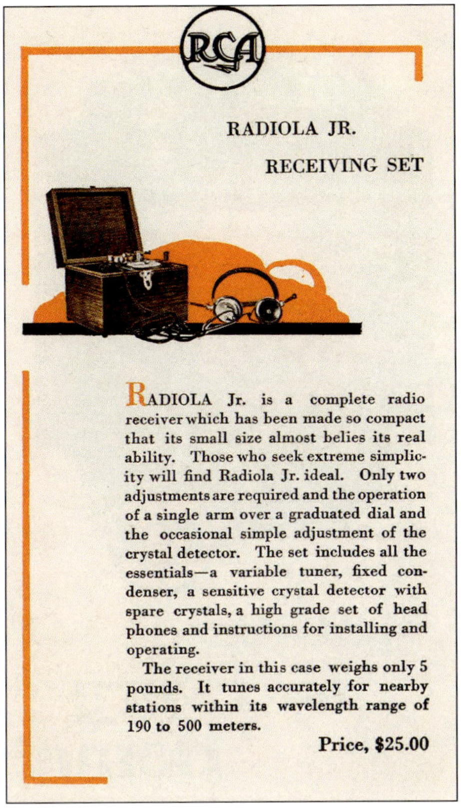

The Aeriola Jr. name was changed to Radiola Jr. shortly before it was discontinued at the end of the 1922 season. (RCA brochure *Radio for Every Home*, ND)

to patent 1,499,331 filed by Frank Conrad on July 22, 1922, the set was designed to increase selectivity without sacrificing sensitivity by maximizing inductance in the antenna circuit while at the same time minimizing capacity. It covered the broadcast band with wavelengths of 190 to 500 meters using two different antenna terminals, one for wavelengths below 350 meters and one for wavelengths between 300 and 500 meters.

For the entire 18 months that the Aeriola Jr. was on the market, it was advertised for $25 with earphones and $32.50 as a complete set including a $7.50 antenna kit. In the summer of 1922, RCA changed the name of all of its receivers to include the Radiola brand name, and while it is not generally known, RCA also changed the Aeriola Jr. to the Radiola Jr.—often spelling it out as Radiola Junior. The Radiola Junior name appeared in a few distributor catalogs and in an RCA display at a 1922 radio show pictured on page 167. It also appeared most publicly as a Radiola Jr. in the February 3, 1923 issue of the *Saturday Evening Post* along with the entire 1923 line. The Radiola Junior name is not well known because RCA suddenly decided to discontinue the set at the end of 1922 before they had time to prepare and implement a new label with the Radiola name. Thus, there are no known crystal sets of this type bearing a Radiola Junior label. Nevertheless, for a short period of time this set was known and marketed as the Radiola Junior—and also the Radiola Jr., as evidenced by the RCA sales brochure shown on this page.

According to Douglas, RCA sold 24,249 sets,[58] but it is likely this number does not include the sets sold by Westinghouse before RCA assumed responsibility for its marketing in early 1922. Since there were no serial numbers on this set, the total number sold cannot be independently determined.

Aeriola Jr. Wannabes

Two British crystal sets manufactured in 1922 imitate the Westinghouse Aeriola Jr. in sufficient detail that, if it were not for the tell-tale printed instructions on the lid, they would easily pass for the real thing—at least at first blush.[59] Upon closer inspection, most of the individual parts of these sets, both exterior and interior, are slightly different from each other and from those in the Aeriola Jr.

The Western Electric No. 44001Crystal Set (left) and the Metropolitan-Vickers Cosmos Radiophone (right)–manufactured and distributed by two different British companies–appear at first glance to be identical to the Aeriola Jr.

The two British sets are the Cosmos Radiophone manufactured by Metropolitan-Vickers Electrical Company, Ltd. (Metrovick), and "The Crystal Set" with the designation No. 44001, manufactured by Western Electric Co. Ltd. Metrovick had access to Westinghouse technology through a long-standing relationship with Westinghouse dating from the time George Westinghouse created British Westinghouse in 1899, which continued even after it became independent in 1916, and was later purchased by Vickers in 1919. Western Electric had access to Westinghouse technology through cross-licensing agreements between Western Electric and Westinghouse.

The two British sets, like the Aeriola Jr., were contained in a dark-stained wood box of similar dimensions and appearance with a hinged lid, had a black wood panel, and used similar Perikon crystal detectors. The two British sets used aluminum dials versus the silvered brass dial on the Westinghouse Aeriola. While the circuits for all three appear to be the same, the construction of the internal circuit components, including the variometer coil and three capacitors, are all slightly different in appearance.

It should also be noted that there was another crystal set named the Aeriola 10 manufactured by the Crane Bros. Radio & Mfg. Co. in Oakland California. It carries an "Approved Radio News Laboratory" sticker dated 1922, which was just at the time RCA took over marketing responsibilities for the Aeriola Jr. from Westinghouse. It is likely that RCA would have either threatened or actually taken legal action to stop Crane from distributing this crystal set, a conjecture supported by the relative rarity of this set.

Another Aeriola designated the Aeriola 10 was manufactured by the Crane Bros. Radio & Mfg. Co. in Oakland, California.

The Aeriola Sr. (Type RF)

The Aeriola Sr. was a one-tube regenerative receiver introduced by Westinghouse in December of 1921 at a price of $75 including one tube, batteries, a headphone and an antenna kit. Like the companion Aeriola Jr. receiver, the Senior was first distributed by Westinghouse, but only for a month or two before RCA assumed that responsibility in January of 1922. RCA

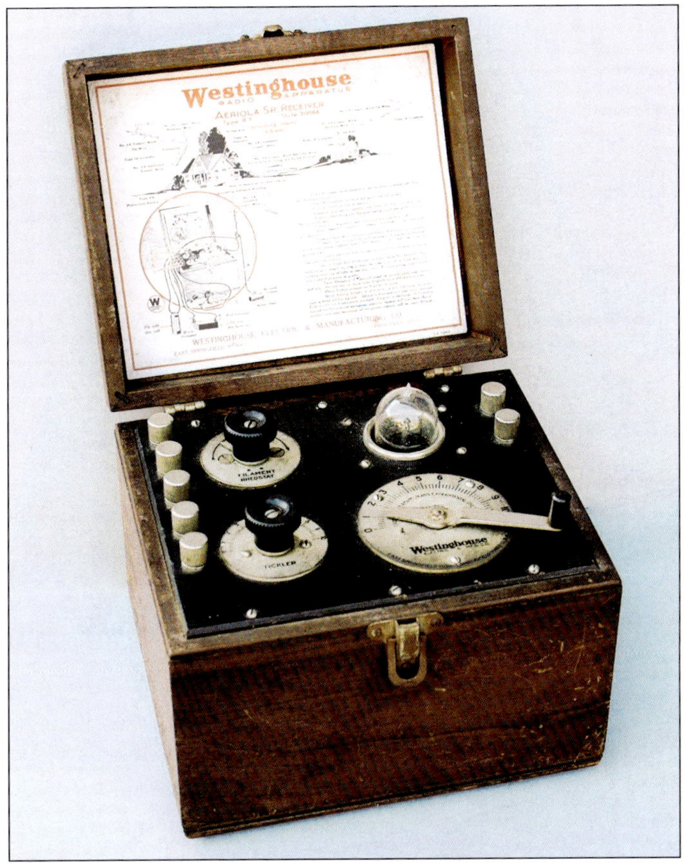

The Aeriola Sr. was similar in appearance to the Aeriola Jr. but it was a more sensitive regenerative receiver equipped with a single WD-11 tube.

The Aeriola Sr. used a unique coil design consisting of an outer coil and two continuously variable inner coils, a design which was used in the RS, the Radiola III, and the Radiola III-A.

Four different styles of instruction sheets were tacked to the interior lid of the Aeriola Sr.—two with the Westinghouse name only and two with both the Westinghouse name and RCA logo at the bottom.

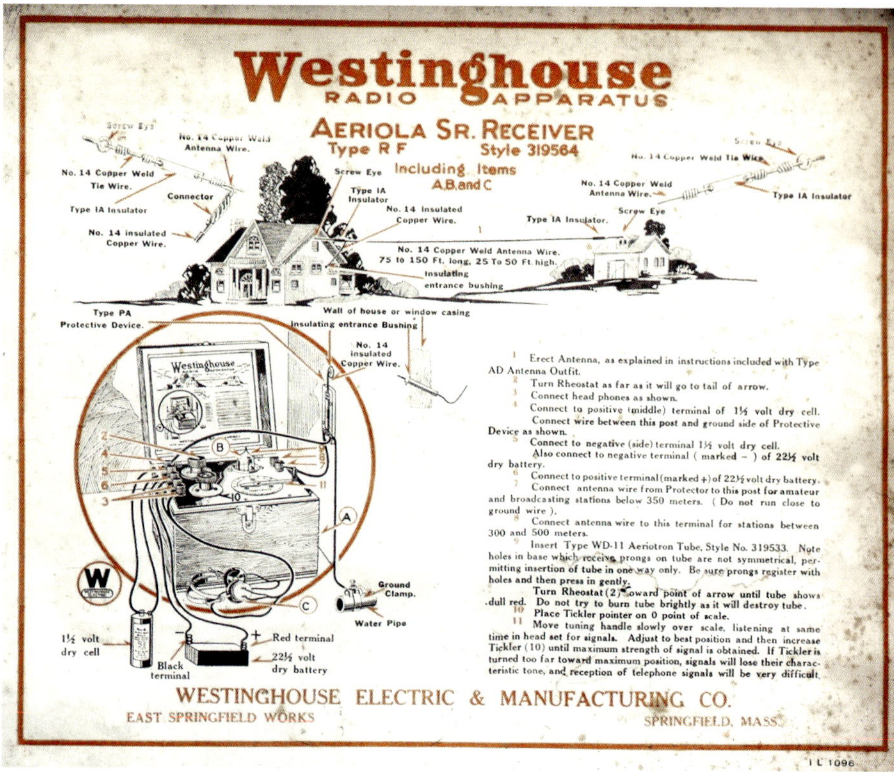

Westinghouse
RADIO APPARATUS
AERIOLA SR. RECEIVER
Type R F Style 319564
Including Items A.B.and C

1 Erect Antenna, as explained in instructions included with Type AD Antenna Outfit.
2 Turn Rheostat as far as it will go to tail of arrow.
3 Connect head phones as shown.
4 Connect to positive (middle) terminal of 1½ volt dry cell. Connect wire between this post and ground side of Protective Device as shown.
5 Connect to negative (side) terminal 1½ volt dry cell. Also connect to negative terminal (marked –) of 22½ volt dry battery.
6 Connect to positive terminal (marked +) of 22½ volt dry battery.
7 Connect antenna wire from Protector to this post for amateur and broadcasting stations below 350 meters. (Do not run close to ground wire).
8 Connect antenna wire to this terminal for stations between 300 and 500 meters.
9 Insert Type WD-11 Aeriotron Tube, Style No. 319533. Note holes in base which receive prongs on tube are not symmetrical, permitting insertion of tube in one way only. Be sure prongs register with holes and then press in gently.
Turn Rheostat (2) toward point of arrow until tube shows dull red. Do not try to burn tube brightly as it will destroy tube.
10 Place Tickler pointer on 0 point of scale.
11 Move tuning handle slowly over scale, listening at same time in head set for signals. Adjust to best position and then increase Tickler (10) until maximum strength of signal is obtained. If Tickler is turned too far toward maximum position, signals will lose their characteristic tone, and reception of telephone signals will be very difficult.

WESTINGHOUSE ELECTRIC & MANUFACTURING CO.
EAST SPRINGFIELD WORKS SPRINGFIELD, MASS.
I L 1096

I.L. 1096

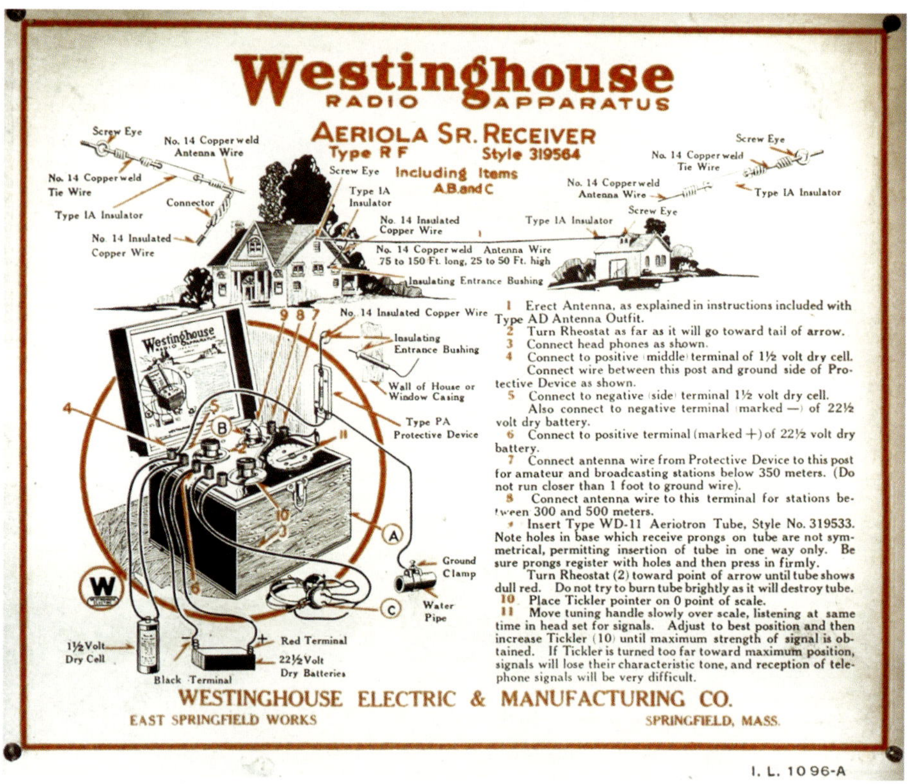

Westinghouse
RADIO APPARATUS
AERIOLA SR. RECEIVER
Type R F Style 319564
Including Items A.B.and C

1 Erect Antenna, as explained in instructions included with Type AD Antenna Outfit.
2 Turn Rheostat as far as it will go toward tail of arrow.
3 Connect head phones as shown.
4 Connect to positive (middle) terminal of 1½ volt dry cell. Connect wire between this post and ground side of Protective Device as shown.
5 Connect to negative (side) terminal 1½ volt dry cell. Also connect to negative terminal (marked —) of 22½ volt dry battery.
6 Connect to positive terminal (marked +) of 22½ volt dry battery.
7 Connect antenna wire from Protective Device to this post for amateur and broadcasting stations below 350 meters. (Do not run closer than 1 foot to ground wire).
8 Connect antenna wire to this terminal for stations between 300 and 500 meters.
9 Insert Type WD-11 Aeriotron Tube, Style No. 319533. Note holes in base which receive prongs on tube are not symmetrical, permitting insertion of tube in one way only. Be sure prongs register with holes and then press in firmly.
Turn Rheostat (2) toward point of arrow until tube shows dull red. Do not try to burn tube brightly as it will destroy tube.
10 Place Tickler pointer on 0 point of scale.
11 Move tuning handle slowly over scale, listening at same time in head set for signals. Adjust to best position and then increase Tickler (10) until maximum strength of signal is obtained. If Tickler is turned too far toward maximum position, signals will lose their characteristic tone, and reception of telephone signals will be very difficult.

WESTINGHOUSE ELECTRIC & MANUFACTURING CO.
EAST SPRINGFIELD WORKS SPRINGFIELD, MASS.
I. L. 1096-A

I.L. 1096-A

marketed the set comparably equipped for $75.90—which consisted of $65 for the set with headphones and tube, $3.40 for batteries, and $7.50 for the antenna kit. The Style No. for all Aeriola Sr.'s was 319564.

According to a patent 1,499,331 filed by Max Bastel on December 11, 1922,

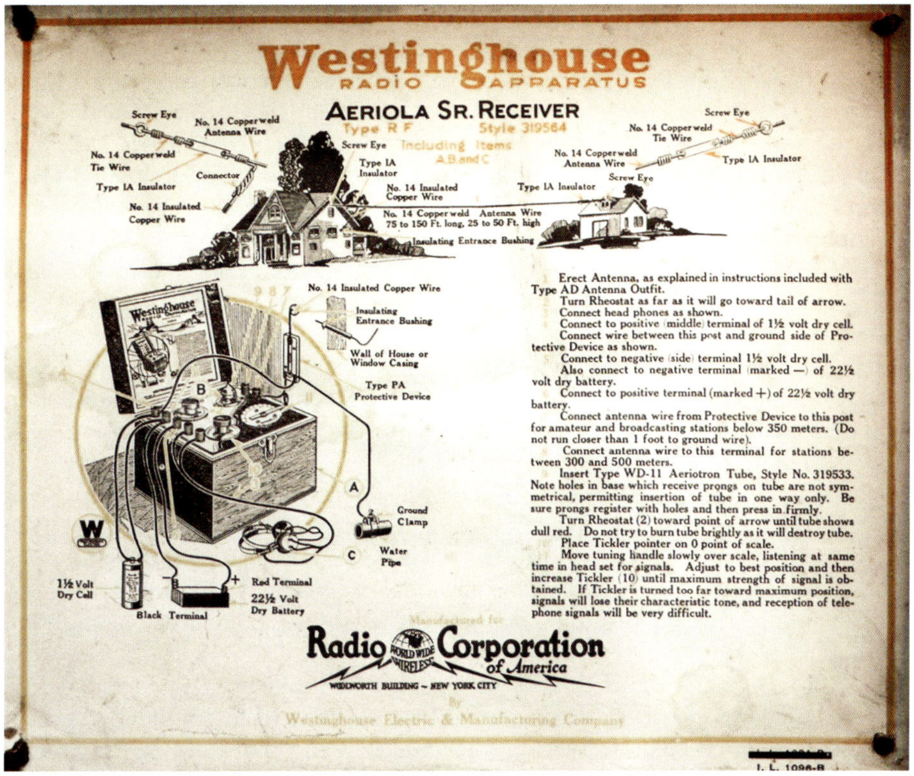

I.L. 1096-B (over stamped)

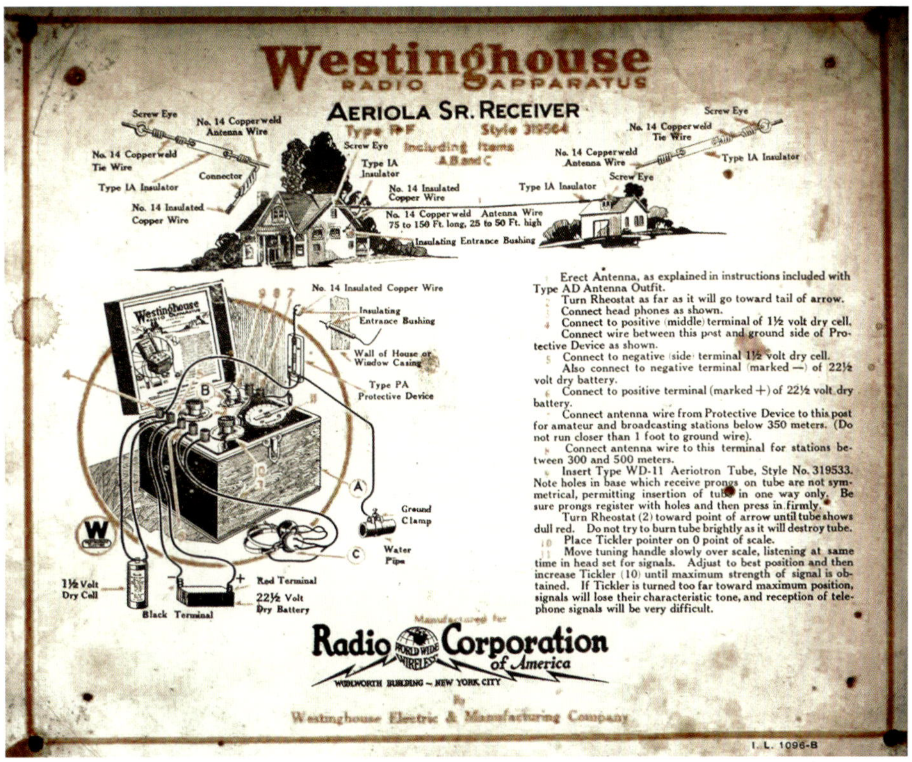

I.L. 1096-B (printed)

the Sr. was designed such that the degree of regenerative feedback was independent of the tuning, thus allowing the tuning and feedback circuits to be adjusted independently. This design objective was similar to that of the RC receiver, but it was achieved with a unique coil design in which an outer coil contained within it two additional continuously-variable tightly-coupled coils. This unique coil arrangement obviated the need for a tuning capacitor such as the one used in the RC design, and was so successful that it was used for the next two model years in the RS and Radiolas III and III-A.

The model distributed first by Westinghouse and then by RCA in 1922 was supplied with a black painted wood panel, a poplar case, and a silvered brass tuning dial, giving it an appearance very similar to the Aeriola Jr. The earliest sets distributed by Westinghouse can be easily distinguished from later sets distributed by RCA by the instruction cards, which had no reference to RCA and bear the identification numbers I. L. 1096 or I. L. 1096-A. RCA replaced this instruction card with one containing the RCA logo, but apparently forgot to assign it a new identification number because the printed number 1096-A was crossed out and replaced by the number I. L. 1096-B stamped directly underneath.

There were several minor modifications to the Aeriola Sr. beginning with the sets bearing instructions I. L. 1096-B, but they were physical changes as opposed to circuit modifications. The physical construction of the tuning coil was changed such that connections to the spiral straps were moved from the ends of the coil to the center. The capacitor paralleling the headphones was changed in style to one with smaller dimensions and moved from its location on the side of the wooden tube support structure directly to the headphone terminals.

In late 1922, the painted wood panel was replaced by a molded bakelite panel, and the poplar case was replaced by a mahogany case. One of the first places the molded bakelite version appeared was in the Andre Radio Supply Catalog No. R-4 circa January 1923. The collector can easily identify a bakelite

A version of the Aeriola Sr. with gold-plated dials and unusual black binding-post caps was made—perhaps as part of a deluxe limited edition or perhaps to memorialize some event. (Alan Douglas Collection)

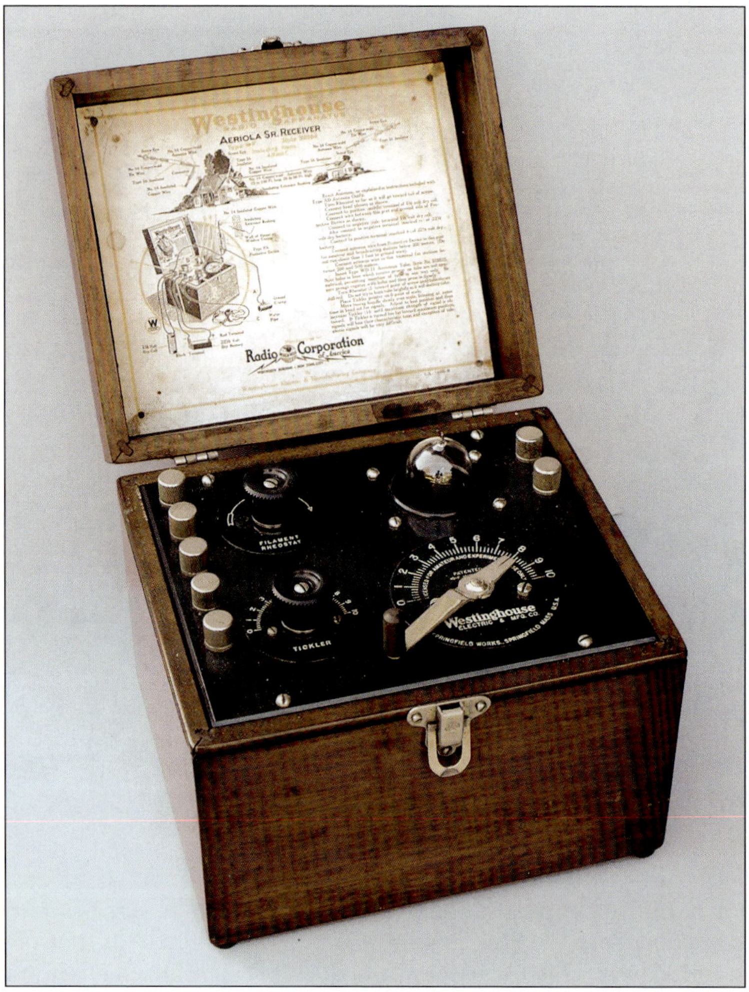

In late 1922 the wood panel of the Aeriola Sr. was changed to a Micarta bakelite base; this set can be easily distinguished at a distance from one with a wood base by the presence of only two screws in the front panel and the RCA logo on the instruction sheet.

panel set at a glance either by the molded black dial plates on the panel in lieu of the nickel-plated brass plates used on the earlier model, or by the reduction in the number of screws used to secure the panel from three to two at the front edge of mahogany box.

The printing error on the instruction card had also been corrected, so the identification number I. L. 1096-B was machine printed. However, the instruction card is not a reliable way to identify this, or any Aeriola receiver for that matter, because of the large number of reproduction cards found on these sets today, most of which are not correct for the set on which they are found.

Douglas states that 55,102 Aeriola Sr. sets were sold through RCA[60], but in addition it is likely that many thousands were sold by Westinghouse through their own distribution system in the months before RCA began to market this set. About half the sets seen on the market today have a Westinghouse label and half have an RCA label, indicating that a substantial number of Aeriola Sr. sets might have been sold by Westinghouse.

The Radiola Senior (Type RF)

Sometime late in 1922, RCA changed the name of the Aeriola Sr. to the Radiola Senior (most often spelling out "Senior") and used the new name in several radio shows including the American Radio Exposition in New York City in December 1922. However, the restyled set bearing the new Style No. 307455 would not be available for several months, and so the old style sets were used at these shows. The Type RF designation remained the same, but the paper instruction sheet on the inside of the lid was replaced by a metal serial number tag, and the sets were assigned serial numbers. Most of the observed serial numbers for the Radiola Senior are between 119,000 and 122,000, although

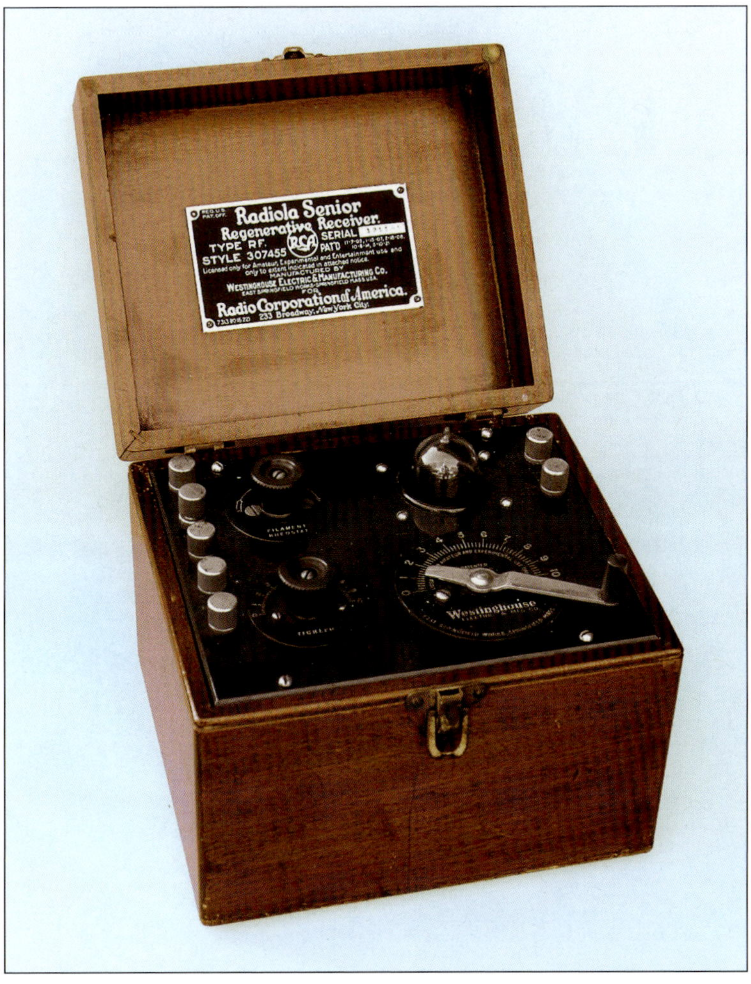

In the spring of 1923, RCA changed the name of the Aeriola Sr. to Radiola Senior, and replaced the instructions in the lid with a name plate stamped with serial numbers.

some sets have been observed with serial numbers in a range around 137,000 and others in a range around 148,000. Based on the serial number evidence, it would appear that about 3,300 Radiola Senior sets were sold by RCA before the set was discontinued by April of 1924 (see Appendix D).

The Radiola Senior had the same circuitry as the Aeriola Sr., and the panel is virtually identical to the latest version of the Aeriola Sr. Towards the end of the production run, the diameter of the hole for the tube was increased from 1-3/8 to 1-5/8 inches to accommodate the shorter WD-11 tubes that RCA began to

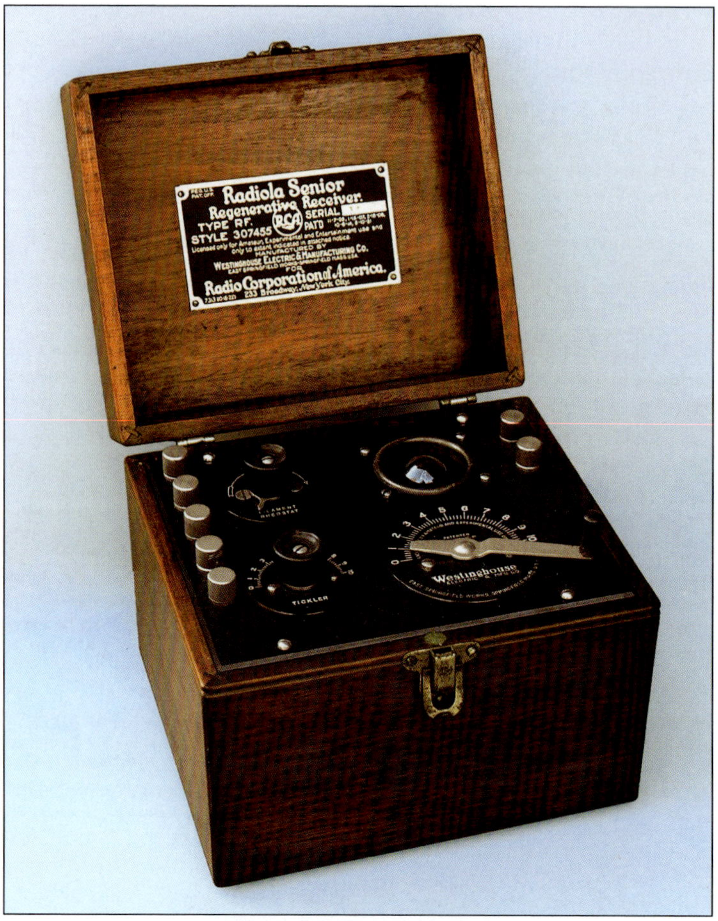

The size of the hole for the tube in the Radiola Senior was enlarged sometime after Westinghouse came out with the smaller version of the WD-11 to allow a better grip on the tube when replacing it, and later a metal bezel was also added.

To accommodate the increase in the hole size on the Radiola Senior panel, the entire tube wood support structure was replaced with an open metal frame.

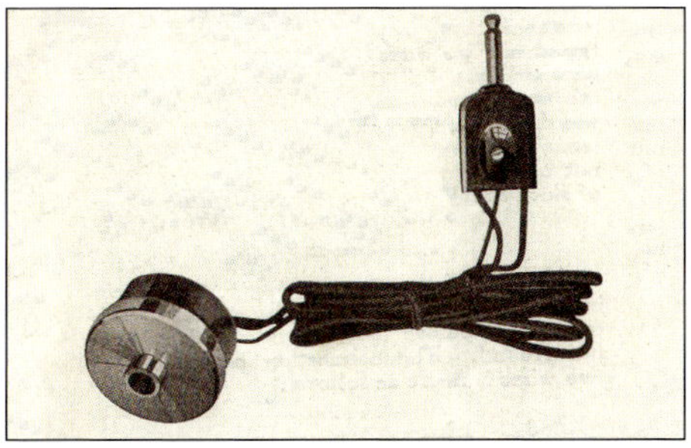

Reproducer attachments for phonographs allowed radios to be played through phonograph speakers, a good alternative for multiple listeners at that time. (RCA catalog *Radio Enters the Home*, June 1, 1922)

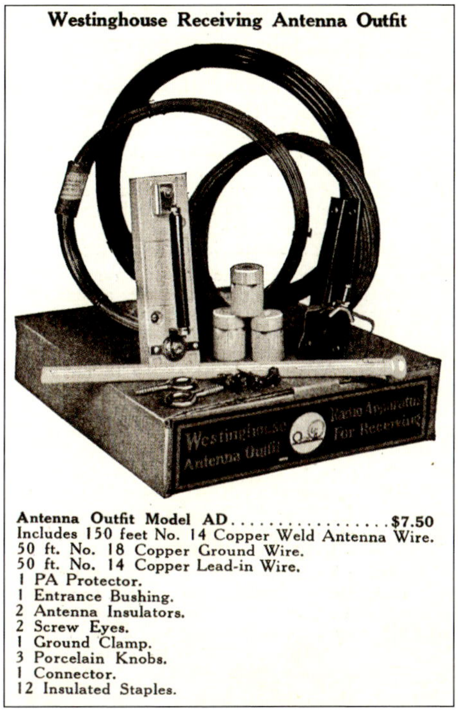

The antenna was an important accessory for early radios, as evidenced by illustrations of a typical antenna installation on virtually all Westinghouse instruction sheets. (*Radio Enters the Home*, June 1, 1922)

A battery charger was a necessary accessory for early radios with tubes that drew sufficient current to require storage batteries.

market. The newer WD-11 tubes were almost an inch shorter than the older brass based tubes, making them difficult to grasp for removal with the smaller hole. Some of the later Radiola Seniors also have metal bezels, or rings, around the tube sockets similar to those found on the earlier Aeriola Sr. sets. Sets with this modification have serial numbers of about 130,000 and higher.

Accessories Complement the Receivers

Westinghouse clearly understood that the average broadcast radio listener wanted to purchase a complete radio set with components that were designed to work together rather than having to select and purchase components separately. They supplied all the accessories required for their early receivers including those manufactured by Westinghouse such as reproducer attachments, battery chargers, and antennas, as well as those manufactured by others such as headphones and batteries. These accessories were sold both individually and in selected combinations as part of a complete set. The accessories addressed here are those that Westinghouse manufactured or marketed in 1921 specifically for their own line of receivers, principally a reproducer attachment to allow radio signals to be acoustically amplified by phonograph speakers, the AD antenna kit, and Rectigon battery chargers. These and other accessories including batteries and headphones were pictured in a Westinghouse Circular 1641 published circa mid-1921.

LS Reproducer Attachment

Westinghouse marketed two reproducer attachments, one designed to replace the reproducer on Victor Victrolas and the other on Columbia Grafonolas, thus allowing the radio to be heard through the phonograph horn. According to the early Westinghouse catalog Circular 1641, the attachment replaced the reproducer mechanism on the phonograph arm holding the needle so that "*Signals which can be heard by but one person using the ordinary head set, can be so amplified as to be heard by a large number when using this attachment.*" Reproducers designated Type LS were designed for the Victor phonograph (Style No. 307262) and the Columbia phonograph (Style No. 307263), which were first advertised in the October, 1921 issue of *The Wireless Age* at $15 each. They were on the market for a number of years, finally discontinued at the end of the 1923 season in early 1924.

Type AD Antenna Kit

An external long-wire antenna was one of the most important components of a radio set in the early years of broadcast radio. Westinghouse first advertised the Type AD antenna kit in conjunction with their radio equipment circa October 1921 for $7.50. The kits consisted of 150 feet of copper antenna wire, copper lead-in and ground wire, a lightning protector, insulators, clamps, porcelain knobs and hardware. There were ample instructions and many articles written on how best to install and maintain a safe and effective antenna. The importance of the outdoor antenna was punctuated by the illustration of a typical outdoor installation that appeared on virtually all of the instruction sheets accompanying the early Westinghouse receivers.

Rectigon Battery Charger

Rechargeable batteries were required as an adjunct to early vacuum tube radios because of the relative large filament currents required. Westinghouse had been manufacturing a line of Rectigon battery chargers for automobile batteries before the broadcast era, and offered two charger styles in 1921 as accessories for radios. The Westinghouse Circular 1641 pictured two Rectigon styles: Style No. 282395 with a 2.5 ampere output for 40 ampere-hour storage batteries (shown here), and Style No. 285168 with a 4 ampere output for larger storage batteries. Westinghouse priced the two at $18 and $28 respectively in an October 1921 ad in *The Wireless Age*.

RCA and General Electric Miss the Boat

Both RCA and General Electric were totally unprepared for the broadcast era ushered in by Westinghouse and KDKA in late 1920, and were unable to satisfy the resulting huge demand for broadcast radio receivers. General Electric engineer William White in his *Early History of Electronics in the General Electric Company* put it this way:[61]

> *"It is only natural that many of us in G.E. at that time felt we "missed the boat." We, at the Works, had radio-telephone transmitters, antenna structures and so had no need for even an active amateur station. All we lacked was the wisdom and initiative to have gotten started in this broadcasting activity. We knew our electronics but not human nature."*

To be sure, RCA and GE were preoccupied with transoceanic and marine communication throughout 1920 and 1921, but there were certainly those in the RCA organization that had the foresight and vision to see the potential for broadcast radio and the huge market in broadcast receivers. David Sarnoff, Commercial Manager of RCA, with the support of GE engineers, sent an extensive letter dated January 31, 1920 to RCA Chairman Owen Young (who was also the President of GE) at his request detailing the potential for RCA in the prospective radio business.[62] One of the major items in this memo was the "Radio Music Box" concept, a concept which was nothing less than broadcast radio.

Sarnoff also wrote Dr. Alfred Goldsmith, head of the Research Department of RCA, regarding the Radio Music Box concept. In Goldsmith's response dated February 26, 1920, Goldsmith first referred to the music box as a "Radiola," a name that would be trademarked and used by RCA to market radio receivers for decades to come.[63] On May 28, 1920, Mr. E. P. Edwards of General Electric wrote the following letter to Sarnoff estimating that $2000 would be necessary to develop the Radio Music Box:

Schenectady, May 28, 1920

Mr. David Sarnoff
Commercial Manager,
Radio Corporation of America,
233 Broadway,
New York City.

Dear Mr. Sarnoff:

SUBJECT: RADIOLA:

With reference to your letter of May 25th, wish to advise that based on Mr. Baker's specifications, a copy of which was handed to you on May 27th, we estimate that $2000 will be necessary to build up the model.

We further estimate that a model built in accordance with these specifications could be produced in from four to six weeks after receipt of authority to proceed.

As pointed out in Mr. Baker's specifications we are now bringing through our factory some 3½ KW telephone sets which would give us an ideal set to test the Radiola with.

Our Mr. Baker has been doing some experimental work in conjunction with loud speaking apparatus, and to date he has been able to get much better articulation than we have heretofore been able to get with any of the loud speaking devices on the market. Mr. Baker feels that further experimentation along this line should be carried on with the Radiola. Work on the Radiola can be carried on in conjunction with the other models we have already started for you and we are preparing to start work as soon as you give us the word to go ahead.

Yours very truly,
E. P. Edwards
Asst. Mgr., Lighting Dept.,
By: F. R. Deakins

RCA records show that a committee of the RCA Board of Directors approved the $2000 expenditure for the Radiola referred to in this letter, which GE presumably received during the summer of 1920. However, according to testimony of GE VP Albert G. Davis at the Federal Trade Commission hearings, these funds were cancelled because GE could not make a satisfactory receiver without a license under the Armstrong feedback patent:

> *"Now, prior to that date [the date of the Westinghouse/GE cross-licensing agreements of July 1, 1921] and after the formation of the Radio Corporation, our people thought it would be advisable to design a receiving set for use in the home."*[64]

> *"At that time most of these sets were made by the users, who would buy coils and so forth and put them together, and they [the GE engineering staff] asked the Radio Corporation for an appropriation for development and they were unable to go ahead because they found in the state of the art then existing they could not make a satisfactory set without a license under the Armstrong feed-back patent and so the matter was dropped for a while and we concentrated our attention on the transoceanic developments and things of that sort."*[65]

> *"I remember particularly, of course, the Armstrong patents which were regarded as exceedingly important, covering, as they did, not only the feedback regeneration which was in the beginning of the broadcasting reception regarded as essential in receiving sets. I think I testified before about appropriations to develop the broadcasting receiving sets at Schenectady which were cancelled because we were not able to get a satisfactory set without that patent."*[66]

Despite misgivings about the prospects for marketing broadcast radios without the Armstrong patent and the cancellation of the appropriations at GE, the RCA Research Department developed at least two broadcast radio prototypes between mid-1920 and mid-1921. The first was an RD 1 prototype with a name plate bearing the RCA Research Department name and a date of September 1920—which fits neatly into the time frame for the broadcast receiver development described in the May 28, 1920 letter from GE.

The Type RD 1, believed to be the first regenerative receiver developed by the RCA Research Department for broadcast radio, is now in the AWA Museum in New York. (Courtesy of AWA Museum)

The name plate affixed to the RD 1 clearly shows it was developed by the RCA Research Department in September of 1920 shortly after the Department was formed. (Courtesy of AWA Museum)

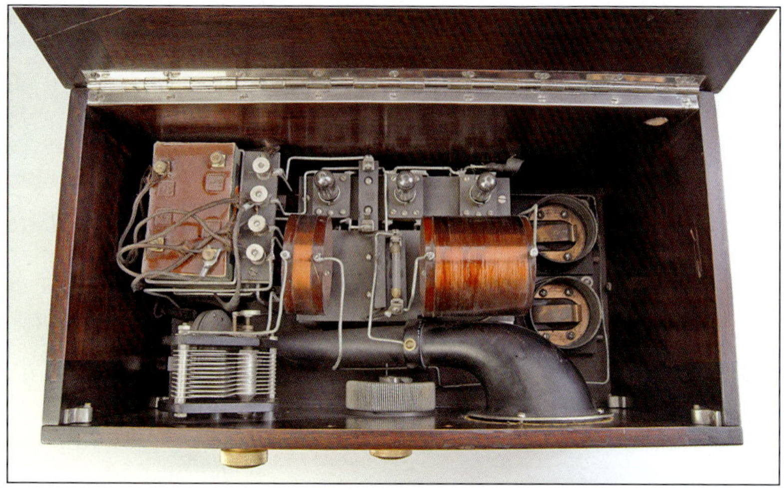

The Research Department of RCA developed a self-contained Radiola with batteries and a speaker that fulfilled Sarnoff's early vision of his Radio Music Box. The Radio Music Box used three Western Electric peanut tubes with low power requirements, thereby permitting the use of self-contained dry cell batteries. (American History Museum, Smithsonian Institution)

The RD 1 prototype has three Navy standard tube sockets which undoubtedly were designed to accommodate the UV-200 and UV-201 tubes. The batteries and speaker, if in fact it was designed to drive a speaker, were external to the unit. This set was in the personal collection of RCA historian George Clark, and is now in the AWA Museum.

There is no record of how well this unit was received by RCA management, but it certainly does not fit Sarnoff's original vision of a self-contained Radio Music Box. In sharp contrast to this set, the second prototype dated 1921 was a totally self-contained receiver with three AT&T 215A peanut tubes, batteries, and a loudspeaker. The peanut tubes were selected because of their low filament voltage and current requirements, thus allowing for the use of dry cell batteries, which fit within the receiver case. This radio is prominently labeled with the "RADIOLA" trademark, and identifies the manufacturer by the name plate inscribed with "Made by Research Department Radio Corporation of America." It is now on display at the American History Museum of the Smithsonian Institution.

Neither RCA nor GE capitalized on these prototypes, however. In June of 1921, Westinghouse joined the Radio Group and dominated RCA receiver products for almost two years thereafter. GE's first entry into the regenerative receiver market in mid-1922 was based on their marine receivers rather than either of these prototypes. It would not be until 1923 when GE produced the Radiola IV that RCA offered the first self-contained receiver (including batteries and speaker) for sale to the public, thereby fulfilling the original vision of David Sarnoff's Radio Music Box.

Thus, while RCA and GE had the resources and staffs with the vision to enter broadcast radio in early 1920, the GE/RCA decision makers clearly were not risk takers. This is in sharp contrast to the Westinghouse decision makers, who clearly were. Westinghouse elected to manufacture and sell the RC receivers using Moorhead tubes despite the RCA patent position on the triodes, which according to Patent Counsel Le Roi J. Williams was done *"...with full knowledge that it had not any right to use those tubes in those sets, and with a full knowledge that the seller probably had no right to dispose of those tubes to the Westinghouse Company for that purpose."*[67] It was their hope that the patent situation would resolve itself before they ran out of Moorhead tubes, and their gamble paid off handsomely.

RCA's First Radio Catalog for Amateurs and Experimenters

According to an H. Gernsback editorial in the December 1921 issue of *Radio News*, *"it was apparent to the careful observer that the country was in the midst of a revolution, as far as radio and the great public are concerned,"* attributing most

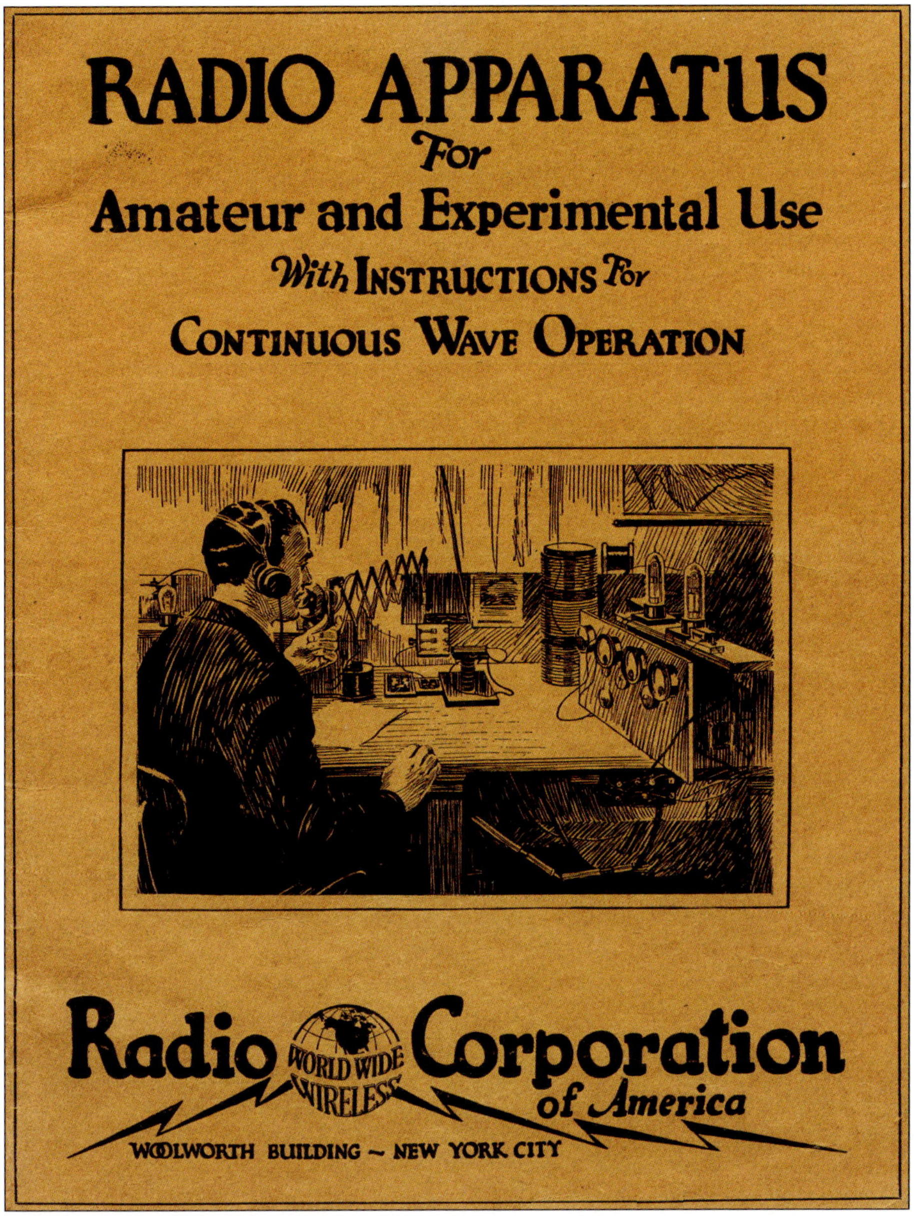

RCA published its first radio catalog directed at the amateur and experimentalist, a catalog which is notable for failing to even mention the words "broadcast" or "broadcast radio" at a time when broadcast radio was entering the mainstream. (RCA catalog *Radio Apparatus for Amateur and Experimental Use*, September 1, 1921)

of this interest to Westinghouse and their multiple broadcasting stations.[68] Clearly, RCA was well aware of the huge potential for broadcast radio sales, and it was no doubt a major factor in prompting RCA to publish its first-ever catalog for radio amateurs and experimenters, dated September 1, 1921.[69]

This catalog contains virtually all of the amateur radio products available to RCA at the time—but exclusive of Westinghouse products, which were not available to RCA at the time this catalog went to press circa May 1921—well before the cross-licensing agreements with Westinghouse were signed on July 1, 1921. What catches the eye is that the catalog contains little of interest to the mass market for broadcast radio, and despite the impending boom in broadcast radio, the catalog never even mentions the word "broadcast radio," much less the term "broadcast." The absence of substantive items relating to broadcast radio is certainly not an oversight or lack of vision, but rather a reflection of the fact that RCA was simply not prepared for the broadcast radio market.

The catalog was neatly divided into two parts. The first part featured radio circuits and components manufactured for RCA by RCA/GE, while the second part featured "*receiving apparatus of quality*," manufactured for RCA by

Wireless Specialty Apparatus Company (WSA), a signatory to the RCA cross-licensing agreements with United Fruit Co. dated February and March, 1921—well before the RCA catalog was prepared. Of the 40 pages of radio circuits and components, 37 were devoted to continuous wave transmitters, while only three pages were devoted to receivers. In these three pages, RCA offered many of the components required to construct a receiver—vacuum tubes and sockets, rheostats, audio transformers, grid leaks, etc. Noticeably absent, however, were the antenna coupling circuits including tuning and tickler coils that were critical to receiver performance in terms of sensitivity and selectivity.

The second part of the catalog listing "apparatus of quality" featured several very expensive receivers that had been designed by the Navy during WWI for marine applications, and were later adapted for commercial marine applications by WSA. The two WSA receivers listed in the RCA catalog were the IP-500 and the IP-501, both of which were used by the United Fruit Company in their marine stations and on their bananas boats in the 1920's and 1930's. With prices between $550 for the least expensive crystal set with detector only, and $750 for the most expensive regenerative receiver without batteries, tubes, earphones or antenna, the cost of these receivers alone was equivalent to about a half a year's wages for the average working man in the early 1920's. Further, the sets were sufficiently complex to operate that they were the subject of course material at radio schools for wireless operators, such as the RCA Institutes, Inc. Needless to say, not many of these sets were purchased by the public for purposes of listening to broadcast radio. The Radio Sales Division of RCA discontinued these receivers and accessories by August of 1922, although the RCA Marine Division continued to supply WSA receivers and accessories for commercial communication applications for many years.

IP-500 Receiver with Detector and One-Step Triode Amplifier, Triode A

The IP-500 is the commercial version of the SE 1220, which was designed by the Radio Test Shop of the Navy Yard towards the end of 1918 and manufactured for them by WSA, among others. This set used inductive coupling for the first time, and had both tuned and untuned secondary coils, either of which could be selected by a double-pole, double-throw (DPDT) switch on the front panel. The untuned circuit provided close coupling between primary and secondary, and was used to receive a broad range of frequencies for initial signal acquisition.

The SE 1220 receiver was designed for medium wavelengths of 300-6800 meters, while the IP-500 was designed for wavelengths of 150 to 6800 meters. Binding posts were provided in the primary, secondary, and tickler circuits which loaded these receivers for operation at longer wavelengths. Separate loading coils

The Wireless Specialty IP-500, first offered to the public by RCA in 1921, was an expensive, if not impressive looking receiver with a number of controls that required substantial skill on the part of the operator. Based on the Navy's SE 1220, it was the first set designed by the Navy with inductive tuning in which either a tuned or untuned secondary coil could be selected by a switch on the panel.

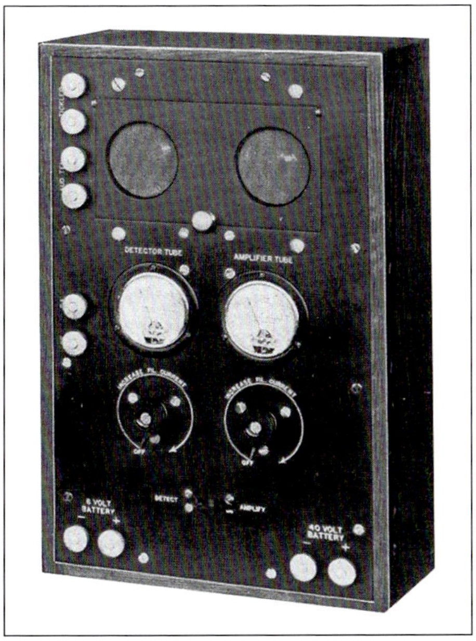

A separate Triode A "control box" containing a vacuum tube detector and a one-step amplifier for regenerative reception was pictured with the IP 501 in the WSA catalog, but the two were shown separately in the RCA catalog without any cross reference, presumably to avoid patent infringement issues. (RCA catalog *Radio Apparatus for Amateur and Experimental Use*, September 1, 1921, p. 42)

and tuning condensers were available for a wide range of wavelengths.

The primary and secondary circuits were housed in the "receiver" cabinet, which was separate from the cabinet containing the detector and amplifier circuits, a practice that was customary at that time. The tickler circuit for the regenerative function was contained in the receiver unit. The receiver could be used with an external crystal detector, or with the flip of a switch it could be connected to a separate "control box" containing a vacuum tube detector for regenerative reception and one stage of audio amplification. In their 1919 catalog, WSA recommended the IP-200 detector for use with the IP-500 receiver.[70]

The WSA catalog also pointed out the benefit of having the two units in separate housings: "*This permits the radio frequency unit to be furnished for commercial use independently of the amplifier unit.*" This is an obvious reference to the fact that WSA could not build regenerative sets for commercial use at that time without fear of patent infringement litigation. RCA had a similar problem at the time their catalog was prepared in the spring of 1921 because Westinghouse, who held the Armstrong regeneration patent, had not yet become a part of the RCA Radio Group. Cognizant of the patent situation, RCA completely separated the IP-500 and Triode A control units in their catalog, and made no cross references connecting the use of one in conjunction with the other. RCA did not even mention the word regeneration or regenerative receiver with regard to the receiver offerings.

The control box for the SE-1220 was designated Type SE-1071 Audion Control Box, while the control box for the commercial Type IP-500 was designated Amplifier Control Box, Type Triode A. RCA changed the designation for this unit in their catalog to Detector and One-Step Amplifier, Triode A, making clear that it contained a detector and a single stage of audio amplification. Binding posts on both units were conveniently co-located so they could easily be connected together with short lengths of wire. RCA offered type UV-200 and UV-201 tubes for the detector and amplifier functions. The IP-500 was priced at $595 and the Triode A control box was priced at $190, so the two-tube regenerative receiver configuration cost a total of $785 plus tubes, batteries, and earphones.

IP-501 Receiver with Two-Step Amplifier, Triode B

The IP-501 was the commercial version of the SE-1420, which was designed by the Navy in 1918 for use with direction-finding equipment. Hazeltine, a consultant to the Navy, devised his method of neutralizing undesired oscillations, and embodied his neutrodyne method of reception in this receiver. The Navy was producing the SE-1220 at the time, and decided to halt production of the SE-1220 to produce the SE-1420, which the Navy did in great

The IP-501, based on the Navy's SE-1420 was a self-contained regenerative receiver which was also first offered to the public by RCA in 1921. The primary and secondary coils of the IP-501 were separated such that a copper plate attached to the cabinet would extend between the two for electrostatic isolation, and the detector tube was integrated into the receiver rather than locating it in a separate controller box as had been the custom.

RECEIVER, IP-501

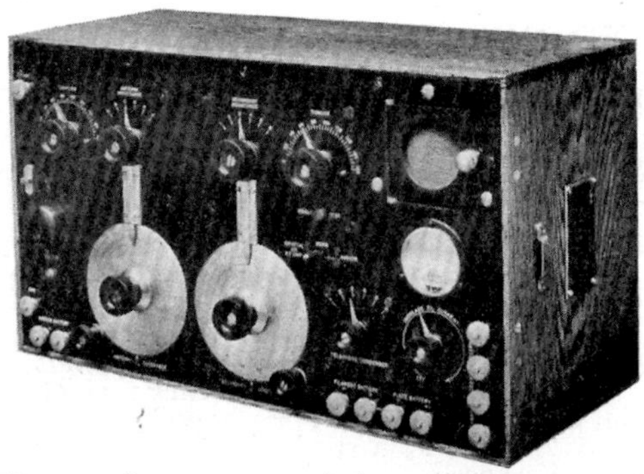

THE **IP-501** Receiver shown in the accompanying illustration is a compact unit containing the radio frequency and detecting circuits in a single case. Its range is 250 to 8,000 meters. It is particularly suitable for installations where cost and space factors are of great importance.

The receiver is similar in mechanical design to the **IP-500,** with the untuned circuit omitted. The capacity coupling between primary and secondary circuits is eliminated in this type by heavy sheet copper boxes separately enclosing the two circuits.

The panel is of Bakelite-dilecto. The coils are bank-wound inductances, of high frequency cable wound on threaded Bakelite-dilecto tubes, impregnated and baked. The condensers are of the self-balanced plate type. The wiring is of rigid copper encased in varnished cambric tubing. All switch points are located beneath the panel. The detector is mounted on a shock-proof support. The feature of condenser dial pointer control by inductance switch is included in the design.

RECEIVER, IP-501...**$550.00**
Overall dimensions: 20 in. x 11 in. x 9 in. Shipping weight: 55 lbs.

The set labeled IP-501 in the RCA catalog is actually an SE 1420, which can be distinguished from the IP-501 by the absence of binding posts for loading coils along the top of the SE 1420 panel. (RCA catalog *Radio Apparatus for Amateur and Experimental Use*, September 1, 1921, p. 41)

quantities. It was considered to be the Navy standard receiver at that time, and was later described in a Navy document as "*the best receiver for general radio reception that the service, and probably the world, has ever seen.*"[71]

The SE-1420 was a medium-wave receiver with a range of 250-6800 meters, while the IP-501 had a range of 250 to 8000 meters. WSA manufactured the Long Wave Loading Unit, IP-503 to extend the range of the IP-500 and IP-501 to 19,000 meters, which covered the wavelength of the highest commercial station at that time.[72] The loading unit was contained in an oak cabinet with binding posts at the bottom which lined up with similarly spaced binding posts located at the top of the receiver front panel. The IP-503 loading coil was not necessary for broadcast reception, and was not offered in the 1921 RCA catalog.

As an aside, it is interesting to note that the set pictured in the RCA catalog as an IP-501 is actually an SE-1420.[73] The major visual difference between the two sets are the six binding posts along the top of the front panel on the IP-501 for loading coils versus the absence of binding posts along the top of the SE-1420 panel. Clearly, there are no binding posts along the top of the panel on the set depicted in the RCA catalog as an IP-501. The set pictured in the RCA catalog is identical in every respect–including the wood grain on the cabinet–to the set correctly identified by Sterling as the SE-1420.[74] The most likely explanation for this error is that Wireless Specialty had not yet actually manufactured a production IP-501 at the time the RCA catalog had been prepared.[75]

The SE-1420/IP-501 was similar to the SE-1220/IP-500 except for several notable changes: 1) the neutrodyne method of reception was implemented, 2) the case was lined with copper and had a center plate extending between the primary and secondary coils for electrostatic isolation, 3) the untuned circuit in the secondary used for pickup work was eliminated, 4) the detector tube was placed within the receiver enclosure, and 5) the control box was replaced with a two-stage audio amplifier, which for the IP-501, WSA designated the Two-Step Amplifier, Type Triode B. The IP-501 with its own internal vacuum tube detector and a tickler circuit was a regenerative receiver, and clearly infringed on the Armstrong regeneration patent. It is not surprising that neither Wireless Specialty or RCA in their respective catalogs ever mentioned

The Triode B (left) manufactured for RCA by Wireless Specialty was a two-stage audio amplifier offered by RCA to the public for general use, but was also used later by RCA with the IP-501 in commercial work. The Triode E (right) was identical to the Triode B except that it was mounted on a breadboard to facilitate experimentation. (*Radio Telegraph and Telephone Equipment*, WSA catalog, 1919, p. 42)

the fact that it was a regenerative receiver, and Wireless Specialty went so far as to recommend the Type IP-202 combined silicon and galena detector for use with the IP-501—notwithstanding the fact that it had its own much more sensitive vacuum detector.

RCA priced the IP-501 at $550 and the two-stage amplifier at $95, for a total cost of $645 for the regenerative receiver. However, WSA had advertised the IP-501 as a complete receiving set for only $348 at retail in the November, 1919 *QST*, and charged RCA $360 for it in the fall of 1921. Thus, the public had to pay almost 60% more for the IP-501 from RCA than it had paid from WSA a year and a half before. The receiver was discontinued for sale to the public in 1922, but RCA used it for many years in their commercial Marine Communication operations.

Two-Step Amplifier, Triode E

The Triode E was identical to the Triode B except that it was mounted on a breadboard to facilitate experimentation. The WSA closed-core audio transformers, Type P, are prominently displayed at the center of the board. The Triode E was displayed in the WSA catalog and appeared in advertisements of the day with the Marconi tubes, but it was displayed without tubes in the 1921 RCA catalog.[76]

IP-501A Receiver

The 501A receiver is essentially identical to a combination of the IP-501 and the Triode B amplifier, except that both units are mounted in a single cabinet for ease of use and economy of space. This receiver was made by WSA for RCA in November of 1922, but did not appear in any of the RCA radio catalogs for broadcast radio. However, it was used in RCA's commercial marine business for many years. The set was equivalent to the Navy SE 1440.

The set used three vacuum tubes which were accessible though a metal door in the upper right corner of the front panel. Telephone jacks were placed on the front panel, which were arranged such that power was connected to the filaments of just those tubes which were in use—the detector tube only, the

The IP-501A, a combination of the IP-501 and Triode B amplifier in one unit, did not appear in RCA catalogs for the public, but was used extensively in RCA's marine communication operations. (John Jenkins, Sparkmuseum)

detector and one amplifier tube, or all three tubes. As in the two other sets, the panel contained a switch to allow the receiver to be connected to the vacuum tube detector or a crystal detector.

IP-503 Long Wave Loading Unit

The IP-503 Long Wave Loading Unit was designed to extend the range of the IP-500 and IP-501 receivers to 19,000 meters–slightly above the longest wavelength used by commercial stations at the time. Later, the IP-501A was designed so that it, too, could be used with the IP-503. The Loading Unit was positioned on top of the receiver cabinet such that the binding posts along the bottom would line up with the respective binding posts along the top of the receiver, and the two units were interconnected with short bus wires.

The IP-503 Long Wave Loading Unit used to extend the range of the IP-500 and IP-501 receivers to 19,000 meters was designed so that the 3rd and 4th binding posts from the left aligned with posts on the IP-500, while the 4th and 5th posts aligned with posts on the IP-501 and IP-501A.

The Eaton Oscillator was used either to generate radio-frequency signals for laboratory work, or as part of a regenerative receiver circuit.

Eaton Oscillator, Triode C

The description of the Eaton oscillator in the RCA catalog is most remarkable for the lack of information on what it is and how it is used. The catalog reads: *"A new feature in radio practice is incorporated in this design, in that the wiring diagram of the complete external circuit is engraved upon the cover, so that no instruction is necessary for its use. The external circuit is engraved in white, the internal circuit in red. On the left-hand side is engraved the oscillating circuit, on the right-hand side the tube circuit."* While the entire description was taken almost verbatim from the WSA catalog, the informative portion from that catalog was completely omitted in the RCA catalog. It would have been impossible for the average man to understand what this device was or how it was to be used.

The Eaton Oscillator could be used either to generate radio-frequency signals for laboratory work, or as part of a receiver circuit. An example of the oscillator used in a regenerative receiver is shown in a book by M. B. Sleeper, where it connected to a loose coupler, a tuning condenser and a triode with batteries—the only components needed in addition to the Eaton Oscillator to form a complete receiver.[77] The oscillator takes the place of the tickler coil and associated feedback circuitry, and allows reception from 250 to 25,000 meters, with the advantage that the only adjustments required are in the tuning circuit.

RCA priced the Eaton Oscillator at $13.50, which compares favorably to the $14.90 price tag advertised by WSA in 1919.[78] Even at that low price, however, the device must not have sold well because it was dropped from the RCA line by April of 1922, well before any of the other WSA devices in the 1921 RCA catalog were discontinued.

Endnotes and References for Chapter 3

1 Wireless telephony was distinguished from wireless (short for wireless telegraphy) in that wireless telephony utilized voice communication while wireless utilized Morse code.

2 Donald G. Little, "Dr. Conrad founds KDKA," *American Heritage,* (Vol. VI, No. 5, August, 1955) p. 71.

3 H. P. Davis, "The Early History of Broadcasting in the United States," in *The Radio Industry: The Story of Its Development* (Chicago/New York: A. W. Shaw Company, 1928) p. 192; reprinted by Arno Press in 1974.

4 *Radioana*, "The Story of Conrad and Station KDKA," Dictated to H. G. Clark by Dr. Conrad and W. W. Rogers in Pittsburgh, about 1940, SRM 4 1559.

5 *Radioana,* "Additional data re RC, Aeriola Jr., Aeriola Sr., Aeriola Sr.," Told to H. G. Clark by Dr. Frank Conrad, about 1940.

6 Frank Conrad, U. S. Patent 1,515,186, "Aperiodic Receiver System," filed June 3, 1920 issued November 11, 1924; U.S. Patent 1,502,848 "Tuning System of Antenna," filed July 7, 1920, issued July 29 1924.

7 "Westinghouse Company Enters Wireless Field," *Electrical Review*, October 16, 1920, p. 615.

8 Donald G. Little, *American Heritage*, August 1955, p. 72.

9 H. P. Davis, *The Radio Industry: The Story of Its Development*, p. 193.

10 *Radioana,* "The Story of Conrad and Station KDKA," dictated to H. G. Clark by Dr. Conrad and W. W. Rogers in Pittsburgh, about 1940, SRM 4 1559.

11 The designation *"Special Land Station"* was an umbrella term comprising three separate license categories–experimental, technical and training school, and special amateur. The number in the call sign corresponded to the district in which the station was located, while the first letter identified its license class–X for experimental, Y for technical and training school, and Z for special amateur. Special land stations could operate at higher powers and on slightly different wavelengths than regular amateur licenses.

12 The reference to 3WE is at odds with Westinghouse Vice President Davis' recollection that the station was designated 2WE. Also, Conrad seems to be confused about the 3 referring to the Third Naval District; Pennsylvania was never in the Third Naval District–or the Second District for that matter. Perhaps Conrad was thinking about his Special Land Station license 8XK, where the first number in the callsign designated the Radio Inspection District in which it was located.

13 Hugo Gernsback, "Amateur Radio Restored," *The Electrical Experimenter*, June, 1919, page 131.

14 Thomas H. White, who has researched the history of U. S. Special Land Stations extensively, found that the first record of any Special Land Station license being issued after the war appeared in the August 1, 1919 edition of the Radio Service Bulletin #28 covering additions in July 1919–although 8XK did not appear: Thomas H. White, *United States Early Radio History*, www.earlyradio historyus.com, Chapter 24 Original Articles, "U.S. Special Land Stations: 1913-1921 Recap."

15 Gleason L. Archer, *History of Radio to 1926*, (The American Historical Society, Inc. New York, 1938) p. 199.

16 In fact, almost half of the entries in the May 1, 1920 issue are duplicates of earlier entries—which are clearly in error—so it would be no surprise if the entry for 8XK in this issue was also a duplicate of an earlier entry. However, since there is no earlier entry for 8XK, it is likely that the license for 8XK was issued earlier but inadvertently omitted from the appropriate earlier issue of the *Radio Service Bulletin*.

17 These words appeared in documents published periodically by Westinghouse celebrating the anniversary of the East Pittsburgh Division founded in 1894. For example: *Westinghouse 1894-1969 East Pittsburgh Divisions' 75th Anniversary* (Westinghouse, 1969); *East Pittsburgh, Pennsylvania Diamond Jubilee 1895-1970* (Westinghouse, 1970); East Pittsburgh 1895-1995 (Westinghouse, 1995).

18 "8XK, Pittsburgh, Pa.," *QST*, September, 1920, pp. 32-34; "8XK," *QST*, November 1920, p.8; "Strays," *QST*, December, 1920, p.48; "Who's Who in Amateur Wireless," *QST*, February, 1921, pp. 39, 45, 48.

19 H. P. Davis, *The Radio Industry: The Story of Its Development,* p. 194.

20 At that time, the license for a commercial broadcast station not only had to be issued, but it had to be in the physical possession of the station.

21 Donald G. Little, *American Heritage*, p. 73.

22 See for example George H. Douglas, *The Early Days of Radio Broadcasting*, (McFarland & Co., Inc. Jefferson, North Carolina, and London, 1934), p. 20; also see www.kdkaradio.com or http://kdkaradio.com/historicphotos.shtml.

23 *QST*, September, 1920, pp. 32-34.

24 "Election Returns Flashed by Radio to 7,000 Amateurs," *The Electrical Experimenter*, January, 1917, p. 650; "A Concert by Wireless," *QST*, January, 1917, p. 26; "News and Music to be Transmitted by Wireless Telephone," *Electrical Review and Western Electrician*, November, 18, 1916, p. 880.

25 "An Almost Unlimited Field for Radio Telephony," *Radio News*, February, 1920, p. 434; "Washington Radio Amateurs Hear Radio Concert," *Radio Amateur News*, May, 1920, p. 646; "Radio Music. *Scientific American Monthly*, October, 1920, p. 178.

26 "A 'Wireless' Concert," *The Electrician*, June, 18, 1920, p. 658; "Melba Entertains Europe by Wireless Telephone," *Telephony*, July 10, 1920, p. 19; "Radio Concerts," *Radio News*, September, 1920, p. 133.

27 "Send Election Returns by Wireless Telephone," *Electrical Review*, November 6, 1920, p. 733.

28 "Westinghouse to Cover Country with Radio Entertainment," *Electrical Review*, December 10, 1921, p. 887.

29 George H. Douglas, *The Early Days of Radio Broadcasting*, pp. 23-29.

30 The basic data for the number of broadcast stations presented in this paragraph are contained in the Radio Service Bulletins issued monthly by the Bureau of Navigation, Department of Commerce. Because of the difficulty of interpreting the data in these bulletins, there are small discrepancies in the number of broadcast stations reported by various authors. The data here were taken from these bulletins, consistent with the interpretation of Thomas H. White, "United States Early Radio History," Internet site (http://earlyradiohistory.us/index.html), http://earlyradiohistory.us/pion622.htm), January 1, 1998. This document also contains a thoughtful discussion of these discrepancies.

31 Frank Conrad, "Radio Receiving Equipment," *IRE Proceedings*, Vol. 10, No. 6, Dec. 1922, pp. 426-439.

32 *Radioana,* "Additional data re RC, Aeriola Jr., Aeriola Sr., Aeriola Sr.," Told to H. G. Clark by Dr. Frank Conrad, about 1940.

33 Ibid.

34 *QST*, Vol. IV, No. 9, April 1921, p.120.

35 Frank Hill interview with Donald G. Little, "KDKA's First Broadcast—How We Did It," *Radio Age*, Vol. 21, No. 11, November 1996, p. 4-6.

36 *The Wireless Age*, Vol. 8, No. 12, September 1921, p. 7; *The Wireless Age*, Vol. 9, No. 1, October 1921, p. 7.

37 *Radio News*, Vol. 3, No. 6, December, 1921, p. 517.

38 The range specified in RCA literature and distributor catalogs was 1600-2800 meters, but the range specified on the Type CB box manufactured by Westinghouse was 1600-2700.

39 *Radio Merchandising*, Vol. 2, No. 6, March, 1923, pp. 36-37; *The Wireless Age*, Vol. 10, No. 6, March, 1923, p. 1.

40 Alan Douglas, *Radio Manufacturers of the 20's, Vol. 3 – RCA to Zenith*, (The Vestal Press, Ltd, Vestal, New York 13851, 1991) p. 5.

41 Alan Douglas, "Radio Corporation of America," *Radio Age*, Vol. 8, No. 9, November, 1982, p. 5.

42 *RCA Victor Service Notes for 1923-1928*, (RCA Victor Company, Inc. Camden N.J, N.D.), see section entitled Radiola RT.

43 Vincent Morley, "Adapting Dry Cell Tubes to the Radiola RC," *The Wireless Age*, Vol. 4, No. 1, October 1923, pp. 46-47.

44 Lauren A. Peckham, "The Westinghouse RC – Part II," *Antique Radio Classified*, Vol. 8, No. 6, June 1991, p. 7.

45 *Radioana*, "Minutes of the General Sales Committee Meeting Held July 10th, 1923," Vol. 1, No. 19, p. 2

46 *The New York Times*, July 23, 1923, p. 7.

47 *The New York Times*, July 29, 1923, p. 8.

48 *Chicago Salvage Stock Store Catalog* (Chicago, IL, 1926) p. 42.

49 Tor van der Lande, "Radioer Jeg Har Mott (Radios I have Met)," *Hallo Hallo*, NR. 53 (1/96), Mars 1996, pp. 36-38 (Norwegian periodical, Tore Moe, ed., Damfaret, 58, N-0682, Oslo, Norway). Kindly provided by Tolbert V. Prowell with a translation.

50 Radiola Guy (Sonny Clutter) web site: (http://www.radiolaguy.com/Showcase/RC.htm); Radiomuseum web site (http://www.radiomuseum.org/r/telefunken_telefunkon_d_type_e3 04. html).

51 Private communication with Sonny Clutter who kindly supplied a copy of the publication containing the picture and technical description, and a second document consisting of the schematic diagram.

52 H. Winfield Secor, "Radio Detector Development," *Electrical Experimenter*, Vol. IV, No. 9, January, 1917.

53 *Radio Receiving Apparatus for Short Wave Telegraphy and Telephony*, Westinghouse Circular 1641, N.D. (circa 1921), p, 4.

54 Alan Douglas, *Radio Manufacturers of the 1920's*, p. 6.

55 The production date on the name plates of these sets is partially concealed with a rivet, so only the day and year (?-26-21) are visible.

56 *The Wireless Age*, Vol. 8, No. 10, July 1921, p. 4.

57 *Radio Enters the Home*, Radio Corporation of America catalog, June 1, 1922, p. 12.

58 Alan Douglas, *Radio Manufacturers of the 1920's*, Vol. 3, p. 6.

59 Alan Douglas, Technology Crosses the Atlantic, *Radio Age*, Vol. 16, No. 3, June 1990, pp. 4-5.

60 Alan Douglas, *Radio Manufacturers of the 1920's*, Vol. 3, p. 6.

61 William C. White, *"The Early History of Electronics in the General Electric Company,"* (William C. White Papers Collection, Schenectady Museum, Museum & Suits-Bueche Planetarium, Schenectady, New York, prepared 1953-1955), Chapter V, p. V-6.

62 *Radioana*, David Sarnoff, "Report on prospective Radio Business," letter from David Sarnoff to C. D. Young, Chairman of the Board of General Electric dated January 31, 1920.

63 Gleason L. Archer, *Big Business and Radio*, (The American Historical Company, Inc., New York, 1939) p. 16.

64 *Federal Trade Commission vs. General Electric Company, et. al., Radio Case, Brief of Facts*, (Sidney C. Ormaby Company, 217 Broadway, New York, 1928) Vol. 4, p. 1891 (testimony given on December 12, 1926).

65 Ibid., Vol. 4, pp.1891-1892 (testimony given on December 9, 1926).

66 Ibid., Vol. 6, p. 2823 (testimony given on April 26, 1927).

67 Ibid., Vol. 7, p. 3491-2 (testimony given on November 16, 1927).

68 H. Gernsback, "The Radiotrola," *Radio News*, Vol. 3, No. 6, December 1921, p 479.

69 *Radio Apparatus for Amateur and Experimental Use with Instructions for Continuous Wave Operation,* (Radio Corporation of America, New York, September 1, 1921).

70 *Radio Telegraph and Telephone Equipment*, (Wireless Specially Apparatus Company, Boston, Mass, 1919), pp. 32, 37.

71 *History of the Bureau of Engineering Navy Department During the World War*, under direction of Edwin Denby, Secretary of the Navy (Government Printing Office, Washington, D. C., 1922) pp. 117-118.

72 Station LY was a commercial station in Bordeaux owned by the French Government that operated at 18,900 meters.

73 *Radio Apparatus for Amateur and Experimental Use with Instructions for Continuous Wave Operation,* RCA catalog, September 1, 1921, p. 41; this error is repeated in the 1922 RCA catalog *Radio Enters the Home*, (Radio Corporation of America, Marketing Department, New York City) June 1, 1922, p. 64.

74 George Sterling, *The Radio Manual*, (New York: D. van Nostrand Company, Inc., 1928) p. 465.

75 Eric P. Wenaas, "Wireless Specialty Apparatus Nomenclature Gone Awry," *The AWA Review*, Vol. 18, 2005, pp. 25-48.

76 *Radio Apparatus for Amateur and Experimental Use with Instructions for Continuous Wave Operation,* RCA catalog, September 1, 1921, p. 42.

77 M. B. Sleeper, *How to Make Commercial Type Radio Apparatus*, (The Norman H. Henley Publishing Co. New York, 1922) pp. 133-135.

78 *QST*, Vol. III, No. III, October 1919, inside cover.

Both radio and RCA entered the home in 1922 with an elaborate catalog for "*those who desired to be entertained.*" (*Radio Enters the Home*, RCA catalog dated June 1, 1922)

Chapter 4

Radio Enters the Home

1922

RCA entered the home with a flourish in 1922, implementing many of the recommendations made by Sarnoff in his 1920 letter to RCA Chairman Owen D. Young, including publication of equipment catalogs and receiver brochures, advertising in national publications, and the formation of a nationwide network of distributors and dealers. The statistics reported to the shareholders in the 1922 RCA Annual Report were impressive. The Commercial Department, which prior to 1922 had been selling radio components and marine receivers, increased their staff from 14 to approximately 200 in order to market the broadcast receivers. A comprehensive catalog was published in June 1, 1922 by the Sales Department to advertise the entire RCA line of radio equipment for the broadcast radio listener and the experimentalist.[1] A distribution system was set up comprising 146 wholesale distributors with 196 warehouses and up to 15,000 retail dealers. Commercial Department revenues increased from $1.5M in 1921 to $14.4M in 1922, most of that attributable to the sale of broadcast receivers and accessories.

The Annual Report went on to tell RCA shareholders that, *"Prior to 1922, no wireless receiving apparatus had been developed which lent itself to quantity production, and the impatient demand for such sets reached its climax in the early part of 1922, when due to the energies of your scientists and engineers, radio receiving apparatus embodying the latest improvements and of a character suited for general use was developed and manufactured in large quantities."*[2] It seems that the RCA management who prepared and/or approved this Annual Report conveniently overlooked the Westinghouse receivers including the RA crystal receiver, the RC regenerative receiver, and the Aeriola Jr. crystal set, all of which were developed, manufactured in large quantities, and sold by Westinghouse prior to 1922. In fact, these sets were actually designed and developed by Westinghouse engineers prior to the agreements between RCA and Westinghouse signed in mid-1921. This example is just one of the numerous overreaching statements that RCA would make in advertisements and press releases throughout the twenties.

RCA Offers its First Broadcast Receivers

RCA introduced its first line of broadcast radio receivers to the public at the New York Amateur Radio Show held at the Pennsylvania Hotel during March 7-11, 1922. Billed by RCA's house organ, *The Wireless Age*, as *"The Most Spectacular Mystifying and Gripping Convention and Exhibition which New York Has Ever Witnessed,"* one might have expected that the RCA radio exhibit would be nothing short of breathtaking.[3] However a close inspection of the exhibit reveals that the receivers billed as "new outfits" were for the most part either engineering models that would undergo further changes before reaching the market, or older Westinghouse models that still sported instruction sheets with Westinghouse logos. Worse yet, the receiver line was essentially a

RCA introduced its fist line of broadcast radio apparatus to the public at the New York Amateur Radio show held March 7-11, 1922. (George H. Clark Collection, Smithsonian Institution)

collection of overlapping and competitively priced models manufactured by three different organizations, each with their own name stamped either on the receiver panel or instruction sheets tacked to the lid.

Three months after this show, RCA issued its first broadcast radio catalog dated June 1, 1922 entitled *Radio Enters the Home,* with the subtitle *"For those who desire to be entertained with concerts, lectures, dance music—as well as for the Radio amateur."* Sections of this catalog were similar to the 1921 catalog addressing the interests of the amateur radio operator and experimenter, yet there was an obvious focus on broadcast radio in the new catalog. The catalog

Table 4-1. Broadcast receivers offered by RCA in 1922

Receiver Sets Offered in 1922	Westinghouse Sets	General Electric Sets	Wireless Specialty Sets
Less expensive crystal sets[a]	Aeriola Jr. $32.50	ER-753[b] $25.50	AR-1375 $47.50
More expensive crystal sets[a]	RA + DB detector $90	AR-1300 $65.50	AR-1382 $85.50[c]
Less expensive regenerative sets[d]	Aeriola Sr. $75.90	None	None
More expensive regenerative sets[e]	RC $261.75	AR-1300 + AA-1400 $250.25	None
Deluxe regenerative sets[f]	Aeriola Grand $409.50	None	None

[a] Included earphones and antenna kit

[b] RCA replaced the ER-753 with the ER-753A in mid-year comparably equipped at $32.50

[c] Earphones and antenna kit @$15.50 added to AR-1382 @ $70 to complete the set

[d] Included earphones, batteries, tube, and antenna kit

[e] Included headphones and external Vocarola speaker, batteries, battery charger, tubes, and antenna kit (RC also include CB loading coil)

[f] Included stand, internal speaker, tubes, batteries, battery charger, and antenna kit

Table 4-2. The 1922 season lineup for RCA broadcast receiver apparatus

	Radiola Model	Apparatus Type	Cabinet Style	Tubes	Intro. Date	Discon'd. By	List Price[a]	Mfg. By
1922 Season — New Models	ER-753	Crystal Set	EZKASE	0	Apr. 1922	Aug. 1922	$18[b]	GE
	ER-753A Radiola I[c]	Crystal Set	Wood Box	0	Aug. 1922	Mar. 1924	$25[b]	GE
	AA-484	Amplifier	Breadboard	2	Apr. 1922	Aug. 1922	**$45**	WSA
	AA-1375	Crystal Set	Metal or Wood Case	0	Apr. 1922	May 1923	$40[b]	WSA
	AR-1382	Crystal Set	Metal or Wood Case	0	Aug. 1922	Feb.1923	$70	WSA
	AR-1300	Crystal Set	EZKASE	0	Apr. 1922	Feb. 1923	$50	GE
	AA-1400	Detector and 2-Stage Amp	EZKASE	3	Apr. 1922	Feb. 1923	**$75**	GE
	Aeriola Sr. Amp.(AC)	2-Stage Amplifier	Wood Box	2	Sep. 1922	Mar. 1924	$68	W
	AA-1520	3-Stage URF Amplifier	EZKASE	3	Aug. 1922	Mar. 1924	**$80**	GE
	AA-485	Detector and RF/AF Amp	Wood Cabinet	6	Aug. 1922	Mar. 1924	**$160**	WSA
	Aeriola Grand (RG)	Regenerative Receiver	Wood Cabinet	4[d]	Mar. 1922	Dec. 1922	$325	W
	Vocarola (LV)	Horn Speaker	None	0	Jan. 1922	Sep. 1923	$30	W
	AG-788	Antenna kit	N/A	0	Apr. 1922	Mar. 1924	$7.50	GE
	AG-1380	Loop Antenna	N/A	N/A	Jun. 1922	Feb. 1925	$25	WSA
Holdover Models	Aeriola Jr. (RE)	Crystal Set	Wood Box	0	Jun. 1921	Dec. 1922	$25[b]	W
	Aeriola Sr. (RF)	Regenerative Receiver	Wood Box	1	Dec.1921	Mar. 1924	$65[b]	W
	RA	Tuner	Wood Case	0	Dec. 1920	Jul. 1923	$68	W
	DA	Detector and 2-Stage Amp	Wood Case	3	Dec. 1920	Jul. 1923	**$70**	W
	DB	Crystal Detector	None	0	Aug. 1921	Mar. 1924	$6.50	W
	RC	Regenerative Receiver	Wood Case	3	Dec. 1920	Jul. 1923	**$132.50**	W
	LS	Loud Speaker Attachment	N/A	0	Not known	Mar. 1924	$18	W
	AD	Antenna kit	N/A	0	Oct. 1921	Sep. 1923	$7.50	W

[a] All sets with prices in bold font did not include tubes; no antenna kits were included in any sets
[b] Included earphones
[c] Name change made in late summer of 1922
[d] Required four WB-800 ballast tubes in addition to four WR-21 triodes

provided not only general information on broadcast radio, but also offered individual receiver components and complete broadcast receiver combinations—five complete sets manufactured by Westinghouse, three by GE, and one by Wireless Specialty (the AR-1382 did not appear in the catalog).

Ten receivers—all but one of which was displayed at the New York show—are grouped in Table 4-1 to facilitate a direct comparison of competing models. All three manufacturers made an inexpensive crystal set with competitive prices as shown on the first line of the table. In the more expensive crystal set category, Westinghouse and General Electric made competing tuners (the RA and AR-1300 respectively) that could function either as a crystal set, or as a regenerative receiver when used in conjunction with separate vacuum tube detector-amplifier units (the DA and AA-1400 respectively). Wireless Specialty also manufactured a more expensive crystal set, the AR-1382, and while it appeared in RCA distributor catalogs for 1922, it did not appear in the RCA catalog *Radio Enters the Home*. Again, the prices of these three competing receivers were similar.

There were only two receivers in the 1922 lineup that did not have directly competing counterparts, both manufactured by Westinghouse. One was the Aeriola Sr., an inexpensive regenerative receiver offered as a complete set at $75.90, and the other was the Aeriola Grand, a deluxe regenerative receiver offered as a complete set at $409.50 with an enclosed speaker and stand. RCA also continued to offer the line of "special high grade receiving apparatus" introduced the previous year for the radio amateur and experimentalist, but the pricing was not competitive for broadcast radio reception by the general public.

In addition to the models RCA introduced at the March radio show and/or appearing in the June 1922 RCA catalog, they also introduced an ER-753A crystal set later in the year which replaced the ER-753 crystal set, an AC amplifier companion to the Aeriola Sr., and an AA-1520 3-stage RF amplifier companion to the AA-1400 amplifier. RCA changed the name of the ER-753A to the Radiola I just after it went into production, but all the sets they sold for almost a year were labeled ER-753A without any reference to Radiola I on either the front panel or the instruction card attached to the lid. It was not until the following year that GE manufactured sets which actually had the Radiola I name on both the panel and an instruction card dated May 1923.

The totality of all broadcast radio receivers and major accessories offered by RCA in 1922 as reflected in their distributor price sheets are shown in Table 4-2. The apparatus introduced in 1922 is shown in the upper portion of the table, while the hold-over apparatus introduced in prior years is shown in the shaded portion of the table at the bottom.

RCA begins the Process of Standardization

It is obvious that the RCA offerings in 1922 were an amalgamation of available receivers from different manufacturers rather than a thoughtful selection based on either performance requirements or market demand. The receivers from the different manufacturers not only had a different look, but carried different manufacturer names and model designations. In some cases the manufacturer names were more prominent than the RCA name, while in other cases the RCA name and logo were nowhere to be found.

RCA understood from the outset the importance of standardizing and integrating their radio offerings in terms of both appearance and performance. The task was a daunting one because RCA depended on three different companies to design and manufacture their sets, and because RCA agreed to purchase 60% of its radio equipment from GE and 40% from Westinghouse regardless of consumer demand. Sets manufactured by Wireless Specialty were counted in GE's 60% share of the market.

RCA began by developing a common model designation system and introducing the Radiola trade name as a hallmark for their receivers. RCA then began to develop design requirements, and in some cases developed designs for individual receivers. The RCA Research Department was tasked to test and evaluate all prospective designs from the three manufacturers, and select the models that would be offered by RCA. The process of standardization would continue for another four or five years before RCA fully achieved its goals of offering the same receiver made by two different manufacturers, meeting the required 60%-40%

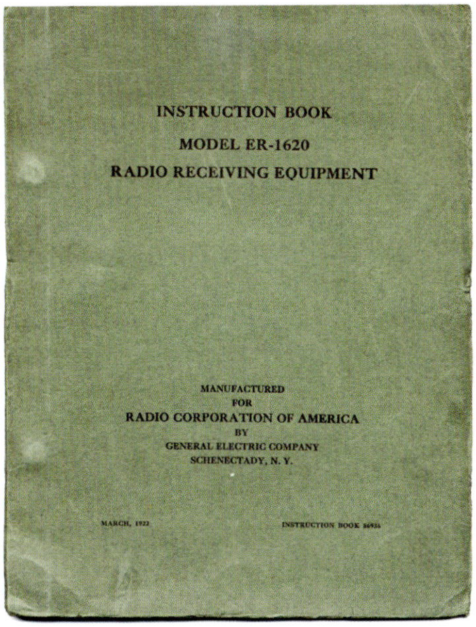

INSTRUCTION BOOK

MODEL ER-1620

RADIO RECEIVING EQUIPMENT

MANUFACTURED

FOR

RADIO CORPORATION OF AMERICA

BY

GENERAL ELECTRIC COMPANY

SCHENECTADY, N. Y.

MARCH, 1922

The ER-1620 Radio Receiving Equipment was one of the first receivers manufactured for RCA to be assigned an ER designation.

proportion between GE and Westinghouse, and assuring that even the most discerning customer could not identify which supplier manufactured the set.

RCA Model Designations

Even prior to the agreements with Wireless Specialty and Westinghouse, RCA had adopted the model designation system used by GE for the radio apparatus that it sold to amateur radio operators and experimenters. The extent and complexity of this designation system was first evident in RCA's June 1921 Radio Apparatus catalog.[4] The model designation system was quite simple in principle: a) two letters, known as "Designation Letters," b) a number, known as the "Serial Number," and c) a letter, known as the "Alteration Letter," used when the apparatus differed slightly from the preceding designs. This system was explained in some detail in a memo prepared by George Clark,[5] which is reproduced in Appendix F.

The first letter of the two designation letters indicated the mechanical form of the apparatus, consisting of one of five letters, P, U, A, S, and E, which stood for piece, unit, assembly, system, and equipment respectively. With regard to receivers, the most relevant classifications were U for units, A for assemblies, and E for equipment. A unit consisted of a device which is complete in itself such as a vacuum tube or transformer. An assembly consisted of an aggregation of units and parts mounted on a common panel or within a common case. An example of an assembly is a tuner or an amplifier mounted in an individual case. Equipment referred to a complete working aggregation which in itself was sufficient for radio transmitting, radio receiving or any major sub-operation thereof. Receiving equipment meant the combination of a) the purely radio apparatus used in receiving, b) the detector, and c) the listening or recording devices. Antennas and ground systems were considered to be assemblies, but were not included in receiver equipment designation.

The second letter of the two designation letters indicated the electrical classification of the apparatus. There were 20 classifications using all letters in the alphabet except for B, E, N, O, and U. The most relevant classifications for radio reception were A for amplifiers, D for detectors, G for antenna and ground systems, P for power equipment, R for tuners, and V for vacuum tubes. All twenty terms are defined in Appendix F.

The serial number assigned to each piece was the main designating feature, and was intended to enable the apparatus to be looked up serially in a catalog. As time passed, sequential serial numbers were also assigned to each apparatus in a production run, so that the numbers appearing in the model designation were referred to as model numbers rather than serial numbers.

The alteration letter was assigned to a redesigned apparatus if the new design interchanged electrically and mechanically with the preceding issue, and where the new electrical or mechanical features could possibly affect "identity of operation or of location." The letter "A" was assigned to the first redesign, "B" to the second, and so on.

Consider for example the ER-1620, the first commercial receiver completely designed by GE for sale by RCA in 1921. The E designates this apparatus as a complete piece of equipment and the R designates it as a receiver. The equipment is made up of several assemblies including an AR-1529 tuned coupler, an AD-1527 vacuum tube detector, two AA-1528 single-stage vacuum tube amplifiers, an AD-692 crystal detector, and two Western Electric type CW-834 headsets. The AR designation was used for the tuner, AD for the two detector assemblies, and AA for the amplifier assembly. Clearly, the E designation category was conceived to account for transmission and reception equipment that consisted of a number of different assemblies.

As the receiver technology evolved, it became customary to house all receiver circuits in a single cabinet, so that the ER designation became redundant with AR. RCA apparently opted to keep the AR designation for complete receivers in lieu of the ER designation after the 1922 season, because the only two broadcast

Rheostats PT-537 (above) and PR-535 (below) are examples of the lowest level of assembly in the RCA designation schema, which were designated with the initial letter P for "Piece" followed by a "T" when designed for a transmitter and an "R" when designed for a receiver.

receivers sold by RCA with an ER designation were the ER-753 and ER-753A Radiola I crystal sets in 1922. The original Radiola III intended for sale in 1923 was also given the designation ER-875 in July 1922,[6] but this receiver was never manufactured. A decal on one Radiola IV has been observed with the designation ER-880, although the official designation was AR-880.

The RCA designation nomenclature was applied to the broadcast sets made for RCA by both GE and Wireless Specialty in 1922 and thereafter, but the nomenclature would not be applied to the Westinghouse sets for another year or two. In 1924, several Westinghouse receivers including the Radiola III and Radiola III-A had both Westinghouse and RCA nomenclatures applied.

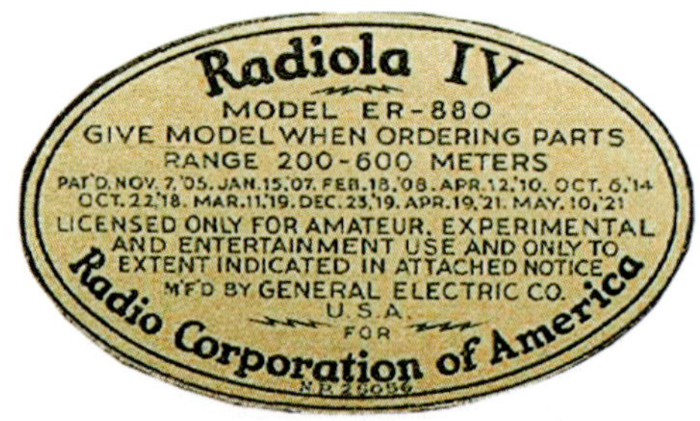

While the Radiola IV was officially given the AR-880 designation, some sets have a decal affixed with an ER-880 designation.

Westinghouse Model Type Designations

Both Westinghouse and Wireless Specialty had developed a model designation system for their receivers prior to 1921, and each were using their own systems at the time they joined the RCA "Radio Group." Westinghouse continued to use its model designation system for several years after joining the Radio Group, and did not fully adopt the RCA system until after the 1924-25 season. Several of the receivers in the 1924-25 season were actually assigned both RCA and Westinghouse designations.

The Westinghouse model designation system for their broadcast receiver apparatus consisted of two "Type" letters. The first letter they used for most of their radio apparatus was either an R, D, A, C, L, P, or I which identified the equipment as being either a receiver, detector, amplifier, coil, loudspeaker or reproducer, antenna protector, or antenna insulator respectively. The second letter was generally assigned in the relative order in which the apparatus was manufactured beginning with the letter A, but there were exceptions. As an example of this Type system, consider the Westinghouse broadcast receiver as a class that all began with the Type letter R. The first receiver to be assigned a Type designation was the RA tuner, which was also a crystal receiver when used with the external DB detector. When the RA tuner was combined with the first Type DA detector/amplifier in a single cabinet, the resulting receiver was given the designation Type RC rather than Type RB. The letter B had been assigned to the RB receiver, a commercial set manufactured by Westinghouse in late 1920 or early 1921,[7] presumably after the RA was assigned its designation but before the RA and DA were combined into a single cabinet and designated the RC.[8] Westinghouse receivers with the two-letter Type designations are shown in Table 4-3. The first letter was invariably R, while the second letter was for the most part sequential—except for certain letters which appear to have been skipped or perhaps were assigned to engineering models that were never produced (e.g. Type RD). For no apparent reason, the RS and RT designations are out of order chronologically.

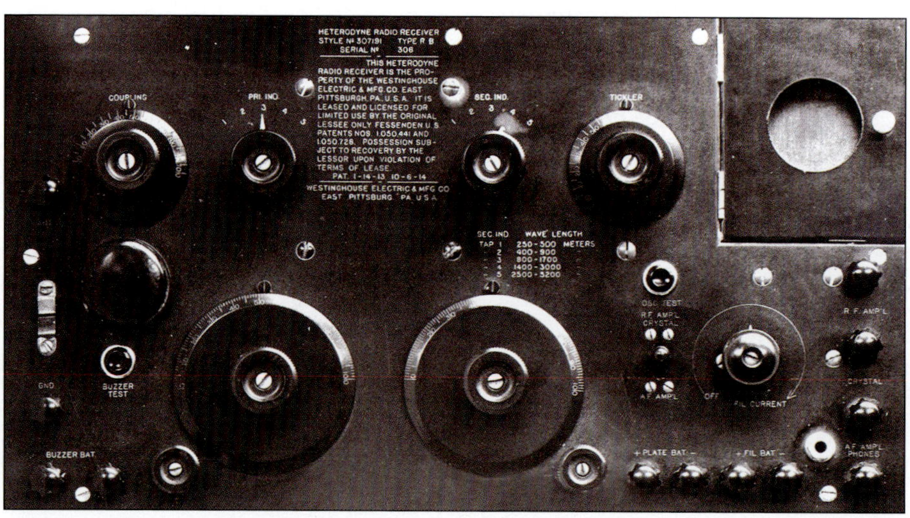

The RB receiver was manufactured by Westinghouse for commercial applications such as reception of radio transmissions at Westinghouse broadcast radio stations. (Courtesy of Alan Douglas)

Table 4-3. Westinghouse used a two-letter system to designate the receiver "Type"

Receiver Type Designation	Release Date	Other Designations	Comments
RA	Dec. 1920	RA Tuner RA Receiver	RA receiver is RA tuner with DB detector
RB	Jan. 1921[9]	Medium Wave Heterodyne Receiver	Commercial set used in broadcast stations
RC	Just after the RB	Radiola RC	RA and DA combined in one cabinet
RE	Jul. 1921	Aeriola Jr.	Later marketed as Radiola Jr.
RF	Dec. 1921	Aeriola Sr. Radiola Sr.	Name change only; same Type designation
RG	Mar. 1922	Aeriola Grand, Radiola Grand	Both assigned same Type designation
RI	Feb. 1924	Radiola III AR-805	Assigned both WH and RCA designations
RJ	Apr. 1924	Radiola Rengenoflex AR-817	Assigned both WH and RCA designations
RK	Mar. 1924	Radiola X AR-816	Assigned both WH and RCA designations
RL	Mar. 1924	Radiola III-A AR-806	Assigned both WH and RCA designations
RS	Jan. 1923	Radiola RS	No RCA designation
RT	Jan. 1923	Antenna coupler	No RCA designation

The Westinghouse spare crystal detector of the pressure type was designated Type DE as well as Style 307450.

Westinghouse manufactured significantly fewer detectors than receivers for broadcast radio. The first two detectors were the DA detector with a 2-tube amplifier and the DB crystal detector. Apparently, the next letter in the alphabet, C, was not used, quite possibly, because the combination DC was in common use as an abbreviation for "direct current." The letters DD and DE were used for spare crystals. The only other known radio detector manufactured by Westinghouse is the DR, a resistance coupled version of the DA. The letter R may have been chosen to connote the resistance-coupled nature of the detector.

Westinghouse manufactured only three broadcast amplifier units for RCA including the AC amplifier designed for use with the Aeriola Sr. and Radiola Sr. receivers, the AP balanced amplifier (also known as the Radiola Balanced Amplifier) designed for use with the Radiola III, and the AR radio-frequency amplifier designed for use with the RC. However, it is likely there were AA and AB amplifier units as well, perhaps for commercial applications. For example, there was a companion amplifier to the RB unit referred to in a Westinghouse letter to RCA which almost certainly had a two-letter designation—perhaps either AA or AB. Both AA and AB units—if indeed they existed—were most likely amplifiers designed for commercial use. It is doubtful that the gap between AC and AP was filled with twelve amplifiers, so following the same line of reasoning that the R in the DR detector alluded to "resistance coupled," it is quite possible that the letter P in the AP designation for the balanced amplifier referred to "push-pull," the type of output circuit used in that amplifier. Perhaps the R in the AR referred to the radio-frequency nature of the amplifier.

Other designations used for receiving equipment were CB for the RC load coil, LS for the reproducing loudspeaker attachment for phonographs, LV for the Vocarola loudspeaker, and AD for a long-wire antenna kit. However, the FH horn speaker with the initial letter F in lieu of the letter L did not follow the pattern for other Westinghouse loudspeaker apparatus.

Wireless Specialty Model Designations

Wireless Specialty used the letters IP (or I-P) followed by a dash and two or three numbers to designate much of their equipment designed before the RCA-WSA agreements, such as the IP-500 and IP-501 receivers. Unlike Westinghouse, WSA began using the RCA nomenclature system for all new broadcast-radio receiving apparatus designed and manufactured for RCA immediately following the RCA-WSA agreements.

RCA Introduces the Radiola Trade Name

The Radiola name first surfaced in the now-famous Goldsmith letter to Sarnoff dated February 26, 1920, in which Goldsmith first referred to Sarnoff's Radio Music Box as a Radiola. It was October 29, 1921 before the potential value of the Radiola name in commercial trade was realized, or at least that is when the name Radiola was filed as a trademark with the U.S. Patent Office.[10] The name was registered on March 27, 1922, and the records show it was first used in commerce on September 27, 1921.

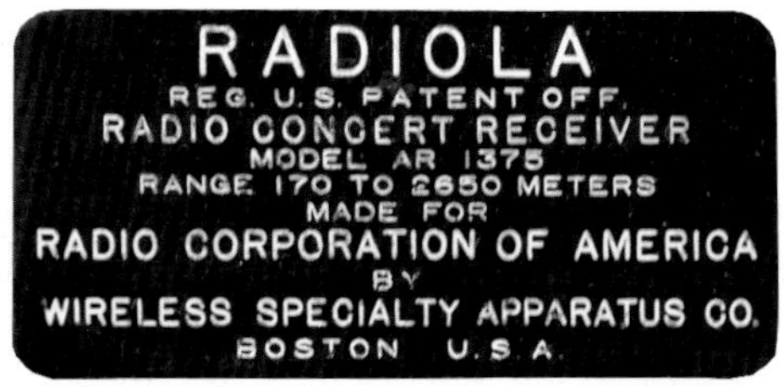

The AR-1375 was the first mass produced RCA receiver to carry the Radiola trade name.

While the registration documents indicate the name was first used in commerce in September of 1921, there is no evidence that any radio distributed by RCA bearing the Radiola name was used in commerce until the spring of 1922. The first radio of record bearing the Radiola name was not the Radiola I (ER-753A), but rather the Wireless Specialty AR-1375, which appeared on the official RCA Price List as early as April 1, 1922. The Radiola I did not appear on the RCA price list until August of 1922, but even then the Radiola name did not actually appear on the ER-753A panel or instruction card in the first two production runs. It was not until 1923 that an ER-753A was produced with the Radiola I designation. There were other Wireless Specialty sets carrying the Radiola name that predated the Radiola I as well, most notably the AA-485 amplifier.

RCA made a decision in mid-to-late summer of 1922 to use the name Radiola on virtually all its receiver apparatus–including the RC receiver and the entire line of Aeriola sets. However, this change did not actually appear on the panels or instructions of either the RC or Aeriola Jr. because a sufficient number of sets had already been produced to satisfy the demand up to the time the set was discontinued.

RCA heavily advertised the Radiola name, and within a short period of time made the Radiola a household word. Radiola was used almost exclusively for RCA radio apparatus in the decade of the twenties, and to a lesser extent for many decades after the RCA-Victor merger in 1929. It was used on selected sets through the era of the transistor radio in the 1960's and beyond, and was one of the most successful trade names of all time.

It is interesting to note that RCA was not the first to use the Radiola trade name. There were at least three radios manufactured or advertised in the U.S. with the Radiola name before April of 1921.[11] RCA must have taken action against these companies immediately after the trade name was registered on March 27, 1922, because shortly thereafter, all advertising for Radiolas other than those marketed by RCA abruptly ceased. The Radiola name was also used in several other countries including France, Australia, England, Sweden, and Canada for many years thereafter. In many cases, the use of the Radiola name abroad was sanctioned by RCA (see Appendix H).

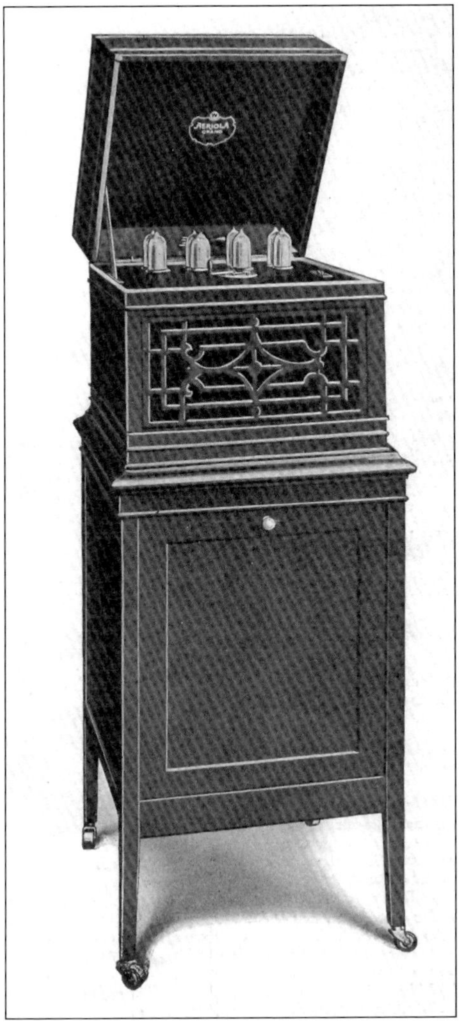

The Aeriola Grand was sold with the battery stand pictured here for $350, or for $325 without the stand. (*Radio Enters the Home*, RCA catalog dated June 1, 1922, p. 29)

Westinghouse Offerings for 1922

RCA introduced three new broadcast radio products in the 1922 season manufactured by Westinghouse including the Aeriola Grand, the AC amplifier offered as an adjunct to the Radiola Senior, and the Vocarola speaker. RCA also held over virtually the entire line of broadcast radio products introduced by Westinghouse the previous year, including the RA/DA/RC receivers, the Aeriola Jr. crystal set, the Aeriola Sr. regenerative receiver, the loudspeaker attachments for phonographs, and the long-wire antenna kit. Once again, Westinghouse dominated the RCA line in 1922, leaving both General Electric and Wireless Specialty in a distant second and third place respectively.

Aeriola Grand (Type RG)

The Aeriola Grand was first introduced to the public at the Amateur Radio Show in New York held on March 7-11, 1922, and first appeared in RCA's Price List for Radio Apparatus dated April 1, 1922. It also appeared in the RCA catalog *Radio Enters the Home* published in June of 1922, and in national ads first appearing circa August 1922.[12] It was the first radio offered by RCA with a self-contained speaker in an attractive polished mahogany cabinet, evocative of the Radio Music Box first described by David Sarnoff many years earlier—except that the batteries were contained in a separate compartment in the optional stand offered with the receiver.

The design goal was to make a high-performance radio with as few controls as possible so that, according to the RCA catalog *Radio Enters the Home* "*any man, woman or child can easily operate it without the slightest technical knowledge*." There were basically two controls, an on-off switch and a tuning dial. The usual filament control rheostats were replaced by four type WB-800 ballast tubes to assure "automatic and uniform heating" of the filaments of the three WR-21-A amplifier tubes and one WR-21-D detector tube used in this set. A special plug in the shape of a vacuum tube could be used to reduce the volume by replacing the last amplifier tube when the volume was too loud. A specially designed stand could be purchased for $35 as an adjunct to this set to hold the batteries.

The Aeriola Grand had two different cabinet styles, both of which were described and pictured in the patent 1,655,985 filed on April 18, 1922 by Frank Conrad.[13] The first was a boxy cabinet with a geometric style grill covering the speaker opening, while the second was a softer style with curves in the lid and a grill with a lyre design. The geometric style pictured to the left was featured in virtually all of the contemporaneous catalogs and ads, and appears to be the more common of the two. The set with the softer style and lyre grill design is shown on the next page. The model designation for both styles is Type RG, but the geometric models viewed by the author bear the Style No. 319529 B, while the lyre model bears the Style No. 319531.

Two types of name decals have been observed on the inside lid of the Aeriola Grand, one with a shield-like appearance in multiple colors and the other with a rectangular shape in bright gold with black lettering. Since both types of decals have been observed on the geometric style receiver in ads during mid-1922, it does not appear that one type of decal was reserved for one style of receiver.

This radio was prominently advertised in RCA catalogs and in magazine ads featuring no less than Guglielmo Marconi boasting that *"The Aeriola Grand is at present the supreme achievement in designing and constructing receiving sets for the home..."*[14] However, it soon became evident that there was something amiss with the Aeriola Grand design. The advertising disappeared by October 1922—only seven months after its introduction in March—and a completely redesigned receiver, the Radiola Grand, was announced in an ad in *The Wireless Age* in December 1922.[15] A memo to RCA Wholesale Distributors announcing the introduction of the Radiola Grand in early 1923 identified the problem as one of insufficient range.[16] Indeed, the

One version of the Aeriola Grand was equipped with grille covering evocative of a lyre and a softer style cabinet with curves in the lid such as the one shown here (left). The Aeriola Grand was designed with only one control to simplify operation—a station selection knob (right).

The resistance-coupled design of the Aeriola Grand is readily apparent from the absence of audio transformers and the presence of large coupling resistors.

RCA catalog states that the receiver was designed for use near high-power broadcast stations not more than 50 miles away. Such performance was incredibly poor for a four-tube regenerative receiver of that era. Worse yet, at $325 without stand or accessories, or $409.50 complete with batteries,

In one of the first ads for the Aeriola Grand, Marconi boasted that *"The Aeriola Grand is at present the supreme achievement in designing set for the home…"* (*Radio Broadcast*, September 1922, p. 451)

SEPTEMBER, 1922 THE WIRELESS AGE 1

When Marconi heard the AERIOLA GRAND

"IT comes closest to the dream I had when I first caught the vision of radio's vast possibilites. It brings the world of music, news and education into the home, fresh from the human voice. It solves the problem of loneliness and isolation.

"The Aeriola Grand is at present the supreme achievement in designing and constructing receiving sets for the home—a product of the research systematically conducted by scientists in the laboratories that constitute part of the R C A organization."

G. Marconi

© UNDERWOOD & UNDERWOOD

RCA

Look for this trade mark at your dealer

The importance of the Symbol R C A

CRUDE radio apparatus of a kind can be made even by embryonic organizations. But the vitally important inventions that have made radio the possession of every man, woman and child are those protected by patents owned by the Radio Corporation of America and developed as the result of costly research conducted in the engineering laboratories of the Radio Corporation of America.

The name-plate of a Radio Set is all-important in the purchase of radio apparatus. If it bears the letters "R C A" the public and the dealer are assured that at the time of its introduction it is the highest expression of the advancing art of radio.

In tone quality, in simplicity of manipulation the Aeriola Grand is unrivalled. A child can snap the switch and move the single lever that tunes the Aeriola Grand and floods a room with song and speech from the broadcasting station.

Any R C A dealer will be pleased to show you the Aeriola Grand and to let you judge its wonderful tone quality for yourself.

There is an R C A set for every purse—Prices range from $18 to $350.

Aeriola Grand with stand $350

Radio Corporation of America — World Wide Wireless

Sales Dept. Suite 2064, 233 Broadway, New York City

District Office, 10 South La Salle St., Chicago, Ill.

When writing to advertisers please mention THE WIRELESS AGE

Two types of decals appeared on the Aeriola Grand lid including one in multiple colors in the shape of a shield, and one in gold with a rectangular shape.

charger, and antenna, the Aeriola Grand was the most expensive and deluxe set in the RCA lineup for 1922.

There were several design features that contributed to the performance problems and poor acceptance by the public. First and foremost, the audio circuits were resistance coupled—similar to the ill-fated DR detector-amplifier—and therefore did not have the gain that would have resulted from transformers, had the audio circuits been transformer coupled. While the WR-21 had twice or so the gain of the WD-11 tube then used on other Westinghouse models, it could not make up for the loss in amplification of 5 to 10 possible from *each* audio transformer. As a result, three stages of audio amplification with three tubes were required to do what two stages did on transformer coupled sets—with the remaining tube used as a regenerative detector. Second, in order to simplify operation, regeneration was accomplished by a fixed coil with taps and a switch as opposed to a continuous tickler adjustment used on most regenerative receivers of the day. The idea was to preset the switch for a given location, thereby reducing the number of controls required to operate the set, but this approach also assured non-optimal operation under most conditions. Third, in order to further simplify operation, the usual filament rheostats used to compensate for battery drain were replaced by the four ballast tubes. This simplification—requiring four additional ballast tubes—not only made the set more expensive to purchase, but the four additional tubes also drained the batteries more rapidly.

While the set was not recalled by RCA, it was replaced free of charge upon request with a Radiola Grand beginning in late December or early January 1923. The returned Aeriola Grand sets were either destroyed or converted into Radiola Grand sets, thus accounting for the rarity of these sets. Only a few sets are known to have survived intact, one of which is pictured here.

It is not known for certain how many of these receivers were made or how many different style numbers exist. An insufficient number of sets remain to estimate the total number from the serial numbers. Three of the sets known to exist have serial numbers of 11162, 46580, and 46738 and a "production date" of (10-27-21). Serial number evidence presented in Appendix D indicates that perhaps as many as 5,200 Aeriola Grand's were manufactured, although very few have survived.

Aeriola Sr. Amplifier (Type AC)

The Aeriola Sr. Type AC amplifier first appeared in the RCA Distributor Price List dated August 21, 1922, and it was first advertised in national magazines

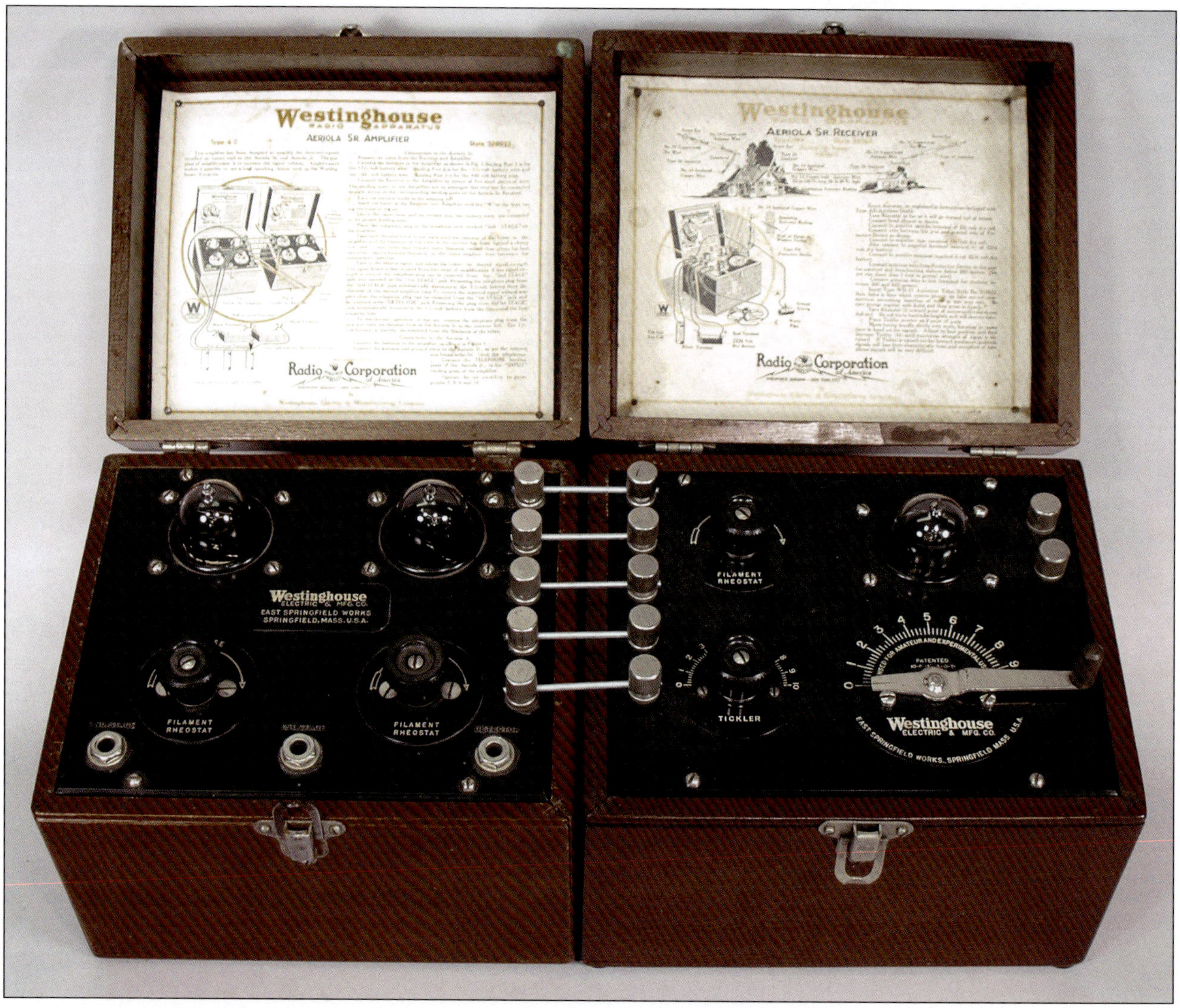

The Aeriola AC had an appearance similar to the Aeriola Sr., with input terminals that lined up with the output terminals of the Aeriola Sr.

The Aeriola AC amplifier was generally advertised in conjunction with the Aeriola Sr. for which it was designed, and was often accompanied by a Vocarola speaker to make the point it could be used for loudspeaker operation. (*The Wireless Age*, October 1922, p. 3)

The transformer-coupled design of the AC was evident from the two audio transformers which dominated the interior.

circa October 1922 for $68 including tubes.[17] George Clark identified the official release date as September 1922. This apparatus is a two-stage audio-frequency, transformed-coupled amplifier that was designed specifically for use with the Aeriola Sr. The sets consisted of two WD-11 tubes with two sockets, two audio transformers, three phone jacks, and two rheostats mounted below a molded Micarta panel which was mounted in a mahogany box. It has a general appearance very similar to the Aeriola Sr. receiver, with the exception of a Micarta panel in lieu of the wood panel found on the Sr. In fact, it is likely that the introduction of the companion AC amplifier with a Micarta panel provided the impetus for replacing the wood panel on the Sr. with the Micarta panel, which in fact, occurred in this time frame—perhaps coincident with the introduction of the AC. This amplifier is very similar to the audio amplifier circuit found in RC receivers designed years earlier, and on balance is quite unremarkable.

The RCA logo was prominently displayed on the instruction sheet tacked to the inside lid of the Aeriola AC, which had the identification number IL 1198 in the lower right corner. The instruction sheet identified this set as Style No. 326032, but these sets were not assigned serial numbers. There is no estimate for the number of sets manufactured.

The Aeriola AC instruction card bearing the identification IL 1198 had a visual appearance very similar to the instructions of the Radiola Sr.

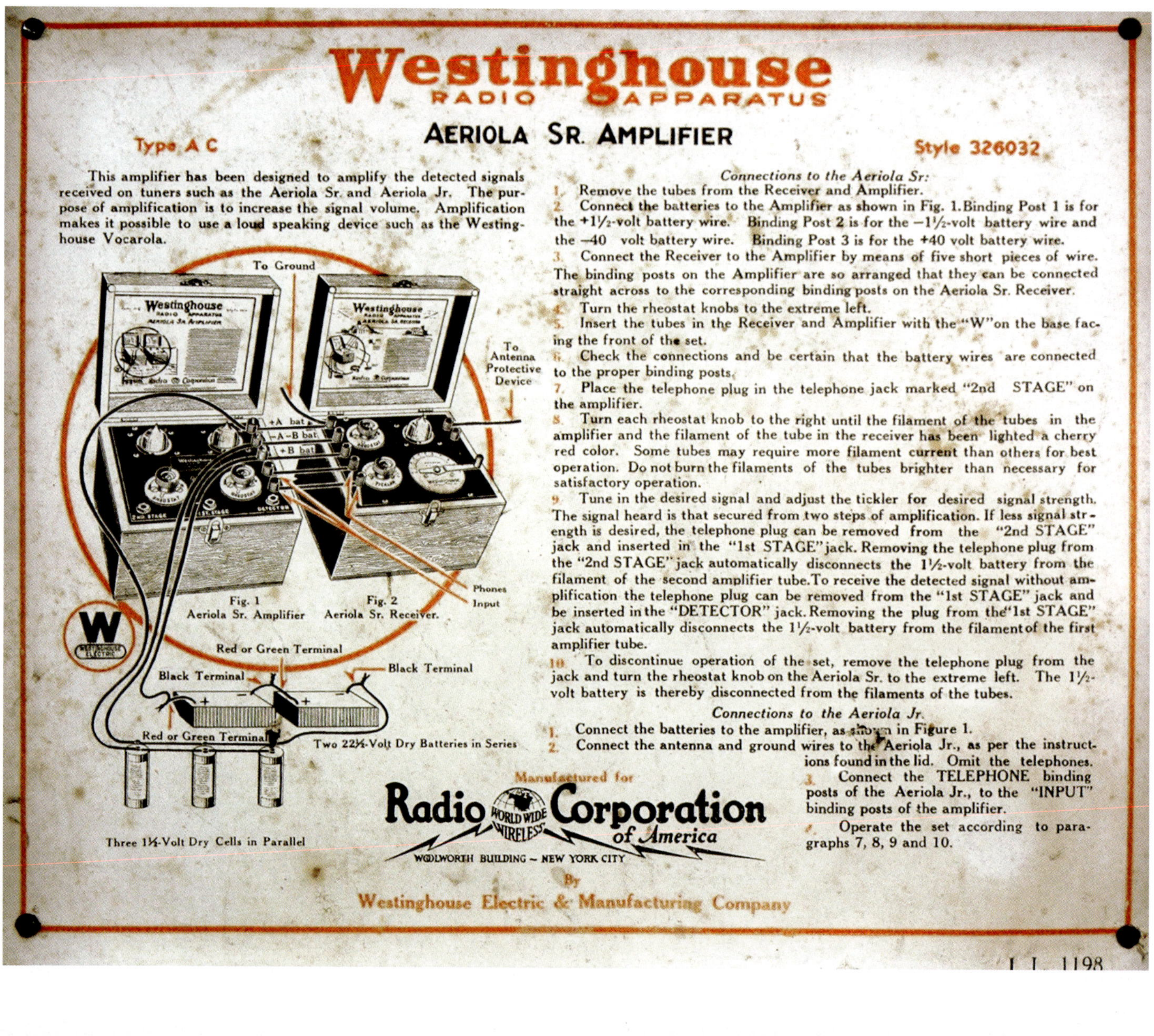

Westinghouse RADIO APPARATUS

AERIOLA SR. AMPLIFIER

Type A C

Style 326032

This amplifier has been designed to amplify the detected signals received on tuners such as the Aeriola Sr. and Aeriola Jr. The purpose of amplification is to increase the signal volume. Amplification makes it possible to use a loud speaking device such as the Westinghouse Vocarola.

Connections to the Aeriola Sr:

1. Remove the tubes from the Receiver and Amplifier.
2. Connect the batteries to the Amplifier as shown in Fig. 1. Binding Post 1 is for the +1½-volt battery wire. Binding Post 2 is for the −1½-volt battery wire and the −40 volt battery wire. Binding Post 3 is for the +40 volt battery wire.
3. Connect the Receiver to the Amplifier by means of five short pieces of wire. The binding posts on the Amplifier are so arranged that they can be connected straight across to the corresponding binding posts on the Aeriola Sr. Receiver.
4. Turn the rheostat knobs to the extreme left.
5. Insert the tubes in the Receiver and Amplifier with the "W" on the base facing the front of the set.
6. Check the connections and be certain that the battery wires are connected to the proper binding posts.
7. Place the telephone plug in the telephone jack marked "2nd STAGE" on the amplifier.
8. Turn each rheostat knob to the right until the filament of the tubes in the amplifier and the filament of the tube in the receiver has been lighted a cherry red color. Some tubes may require more filament current than others for best operation. Do not burn the filaments of the tubes brighter than necessary for satisfactory operation.
9. Tune in the desired signal and adjust the tickler for desired signal strength. The signal heard is that secured from two steps of amplification. If less signal strength is desired, the telephone plug can be removed from the "2nd STAGE" jack and inserted in the "1st STAGE" jack. Removing the telephone plug from the "2nd STAGE" jack automatically disconnects the 1½-volt battery from the filament of the second amplifier tube. To receive the detected signal without amplification the telephone plug can be removed from the "1st STAGE" jack and be inserted in the "DETECTOR" jack. Removing the plug from the "1st STAGE" jack automatically disconnects the 1½-volt battery from the filament of the first amplifier tube.
10. To discontinue operation of the set, remove the telephone plug from the jack and turn the rheostat knob on the Aeriola Sr. to the extreme left. The 1½-volt battery is thereby disconnected from the filaments of the tubes.

Connections to the Aeriola Jr.

1. Connect the batteries to the amplifier, as shown in Figure 1.
2. Connect the antenna and ground wires to the Aeriola Jr., as per the instructions found in the lid. Omit the telephones.
3. Connect the TELEPHONE binding posts of the Aeriola Jr., to the "INPUT" binding posts of the amplifier.
4. Operate the set according to paragraphs 7, 8, 9 and 10.

Manufactured for Radio Corporation of America

WOOLWORTH BUILDING ~ NEW YORK CITY

By Westinghouse Electric & Manufacturing Company

I L 1198

Radiola AC Amplifier (Type AC)

RCA changed the name of the Aeriola Sr. Amplifier to the Radiola AC in mid-1922, but sets were not produced with this name until late 1922. The new name appeared on a metal tag tacked to the inside lid bearing the date (10-16-22), replacing the paper instruction sheet used in the Aeriola AC. The new Radiola AC name appeared in RCA's first catalog for 1923 and in advertisements as early as February for $68, the same price as the Aeriola Sr. Amplifier. Unlike the Radiola Sr., the Radiola AC did not undergo a change in Style number when the name was changed.

In contrast to the Aeriola AC, the Radiola AC was assigned serial numbers which were stamped on the metal name plate in the lid. Serial numbers as low as 112,000 and as high as 143,000 have been observed, but not all serial numbers between these two extremes were assigned to the Radiola AC. The tube-socket holes in sets with the lower serial numbers were smaller in size, but sufficient to accommodate replacement of the taller brass Westinghouse WD-11 tubes. However, the tube-socket holes in sets with higher serial numbers were enlarged in a manner similar to the Radiola Senior to accommodate removal of the shorter WD-11 Radiotrons.

Coincident with the change in tube-socket hole size, new tube sockets were designed with a battery holder between the two sockets to accept a 1.5 volt "C" (grid bias) battery. The bias battery was needed to reduce distortion which became noticeable when the plate voltage on the WD-11 tubes exceeded 45

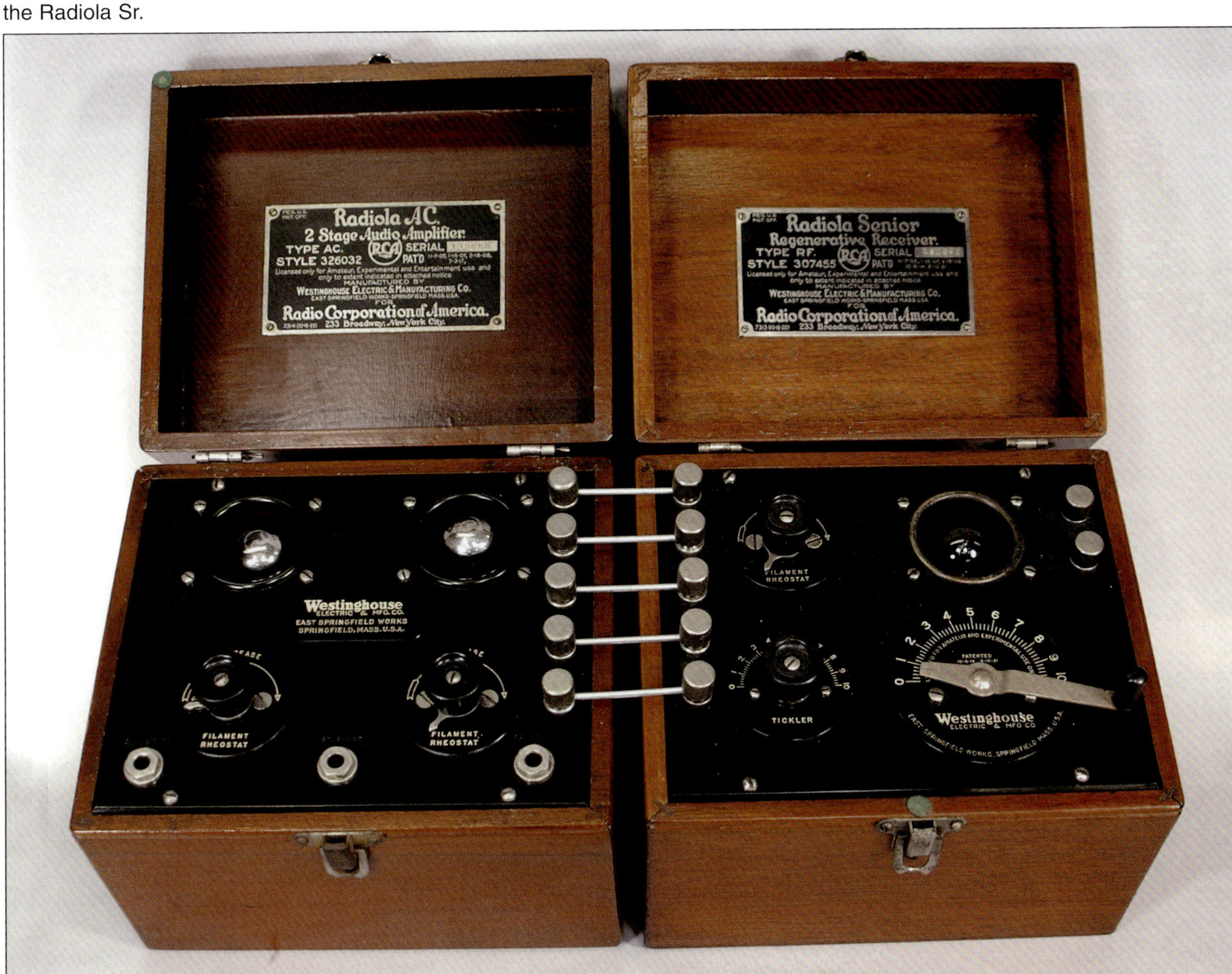

The Radiola AC was designed to match the Radiola Sr.

Coincident with the enlargement of the tube-socket hole for smaller WD-11 Radiotrons, the tube sockets were redesigned to accommodate a new "C" battery to provide a grid bias.

volts—the increased plate voltage being required to obtain better speaker performance. RCA first encountered this distortion problem in early 1923 when they substituted the WD-11s for the UV-200 and UV-201 in the RC receiver. In fact, they published an article in *The Wireless Age* in October 1923 showing the public how to modify the RC by adding a 1.5 volt "C" bias battery to improve the performance of RC's equipped with WD-11's.[18] The decision was made to locate the battery inside the cabinet rather than externally because—as the new instruction book stated—the "C" battery is not required to produce any current and therefore should last 12 to 18 months.

The Radiola AC was offered until the end of the 1923 season circa March 1924. There is no definitive information on the number of Radiola AC's made, but serial number evidence indicates that about 5,800 were manufactured.

The first Vocarola ad was placed by Westinghouse just before RCA assumed responsibility for marketing and distributing this early horn speaker. (*The Wireless Age*, January 1922, p. 1)

Vocarola Loudspeaker (Type I and LV)

From the very beginning of broadcast radio, there was a strong desire for families and others to listen to radio as a group. This need for a collective listening experience is well documented by the numerous photographs of families huddled around a radio with multiple sets of earphones that were taken in the early 20's. The Vocarola loudspeaker was designed and manufactured by Westinghouse specifically to meet this need. It was first advertised in January of 1922 for $30, but it had been reduced to $18 by May 1, 1923, the last time it appeared in an RCA price list.

The evolution of the Vocarola speaker was recounted by Frank Conrad to historian George Clark as follows:[19]

> *"When the idea of broadcasting suddenly blossomed at Pittsburgh, there was no loudspeaker available on the market, so Frank Conrad hastily tried out some on-the-minute ideas for making one of these devices. H. P. Davis's automobile happened to be within sight, and the horn suddenly gave Conrad an inspiration. If this spread the sound of a noise maker actuated by a rubber bulb, how would it work with a telephone unit? Hurriedly he removed the horn, and fitted it to a telephone reproducer. It worked so well the first time he tried it that he stopped right there, and thus the Vocarola was born."*

The Vocarola is a painted brass horn similar to those used on automobiles of the day, with a modified telephone receiver specifically selected to drive the horn.

The Vocarola horn was produced with two different decals, the most common being rectangular in shape with the Type LV designation.

The decal on the earliest Vocarola horn was heart shaped with the designation Type I, a designation which can be discerned (barely) from a photograph appearing in the June 1922 issue of *Radio Broadcast* on page 184.

Conrad's statement that there were no loudspeakers on the market at the time the idea of broadcasting blossomed at Pittsburgh (circa September 1920) is a bit of a stretch. The loudspeaker was of interest to the both the phonograph and wireless community well before broadcast radio, and more than one company was actively marketing loudspeaker devices by September 1920. Magnavox developed loudspeaker devices as early as 1915,[20] and was actively marketing two versions of a "Radio Telemegafone" in the year before broadcast radio blossomed in Pittsburgh—a smaller horn priced at $75 and a larger horn priced at $95.[21] In 1919 and 1920, the G. Boissonnault Company was marketing a Model B horn for $12 described as a "Loud Talking Receiver."[22] The use of this horn by amateur radio operators was chronicled in articles published by *Radio Amateur News* in 1920.[23] It seems obvious that Westinghouse developed the Vocarola, not because there was no other speaker available, but so that it could offer a complete receiving set with all pertinent accessories under the Westinghouse name.

The Vocarola essentially consisted of a brass horn similar to those used on automobiles of that era fitted with a telephone receiver to drive the horn, in this case a Baldwin receiver. The standard Baldwin mica diaphragm was replaced with a special metal diaphragm capable of withstanding abuse in the event of being overdriven.[24] Westinghouse engineers claimed that the Vocarola *"furnished as high a quality of reproduction as can be obtained except through the use of a very elaborate sound chamber such as is found in high-priced talking machines."* This claim was also a bit of a stretch.

The horn itself was made from polished brass and painted black. It was attached to a metal plate so the speaker could be placed either on a flat surface or hung from the wall. The paint on many units has been removed by unsuspecting collectors who believed that the underlying brass must have been the original finish. In doing so, they also would have scraped off the very desirable decal which is absent from many Vocarolas today.

According to the decal labels, there were two different types of Vocarolas including Type I with a shield-shaped decal similar to one appearing on the Aeriola Grand, and Type LV with a rectangular decal. Type I appears to be the earlier of the two, not only because it appeared first in advertisements, but also because the Type LV Vocarola appeared in the RCA price sheets up until the time the Vocarola was discontinued circa September 1923. Consequently, the Type I could not have succeeded the Type LV.

Since there is no obvious difference between the two types, the reason for this change in Type designation is unknown. Perhaps it was changed as part of the scheme to develop a uniform designation system. The phonograph attachment had already been designated Type LS, so it would have been logical to designate the Vocarola Type LV. However, this conjecture does not explain why Westinghouse initially used the single letter Type I designation in contrast to their other radio apparatus, all of which had two-letter Type designations.

The Vocarola was featured prominently in many ads with several RCA receivers—most notably the RC—from January 1922 until the latter part of 1923, when RCA came out with its first upright horn speaker. It is not known how many Vocarolas were sold during this period, but an internal RCA memo dated July 10, 1923 documented the need for disposing of an excess inventory of 3000 Vocarolas.[25] One can only presume that a significantly greater number had been sold prior to this date.

General Electric Offerings for 1922

General Electric had been ill-prepared to meet the demand for broadcast receivers in 1921—and their offerings in 1922, while substantial compared with the previous year, still paled in comparison to those of Westinghouse. GE basically had three new offerings: 1) the ER-753 followed quickly by the ER-753A crystal set that competed directly with the Westinghouse Aeriola Jr.,

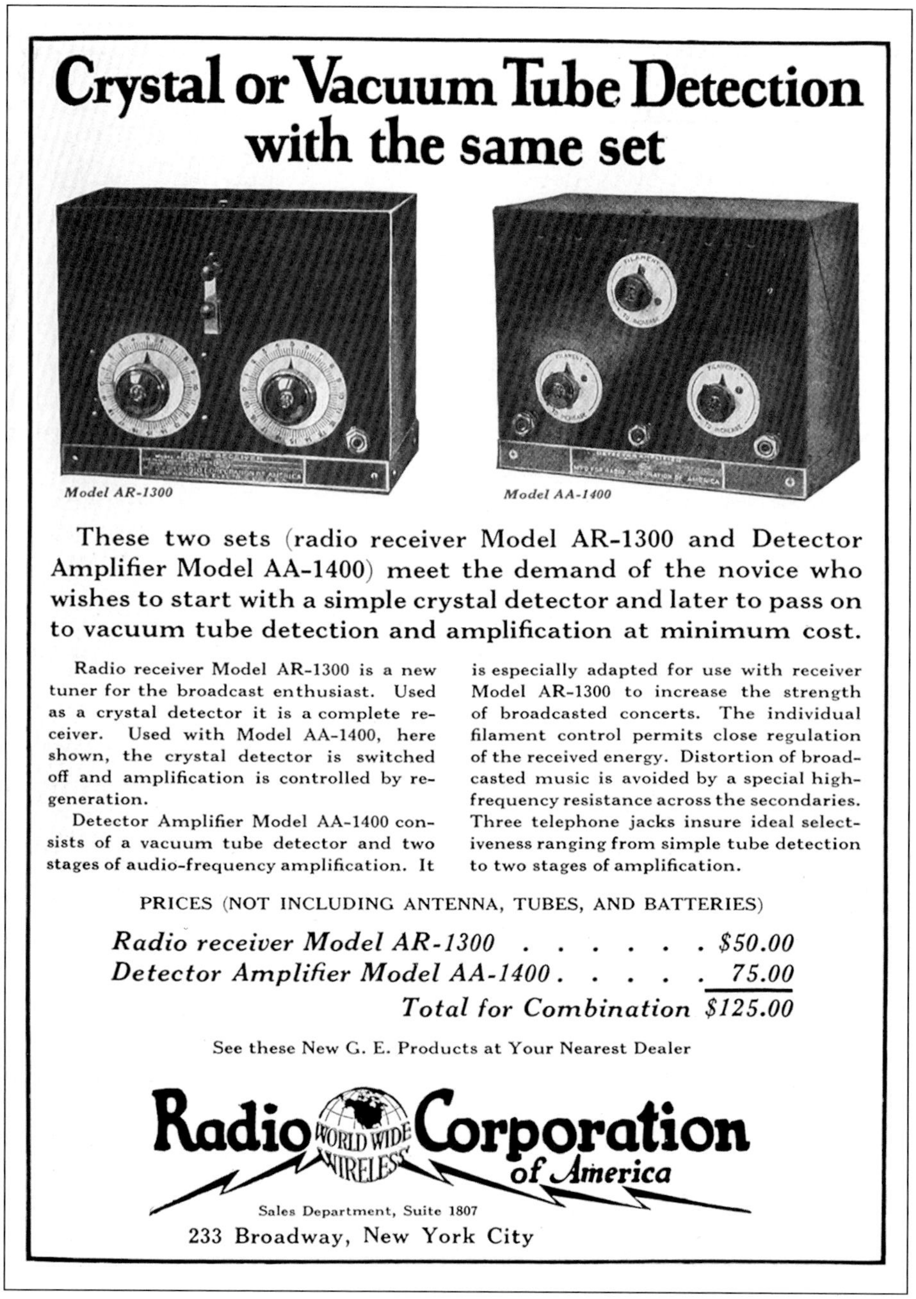

Crystal or Vacuum Tube Detection with the same set

Model AR-1300 *Model AA-1400*

These two sets (radio receiver Model AR-1300 and Detector Amplifier Model AA-1400) meet the demand of the novice who wishes to start with a simple crystal detector and later to pass on to vacuum tube detection and amplification at minimum cost.

Radio receiver Model AR-1300 is a new tuner for the broadcast enthusiast. Used as a crystal detector it is a complete receiver. Used with Model AA-1400, here shown, the crystal detector is switched off and amplification is controlled by regeneration.

Detector Amplifier Model AA-1400 consists of a vacuum tube detector and two stages of audio-frequency amplification. It is especially adapted for use with receiver Model AR-1300 to increase the strength of broadcasted concerts. The individual filament control permits close regulation of the received energy. Distortion of broadcasted music is avoided by a special high-frequency resistance across the secondaries. Three telephone jacks insure ideal selectiveness ranging from simple tube detection to two stages of amplification.

PRICES (NOT INCLUDING ANTENNA, TUBES, AND BATTERIES)

Radio receiver Model AR-1300 . . . . $50.00
Detector Amplifier Model AA-1400 . . . . . . 75.00
Total for Combination $125.00

See these New G. E. Products at Your Nearest Dealer

Radio Corporation of America

WORLD WIDE WIRELESS

Sales Department, Suite 1807
233 Broadway, New York City

Advertisements for the AR-1300 receiver and AA-1400 detector-amplifier began to appear in the summer of 1922. (*Radio Broadcast*, June 1922, p. 169)

2) the AR-1300/AA-1400 regenerative receiver that competed directly with the Westinghouse RA/DA/RC, and 3) a new AA-1520 RF amplifier designed for use with the AA-1400. However, the Westinghouse offerings had been introduced to the public a year or more before the GE offerings, and outsold the GE sets by a wide margin.

AR-1300/AA-1400 Regenerative Receiver

The AR-1300 receiver and AA-1400 detector-amplifier combination, one of GE's first entries in the broadcast radio market, made its first public appearance at the New York Amateur Radio show in March of 1922, and also appeared in the April 1, 1922 RCA distributor price sheets. The first instruction cards are dated May 1922, a month before national ads began to appear circa June 1922.

The AR-1300 was sold as either a "Crystal Radiophone Receiver" or a "Regenerative Tuner" for $50, while the AA-1400 was sold as a detector and 2-stage amplifier for the AR-1300 for $75 without tubes—the UV-200 detector

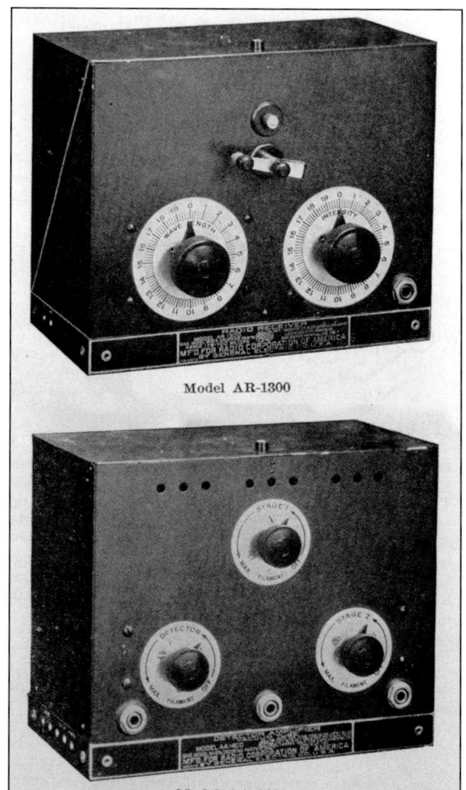

The early AR-1300 receivers and AA-1400 detector-amplifiers were equipped with a lid release mechanism consisting of a circular button on top similar to the ones GE used on marine receivers. (*Radio Equipment and Supplies*, Radio Catalog No. 22, 1922, pp. 21-22)

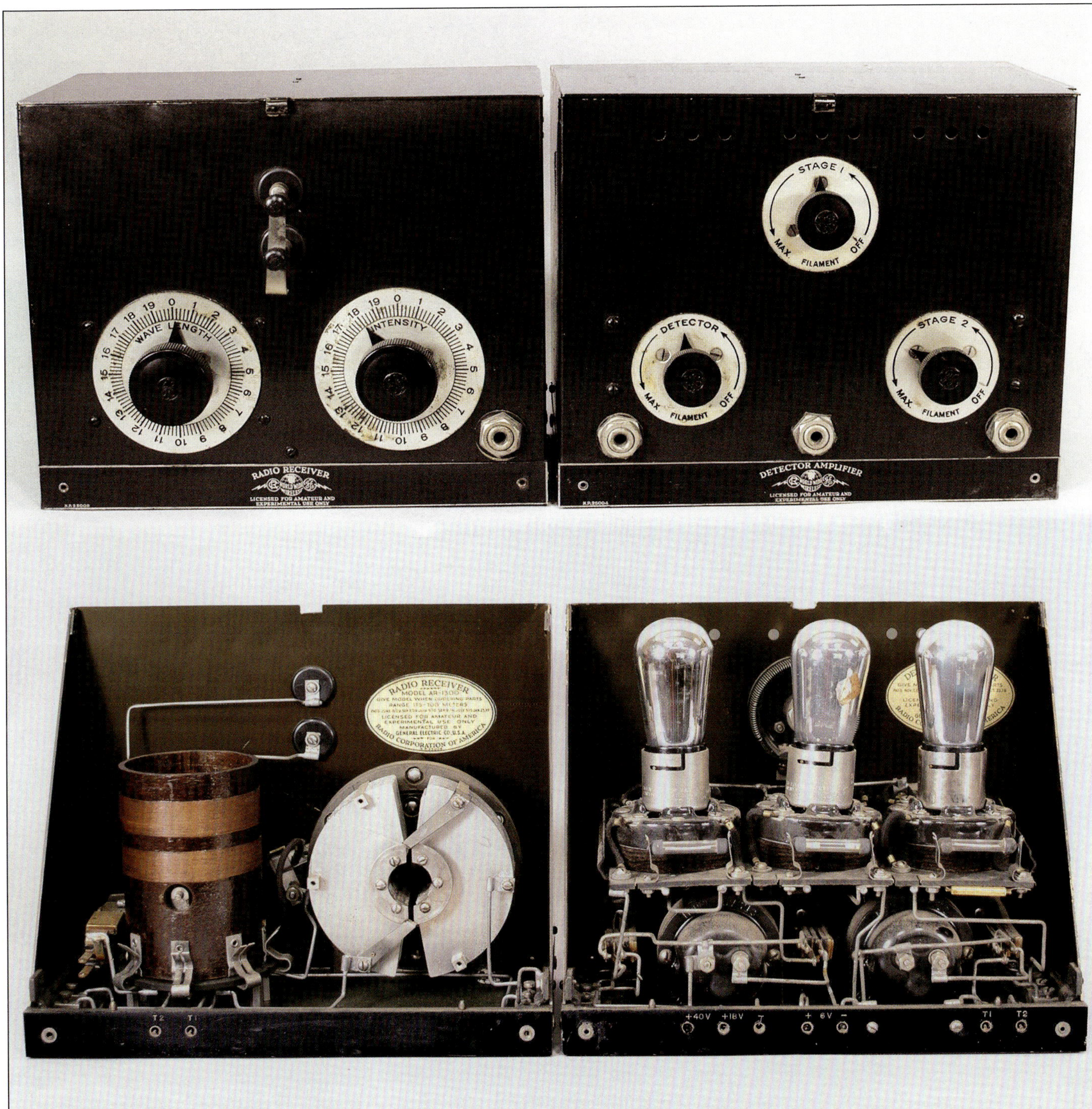

The AR-1300 receiver and AA-1400 detector-amplifier were housed in dark green EZKASES similar to the ones GE used on marine sets.

tube and two UV-201 amplifier tubes were an extra $18. A complete regenerative receiver including the 1300/1400 units, a Vocarola speaker, headset, batteries, battery charger, and antenna was priced at $250.25. The price was undoubtedly set to compete favorably with a similar Westinghouse package featuring the RC receiver priced at $261.75.

The two units were housed in green-enameled metal EZKASE cabinets, obviously direct descendents of the ER-1620 marine receivers designed for use aboard ships. In fact, the AA-1400 consisted of two AA-1528 amplifiers and an AD-1527 detector combined within a single case, and was used for both broadcast and commercial marine communications. The AR-1300, a single-circuit tuner enclosed in a matching green case, was designed specifically for

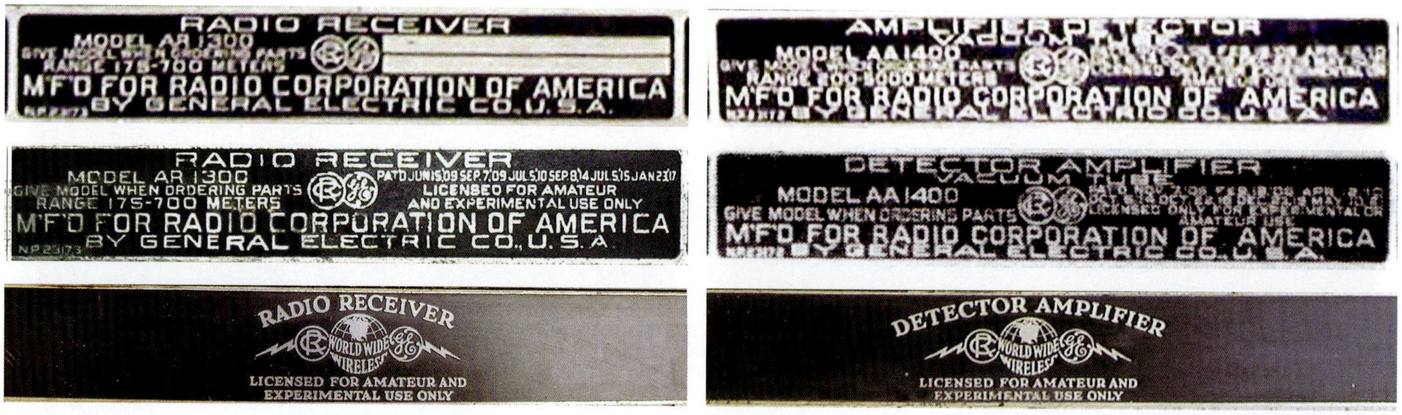

Three sets of name plates have been observed for the AR-1300 (left column) and the AA-1400 (right column), one believed to be associated with prototypes (top), one with the first production run (center), and one with a later production run (bottom).

broadcast radio with operating wavelengths of 180 to 700 meters.[26] A long wave coil system was also available to change the wavelength range of the AR-1300 to cover 650 to 2800 meters, although this option was not widely advertised. A sensitive Perikon detector was mounted on the front panel so it could be used as crystal receiver by itself, or by switching the crystal detector off, it could be used as a regenerative tuner in conjunction with the with the AA-1400 detector and two-step amplifier. The tuner sold for $50, and the combined price for both without tubes was $125, slightly less than the Westinghouse RC priced at $132.50 without tubes. At some point, a black wood base and top designed to frame the two units was also offered for sale.

Variations among the AR-1300 and AA-1400 units with green cases have been observed in both existing antique collections and contemporaneous catalogs. Early sets appearing in catalogs, RCA ads, and the instruction sheets had a round button on top of the set for releasing the lid—obviously copied from the mechanism used on the ER-1620 marine receiver units. Later sets like the pair shown here hade a distinctly different clasp arrangement consisting of a small bar at the front edge to release the lid. This change was made most likely to reduce cost for the consumer market. However, it does not appear that green-cased sets with the bar release mechanism ever appeared in any contemporaneous ads or catalogs. There were other more subtle differences as well. For example, the filament control on the AA-1400 sets appearing in early RCA ads such as the one shown here do not have an "off" position marked on the dial, while the AA-1400's appearing in the 1922 RCA catalog and on many existing sets do.

Three distinctly different name plates for both the AR-1300 and the AA-1400 have been observed in catalogs, photographs, and radio collections. Two of the three are clearly associated with production models while the third in each case is almost certainly associated with a prototype. The name plates on the AR-1300 with the two long blank spaces and on the AA-1400 with the designation "Amplifier Detector" are believed to be associated with prototypes (production versions of the AA-1400 were designated "Detector Amplifier"). The plates which are dominated by block-style print are believed to be associated with the first production units, while plates dominated by the world wide wireless logo with lightning flashes are associated with a later production run. A 1300/1400 pair with the block-print style name plates is shown in several Alan Douglas publications,[27] while a pair with the world wide wireless logo is shown here.

It is not known how many of these sets were sold, but judging by their relative rarity today, it was probably only a few thousand in number. This receiver had to compete with the entrenched RC receiver, and the relative crude green metallic cases could not have been very endearing to the average homemaker. They were on the market only until the end of 1922, when they were replaced by the Radiola V, a dressed up version of the AR-1300 and AA-1400.

A long-wave coil covering 650-2800 meters was available for the AR-1300, which replaced the coil for the broadcast band by simply lifting it off of the base where it was held by seven s-shaped friction clips. (Stephen Bown Collection)

AR-1300-A Two-Circuit Tuner

Towards the end of 1922 General Electric modified the AR-1300 by adding a series capacitor to the intensity coil and rewiring the coil connections so that they would mate up with the antenna and ground input connections of the AR-1300. The resulting combination of the AR-1300 and the AR-1300-A formed a two circuit tuner to provide greater selectivity. As the accompanying instruction manual dated February 1923 explained, this unit (the AR-1300 and AR-1300-A) could be used in any of the following combinations: 1) as a single or two circuit crystal receiver, 2) as a two circuit tuner with a detector-amplifier such as the Model AA-1400, or 3) as a two circuit tuner with a radio-frequency amplifier and detector-amplifier such as the Radiola VI.

The AR-1300-A is evocative of the Westinghouse RT tuner which Westinghouse used to convert their RA crystal receiver to a more sensitive two circuit receiver. Since both the RT and the AR-1300-A appeared at about the same time, one cannot be certain whether it was GE or Westinghouse who was first with the idea of converting the basic one circuit receiver to a two circuit receiver by adding a companion unit. The existence of the AR-1300-A—which apparently has not here-to-fore been chronicled—is most interesting because it means that GE with its AR-1300, AR-1300-A and Radiola VI had designed a lineup similar to the well-known Westinghouse RA-DA-AR-RT lineup.

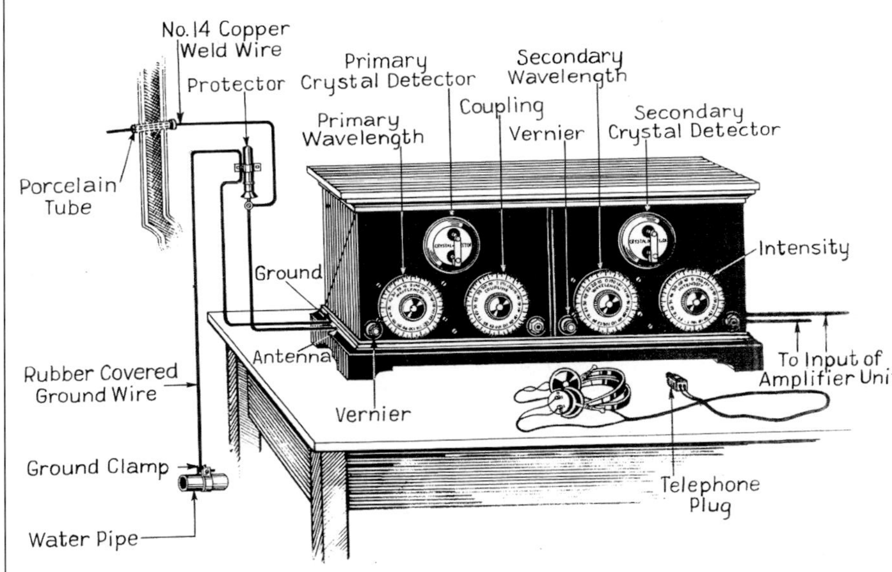

A GE instruction manual for a two-circuit tuner dated February 1923 shows the AR-1300-A (left) and AR-1300 (right) mounted together in a wood tray and capped with a wood top. (George H. Clark Collection, Smithsonian Institution)

According to the picture in the instruction manual, the AR-1300 and AR-1300-A were placed on a wood base (with the AR-1300-A on the left) and capped with a wood top, giving an appearance similar to the Radiola VI. The four units operating together would have appeared like two Radiola VI units positioned side-by-side. The right knob on the AR-1300-A became a "coupling" knob, and was labeled as such. The coupling knob on a high setting resulted in a broad tuning with the signal appearing over a large portion of the tuning scale of the AR-1300. The coupling knob of the AR-1300-A was successively decreased to sharpen tuning while at the same time the tuning control was successively readjusted to maximize signal strength. There is no record that this set was ever marketed by RCA, which is not surprising because the similar Westinghouse RT introduced in January of 1923 was certainly not a commercial success.

AR-1520 Radio Frequency Amplifier

The AA-1520 three-stage untuned radio-frequency (URF) amplifier was introduced in August of 1922 but the first ads did not appear until circa January 1923.[28] It was specifically designed for operation with the AA-1400 detector-amplifier in conjunction with an indoor loop antenna such as the AG-1380, but it could also be used with an outside antenna by adding the AR-1300 tuner or equivalent. Company employees derisively referred to the three units together as the "freight train."

In addition to eliminating the need for a large outside antenna, the AA-1520 was also touted as having better sensitivity, allowing reception at long distances. The AG-1380 loop antenna manufactured by Wireless Specialty was recommended for use with the AA-1400 and AA-1520. This antenna was supplied with the UC-1820 variable condenser which could be used to tune the radio to the desired station.

The AA-1520 was housed in a green metallic EZKASE that matched the AR-1300 and AA-1400. The first national ad placed by RCA showed a unit with the new bar-type lid release mechanism rather than the button first used on the AR-1300 and AA-1400. The set was designed for three UV-201 tubes, but it was sold without the tubes for $80. The AA-1520 was sold as a stand-alone unit in 1922, but in 1923 it was repackaged in a faux-mahogany EZKASE and

One of the first advertisements for the AA-1520 appeared in the January 1923 issue of *The Wireless Age*. (*The Wireless Age*, January 1923, p. 1)

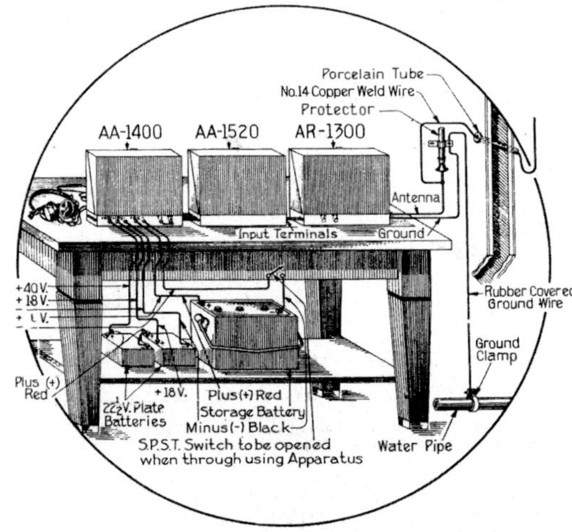

To increase sensitivity, the AA-1520 RF amplifier could be used with the AA-1400 detector-amplifier and the AR-1300 tuner as shown in the connection diagram from a GE instruction card.

The AA-1520 was spruced up for the 1923 season by changing the EZKASE from dark green to a faux-mahogany finish and replacing the knobs with a new design.

The interior of the AA-1520 is similar to that of the AA-1400, the major difference being three UV-1714 RF transformers in lieu of two audio transformers.

A Perikon detector similar to the one found on the ER-753 was used on all broadcast radio crystal sets manufactured by GE for RCA.

sold both as a stand-alone unit, and in combination with the AA-1400 as the Radiola VI. It is not known how many of AA-1520 sets in green cases were made, but judging by the number on the current auction market, it must have been a small number. At least 2774 sets in the faux mahogany case were manufactured for use in Radiola VI sets.

ER-753 Crystal Set

The ER-753 crystal set was designed by GE engineer Arthur Van Dyck and advertised as "*a simple, compact, inexpensive receiver [which] is available for everyone located within reasonable distance of a broadcasting station.*" The set could be tuned from 180 to 700 meters in two ranges (180-400 meters on the lower binding post and 300-700 meters on the upper binding post) by varying the inductance of a flat coil mounted against the front panel with a copper plate rotating over the exposed surface facing the interior. The same Perikon detector used on the Type DB detector and the Aeriola Jr. was used on the ER-753.

The ER-753 vertical wedge with a more delicate engraved style print filled with white ink appeared most often in advertisements and on the instruction card. The ER-753 was also produced in a horizontal wedge configuration which has been observed with an embossed block-style print with no ink fill, but most likely came with the more delicate white-ink fill as well.

The first ads for the ER-753 began to appear circa May 1922. (*The Wireless Age*, May 1922, p. 4)

The instruction cards for the ER-753 were not attached to the set, and as a result are now quite rare.

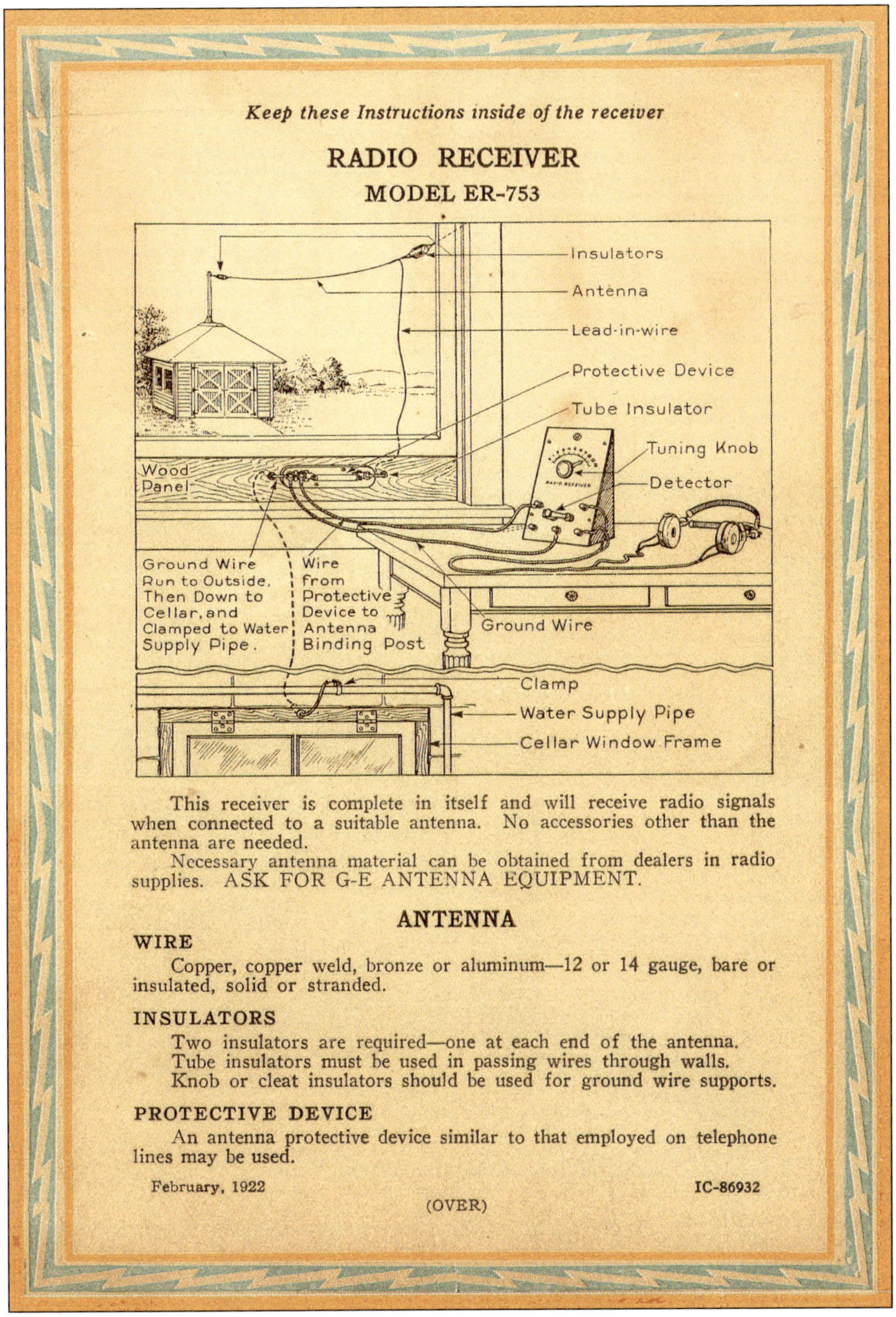

Keep these Instructions inside of the receiver

RADIO RECEIVER

MODEL ER-753

This receiver is complete in itself and will receive radio signals when connected to a suitable antenna. No accessories other than the antenna are needed.

Necessary antenna material can be obtained from dealers in radio supplies. ASK FOR G-E ANTENNA EQUIPMENT.

ANTENNA

WIRE

Copper, copper weld, bronze or aluminum—12 or 14 gauge, bare or insulated, solid or stranded.

INSULATORS

Two insulators are required—one at each end of the antenna.
Tube insulators must be used in passing wires through walls.
Knob or cleat insulators should be used for ground wire supports.

PROTECTIVE DEVICE

An antenna protective device similar to that employed on telephone lines may be used.

February, 1922 IC-86932

(OVER)

The interior of the ER-753 reveals a very simple tuning mechanism consisting of a copper plate rotating over a flat coil to change its inductance.

Incredibly, GE selected one half of an EZKASE enclosure for the ER-753, giving it the appearance of a wedge.[29] Clearly, GE was not invested in the aesthetics of radio cases in the beginning. The set was produced in two styles, a vertical wedge and a horizontal wedge. The only difference between the two was the side on which the metal feet were stamped. On the vertical wedge, the feet were stamped on the short side, while on the horizontal wedge, the feet were stamped on the long side. The horizontal wedge was designated "Design No. 1," while the vertical wedge was designated "Design No. 2." In both cases, the front panel was easily removed so that the headphones, which were sold with the set, could be easily stored out of the way. The front panel was produced in at least two styles, one with block-style embossed print without any added coloring, and the other with more delicate embossed print filled with white ink.

It is not known how many sets were manufactured, but it could not have been very many—even though the price was only $18 including headphones—

because it was on the market for only a short while. GE was encouraged by RCA to improve the design, and by August 1922 the set had been replaced by a redesigned ER-753 designated Model ER-753A.

ER-753A Crystal Set

The ER-753A was introduced in mid-1922 for $25 including earphones (or $32.50 with an antenna kit) as a replacement to the ER-753 crystal set. According to the date appearing on the earliest instruction sheet, the ER-753A was introduced in July of 1922, too late to appear in the June 1, 1922 RCA catalog. Many collectors believe this set was the Radiola I, but that is not the

The ER-753A *"opened like a book"* to expose the front panel, with a handle on top so it could be *"carried like a satchel."*

The interior of the ER-753A was dominated by a variometer used for tuning that replaced the flat coil in the ER-753.

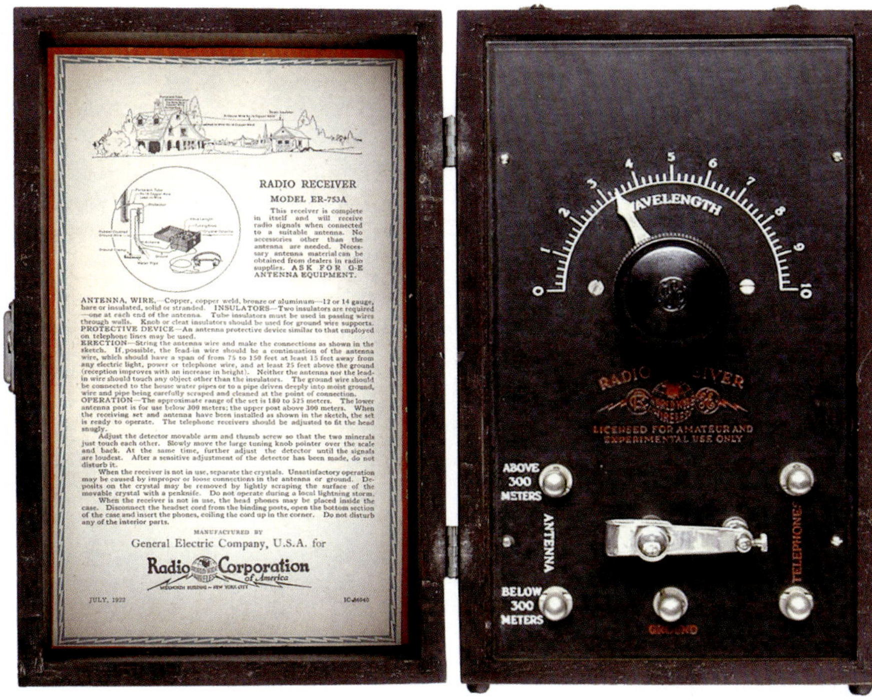

The earliest ER-753A sets, with an instruction card dated July 1922, were manufactured with an engraved-style print on the front panel filled with either white ink, or a combination of both red and white ink as shown here.

A second production run of the ER-753A, with an instruction card dated August 1922, was manufactured with a block-style print on the front panel without any ink fill.

case—at least not at the time it was first produced. Early sets were identified as the ER-753A on the instruction sheet and on a gold seal in the rear compartment, both without reference to Radiola I. An official GE photograph of this receiver was clearly labeled ER-753A without reference to Radiola I.

It was not until late summer of 1922 that RCA decided to change the name of the ER-753A to Radiola I, well after production began. The first known reference to Radiola I was in the August 22, 1922 RCA price sheet, but, GE must have made a relatively large number of ER-753A sets during the first two production runs before this decision because it was not until May of 1923 that sets were actually manufactured with the Radiola I name. Both the Radiola I name and ER-753A model number appeared on sets with instruction sheets dated May 1923, and the Radiola I name was also engraved into the front panel. The Radiola I retained the original GE model designation ER-753A—much as the Radiola Sr. retained the original Westinghouse Type RF designation of the Aeriola Sr. after the name was changed. It is curious that the Aeriola Sr. and Radiola Sr. have kept their separate identities among collectors and historians, whereas the ER-753A and Radiola I have not—although they have an almost identical evolutionary relationship as the two Senior sets.

The ER-753A replacement for the ER-753 wedge was electrically similar with the notable exception that the flat coil was replaced by a simple variometer. As a result, the upper operating range was reduced from 700 meters to 525 meters, still sufficient to cover the broadcast band as it existed then. The labeling on the two binding posts was changed accordingly to "above 300 meters" and "below 300 meters." The major physical difference was the attractive walnut cabinet used to enclose the set, which opened to the front to expose the operating panel, and opened to the rear to access a compartment for storing the headphones. A leather handle was situated at the top of the cabinet so that, as later ads stated, it could be "*Opened like a book*" and "*Carried like a satchel.*"

The front panel arrangement of the ER-753A was also similar to the ER-753, with the most noticeable differences being in the labeling. There are two different panel styles, and each panel style was accompanied by a slightly different set of instructions tacked to the inner lid. The earliest panel was embossed with "Radio Receiver" which was filled with both red and white ink on one version, and with all white ink on another version. The accompanying instruction sheet dated July 1922 bears the identification number IC-86940. The second version of the panel is also embossed with "Radio Receiver" but has no ink fill. The revised instruction sheet accompanying this version was dated August 1922 and bore the number IC-86940-A. Both versions were identified as an ER-753A, but neither version was marked with the Radiola I logo, on either the panel, the instructions or the gold seal in the rear.

A second version of the early ER-753A panel was filled with white ink. (Mike Bell collection)

Radiola I Crystal Set (ER-753A)

The first sets marketed as a Radiola I were actually ER-753A sets manufactured by GE before RCA made the decision to change the name of all their receivers to include the Radiola trade name. While the Radiola I name first appeared in RCA price sheets dated August 21, 1922 at a price of $25 including headphones, the first national ads did not begin to appear until circa November 1922.[30] It should be noted in this ad that the ER-753A name is given prominent billing on a par with the Radiola I name—something very unusual for sets with the Radiola name. Clearly, this reference was included in ads for the consumer because the sets RCA sold at that time had been manufactured with the ER-753A name on the instructions—not Radiola I.

The first ER-753A bearing the Radiola I name appeared circa May 1923, the date printed on the instruction label accompanying this set which bore the number IC-86940-B. The Radiola I name appeared on both the instruction sheet and on the front panel, which was embossed with "Radiola I" without any ink fill. The Radiola I name was also added to a gold seal located in the rear compartment used to store the earphones.

In addition to the variations among front panels, instruction sheets, and gold seals, there are other minor variations to ER-753A sets. For example, two different types of door clasps have been observed, the more common form seen in most advertisements and publications with one flat side, and a rarer form with an oval shape.

It has often been stated or implied that the Radiola I was the first production receiver marketed by RCA with the Radiola name,[31] but this is not the case. This distinction belongs to the AR-1375 Radio Concert Receiver manufactured

While the Radiola I appeared in RCA price lists in August, advertisements did not begin to appear until November 1922. (*The Wireless Age*, November 1922, p. 1)

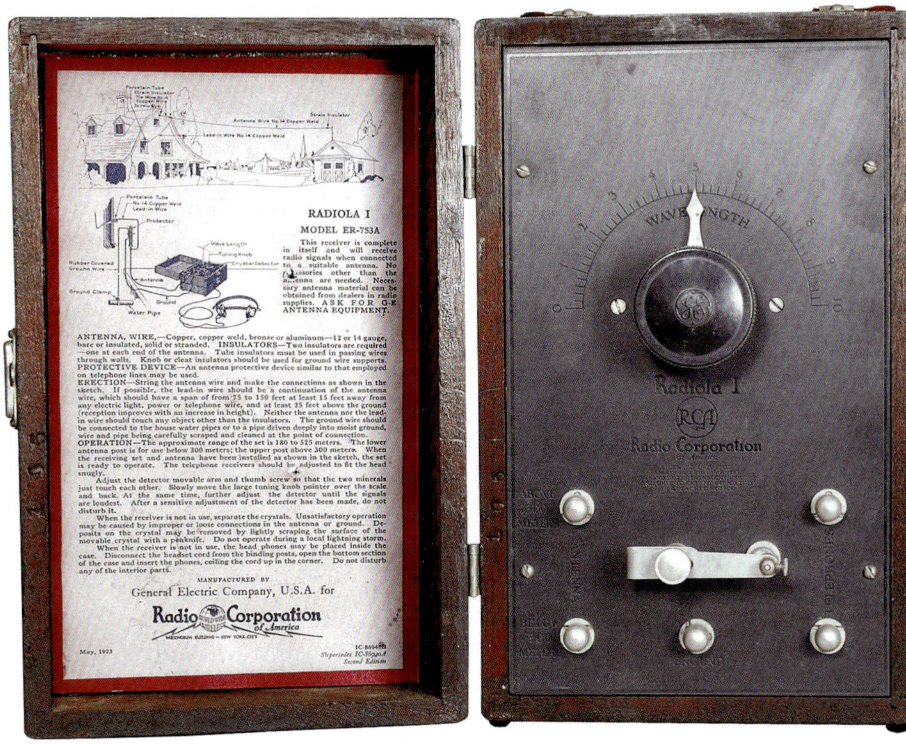

The Radiola I was manufactured with an instruction card dated May 1923, and the Radiola I name appeared on the front panel in a block-style print without any ink fill.

Two different styles of door clasps have been observed, the more common one with one straight edge (top), and a rarer one with an oval shape (bottom). (Courtesy of Sonny Clutter)

by Wireless Specialty Apparatus. This receiver appeared in the New York Amateur Radio Show in March, 1922, on the RCA Price List dated April 1, 1922 and in the RCA catalog dated June 1, 1922, well before the appearance of either the ER-753A or Radiola I. While the Radiola name does not appear in the text of the June 1922 catalog, it clearly appears on the AR-1375 receiver itself, which is pictured in the RCA catalog on pages 16-17.

According to Douglas, RCA records show that either 7826 or 8064 sets were made.[32] However, the *Minutes of the General Sales Committee for July 10th, 1923* showed that RCA was looking into a means of disposing of 6000 Radiola I sets in their inventory as of July of 1923.[33] Therefore, if the numbers Douglas reported are correct, there would have been only about 2000 Radiola I sets sold to the public in the one year period from August 1922 to July 1923, the remainder being disposed of at discount. Such a small number seems unlikely,[34] so could it be that the RCA records actually reflected that 7826 of the ER-753A and 8064 of the Radiola I sets were made?

AG-788 Antenna Outfit

Virtually all of the early inexpensive receivers—crystal sets in particular—depended on a long-wire antenna in order to collect sufficient power to provide audible signals. RCA marketed two virtually identical long-wire antenna kits to meet this need, the AG-788 manufactured by GE and the AD manufactured by Westinghouse, both selling for the same price of $7.50. RCA bundled the GE antenna kit with the GE receivers and the Westinghouse antenna kit with the Westinghouse receivers.

The GE AG-788 kit introduced in the April 1, 1922 RCA price list consisted of 175 feet of #14 copper wire, lead in and ground wires, a lightning protector, and miscellaneous hardware. RCA continued to sell this kit until circa March 1924 when the new radio line was introduced.[35] It is not clear why it was dropped at this time because, according to the instructions for a majority of sets in the new line, an external antenna was required. Perhaps it was dropped because the antenna kit at $7.50 was not selling well. The almost-identical Westinghouse AD antenna kit had been dropped from the RCA line earlier in September of 1923.

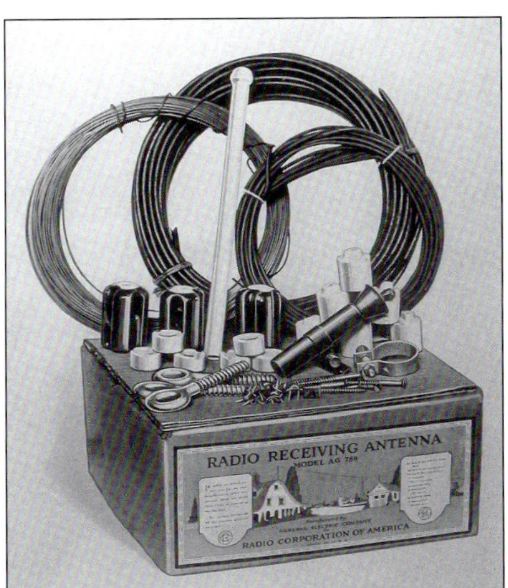

The AG-788 antenna outfit was offered by GE to compete with the Westinghouse AD antenna outfit. (George H. Clark Collection, Smithsonian Institution)

Wireless Specialty Offerings for 1922

According to the RCA Distributor Price lists, RCA offered four Wireless Specialty products in 1922 including an AA-484 amplifier, an AA-485 amplifier, an AR-1375 crystal set and an AG-1380 loop antenna. The AR-1382 receiver, while never listed in the price lists or in RCA catalogs, was offered in RCA distributor catalogs as a companion to the AA-485 amplifier. Two additional Radiola amplifier prototypes developed by WSA for RCA were documented in contemporaneous publications, one designated AA-1376 and the other AA-1381, but apparently neither was produced in quantity. WSA documents also refer to a heterodyne unit assigned the type number AD-1377, which was designed to produce oscillations between 130 and 1200 meters. This set, which was mounted on a panel containing a socket and filament rheostat, a variable condenser, and a variometer has not surfaced in the collecting community.

In the same way that Westinghouse used the Aeriola name for their early broadcast radio apparatus, Wireless Specialty used the name Radio Concert for the early broadcast radio apparatus they made for RCA. The AR-1375 and AR-1382 were designated Radiola Radio Concert Receivers, the AA-485, AA-1376, and AA-1381 were designated Radiola Radio Concert Amplifiers, and the AG-1380 was designated Radiola Radio Concert Loop Antenna. These names appear prominently at the top of the front panel of cabinet-style sets, and on name tags affixed to the AR-485 amplifier and AG-1380 loop antenna.

None of the Wireless Specialty receivers or receiving apparatus distributed by RCA was widely advertised, and as a result, the sales were quite limited. Consequently, very few of these items survived, and they are now highly sought after by collectors.

Radiola Radio Concert Receiver (AR-1375)

The AR-1375 Radio Concert Receiver was first introduced to the public at the New York Amateur Radio Show in March of 1922, and listed in the April 1, 1922 price list. The radio on display at the show appears to be an engineering model because it is noticeably different from the one that appeared in the 1922 RCA

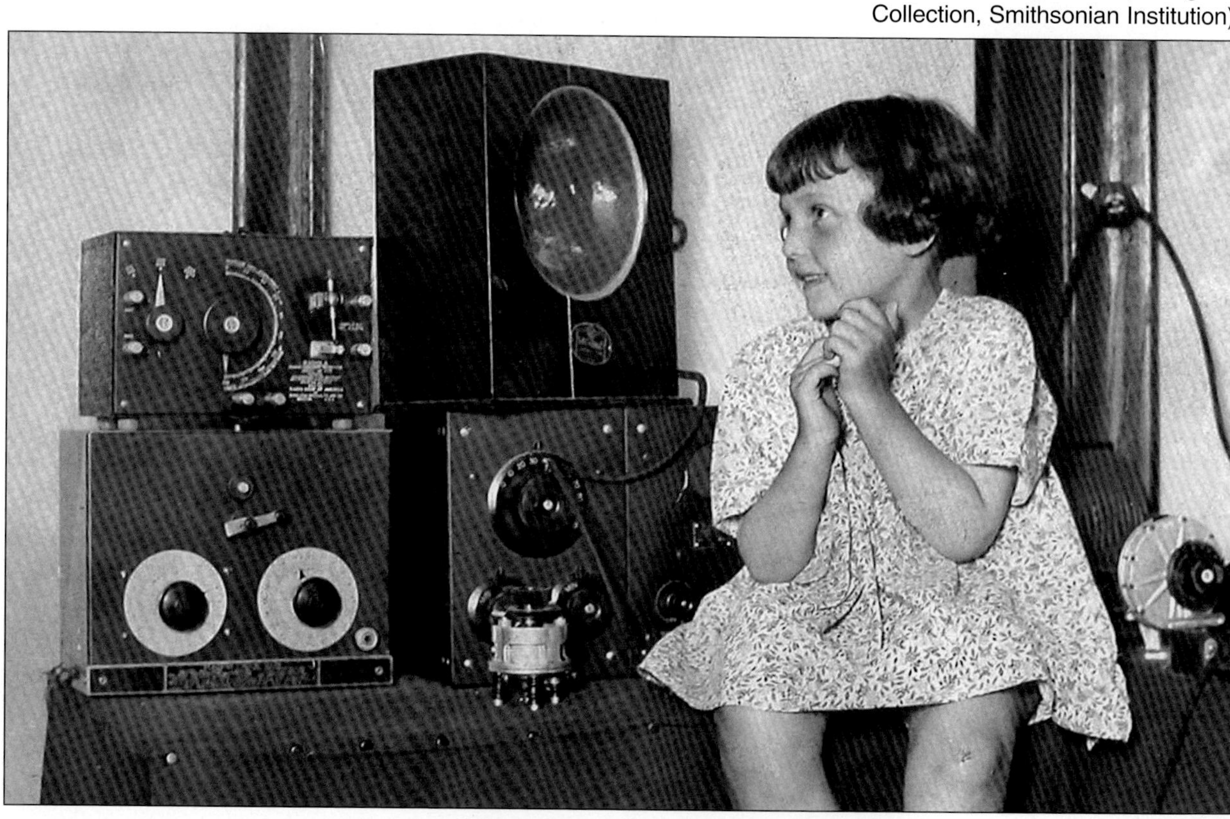

The early "horizontal version" of the AR-1375 similar to the one displayed at the New York Amateur Radio Show in March 1922 can be seen in the upper left corner of a photograph which appeared in the August 12, 1922 edition of Radio Digest Illustrated Magazine. (George H. Clark Collection, Smithsonian Institution)

The AR-1375 was the first receiver distributed by RCA with the Radiola name on the front panel. (Courtesy of AWA Museum)

The AR-1375 was a relatively simple crystal set using a variometer for tuning and three capacitors, two of which are included together in a single assembly directly to the left of the switch. (Doug Furney collection)

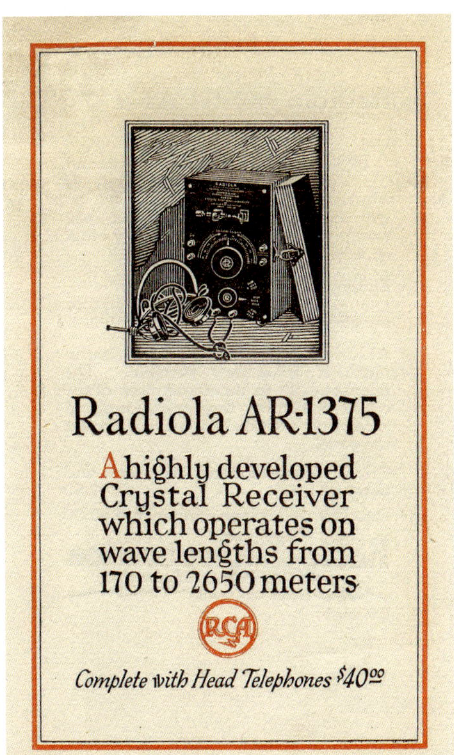

catalog where it was offered for sale to the public for the first time. The set appearing in the catalog is oriented with the long dimension vertical, while the set at the show was oriented with the long dimension horizontal. This horizontal version of the set also appeared in a photograph in the August 12, 1922 edition of *Radio Digest Illustrated* in a home setting–suggesting that at least a few of the early AR-1375 receivers were sold to the public.[36] However, the young girl in the photograph appearing in the *Radio Digest Illustrated* (a portion of which is reproduced on the opposite page) has been identified as Marilyn Bucher, daughter of RCA engineer Elmer Bucher. Therefore, it is likely to be an engineering model rather than a production unit.

This Concert Receiver was the first broadcast receiver made by Wireless Specialty for RCA, and was also the first to bear the name Radiola, predating the ER-753A forerunner to the Radiola I by several months, and predating the actual Radiola I by almost a year. The AR-1375 described in the June 1922 RCA catalog as *"a high grade crystal receiver of superior workmanship,"* was designed for operation between 170 and 1650 meters in three steps: 170-410, 350-965, and 925-2650 meters. The latter range was used to receive Navy stations including time signals from Arlington, weather reports, and other stations operating on these wavelengths. It had a removable metal lid at the top enclosing a compartment for the headphones, and a metal cover to protect the front panel. The set had provisions to connect to a vacuum tube amplifier unit, and in fact the AA-1376 audio amplifier was designed for use with this set, but apparently it never went into production.

According to RCA records,[37] this set also came in a mahogany cabinet with gold-plated parts and headphones, although no receivers or photographs with this configuration could be found. This cabinet must have been similar to the wood cabinets found on the WSA AA-1381 amplifier on page 163,[38] and the Radiola Special which is pictured on page 197. The total cost to RCA was $15 including earphones for the Radio Special in a metal case, and $19.15 for the same apparatus with a mahogany case.

The set was on the market for about 12 months, but very few sets were actually sold, in part because of the relatively high price when compared to other RCA crystal sets, and in part because it was not well advertised. At $40 including headphones, it had to compete directly with the lower-priced Aeriola Jr. at $25, the ER-753 at $18, and later the Radiola I (ER-753A) at $25—all with headphone included, as well. The set was withdrawn from the market during the first few months of 1923, modified into a regenerative receiver by adding a UV-199 tube, given a new look with a new overlay panel plate, and reintroduced to the market as the Radiola Special (see Chapter 5). As a result, only a very few sets are known to exist without major restorations.

AA-1376 Audio Amplifier

The AA-1376 Radio Concert Amplifier was designed as a companion to the AA-1375 crystal set to increase its responsivity and therefore its range, and to drive a loudspeaker. The set never appeared in any RCA price list or catalog, and apparently never went into production. However, this amplifier does appear in the instruction manual distributed with the AA-1375 receiver. It is not known if any models still survive.

The AA-1376 Radio Concert Amplifier was pictured in the instruction manual for the AR-1375 receiver, but was apparently never distributed by RCA.

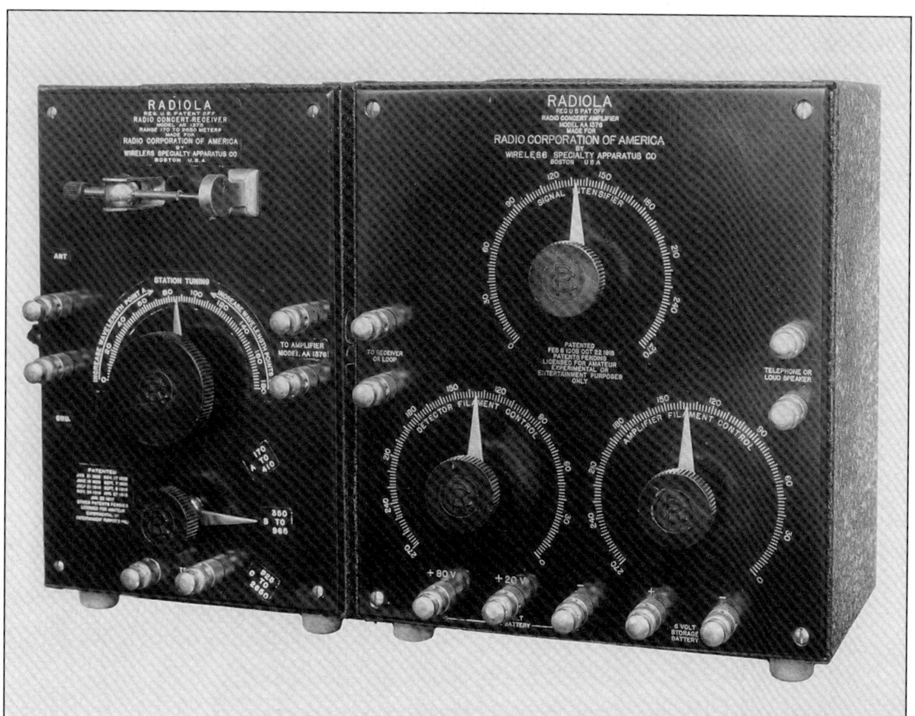

The amplifier consisted of three stages, a detector and two transformer-coupled audio stages, presumably designed around the UV-200 and UV-201 tubes. The binding posts at the left of the amplifier line up with the binding posts at the right of the crystal set as shown in the photograph. Alternatively, a loop antenna such as the Wireless Specialty AG-1380 was connected to the input terminals of the AA-1376 in place of a crystal set, in which case the UC-1820 variable capacitor mounted on the base of the AG-1380 antenna was used to select desired stations. This tuning arrangement is similar to the one used by GE in the Radiola VI, and since GE used the Wireless Specialty loop, it is likely they copied the idea from Wireless Specialty.

AA-484 2-Stage Audio Amplifier

The two-stage audio amplifier AA-484 was a breadboard-style instrument designed for the advanced amateur to carry on experiments in audio amplification. It first appeared in the April 1, 1922 RCA Price List at $45, and was first pictured on page 52 of the June 1, 1922 RCA catalog. It is mislabeled as the model AA-485 on this page of the catalog, but it was correctly listed as the AA-484 in the price list on page 112.

The AA-484 appearing in the 1922 RCA catalog is believed to be a prototype, which is easily distinguished from a production unit by the absence of the Wireless Specially name plate with a serial number. (RCA catalog *Radio Enters the Home* dated June 1, 1922, p. 52)

The production units of the AA-484 amplifier differed from the prototype by the presence of a WSA name plate and a more symmetric wiring layout made possible by a slightly different placement of holes in the board.

The name plate on the AA-484 amplifier identified it as a Wireless Specialty Apparatus Model AA-484, 2-Stage, Low Frequency Amplifier made for RCA.

The photograph appearing in the catalog is a prototype or preproduction unit which is slightly different from the production unit, a rare example of which is shown here. The production unit can be distinguished from the prototype by the presence of a Wireless Specialty name plate bearing a serial number and a slightly different wiring layout dictated by a slightly different placement of holes in the board which resulted in a more symmetric wiring layout.

The AA-484 has a strong resemblance to the Wireless Specialty Two-Step Amplifier, Type Triode E reproduced in Chapter 3 from the 1919 Wireless Specialty catalog. The descriptions of the two sets are similar, right down to the particular type of board used in both—a "bakelite dilecto." It is likely that RCA did not want to market a set with exposed components clearly labeled with the Wireless Specialty name, and instead, asked WSA to redesign the set with GE components. According to the WSA catalog, the Triode E was electrically equivalent to the Triode B, and therefore it is likely that the AA-484 is electrically equivalent to the Triode B as well, a set which is also described in Chapter 3.

The AA-484 pictured here was restored using an original board and correct GE and WSA components. Very few, if any examples may exist in the original factory condition because the purchaser was encouraged to *"carry on experiments in amplification and thus study at first hand the action of various forms of radio receiving circuits."* Further, a number of these sets may have been sold in kit form. In a letter dated January 13, 1922, Wireless Specialty offered the AA-484 to RCA in three configurations: a "wired" amplifier for $20.18, an "unwired" amplifier for $17.25, and a configuration described as "Panels Only Drilled with Feet" for $3.67.[39]

RCA marketed the AA-484 for a short five-month period between April 1 and August 21 of 1922, but with little or no advertising, very few were sold. Consequently, very few of these sets were made—probably only a few hundred—judging from the serial number of the set pictured here.

AA-485 Six-Tube Radio-Audio Amplifier

The Radiola AA-485 RF/audio amplifier was introduced to the public at the Amateur Radio Show in New York in March of 1922, although the model pictured there was an engineering model with no nameplate, and had knobs which were of a different style than found on the production models. An engineering model as it appeared at the radio show is pictured in more detail in the June 1922 issue of *Radio Broadcast* magazine.[40]

The AA-485 was one of the first apparatus distributed by RCA sporting the Radiola name which was prominently displayed on the name plate attached to the front panel.

The interior of the AA-485 is dominated by the two audio transformers, two rheostats, and five RF transformers.

The AA-485 appeared in the RCA Price List for Radio Apparatus dated April 1, 1922, but did not appear in the June 1, 1922 RCA catalog or in the RCA price list of the same date—most likely because there was a delay in the release of this unit. However, it did reappear in the price list dated August 21, 1922 at $160 without tubes, and in subsequent price lists at the same price until the end of the 1923 model season. One of the few published photographs of the completed model as it was offered for sale appeared in selected RCA distributor catalogs for 1922 such as the one published by Robertson-Cataract Electric Company.[41]

This set was advertised as one of the most sensitive devices ever developed for the reception of broadcast signals. This claim was likely to have been true at the time, considering the fact that six vacuum tubes were used to create five steps of RF amplification, a detector stage, and two steps of transformer-coupled audio amplification. The last two tubes were "reflexed" back through the third and fourth tubes to achieve the equivalent of five stages of RF amplification with three tubes. Five hard-vacuum UV-201 tubes were recommended for the amplifier circuits, and the soft-vacuum UV-200 tube was recommended for the detector. The amplifier with an operating range of 200-500 meters was designed for use in conjunction with either a loop antenna equipped with a tuning condenser such as the WSA AG-1380, or with a standard tuned-circuit receiver such as the AR-1382 described below.

The AA-485 rivals the Aeriola Grand as being the rarest of early Radiolas, and it certainly has an aura that is unrivaled by any other Radiola including the Aeriola Grand. With its six tubes in a row prominently jutting out from atop the attractive hardwood cabinet, and with the reflex circuitry used for the first time on RCA receivers, it is unique both esthetically and electrically. This set, with the Radiola name clearly affixed to the name plate, shares the distinction with the AR-1375 of being one of the first two RCA sets to bear the Radiola name.

Based on the serial number evidence, there may have been as many as 500-1000 sets made. However, given the rarity of this set today, it is likely that few were actually sold, and a large number may have been disassembled by the manufacturer for parts or destroyed. The set continued to appear in the RCA price sheets until the end of the 1923 model year circa February 1924, at which time it was dropped.

AR-1382 Crystal Receiver

The AR-1382 crystal receiver never appeared in an RCA price list or RCA catalog, but it did appear in the 1922-1923 catalogs of several RCA distributors at $70 without a headset.[42] It had to compete with a number of other crystal sets offered by RCA, including WSA's own AR-1375 priced at only $40 with a headset. It was never advertised nationally by RCA, and only a very few sets must have been sold.

The receiver was designed with a two-tuned antenna circuit and a wide range of coupling to allow for greater selectivity. It was particularly suited for use with an RF amplifier such as the AA-485 amplifier, with which it could be used by lifting the cat's whisker and connecting to the terminals marked "amplifier." The detector on this set appears to be the same one used on the AR-1375. The set can also be used with an audio amplifier by connecting to the terminals marked "telephones." It had an operating range of 250-750 meters.

A panel also labeled AR-1382—which is somewhat different than the one pictured here—has surfaced in the collecting community. This panel differs from the one pictured here in the following ways: 1) there is no crystal detector, making the set a tuner rather than a receiver, 2) there are no binding post at the bottom right for headphones, and 3) the binding posts at the center right of the panel are labeled "To Detector Amplifier" rather than "To Amplifier." Apparently, more than one type of set was contemplated for the AR-1382 model number.

Model AR-1382

While the AR-1382 never appeared in an RCA price list or catalog, it did appear in RCA distributor catalogs. (*Radio Supplies and Equipment,* Robertson-Cataract Electric Company Catalog No. 22, 1922-1923, p. 22)

AA-1381 Amplifier

The AA-1381 is a one-tube amplifier which was designed for operation with a tuning receiver such as the AR-1382, or the AG-1380 loop antenna with a tuning condenser. Very little is known about this set except what can be gleaned from a physical inspection of the panel and the interior. Since this set

A photograph identified as a Wireless Specialty AA-1381 with four tubes, at least three RF transformers and one audio transformer was found in the *Radioana Collection.* (George H. Clark Collection, Smithsonian Institution)

An artist's depiction of the AA-1381 provides detail on an unusual detector which appears to be unique among all broadcast receivers marketed by RCA. (George H. Clark Collection, Smithsonian Institution)

A version of the AA-1381 with a wood cabinet appeared in several issues of the Old Timer's Bulletin. (*The Old Timer's Bulletin*, June 1992, front cover)

does not appear in any RCA price list or catalog, nor was it ever advertised by RCA, it is likely that it never went into production.

Two slightly different versions of this set have been photographed, one with a metal case and one with a mahogany case. The version with a metal case was photographed at a radio show circa 1922,[43] while a version with the wood case appeared twice in *"The Old Timer's Bulletin"*.[44] A third image of this set, also reproduced here, was sketched by an artist for RCA and reproduced later in an issue of *Radio-Craft* magazine in 1938.[45] The detector appearing in the *Old Timer's Bulletin* does not match the detector in the RCA files or the one in the photograph at the unidentified radio show circa 1922, but instead appears to match the detectors used on the AR-1382 and AR-1375 sets.

A photograph of an alternative version of the AA-1381 amplifier has been found in the *Radioana Collection* with four tubes. The set has at least three RF transformers and one audio transformer similar to the ones found in the WSA AA-485, and it may or may not have been reflexed in a manner similar to the AA-485. It does not appear to have a tuner, so it was likely designed to be used with either the AR-1382 tuner or perhaps the WSA AG-1380 antenna and UC-1820 tuning condenser (often mounted to the base of the antenna).

AG-1380/1385 Collapsible Loop Antenna

The AG-1380 collapsible antenna was the first loop antenna sold by RCA for broadcast reception. In the 1922 catalog, RCA explained that the loop antenna *"is found advisable for use in congested areas where interference from a number of transmitting stations operating at the same time, is experienced."* Congested areas were generally synonymous with large cities—where it was often impractical to string long-wire antennas for hundreds of feet—so a second advantage was that it obviated the necessity to string a long-wire antenna. In these situations, selectivity could be enhanced by pointing the loop antenna in the direction of the desired station, but the loop antenna with its smaller

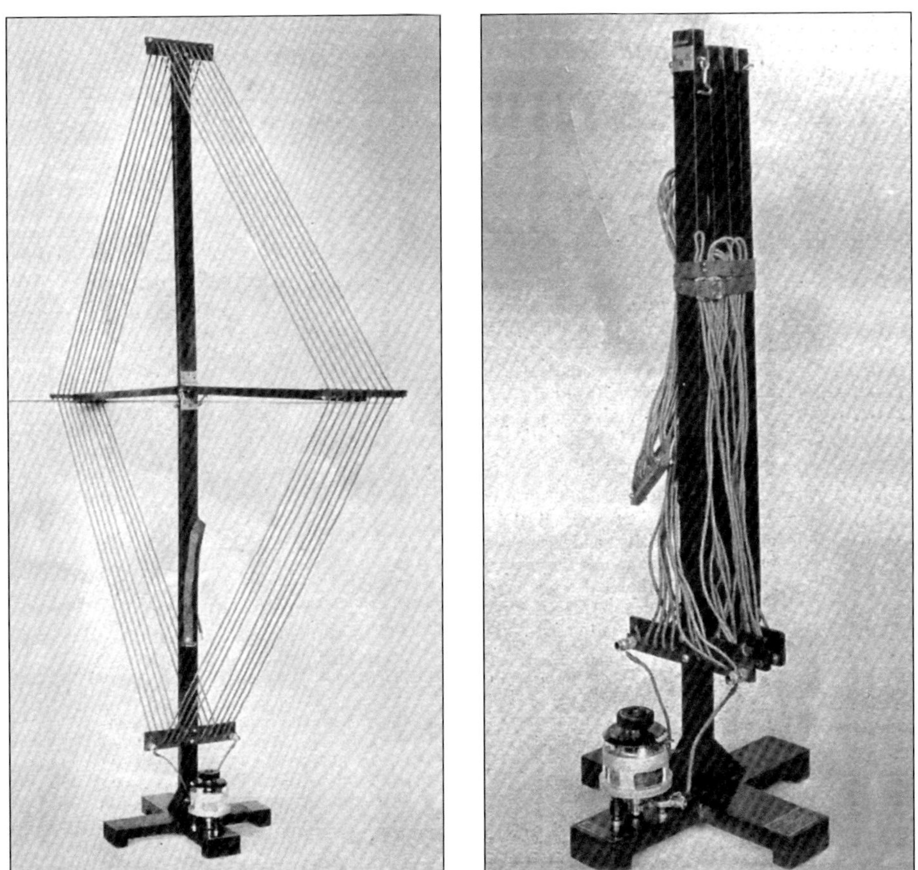

The AG-1380 loop antenna (with the UC-1820 tuning condenser mounted on the base) had an erected span of almost five feet, requiring a significant amount of space. The AG-1380 loop antenna with an open span of five feet could be readily collapsed to save space when not in use. (George H. Clark Collection, Smithsonian Institution)

dimensions collected less power at broadcast wavelengths than a long wire antenna, so it was necessary to add additional amplification, preferably radio frequency amplification.

The loop was designed specifically for several WSA sets, most notably the AA-485 RF/audio amplifier, the AA-1381 amplifier, and the AA-1376 amplifier. However, none of these sets appeared in the June 1922 catalog where RCA first featured this antenna. In fact, this antenna was not paired with any of the receivers or amplifiers offered by RCA in the catalog, so the average broadcast radio customer must have wondered why this antenna was in the catalog. It was most likely put in the catalog for use with AA-485 amplifier, which apparently was deleted because it was not ready for release. It was not until August 1922 that RCA marketed the AG-1380 with both the WSA AA-485 amplifier and GE's new AA-1520 radio frequency amplifier. It was also marketed with the Radiola VI in 1923.

The frame of the antenna itself is "*artfully finished in mahogany and completely collapsible*," and with good reason, because the open antenna was rather large with dimensions of about five feet in height and width. The antenna first appeared in the June 1, 1922 RCA price list for $25 without the tuning condenser, or for $32.50 with the companion UC-1820 tuning condenser that was often mounted on the base. Strangely, the price was reduced almost immediately to $24 in the August 21, 1922 price list.

The original antenna introduced in June 1922 had a fixed base, but in the September 15, 1923 RCA Price list, an AG-1385 model was introduced with a rotating base. The AG-1385 also appeared in a few distributor catalogs.[46] Given the size of the antenna and the need to rotate the antenna to receive different stations, this feature would have been most welcome. These loops were sold for several years, and discontinued by April 15, 1924.

The AG-1385 loop antenna was similar to the AG-1380 except that it contained a mechanism allowing the antenna to rotate without having to rotate the base. (Courtesy of Stephen Brown)

Endnotes and References for Chapter 4

1 *Radio Enters the Home*, (Radio Corporation of America catalog, Marketing Department, New York City, June 1, 1922).

2 RCA Annual Report for 1922, (RCA Corporation, New York, 1923) p. 21.

3 "New York's Amateur Show," *The Wireless Age*, Vol. 9, No. 7, April, 1922, p. 21.

4 *Radio Apparatus for Amateur and Experimental Use with Instructions for Continuous Wave Operation*, (Radio Corporation of America, New York, September 1, 1921).

5 *Radioana,* George W. Clark, "The Radio Corporations Model Designation," undated notes GWC-22-6A.

6 Memo from W. M. Derrick to A. D. Mc Kenzie and P. H. Boucheron dated July 25, 1922 assigning the number ER-875 to the Radiola III (kindly provided by Alan Douglas).

7 The RB receiver can been seen in many photographs of Westinghouse broadcast stations such as the one taken at WJZ appearing in the following book: Austin C. Lescarboursa, *Radio for Everybody*, (Scientific American Publishing Company, Munn & Company, New York, 1922) p. 51.

8 It should be noted that virtually all the early units with serial numbers below 500 were housed in a single cabinet and sold as either as an RA or DA, indicating that the RAs and DAs were indeed released before the RCs. Of all the RCs observed to date, at least one of the two units has a serial number above 500, indicating that the RCs may not have been produced until about a third of the way through the first production run of about 1600 RAs and DAs.

9 *Radioana,* letter from E. A. Tilden, Westinghouse Supply Division to E. E. Bucher, RCA Corporation re: "Data Required for Our New Catalog of Commercial Apparatus Prop. R-5786," March 6, 1922. An attachment to this letter dated January 14, 1921 announces the RB, and gives its technical specifications.

10 "Radiola" Word Mark, (United States Patent and Trademark Office, Washington D.C., March 7, 1922), Registration Number 71154767.

11 See Appendix H by a description of Radiolas made for companies other than RCA.

12 *The Wireless Age*, Vol. 9, No. 11, August 1922, p. 11.

13 Frank Conrad "Wireless Receiving Cabinet," (United States Patent and Trademark Office, Washington D.C., January 10, 1928; filed April 18, 1922) No. 1,655,985.

14 *Radio Broadcast*, Vol. 1, No. 5, September 1922, p. 451.

15 *The Wireless Age*, Vol. 10, No. 3, December 1922, p. 92.

16 "Recently Discovered," *Radio Age*, Vol. 18, No. 7, July 1993, p. 7. (This article contains only the first page of a five-page undated memo from the manager of the RCA sales department to wholesale distributors.)

17 *The Wireless Age*, Vol. 10, No. 1, Oct. 1922, p. 3.

18 Vincent Morley, "Adapting Dry Cell Tubes to the Radiola RC," *The Wireless Age*, Vol. 4, No. 1, October 1923, pp. 46-47.

19 Floyd A. Paul, *Radio Horn Speaker Encyclopedia,* (Published by Floyd A. Paul, Glendale, CA. 2nd Ed., 1989) p. 20.

20 Magnavox's development of the horn speaker is chronicled by Walt Sanders in a three part article: "Magnavox—The Great Voice! Part I," *Radio Age*, Vol. 14, No. 4, April 1988, pp. 1-2; "Magnavox Part II," *Radio Age*, Vol. 14, No. 5, May 1988, pp. 1-2; "Magnavox Part III," *Radio Age*, Vol. 14, No. 7, July 1988, pp. 3-5.

21 *Radio Amateur News*, April 1920, p. 531. This ad for the Magnavox Telemegafone appeared well before the idea of broadcast radio came to Westinghouse.

22 *Radio Amateur News*, December 1919, p. 256.

23 Frank Frimmerman, "Frank Frimmerman's Station," *Radio Amateur News*, February 1920, pp. 420-421.

24 "The Vocarola for Radiophone Reception," *The Wireless Age*, Vol. 9, No. 9, June 1922, p. 61.

25 *Radioana,* Minutes of the General Sales Committee Meeting held July 10th, 1923, Document No. CWC5 2020A.

26 The instructions placed the lower limit at 180 meters, but other RCA documents such as distributor price lists placed the lower limit at 175 meters.

27 Alan Douglas, *Radio Manufacturer's of the 1920's*, Vol. 3-RCA to Zenith (Vestal Press, Ltd., Vestal, New York, 1991), p. 36-37; a close-up of the pair appears in Alan Douglas, "Radio Corporation of America, Part IV, *Radio Age*, Vol. 8, No. 10, December 1982, p 3.

28 *The Wireless Age*, Vol. 10, No. 1, January 1923, p. 1.

29 RCA sometimes used the spelling "EZ case" in advertising

30 *The Wireless Age*, Vol. 10, No. 2, November, 1922, p. 1.

31 Gleason L. Archer, LL.D., *Big Business and Radio*, (The American Historical Company, Inc., New York, 1939) pp. 15-16.

32 Alan Douglas, *Radio Manufacturers of the 1920's, Vol. 3—RCA to Zenith*, p. 14.

33 *Radioana,* Minutes of the General Sales Committee Meeting held July 10th, 1923, Document No. CWC5 2020A.

34 Such a small number seems unlikely because if so few sets were sold in the eight months between June 1922 and March 1923, RCA would not have ordered a third production of ER-753A sets with the Radiola I name in March of 1923.

35 The AG-788 antenna was first dropped from the April 15, 1924 RCA price list.

36 *Radio Digest Illustrated*, Vol. 2, No. 5, August 12, 1922, p. 16.

37 *Radioana*, letter from A. E. Crocker, Sales Department to Dr. Alfred M. Goldsmith dated January 24th, 1923 describing the unit cost price of the AR 1375 "W. S. A. Receiver in mahogany case, gold plated parts with telephones" at $19.50.

38 *The Old Timer's Bulletin*, Vol. 33, No. 2, July 1992, front cover; *The Old Timer's Bulletin*, Vol. 17, No. 3, December 1976, front cover, p. 9.

39 *Radioana*, "Prices Quoted by Wireless Specialty Apparatus Company to Radio Corporation for New Devices," January 13, 1922.

40 *Radio Broadcast*, June 1922, Vol. 1, No. 2, p. 180.

41 *Radio Supplies and Equipment,* Robertson-Cataract Electric Company Catalog No 22, 1922-1923 (Reprinted by the Vestal Press Ltd., Vestal, New York 13850, 1922) p. 22.

42 Ibid., p. 12.

43 Alan Douglas, *Radio Manufacturers of the 1920's , Vol. 3—RCA to Zenith*, p. 23.

44 *The Old Timer's Bulletin*, Vol. 33, No. 2, July 1992, front cover; *The Old Timer's Bulletin*, Vol. 17, No. 3, December 1976, front cover, p. 9.

45 Hugo Gernsback, "Progress of Radio Receiving Equipment," *Radio-craft*, Jubilee Souvenir Number, March 1938, p. 549.

46 *Radio Supplies*, (Julius Andrae & Sons Co., Milwaukee Wisconsin, September 1923) Catalogue R-6, p. 33.

There's a **Radiola** *for every purse*

THE public has made radio what it is today and the public will dictate what radio will become tomorrow.

In the meantime, the Radio Corporation of America is doing what it can to protect the public—

RCA

This symbol of quality is your protection

By systematically developing Radiolas, which are so simple, so easily manipulated, that no technical knowledge is required to make the most of them.

By research that not merely improves apparatus, but will stabilize the quality of broadcast transmission.

By appointing only responsible distributors who have the same sense of obligation to the public.

The Radio Corporation of America manufactures Radiolas to suit every purse. Prices range from $18 to $350. RCA dealers will frankly advise you in making a selection.

Before purchasing any Radio Set, be sure to buy the book "Radio Enters the Home" at your dealer—Price 35c or write direct to

Radio Corporation of America

WORLD WIDE WIRELESS

Sales Department
233 Broadway
New York City

District Office
10 South La Salle St.
Chicago, Ill.

Chapter 5

There's a Radiola for Every Purse

1923

RCA introduced its 1923 line of radio apparatus to the public in several expositions late in 1922 including the Chicago Radio Show during the week of October 14-20, and the American Radio Exposition in New York City during the week of December 21-30. A photograph taken at what is believed to be the Chicago show focuses on the new Radiola lineup for 1923 with the new RCA slogan *"There's a Radiola for Every Purse."* The slogan must have resonated with the public, because it appeared prominently in RCA advertising not only throughout 1923, but for several years thereafter.

The RCA line for 1923 was captured in a photograph believed to be taken at the Chicago Radio Show in October of 1922. (George H. Clark Collection, Smithsonian Institution)

90 **THE SATURDAY EVENING POST** *February 3, 1923*

Radiolas for

REG. U. S. PAT. OFF.

This symbol of quality is your protection

BACKED by the manufacturing, scientific and research resources of the Radio Corporation of America, the full line of RCA sets and apparatus makes the marvels of radio available to every home and within reach of anyone's purse. Simply tell your dealer you want a RADIOLA.

If you can turn a knob you can operate a RADIOLA, the trade-marked receivers of the Radio Corporation of America. These modern products make experience unnecessary. They are beautifully made, fitting into the decorative arrangement of every home.

For permanent satisfaction, service by reliable dealers, and tested quality, insist upon an RCA set. When you insist upon one of these sets, you know you are purchasing the result of the scientific achievements, research experience and conscientious public service of the Radio Corporation of America, in whose establishment such companies as the General Electric and Westinghouse participated.

The best receiving sets are now here. Be prepared for broadcasting in 1923 which will include the finest programs that can be provided. Every RADIOLA described here has the necessary wave length range to receive the broadcasted programs of established broadcast stations. Don't delay. Order your RADIOLA from your nearest RCA dealer today.

*"There's a **Radiola** for every purse"*

Radiola Grand

One of the DeLuxe instruments in the RCA group. So attractively designed that it harmonizes with the furniture of any room. Especially suitable for those desiring a large volume of reproduction for family dancing and for the entertainment of large groups. Furnishes local and long-distance reception. Four Radiotron tubes (one detector and three amplifiers) and plate batteries. Perfect in tone. Easy to manipulate. **Price $350.**

Radiola IV

The latest addition to the RADIOLA line. A DeLuxe instrument of the console type, completely self-contained, including dry batteries for filament, plate and grid voltage, a new loudspeaker, head-telephone jack, antenna and ground plug connectors. Artistic in design to harmonize with fine furniture. Provides long-distance as well as local reception. An ideal addition to the modern music-room.

Price $275.

Radiola V

Attractive receiver with mahogany case for well-appointed homes. Very sharp tuning possible. Vacuum tube detector with two stages of amplification. Also fitted with a crystal detector. Good for distances up to several hundred miles. **Price $132.50 without Radiotrons or batteries.**

Radiola VI

Receives with an indoor-loop antenna. Six tube set including three stages radio frequency, one detector, and two stages of audio amplification. Designed for long-distance reception (up to 1000 miles under favorable conditions). This Radiola is designed to be operated with Model AG-1380 loop antenna. **Price $162.50 without Radiotrons or batteries.**

Radiola RC

One of the most popular of the Radiola group. More than 25,000 in use. Accepted as the standard universal receiver. Can be tuned to receive stations sending on wave lengths between 175 and 700 meters. Remarkably clear reproduction. **Price $132.50 without Radiotrons or batteries.**

Radio Corporation of America

WORLD WIDE WIRELESS

Sales Department
233 Broadway
New York

District Sales Offices
10 So. La Salle Street
Chicago, Ill.

433 California Street
San Francisco, Cal.

RCA introduced the 1923 lineup in the February 3rd issue of the *Saturday Evening Post*.

THE SATURDAY EVENING POST 91

every home in 1923

This is what a **Radiola** will bring you throughout 1923

The best entertainment and education that the world, serious and at play, can provide—from the local broadcasting stations and distant ones—broadcasting which will be more elaborate than ever in 1923—complete, constantly changing programs that will thrill the family circle during the long winter evenings—

- Music by famous orchestras—
- Songs by famous artists, singing to you in person—
- Complete church services—
- Instantaneous reports of sporting events, play by play, blow by blow—
- Weather reports, crop information, time signals—
- Stock and bond quotations—
- Fashion hints—
- Bedtime stories for the children—
- Statesmen explaining national problems—
- Election returns—

Front seats in the new theatre where life's great drama is played out, day by day, night by night. New programs, new artists, different entertainment every day.

Radiola Sr.

Most popular of single-tube sets. More than 40,000 of these sets in actual use. Wonderfully efficient. Anybody can manipulate it. Receives up to distance of 1500 miles. Can be fitted with amplifier for long distances and with loud-speaker. Operated by dry batteries only. **Price, including one WD-11 Radiotron and head telephones, $65.**

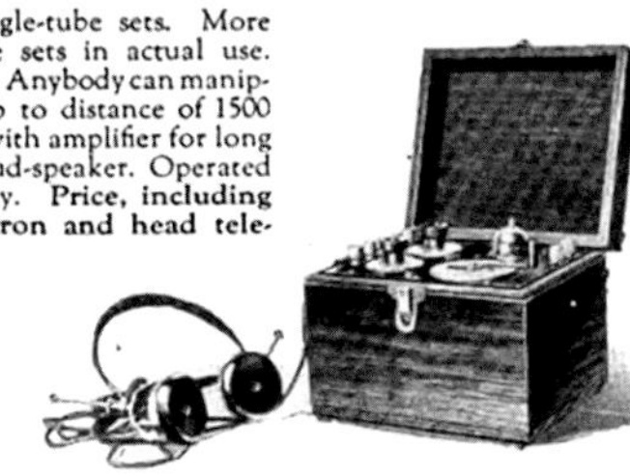

Radiola Sr. Type AC Amplifier.

Combine it with Radiola Sr. and obtain loud-speaker operation of sufficient strength to fill a room with music. Operated by dry batteries. **Price, with two famous Radiotron WD-11 vacuum tubes, $68.**

Radiola Jr.

Complete crystal receiver, easy to set up and operate and good for distances up to 30 miles. All the joy of radio at the lowest price. Mahogany finished case, head telephones and super-sensitive crystal.

Price complete $25.

"There's a **Radiola** for every purse"

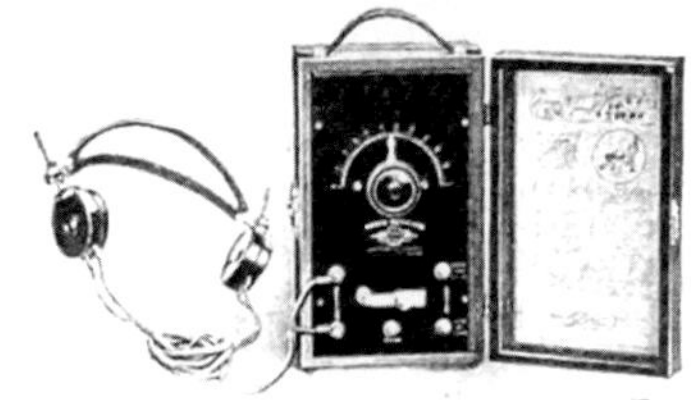

Radiola Combination

A super-selective set, particularly adapted for lectures, class room demonstration and experimenting by the discriminating. Extremely sharp tuning for local and long-distance reception under conditions of severe interference. Comprises coupler RT, tuner RA, radio-frequency amplifier AR and detector-amplifier DA. **Price $263 without tubes or batteries.**

Radiola AR-1375

The finest crystal receiver available. Good for distances up to fifty miles. Receives stations sending on wave lengths of 170 to 2650 meters. Very easy to operate. Perfect tone quality. Compartment for head telephones. Price, with telephones, $40.

Radiola I

An exceptionally well-finished crystal receiver, ideal for the boy or grown-up who wants to listen to nearby stations and to enjoy radio in a modest way at a moderate cost. Receives up to 30 miles. Self-contained, with but one knob to manipulate. Complete with head telephones. **Price $25.**

Radio Corporation of America

WORLD WIDE WIRELESS

Sales Department
233 Broadway
New York

District Sales Offices
10 So. La Salle Street
Chicago, Ill.

433 California Street
San Francisco, Cal.

Table 5-1. Broadcast radio apparatus marketed by RCA during the 1923 season, including holdover apparatus (shaded entries)

	Model	Apparatus Type	Cabinet Style	Tubes	Intro Date	Discon'd. By	List Price[a]	Mfg. By
1923 Season New Models	Radiola II (AR-800)	Regenerative Receiver	Portable Wood Case	2	Dec. 1922	Mar. 1924	$97.50	GE
	Radiola IV (AR-880)	Regenerative Receiver	Table Wood Case	3	Dec. 1922	Mar. 1924	$275.00	GE
	Radiola V (AR-885)	Regenerative Receiver	EZKASE Metal Case	3	Dec. 1922	Mar. 1924	**$132.50**	GE
	Radiola V-A (AR-885A)	Regenerative Receiver	EZKASE Metal Case	3	Mar. 1923	Mar. 1924	$142.50[b]	GE
	Radiola VI (AR-895)	Untuned RF Receiver	EZKASE Metal Case	6	Dec. 1922	Mar. 1924	**$162.50**	GE
	Radiola RT	Coupler	Wood Case	0	Jan. 1923	Mar. 1924	$45.00	W
	Radiola AR	URF Amp	Wood Case	3	Jan. 1923	Mar. 1924	**$80.00**	W
	Radiola RS	Regen. Rcvr.	Wood Box	2	Jan. 1923	Mar. 1924	$85.00	W
	Radiola Grand (RG)	Regenerative Receiver	Wood Cabinet	4	Dec. 1922	Mar. 1924	$350.00	W
	Radiola Special	Regenerative Receiver	Metal Box	1	Feb. 1923	Mar. 1924	$62.50	WSA
	Radiola VII (AR-905)	TRF Receiver	Wood Cabinet	5	Sep. 1923	Jan. 1924	$290.00	WSA
	UD-1430	Crystal Det.	None	0	May 1923	Mar. 1924	$1.60	WSA
	UD-1432	Crystal Det.	None	0	May 1923	Mar. 1924	$2.30	WSA
	UZ-1320	Speaker	None	N/A	Nov. 1923	Oct. 1924	$36.50	GE
	FH	Speaker	None	N/A	Sep. 1923	Feb. 1925	$36.50	W
	AG-1385	Loop Antenna (Rotating)	None	N/A	Sep. 1923	Mar. 1924	$24	WSA
Holdover Models	Radiola I (ER-753A)	Crystal Set	Portable Wood Box	0	Aug. 1922	Mar. 1924	$25.00	GE
	AA-1520	RF Amplifier	EZKASE	3	Aug. 1922	Mar. 1924	$80.00	GE
	Radiola AC	Audio Amp	Wood box	2		Mar. 1924	$68.00	W
	DA	Det. + Amp.	Case	3	Dec. 1920	Jul. 1923	**$70.00**	W
	RA	Tuner	Wood Case	0	Dec. 1920	Jul. 1923	$68.00	W
	Radiola Sr.	Regen. Rcvr.	Wood Box	1	Dec. 1921	4-15-24	$65.00	W
	Radiola RC	Regen. Rcvr.	Wood Case	3	Dec. 1920	Jul. 1923	$142.50[b]	W
	AA-485	RF + Audio Amp	Wood Cabinet	6	Mar. 1922	Mar. 1924	**$160.00**	WSA
	DB	Crystal Det.	None	0	Aug. 1921	Mar. 1924	$6.50	W
	AG-788	Antenna kit	N/A	0	Apr. 1922	Mar. 1924	$7.50	GE
	AD	Antenna kit	N/A	0	Oct. 1921	Sep. 1923	$7.50	W
	Vocarola LV	Horn Speaker	None	0	Jan. 1922	Sep. 1923	$20	W
	LS	Speaker Attachment	None	0	Not Known	Mar. 1924	$18	W
	AG-1380	Loop Antenna	None	N/A	Jun. 1922	Feb. 1925	$20	WSA

[a]All apparatus included tubes except those noted with the price in bold font

[b]Included UV-199 tubes, batteries and headphones

The new radios for the 1923 season displayed at this show were the Radiola IV, Radiola V, Radiola VI, and the Radiola Grand. Other apparatus from the existing 1922 line were also on display: the Radiola AR-1375 crystal set, Radiola Jr., Radiola Sr., Radiola Sr. with AC Amplifier, RC receiver, and the IP-501 receiver. This display was notable because it clearly signaled RCA's intent to brand name all their broadcast radio apparatus with the Radiola name—including the Radiola Jr., the Radiola Sr., and the Radiola AC Amplifier—all of which had been on the market for a year or more with the Aeriola name. The Radiola Jr. was abruptly discontinued in December of 1922, and therefore did not appear in the American Radio Exposition in December 1922 or in the 1923 RCA catalog, although it did appear in a February 1923 magazine ad which was most likely prepared before the decision to discontinue it was made.

In the trade magazine *Radio Merchandising,* RCA announced that it would advertise their entire 1923 line to the public in the February 3rd issue of the *Saturday Evening Post.*[1] The two-page ad in the *Post* was the first ad—and perhaps the only ad—featuring the entire 1923 RCA line together.[2] Prominently placed in this ad was the slogan "*There's a Radiola for Every Purse,*" which RCA began to use extensively in 1923—and with good reason. RCA offered 20 or so distinctly different sets in 1923 with prices ranging from $25 for a crystal set to $350 for the four-tube Radiola Grand regenerative receiver with an enclosed speaker. In fact, the 1923 line contained the largest number of different sets offered by RCA in any one year during the entire decade of the twenties (see Table 5-1).

It is also quite evident from this ad that every set in the entire line had been given the new Radiola brand name—including the renamed Aeriola Jr., Aeriola Sr., Aeriola AC Amplifier, and the RC. From this point on, RCA used "Junior" and "Senior" interchangeably with the abbreviated forms "Jr." and "Sr." in advertising and catalogs. The Jr. and Sr. form continues to be used here because it seems to have been the most common form used by RCA in advertising and catalogs of the day.

The 1923 Season Overview

RCA had originally planned ten new sets for the 1923 season: six numbered Radiolas (Radiolas II through Radiola VII), the Radiola RS, the Radiola Grand, and the Radiola AR and RT adjuncts for the Radiola RC.[3] GE was scheduled to manufacture Radiolas II through VI, but for reasons unknown the 1923 version of the Radiola III was cancelled before production began. Westinghouse was scheduled to manufacture the RS, the AR and RT adjuncts for their RC, and the Radiola Grand—a replacement for the ill-fated Aeriola Grand introduced in 1922. WSA was originally scheduled to manufacture only one new set, the Radiola VII, but early in the 1923 season RCA made a decision to drop the AR-1375 crystal set which had been briefly held over from the 1922 season, and decided instead to convert the existing inventory of AR-1375 crystal sets to a one-tube regenerative receiver and rename it the Radiola Special. As a result, Wireless Specialty actually made two of the new sets introduced in 1923, while GE made only four.

RCA had originally planned to hold over three competing crystal sets: the WSA AR-1375, the GE Radiola I, and the Westinghouse Aeriola Jr.—which had only recently been renamed the Radiola Jr. several months prior. All three of these sets appeared in the RCA ad in the February 3rd issue of the Saturday Evening Post, but RCA made the decision to drop two of three sets, most likely right after the ad had been prepared. The Radiola Jr. was heavily discounted in December for the Christmas season, a sure sign the set was to be dropped. Shortly thereafter, RCA decided to convert the existing stock of unsold AR-1375 crystal sets to one-tube Radiola Specials. Only the Radiola I continued on the market for a while, but it too was discontinued before the end of the year because of poor sales. As mentioned previously, RCA still had 6000 sets in stock as of July of 1923 for which they were seeking a means of quick

Table 5-2. Competing broadcast receiver sets manufactured for RCA in the 1923 season

Apparatus Type	Westinghouse Sets	General Electric Sets	Wireless Specialty Sets
One-tube regenerative sets[a]	Radiola Sr. $65	None	Radiola Special $62.50
Two-tube regenerative sets[a]	Radiola RS $85	Radiola II $97.50	None
Three-tube regenerative sets[a]	RC or RA-DA $142.50	Radiola V $142.50	None
Deluxe regenerative sets with speaker	Radiola Grand $350[b]	Radiola IV $275[a]	None
Three-stage untuned RF amplifier	AR $80	AA-1520 $80	None

Note: Prices are from RCA dealer price sheets dated May 1, 1923
[a] Included headphones, Radiotrons, and batteries
[b] Included a floor stand in which the batteries could be placed

disposition. It was clear that RCA understood the era of the crystal set was over, at least as far as they were concerned.

RCA published their 1923 line in a small 17 page catalog entitled "*Radio Apparatus for Broadcast Reception.*" In contrast to the large comprehensive 128 page catalog for 1922, the new RCA catalog for 1923 was little more than a brochure at 3 x 5 inches. By the time this catalog was published in early 1923, the AR-1375 had been dropped from the line, but oddly, the replacement Radiola Special did not appear in this RCA catalog, or any other distributor catalog for that matter. In fact, the Special did not appear in any RCA distributor price list—even after it was introduced circa May, 1923—nor did it ever appear in any national advertisements.

By June 26, 1923 it had become apparent to RCA management that the receivers for 1924 were going to be delayed, and so they decided to delay the publication of a 1924 catalog, and instead printed 300,000 copies of a revised 1923 catalog in which the Radiola I, RC, RT, and AR were dropped, and the Radiola VII and IX receivers and FH and UZ-1320 horn speakers were added.[4] The two versions of this catalog were nearly indistinguishable from the front cover, but featured a number of distinctly different apparatus. This catalog was in use until the 1924 line was introduced towards the end of February of 1924.

While the complete Radiola 1923 line still contained a number of redundant and competing sets reminiscent of the 1922 season (see Table 5-2), the reasons were more a result of constraints placed on RCA than a lack of vision or discipline. First, at the end of 1922 RCA found itself with a glut of unsold receivers as a result of an unanticipated downturn in sales during the second half of the year, and so they decided to hold over a number of models to clear out the existing stock. For example, RCA had tens of thousands of unsold RC/RA/DA receivers designed in 1920 that were all but obsolete by 1923. Clearly, RCA had to either continue marketing these sets or write them off at a substantial cost. The AR and RT adjuncts were almost surely offered in an attempt to bolster sales of the RC/RA/DA receivers.

Second, RCA was obligated to sell radio equipment according to the 60%-40% agreements with GE and Westinghouse, but since Westinghouse was well ahead of its quota for 1922, RCA had to favor GE in the 1923 season, independent of customer demand. It is likely that the Radiola V and VI, which were no more than dressed up combinations of the thoroughly mundane 1300/1400/1520 receivers from the 1922 season, were included in

The RCA catalog for 1923 was significantly smaller in size and scope than the 1922 catalog.

the 1923 lineup to help increase GE revenue share. RCA heavily advertised the GE sets in comparison to those from Westinghouse and Wireless Specialty, thus encouraging the public to buy the sets that would help RCA satisfy the 60%-40% agreements.

General Electric Offerings

Of the four Radiolas offered by GE in the 1923 season, only the Radiola II and IV were truly new designs. The Radiola I was a hold over from 1922, the Radiola V was a combination of the AR-1300 and the AA-1400 from 1922 with a new coat of paint and new knobs, and the Radiola VI was a combination of the AA-1400 and AA-1520 from 1922, also with a new coat of paint and new knobs. The Radiola III was also to be a new design, but in the end it was dropped from the 1923 lineup. In addition to these receivers, GE also produced a new Radiola horn speaker, the UZ-1320.

Radiola II (AR-800)

According to George Clark, the Radiola II was introduced in December, 1922, but the set did not appear in the December 1922 American Radio Exposition in New York, nor did it appear in the February 3, 1923 *Saturday Evening Post* ad introducing the new RCA line for 1923. Further, it did not appear in RCA price sheets until May, 1, 1923, where it was offered for $97.50 complete, nor did national ads for the Radiola II appear until July 1923.[5] Consequently, it is difficult to reconcile Clark's early release date of December 1922 with the later dates of other reliable indicators.

RCA offered a number of new sets for sale to the public including the Radiola II, RS, Radiola Grand, and Radiola IV at a unique type of radio show held at the Bamberger Department Store in January 1923. (George H. Clark Collection, Smithsonian Institution)

In support of Clark's early release date, the Radiola II did appear in the first RCA catalog for the new 1923 line printed no later than February or March 1923. More to the point, an internal RCA memo dated January 25, 1923 clearly stated that Radiola II sets were sold to the public at a special showing held by Bamberger's Department store in New York City towards the end of January, 1923.[6] Based on these two documents, it appears likely that the set was indeed released in December of 1922, but that it was not widely promoted or advertised—at least not initially—for several possible reasons described in the following paragraphs.

First, according to GE engineer William White, there was a scarcity of UV-199 tubes in early 1923, with about 200 satisfactory tubes being produced per day.[7] Also, according to White, these tubes were in great demand for the Radiola IV, which was in production at that time—but White made no mention of tubes being allocated to the Radiola II. The scarcity of tubes might explain a delay in the introduction of the Radiola II, but by April the GE factory was producing about 3000 a day, more than enough to meet production demands for both the Radiola II and IV. Thus, a lack of tubes was probably not the main reason that advertising for the Radiola II was delayed until July 1923.

The Radiola II was housed in an attractive mahogany cabinet which also contained the dry cell batteries needed to power the two UV-199 tubes used in this receiver.

The most common version of the Radiola II contains a small terminal board in the lower left portion of the rear panel which was used to accommodate two additional 22½ volt batteries needed to drive an external speaker horn speaker, and a "C"-battery to reduce distortion introduced by increased plate voltage from 45 to 90 volts.

A second and more convincing reason that advertising for the Radiola II was delayed was that RCA feared it would adversely affect sales of over-stocked models such as the Radiola Sr. and AC amplifier that were already clogging inventories, and the competing RS. This conjecture is supported by the following observation documented by a salesmen at the Bamberger show:

> *"If we start showing the Radiola II, we will kill the Senior sales completely, and probably the RS only [sic]. The Radiola II has an appearance of a real set, and is complete in every way."*[8]

The fear that the Radiola II would adversely affect sales of other RCA receivers was well founded because it was a portable two-tube regenerative receiver packaged in an attractive mahogany case topped off with a leather handle. It was one of the first to be equipped with GE's new low-current UV-199 tubes designed to operate on dry-cell batteries, which were small enough to fit within the case.

This two-tube set with one detector and one audio amplifier tube was designed for a wavelength range of 200 to 600 meters. GE selected the double variometer arrangement first used by Westinghouse in the Aeriola Sr., although the two variometers are not quite identical in appearance. The set was relatively simple to operate with three controls—a filament control for both tubes, a tuning control to select stations, and an intensity control to adjust feedback. The set was heavily advertised as a portable that could be used in the countryside or on vacations.

While the set was designed to operate as a portable using flashlight batteries for the filaments and miniature "B" batteries for the plate voltage, it could also be used in a fixed location such as the home using longer-lasting external dry cell batteries for the filaments and standard size "B" batteries. The wire harness for the external batteries could be routed through a hole in the bottom of the battery compartment.

The Radiola II was originally designed to operate with two 22½ volt batteries which provided a plate voltage of 45 volts—sufficient for driving earphones but not speakers. There must have been a desire by the public to use the Radiola II with a speaker, because the set was soon modified to accept four batteries at 90 volts, sufficient to drive an external speaker. This modification for the extra batteries was made by adding a small battery terminal board to the rear of the set, a modification which was made to most of the sets produced. It is possible to find Radiola II sets without this board, but they are quite rare.

While early instructions and sales pamphlets referred to two "B" batteries for 45 volts without any reference to four "B" batteries or a "C" battery, later

An early and rarer Radiola II without a terminal board (left) is compared to a later and far more common Radiola II with the terminal board (right), which was added shortly after production began.

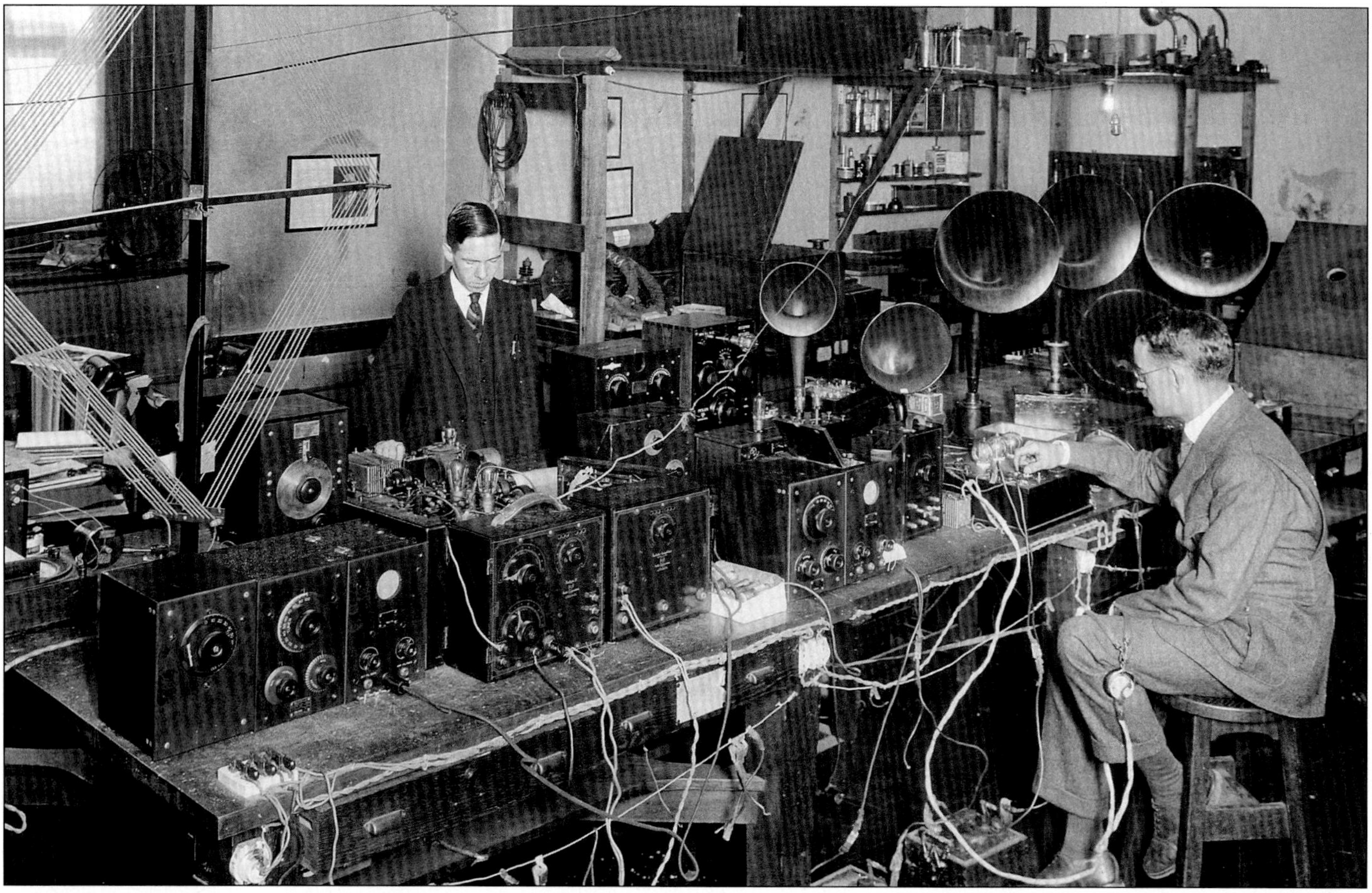

The RCA Research Department developed a prototype amplifier for the Radiola II in a matching cabinet, although it was never sold to the public.

documents refer to four "B" batteries for use with a loudspeaker, and an additional "C" battery which was needed to reduce distortion which would have occurred at the higher plate voltage without a change in the grid bias point. It is interesting to note that early advertisements for the Radiola II focused on earphones, while later ads pictured the Radiola II with a loudspeaker.[9]

Despite the increase in plate voltage, there must have been an even greater demand for increased amplification, because records show that RCA requested its engineering department to prepare a diagram for an additional stage of audio amplification to be provided to the public upon request.[10] A matching prototype amplifier was also developed and tested, although it was never offered for sale to the public. A picture of this amplifier undergoing test and evaluation in the RCA Research Department appeared in the April 1923 issue of *The Wireless Age*.[11] The prototype amplifier is housed in a portable cabinet with a general appearance similar to the Radiola II itself.

According to Douglas, RCA records show that 9594 Radiola II's were made, outselling the competing RS by better than two to one. However, it is not clear whether the Radiola II was a better set or just more heavily advertised.

The prototype Radiola II amplifier was placed in a Radiola II cabinet for test and evaluation on the bench.

Radiola III (ER-875) – (never produced)

It is clear from RCA documents that a Radiola III was approved for inclusion in the original 1923 line, and that it was to be manufactured by GE. This Radiola III—designated the ER-875 according to an RCA memo—was to be a portable regenerative receiver with a detector and two stages of audio amplification using UV-199 tubes, and a loudspeaker, all contained in a carrying case.[12] With the third tube and the enclosed speaker, the Radiola III would have been a step up from the portable Radiola II, but it also would have nearly duplicated the Radiola IV, which also had three UV-199 tubes and an enclosed speaker.

For reasons unknown, the Radiola III was dropped before it went into production. It may well be that the similarity of the proposed Radiola III and the Radiola IV was the reason why the III was eventually dropped, but that would still leave the question unanswered as to why RCA approved the Radiola III for production in the first place. The Radiola III name was skipped in the 1923 line, only to be resurrected in the 1924 line as a relatively simple two-tube regenerative receiver.

Radiola Two-Tube Portable Prototype

While there were many Radiola prototypes produced over the years, the one shown here is of particular interest because it fills a gap in the original Radiola II-III sequence. The Radiola II was a two-tube portable without a speaker, and the original Radiola III was to be a three-tube portable with a speaker—but there was no two-tube portable with a speaker. It is clear from a recently discovered prototype built by GE for RCA that both companies gave serious consideration to a two-tube portable set with a speaker for the 1923 season.

The prototype was designed with a handle, a lid to protect the panel, an enclosed speaker, and two 22½ volt batteries. It had two developmental UV-199 "peanut" tubes as well as a tuning coil similar to the one used in both the Radiola Sr. and the much-different Radiola III introduced in 1924. The GE monogram and the early RC monogram both enclosed in separate circles appeared on the name plate on the front panel. The prototype can be dated to a time no earlier than late 1921 by the UV-199 developmental tubes which first became available at that time, and no later than September 1922 when the RC monogram enclosed within a circle was replaced by the ubiquitous RCA monogram with the lightning flash. This time frame coincides with the time frame when GE and Westinghouse would have been designing sets for RCA's 1923 season.

Based on the experience reported with the Radiola II, RCA must have determined that a minimum plate voltage of 90 volts would be required to drive a speaker with two UV-199 tubes. Since four 22½ volt "B" batteries—rather than the two found with this set—would have been required to produce the 90 volts, the "B" batteries alone would have weighed 20 pounds, and the whole set would have weighted 45 pounds! Thus, the prototype with 45 volts most likely did not have adequate performance, and a similar design using four 22½ volt batteries to produce a total of 90 volts may have been too heavy to mass produce for the public in 1923.

The Radiola III (ER-875) originally planned for 1923 undoubtedly looked something like this prototype, with the addition of a third tube and a second audio transformer for the second stage of audio amplification. The batteries would have been a problem for this set, however. Perhaps RCA intended to use

A prototype two-tube portable receiver with a speaker manufactured by GE for RCA in late 1922 filled the gap between the Radiola II and proposed Radiola III (left). The two-tube prototype used a variometer coil similar to the Radiola III and two prototype UV-199 developmental tubes that became available in late 1921 (right).

four of the small 22½ volt batteries of the type used in the Radiola II, and flashlight batteries for the filaments. In this case, the battery life would have been relatively short.

Radiola IV (Model AR 880)

The Radiola IV was displayed at several radio shows near the end of 1922 at a list price of $275. It was originally designed to be the top-of-the-line receiver for 1923, but the Westinghouse Radiola Grand, an unplanned replacement for

The Radiola IV was the first receiver marketed by RCA that fulfilled David Sarnoff's vision of the self-contained Radio Music Box, complete with batteries and speaker.

The Radiola IV is very similar to the Radiola II except for an additional stage of amplification.

The first ads for the Radiola IV also featured the Radiola Grand, which competed with the Radiola IV at the high-end in the 1923 season. (*The Wireless Age*, February, 1923)

the Aeriola Grand at $350, competed for this position. In fact, the Radiola Grand was often pictured together with the Radiola IV in early ads.[13] The Radiola IV was a three tube receiver with a speaker and batteries enclosed in an attractive cabinet, while the Radiola Grand was a four-tube regenerative receiver with a speaker enclosed in an equally attractive cabinet, but the batteries were not self contained.

According to serial number data, there were approximately 10,700 Radiola IV's and 11,100 Radiola Grand's manufactured, but about half the Radiola Grand's were used to replace the Aeriola Grand's free of charge, so the Radiola IV actually outsold the Radiola Grand by about two-to-one. The relative success of the Radiola IV was undoubtedly due to the smaller, more attractive cabinet with self-contained batteries and a lower price. The IV was on sale for the entire 1923 season from December 1922 to March of 1924.

Electrically, the Radiola IV is similar to the Radiola II except for one additional stage of audio amplification which was needed to drive its enclosed

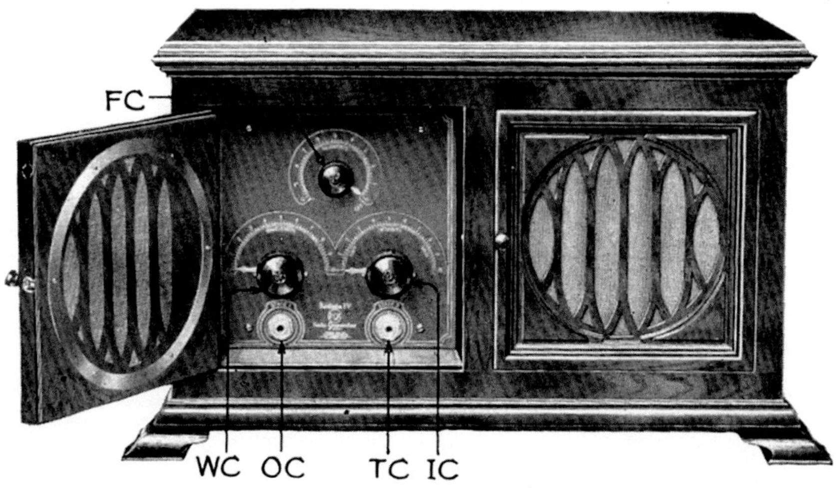

Fig. 3. Front View of Control Panel

The early Radiola IV instruction manuals did not show the vernier dial in the lower center of the panel that was added later to assist in tuning shortly after production began. (Radiola IV Instruction Manual 86945A)

speaker. The RF and detector circuits of the two are virtually identical, both using the same double-tuning variometer coils, which are similar but not identical to those used on the Radiola Sr. Like the Radiola II, the IV has two antenna connections for two wavelength ranges, one for wavelengths between 200 and 400 meters, and one for wavelengths between 350 and 600 meters, slightly different than the corresponding ranges for Radiola II.

Controls on the Radiola IV front panel include Filament, Intensity, Wavelength and Vernier—also identical to the controls on the Radiola II. The vernier control was added soon after production began to allow for a fine tuning of wavelength—first appearing circa March of 1923. The vernier dial was not included in the earliest version of the IV—the one that appeared in most catalogs and advertisements of the time—nor did it appear in the instruction manuals with dates earlier than March 1923. It must have been added in response to customer-reported difficulties in station tuning. It is estimated that only 10-20% of these sets were made without a vernier dial, and so it is a "must" for the serious collector.

Volume in previous sets utilizing earphones had been controlled by selecting the appropriate phone jack on the front panel, but since this approach did not work for sets with a self-contained speaker, a volume control was added to the top of the chassis panel accessible by lifting the lid of the cabinet. The Radiola IV was the first broadcast receiver marketed by RCA to have a separate volume control. The Radiola IV was also the first broadcast receiver manufactured by GE that was assigned individual serial numbers. The serial numbers stamped onto the chassis began with a low number—perhaps one or perhaps 100—and continued in sequence to approximately 10,700.

Radiola V (AR-885)

The Radiola V consisted of the AR-1300 and AA-1400 dressed up with a faux-mahogany finish and topped off with "pants and vest"—a wood base to hold the two sets, and a wood cap on top to balance the base. The set, which was electrically equivalent to the old AR-1300 and AA-1400, did not represent new technology and therefore took very little effort on the part of GE to bring to market. A metallic name tag bearing the name Radiola V was placed on the base. The wood bases and lids were also sold separately for $2.50 without name tags for use with the green AR-1300, AA-1400, and AA-1520 units sold the previous year.

The finish on this set was hyped in ads and catalogs stating: "*the brass cases have been given a mahoganized finish which makes them the equal in beauty of a cabinet made of solid mahogany...*" However, this view was not universally held

Historian George H. Clark thought that the exterior of the Radiola V (identical to the V-A shown here) looked more like a cross section of a diseased liver than faux-mahogany metal panels.

Radiola V ads boasting over 20,000 satisfied customers indicated that it was one of the most popular Radiola sets for 1923. (*The National Geographic Magazine*, October, 1923)

within the RCA inner circles, as evidenced by a poignant excerpt from a memo written by George Clark to David Sarnoff dated January 29, 1923: *"Riding in the sleeper the other evening, I noticed that it was mahoganized steel, but the design was real mahogany, or at least what the man in the street imagines mahogany to look like. The G.E. sets [Radiola V and VI] look more like a cross section of a diseased liver than they do mahogany."* Apparently the man in the street did not agree with Clark because an RCA advertisement appearing just before Christmas of 1923 touted that there were "20,000 satisfied users of the Radiola V."

The Radiola V first appeared on the market in December of 1922 at a price of $132.50 without tubes, or $207.50 complete with one UV-200 tube and two UV-201 tubes, batteries, battery charger, antenna kit, and headphones and plug—close to the price of the units sold in combination in 1922. In March of 1923, RCA offered the V for $142.50 complete with WD-12 tubes, dry cell batteries, headset and plug. By June, RCA offered the V for use with UV-201-A or UV-199 tubes—using UR-589 adaptors to fit the standard Navy base—for the same price of $142.50 (see V-A below). By the end of 1923, the V was offered complete—including a Westinghouse FH speaker—for the same price of $142.50.

Radiola V-A (AR-885-A)

There are a scant few references to a Radiola V-A model in RCA documents and publications. While there appear to be no references to a Radiola V-A in RCA price lists, RCA catalogs, or RCA advertisements, the V-A is explicitly identified on page 8 of the 1923-24 Pettingell-Andrews distributor catalog supplied with the UV-199 tubes. Historian George Clark also refers to a Radiola V-A in his list of release dates, identifying the release date for this set as March 1923. The V-A model can be readily distinguished from the V by the AR-885-A designation on the brass label on the front of the set, or by the AA-1400-A designation on the gold seal inside the detector-amplifier unit.

There were distinct differences in the circuitry of the detector-amplifier unit used in the two sets that caused RCA to distinguish between the two by the letter A. The changes in circuitry were needed to accommodate the newer UV-199 or UV-201-A tubes, versus the older UV-200/201 and WD-11 tubes used in the V. First, filament control rheostats with higher resistances were required to accommodate the new tubes with proportionally higher resistances. Second, a 4.5 volt "C" battery was added to bias the grids of the UV-199 tubes to reduce

Radiola V-A Complete with Radiola Loud Speaker

The Radiola V-A appeared only in the larger RCA distributor catalogs such as the one published by Pettingell-Andrews. (1923-24 Pettingell-Andrews catalog)

The Radiola V-A can be readily distinguished from the Radiola V either by the presence of the "A" on the AR-885-A metal nameplate affixed to the outside of the wood base (above) or the "A" on the AA-1400-A gold seal affixed to the inside panel of the amplifier cabinet (below).

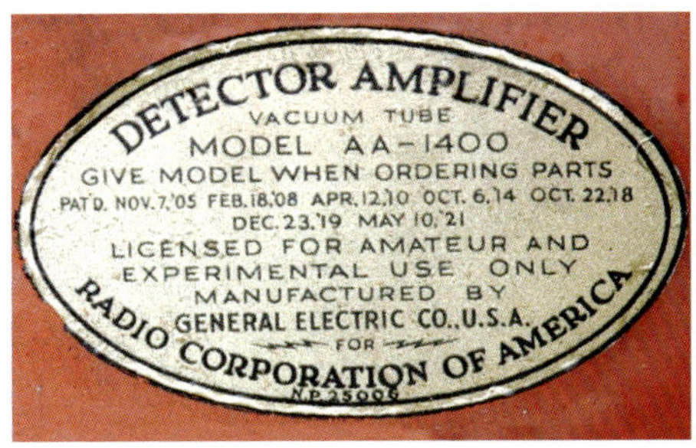

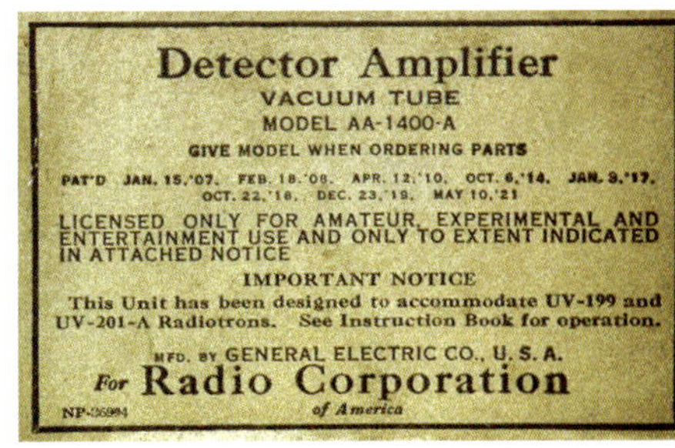

distortion. There were several other minor wiring changes in the A version as documented in the RCA Victor Service Notes.[14]

It is interesting to note that RCA added the A to the model number on the name plate and operating instructions accompanying the set, but did not do so for the model name. Apparently RCA did not wish to call attention in public to the design modifications in the Radiola V by openly advertising the Radiola V-A or adding the A onto Radiola V labels attached to the set, but it is also clear that distributors needed a means of distinguishing between the two versions in order to provide the appropriate tubes at the point of sale, since many of the V and V-A sets were offered by RCA without tubes.

Both the V and V-A sets were discontinued circa March 1924 when the new 1924 line of Radiolas was released. According to an RCA ad appearing just before Christmas 1923 there were "20,000 satisfied users of the Radiola V,"[15] and according to Douglas, RCA records show a total of 21,138 Radiola V sets were made, a number which certainly must have included both versions. It is not known how many V versus V-A models were made, but the number of each appearing in auctions and flea markets today appears to be about equal.

With over 20,000 sets sold at the regular price, the Radiola V was the most popular receiver offered by RCA in the 1923 season, outselling the also-popular Radiola IV by two-to-one. There were actually more RC's sold in 1923 than Radiola Vs, but 20,000 of them were sold for less than RCA's cost in a close out sale to eliminate a stagnant inventory that the public would not otherwise buy.

Radiola VI (AR-895)

The Radiola VI consisted of the AA-1400 and AA-1520 dressed up with a faux-mahogany finish and a wood cap and base, just like the Radiola V. A metal name tag bearing the name Radiola VI was placed in the center of the base. The set was released in December of 1922, and the first ad appeared in the January 1923 issue of *Radio Merchandising*, where it was featured together with the Radiola V.

The set was sold for $162.50 without accessories, but it required a tuning device such as the Wireless Specialty AG-1380 antenna with a variable capacitor—hardly an accessory—in addition to tubes, batteries and other items. The price for a complete set was not advertised, but according to RCA price sheets, it would have totaled more than $300. With a steep price tag and the need for a large indoor antenna, it is not surprising that it was not a best seller.

The Radiola VI was featured with the Radiola V in the first ads appearing early in 1923. (*Radio Merchandising*, January 1923—courtesy of Alan Douglas)

According to Douglas, RCA records show that 17 were sold in 1922, 670 in 1923, and 2087 in 1924. With only 670 sold in all of 1923, it is likely that many of the 2087 sets sold in 1924 were sold at a steep discount to clear inventory at the end of the 1923 season, circa March 1924. The very low number of Radiola VI sets sold was punctuated in a 1923 distributor catalog referring to hundreds–not thousands or tens of thousands–of families: *"Hundreds of families living in apartments in which it is desirable or impossible to erect antenna wires, can now hear the enjoyable, ever-changing programs put on the air throughout the day and night."*[16]

The exterior of the Radiola VI had an appearance very similar to the Radiola V, with the most noticeable differences being the two additional knobs and the absence of the Perikon detector.

While RCA did not provide tubes with the VI, the instructions recommended one UV-200 or UV-201-A detector tube, and five UV-201 or UV-201-A amplifier tubes. However, each distributor was free to make its own offering of tubes, and one catalog gave the following recommendation: *"Any of the various types of tubes now available may be used in this receiver. It is provided with sockets that take standard base tubes such as Radiotrons UV-201-A and WD-12 and, by means of an adaptor, UV-199. If facilities are readily available for the charging of a storage battery, it is suggested that five Radiotron UV-201-A be used as amplifiers and a UV-200 as a detector. Where it is impractical to use a storage battery to light the filament of the tubes, or should the owner prefer dry cell operation, either UV-199 or WD-12 may be used."*[17] It is strange that the Radiola V required different rheostats for the UV-201 versus the UV-201-A but the VI did not—perhaps the distributor was unaware of this subtlety.

Radiola UZ-1320 Horn Speaker

The UZ-1320 was to be the first of several large upright horn speakers offered by RCA. It was originally to be offered with the Radiola VII manufactured by Wireless Specialty in mid-1923, but GE had significant problems delivering the speaker to RCA, so much so that RCA substituted the FH speaker for use with the Radiola VII. RCA communicated this decision to wholesale distributors in an October 9, 1923 memo stating that it was impossible to supply the UZ-1320

The first ads for the UZ-1320 focused on the problems and difficulties involved in developing a loudspeaker, almost as if RCA was trying to justify the delay in bringing the speaker to market. (*Radio Merchandising*, November 1923)

for the Radiola VII, and that the Westinghouse FH speaker was to be substituted.[18] By the time this decision had been made, the instruction manual to accompany the set had been printed with a reference to the UZ-1320 as the speaker *"recommended as being adapted for use with the Radiola VII."*[19]

While the UZ-1320 appeared in the September 15, 1923 RCA dealer price list and in the revised 1923 RCA catalog published in the fall of 1923, it was not available until November. An RCA ad appeared in the November 1923 edition of *Radio Merchandising* introducing the UZ-1320 in conjunction with the Radiola II. The retail price for the speaker was $36.50. Oddly, this and other early ads focused on the problems and difficulties involved in developing a loudspeaker, almost as if RCA was trying to justify the delay in bringing the speaker to market. The UZ-1320 was on the market for about a year until the UZ-1325 horn was introduced circa October of 1924.

The Radiola horn consisted of two major parts, a wood base with a driver unit, and a molded hard-rubber horn consisting of a neck and flared bell. The driver unit consisted of an electromagnet and diaphragm similar to those used in ear phones. The molded horn was manufactured by the Miller Rubber Company of Akron Ohio, as were all the hard rubber horns offered by RCA. A knurled thumb wheel was located in the base to permit an adjustment of the spacing between the diaphragm and pole piece, depending on signal strength. This adjustment was used to decrease spacing for weak signals which increased sensitivity, and to increase spacing for strong signals which prevented the diaphragm from striking the pole piece causing a chattering sound.

UZ-1336 Horn Speaker Prototype

General Electric developed a number of horn speaker concepts in the 1922-1924 time frame, several of which made it only to the prototype stage for test and evaluation by RCA. Floyd Paul reported on two of these in a 1987 article in *The Old Timer's Bulletin*—UZ-1336 and UZ-1344 prototypes manufactured by GE—in which he stated that a brochure for the UZ-1336 dated April 1923 existed, but that no UZ-1336 speaker was known to exist in any collection.[20] The brochure described the UZ-1336 horn as having *"...a base similar to the UZ 1320 base with a locknut and adjusting nut arrangement for diaphragm placement. The top bell appears squattier and not as tall as the UZ-1320."*

The UZ-1320 speaker consisted of two major parts including a wood base with a driver unit, and a molded hard rubber horn consisting of a neck and flared bell.

A knurled thumb wheel was located in the base of the UZ-1320 for adjusting the spacing between the diaphragm and pole piece to compensate for signal strength.

The taller UZ-1336 horn on the left is a prototype, while the squattier UZ-1336 horn on the right appears to be a pre-production or production unit. (Richard Ransley collection) Both have identical nameplates with the RCA and GE logos.

In contrast to the large threaded thumb nut used on the production version of the UZ-1336 to secure the bell to the base, a collet with a tightening screw was integrated into the top of the prototype brass base for this purpose.

Two horn speakers bearing a GE nameplate with the UZ-1336 model number recently surfaced from private collections.[21] One has a large bell resembling that of the UZ-1320 and appears to be a prototype, while the other with a shorter squattier bell appears to be pre-production or production unit. The prototype has a flat brass base with a collet on the neck which is tightened with a screw to hold the bell. In sharp contrast, the production unit has a distinctly canted base which angles the bell neck backward and the bell opening upward, while the bell is attached to the base with the same large finger-nut arrangement used on all other GE production horns. Both have the same locknut and adjusting nut arrangement for adjusting the diaphragm.

Historian George Clark included the UZ-1336 speaker on several lists he prepared identifying apparatus that RCA actually sold during the 1923 year. However, the sale of this loudspeaker to the public cannot be independently confirmed because it did not appear in any RCA price list, catalog, or ad.

UZ-1344 Horn Speaker Prototype

The existence of a UZ-1344 horn was also reported by Paul in the same AWA article.[22] The base of the UZ-1344 is distinctly different from the other horn speakers manufactured by RCA. The large cylindrical base contains equally large horseshoe magnets and an armature mechanism which is unique among all known GE speakers. The horn is described as being of hard rubber and is about two thirds the height of the UZ-1320.

The name plate clearly identifies this speaker as a UZ-1344 made by General Electric (left). The base of the UZ-1344 is unique among all horn speakers marketed by RCA, with an appearance more like the Western Electric 10-D horn than other GE horns (right). (Collection of John L. German)

A letter from Alfred Goldsmith to E.E. Bucher dated December 22, 1922 reports on the test and evaluation of the UZ-1344 speaker, referred to in the letter as the "Hoxie Loud Speaker," obviously a reference to GE engineer Charles A. Hoxie who must have had a hand in its design.[23] The letter states that three prototypes were tested and that there were intermittent problems with all three. However, according to this letter the main reason the speaker was rejected was the high cost of its unique base. The cost of the parts in the base alone was estimated to be a whopping $27.35. Compare this to the *retail* price of the UZ-1320 speaker at $36.50 in 1923 and the UZ-1325 at $25 in 1924. The date of this letter suggests that the numbering scheme for the development of GE horn speakers did not follow chronologically. The UZ-1344 was developed and rejected by the end of 1922, while the UZ-1325 was not introduced until late 1924.

The only known prototype of the UZ-1344 horn has been in the private collection of John L. German of Olathe, Kansas for many years.

FD Horn Speaker Prototype

GE developed a little-known type FD horn speaker that was apparently never seriously considered for production. The main significance of this horn is that the existence of this GE prototype FD in conjunction with the existence of the Westinghouse FH horn indicates that RCA must have had a hand in specifying the letter designations of prototype horns. That would explain why the first letter in the Westinghouse FH speaker designation was an F instead of an L—the L designation being used by Westinghouse for their prior loudspeaker devices.

It may also mean that the FH horn, discussed later in this chapter, was originally a prototype not slated for production until the delay in the production of the UZ-1320 prompted the decision to put the FH speaker into production as a replacement. Certainly, GE dominated the R&D and production of speakers for RCA in the 1920's, and, discounting the Vocarola horn, the FH is the only speaker that was produced independently by Westinghouse for RCA.[24]

One of the prototype speakers that GE developed for RCA was designated the FD horn. (George H. Clark Collection, Smithsonian Institution)

Westinghouse Offerings

None of the Westinghouse receiver offerings for 1923 was either new or innovative. The Radiola Grand was a replacement for the Aeriola Grand with the same styling and regenerative technology, and the RS was basically a Radiola Sr. with the addition of one stage of amplification. The AR radio-frequency amplifier was a knock-off of the GE AA-1520 amplifier from the previous year, and the RT coupler, while useful, was simply an adjunct to the nearly-obsolete RC receiver. RCA had clearly reigned in Westinghouse in 1923 in order to give GE a chance to catch up on sales in the 1923 season.

Radiola Grand (RG)

As discussed in the previous chapter, the Radiola Grand was designed to replace the ill-fated Aeriola Grand. Announcements for the Radiola Grand began appearing in advertisements circa December 1922,[25] and the receivers became available circa January 1923, which was also the earliest date appearing on the instructions accompanying the set. Ads featuring the Radiola Grand at $350 together with the Radiola IV at $275 began to appear circa February 1923.[26] This receiver was on the market for just over a year until the 1924 models were introduced in February and March of 1924.

The Radiola Grand came in the same two cabinet styles as the Aeriola Grand with the geometric style also being the most common—which is not surprising because many of the Aeriola Grand cabinets were used for the Radiola Grand receivers. A tell-tale sign that an Aeriola Grand cabinet was used for the Radiola Grand is a filled hole on the inside of the lid where a bracket was once mounted in the Aeriola Grand to hold the special plug used to replace the final amplifier tube. Another sign of this conversion is the rectangular Aeriola Grand decal on the inside of the lid, which in many cases was simply covered over with a Radiola Grand decal of similar size and shape.

One type of Radiola Grand was made with a boxy look and a geometric-style grill covering the speaker opening virtually identical to the Aeriola Grand.

The Radiola Grand panel had five controls and four tubes versus the single control and four additional ballast tubes found on the Aeriola Grand (left). The right-hand tube socket in later Radiola Grand's was larger than the other three.The interior of the Radiola Grand reveals the presence of two audio transformers, replacing the coupling resistors found in the Aeriola Grand (right).

While the cabinet and general exterior appearance remained the same, the Radiola Grand differed from the Aeriola Grand in a number of ways, as articulated in a memo from RCA to wholesale distributors in August or September of 1922.[27] The most noticeable of these differences are briefly summarized in the following paragraph.

Four WD-11 dry cell tubes were used in place of the four WR-21 storage battery tubes previously used, and the four WB-800 ballast tubes were eliminated in favor of a filament control. A single-circuit tuner was used with a continuously variable tickler to allow the user to control regeneration and sensitivity. An intensity control was placed at the rear of the panel to reduce the volume for local high-power stations. An external jack was provided on the side of the cabinet to accommodate external headphones. The amplifier circuits were transformer coupled, thereby increasing audio amplification capability, and the last two tubes were designed for push-pull operation. A "C" battery was provided for grid bias, presumably to reduce distortion and increase amplification.

Even with such a drastic change in circuitry, Westinghouse and RCA kept the same RG model designation for the Radiola Grand as the Aeriola Grand. The model designation RG was prominently displayed on the serial number tag affixed to the back of the set. Serial numbers are clumped between 132,000-136,000 and 144,000-149,000, with some scattered between 122,000 and 129,000. The number of RG sets manufactured could not be accurately determined from the available serial number data because of the small number of sets sampled to date. Based on serial number data presented in Appendix D, it is estimated that as many as 11,000 sets were made. This receiver was on the market for just over a year until February 1924.

Aeriola Grand cabinets were used in the manufacture of the Radiola Grand, as can be seen from this Radiola Grand decal which was used to cover over the Aeriola Grand decal. (Sonny Clutter collection)

Radiola RS

The RS receiver was essentially a Radiola Sr. with one stage of amplification added. It was contained in a mahogany box with an appearance similar to the Radiola Sr., except with two tubes. A complete set including two WD-11 dry cell tubes, head phones and batteries sold for $85 when it went on the market in January 1923. It competed directly with the Radiola II equipped with two UV-199 tubes and equivalent accessories for $97.50. In fact, the two sets were introduced to the trade together in a July 1923 ad in *Radio Merchandising*.[28] Despite the higher price, the Radiola II outsold the RS by two to one, perhaps because of its portable mahogany case with self-contained batteries—a feature favored by the public at that time—and/or because the Radiola II was more heavily advertised than the RS. The set was discontinued at the end of the 1923 season, and according to Douglas only 4613 were sold.[29]

The Radiola RS was thinly advertised, with national ads appearing many months after it was first sold to the public. (*Radio News*, August 1923)

While the RS was manufactured at the end of 1922 and sold to the public in January of 1923, it was not included in RCA price lists until the May 1, 1923 edition, not introduced to the trade until July 1923, and not nationally advertised until August 1923. The most logical explanation for the delayed advertising of the RS (and the Radiola II) was that RCA had a glut of unsold radios that they feared would be made obsolete by the introduction of the RS and Radiola II.

The controls and binding posts on the RS panel are essentially the same as those on the Radiola Sr., the only difference being one additional binding post on the RS for battery connections. The advertised wavelengths for RS are slightly different than the Sr.: 180-375 meters when the antenna is connected to the "Short Wave" posts, and 320-550 meters for the "Long Wave" posts. A metal name tag was affixed to the inside of the lid bearing the name Radiola RS, the Style No. 358101, and the date of manufacture (10-16-22). A serial number was stamped on the name plate, each bearing a different six-digit number. The first three digits of most serial numbers observed begin with 127, 130, 131, 132,

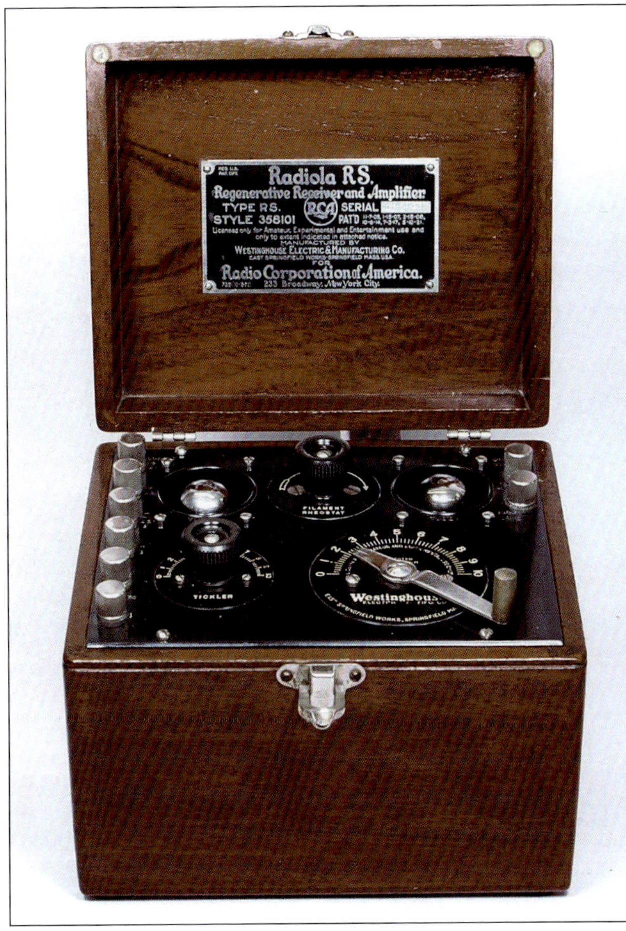

The early RS (left) was designed with small tube holes which easily accommodated the larger brass WD-11 tubes, while the later RS (right) was redesigned with larger holes for easier removal of the new shorter WD-11 Radiotrons.

135, or 136, although there are a few beginning with 111 and a few with 150. The exact number of sets cannot be inferred with any accuracy from the limited data base available, but the upper bound would be 6000, consistent with the 4613 reported by Douglas.

Like several other Westinghouse models using WD-11 tubes, the RS came in two versions, one with the small tube holes designed for the older and taller brass-based WD-11 tubes, and the other with larger diameter holes designed for easier grasping of the newer shorter WD-11 Radiotrons with bakelite bases. The version with the small holes had an enclosed wood tube socket, while the version with the larger holes had a redesigned open base. Versions with the larger holes generally have serial numbers above the 135,000 range.

The interior of the early RS with the small tube holes (left) had a significantly different appearance than the later RS with the larger tube holes (right).

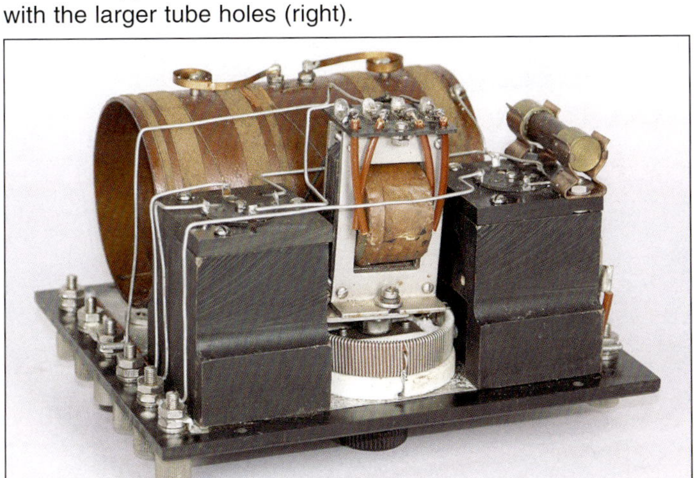

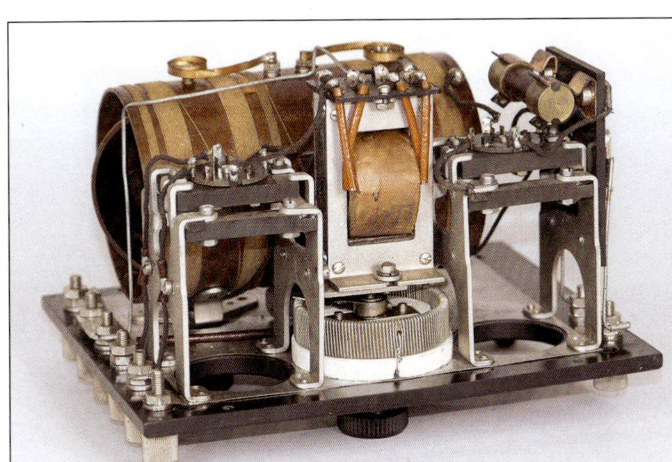

The AR (left) and the RT (right) had an appearance similar to the RA and DA, but unlike the RA and DA, they had name plates bearing the Radiola name.

Radiola AR Radio Frequency Amplifier and RT Antenna Coupler

The AR Radio Frequency Amplifier and RT Antenna Coupler were designed for use with RA, DA, or RC in various combinations to increase sensitivity and selectivity where broadcast reception over a long range was desired. The RT consisted of an adjustable series circuit with a variometer and tuning condenser similar to the circuit found in the RA. It could be used with the RA to form a two-circuit tuner to increase selectivity. The AR was a three-step untuned RF amplifier which could be used to increase sensitivity, a useful feature when used with the RT, which by its very nature introduced a decrease in sensitivity.

The interior of the AR amplifier (left) has the same general appearance as the DA amplifier, while the interior of the RT antenna coupler (right) has the same general appearance as the RA tuner.

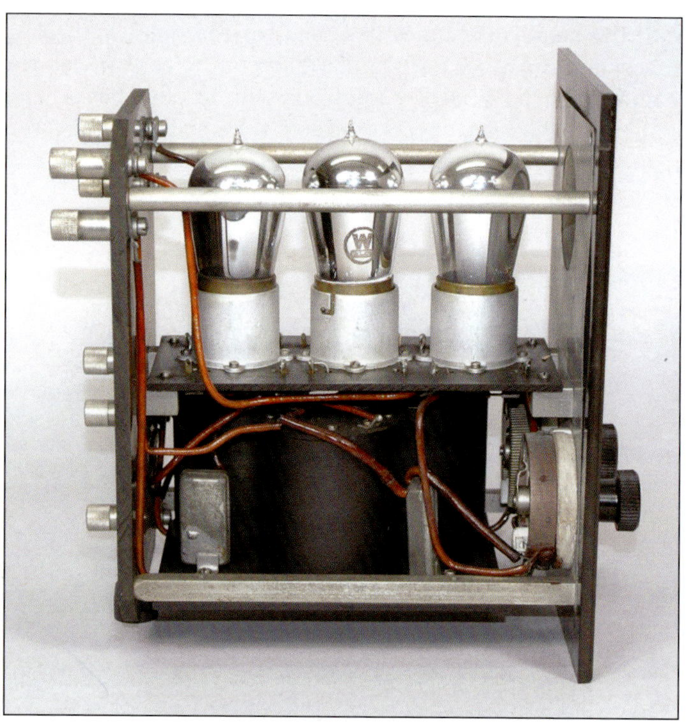

Of the five different combinations of the AR and RT recommended in conjunction with the RA/DA/RC receiver, the RT/RC combination is one of the most interesting because there is no direct electrical connection between the two units. (RT Instruction Manual)

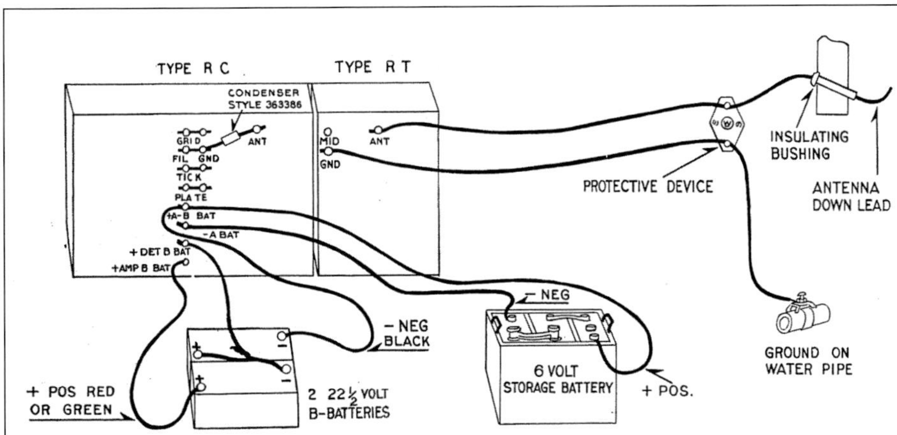

Fig. 2—Radiolas RT and RC

Five combinations were suggested by the manufacturer in the AR and RT instruction manuals including 1) the RT and RC, 2) the RT-RA-AR-DA, 3) the RT and DA, 4) the AR-RA-DA with the AG-1380 loop antenna, and 5) the RT-AR-DA with the AG-1380 loop antenna. Of course, the individual RA and DA units could be used in place of the RC unit in the first three combinations.

The first combination consisting of the RT and RC was described as the best combination for general use. The RT was essentially a tuned circuit with an antenna and ground connection, but without a direct hard-wired connection to the RC. The antenna post of the RC was connected to the ground post though a Style 363386 condenser supplied with the RT to complete the tuning circuit. Sufficient coupling was achieved by setting the RT in physical proximity to the RA unit. The coupled circuit in the RT provided additional selectivity to eliminate nearby stations, and the regenerative circuit in the RC provided good sensitivity.

The AR radio-frequency amplifier was a recommended addition to the RT/RC combination to obtain increased sensitivity where greater ranges were desired. The set had three stages of transformer-coupled radio-frequency amplification using three UV-201 tubes. This set was a direct counterpart to the GE AR-1520 radio-frequency amplifier, which no doubt provided the impetus to Westinghouse for development of the AR unit.

The AR and RT were nationally advertised only in conjunction with the RA/DA/RC sets. (*The Wireless Age*, April 1923)

The four units consisting of the RT/RA/AR/DA when used in combination were derisively known as the "freight train," and the UZ-1320 horn speaker when used with this combination was known as the "caboose."

The RT and DA were mentioned as a possible combination, but lacking regeneration, this combination had a very limited range. The last two configurations used a loop such as the AG-1380 to make use of its directional properties—to minimize the pickup from a strong nearby station, for example—but at the sacrifice of range.

The RT-RA-AR-DA combination (in that order from left to right) is the one that was pictured most often in ads and catalogs of that era including the RCA catalog. George Clark recounts that this particular combination was described by RCA insiders as the "freight train," and when a horn speaker accompanied the four boxes, it was referred to as the "caboose."[30] It is amusing to note that in George Clark's account, he recalls: *"The General Electric Company countered with a similar agglomoration [sic], using its AR 1300, AA-1400, and AA-1520."* However, he has it just reversed—it was Westinghouse who countered GE the following year by introducing the AR and RT.

The instructions recommended that the boxes be positioned in the order shown with the RA and AR in the center, but most listeners had purchased the RC, making it somewhat difficult to achieve this configuration. The instructions went on to say that the listener could remove the AR and DA from their respective cabinets and switch them, which required that the bakelite strips on the rear be removed and switched as well, not to mention the need to drill additional holes in the rear of the AR cabinet to accommodate the more numerous connections on the rear of the DA unit.

The AR and RT were put on the market at the beginning of the 1923 model year, with the AR priced at $80 without tubes or batteries, and the RT at $45. All four units were advertised in April 1923 at $263 without tubes, batteries, antennas, or headphones, making it one of the most expensive sets that RCA marketed in 1923. Not surprisingly, there were very few AR and RT models sold. The serial number evidence suggests there were perhaps as many as 2,500 of each made. The AR units have six-digit serial numbers, most of which begin with 122 or 123, while the RT units have six-digit serial numbers, most of which begin with either 132, 148 or 149, although other numbers have surfaced. The AR has the Style Number 319518, while the RT has the Style Number 359970. Dates on the serial number plates are (10-4-22) for the AR and both (2-11-22) and (10-16-22) for the RT, although there is strong evidence that these sets did not go into production until January, 1923. The AR and RT remained on the market—even after the July clearance sale of all companion RC, RA, and DA sets—most likely just to clear out unsold inventory. These sets were dropped from RCA price sheets when the 1924 models were introduced.

The FH horn speaker generally appeared in ads for the Radiola VII, but was rarely if ever identified or described. (*The Wireless Age*, November 1923)

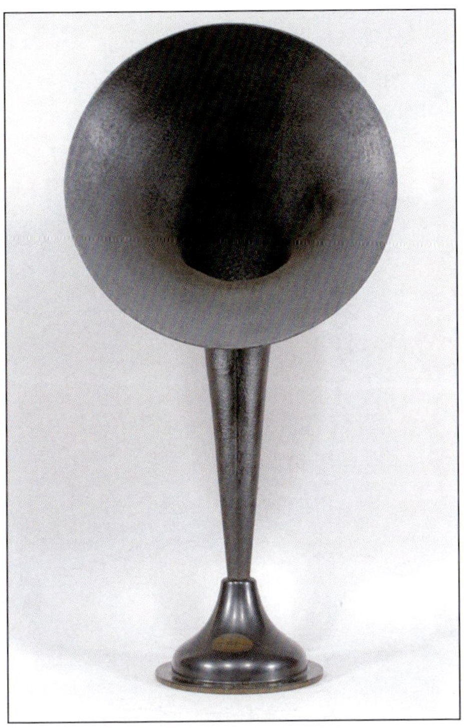

The Westinghouse FH speaker was approximately 2½ inches larger in bell diameter and 4 inches taller than the respective diameter and height of the GE UZ-1320 speaker.

The FH horn speaker can be easily distinguished from GE horns by the rounded shape of the base and the absence of a thumb wheel to adjust to control the driver spacing.

FH Horn Speaker

According to an RCA memo to its distributors, the FH horn was distributed on September 25, 1923 in conjunction with the Radiola VII receiver.[31] It was a last minute replacement for the GE UZ-1320 which was supposed to be distributed with the Radiola VII, but design or production problems delayed the GE speaker to the point where the FH was substituted. The FH horn first appeared in the September 15, 1923 RCA Distributor Price List at $36.50, matching the price of GE's UZ-1320 horn when it finally came on the market in November. Clearly, the two price points were selected so that neither horn would have a price advantage. The earliest ads showing the FH speaker with the Radiola VII receiver began to appear in national magazines circa November 1923, coinciding with the time that the GE UZ-1320 was finally introduced.[32] The FH speaker was rarely if ever advertised by itself—and it was generally not identified or described when pictured in Radiola VII ads. It may well be that if the GE UZ-1320 had been available for the Radiola VII in a timely fashion, the FH may never have been put on the market.

This speaker was not given the typical two-letter designation beginning with a UZ that one might have expected from RCA, nor was it given a two-letter designation beginning with L that might have been expected if it were named by Westinghouse. It appears that RCA assigned the two letter designation beginning with an F and not Westinghouse, because at least one other speaker manufactured by GE was given a similar designation beginning with F, namely the FD horn (see page 188). Perhaps RCA gave the two-letter designation beginning with F to speakers that were being developed for research purposes, but were not seriously being considered for distribution.

The FH horn was made of hard rubber manufactured by the Miller Rubber Company, as were all the upright horns manufactured for RCA. At least one horn made of fiber has been observed, but this is the exception, and is believed to be either a prototype or an early production model.[33] The base is made of metal, but in contrast to the GE horns, it was designed so that no adjustment for the spacing between the pole piece and diaphragm was required.

The FH speaker was on the market for about a year until circa February 1925, shortly after the new UZ-1325 speaker was introduced at $25. The exact number of FH speakers manufactured is not known, but the evidence suggests that there were fewer FH speakers manufactured than the UZ-1320, and certainly far fewer speakers than the more popular UZ-1325 appearing in 1924 priced 30% lower.

Wireless Specialty Offerings

Wireless Specialty was never a major player in the manufacturing of broadcast receivers for distribution by RCA, and 1923 was no exception. Aside from the Radiola Special—a last minute conversion of the AR-1375 crystal set to a one tube regenerative receiver—WSA was commissioned to manufacture only the Radiola VII, a relatively innovative but hopelessly complex receiver for the average man on the street. While almost 10,000 Radiola Specials were sold, very few of the Radiola VII's were sold. WSA was also commissioned to manufacture two inexpensive crystal detectors that competed with the more expensive Westinghouse DB detector.

Radiola Special

RCA planning documents as late as January 1923 indicated that the Wireless Specialty AA-1375 crystal set was to be held over for the 1923 season, most likely because there was a large unsold inventory from the previous year.[34] Early in the year, the decision was made to drop the poorly selling AR-1375 and

The Radiola Special was manufactured with both a wood case and a metal case like the one pictured here.

The AR-1375 ancestry of the Radiola Special can be seen by comparing the Radiola Special panel with the AR-1375 panel which it cleverly covered.

The interior of the Radiola Special reveals how WSA cleverly converted the AR-1375 to a regenerative receiver with the addition of a single UV-199 tube in the upper portion of the set where the earphones were originally to be stored.

retrofit the unsold inventory with a single UV-199 tube to convert the set into a more sensitive regenerative receiver. The crystal detector was removed, a new front panel was added to cover the existing panel, and an instruction sheet was attached to the inside of the front lid. The front panel can be easily removed from the Radiola Special to reveal the true ancestry of this set.

The UV-199 tube was incorporated into the top compartment which was originally intended to store the headphones. The wiring was changed significantly, but many of the original components were retained—including the relatively fragile variometer which is invariably found with its bakelite-type frame fractured in several places. Many of the sets found by collectors today have been modified in some way, so it is difficult for the collector to find a set in its original state.

While this set never appeared in national ads or in RCA distributor price lists, the price can be inferred from several ads appearing in New York City newspapers. An early ad for the Special appeared in *The New York Globe* on June 2, 1923 with a price of $62.50 complete with headphones and a UV-199 vacuum tube.[35] Wireless Specialty documents from the *Radioana Collection* indicate that the set was also sold with an attractive mahogany case and gold lettering on the front panel for an additional $10 or so.

At some point, a two-stage amplifier using two UV-199 tubes was offered specifically for the Special, presumably so that it could drive a loudspeaker. The amplifier was enclosed in a green wood box constructed so that it would fit on the top of the Radiola Special case when the sliding metal lid was removed. It is not known to this author who manufactured this amplifier, but they appear from time to time at swap meets and auctions.

The Radiola Special manufactured with a wood box had an ornate control panel with gold-plated lettering and hardware. (Stew Oliver collection)

Unknown to most collectors, serial numbers were stamped into the top edge of the front panel of the Radiola Special (and/or the AR-1375) which have been observed to be as high as 9895—indicating that perhaps as many as 10,000 Radiola Specials (and/or AR-1375's) were manufactured.

Gimbel Brothers placed an ad in *The New York Times* on November 2, 1923 for a 3-tube "Radiola Special Loud-Speaking Receiving Set" with a special amplifier and a Perfectone loudspeaker for $79.[36] The 3-tube set referred to in the ad was very likely this 2-tube amplifier mounted on top of the 1-tube Special. The outfit was obviously offered "at one extremely low price" to clear out the Special inventory before the new 1924 season models arrived.

Oddly, the Special was never given an RCA designation number such as an "AR" like the other broadcast receivers made by Wireless Specialty for RCA. Because it did not appear in national ads or distributor price lists, it would appear that RCA did not consider the Radiola Special as a serious entry into the market—it was more of an afterthought to liquidate an otherwise stagnant inventory. The trick was to move the inventory without negatively affecting sales of competing GE and Westinghouse models. The Special competed directly with the Westinghouse one-tube Radiola Sr., priced at $65.00, also equipped with a WD-11 vacuum tube and headphones. It is not known how many of these sets were sold, but judging from serial number data and the number of Specials on the market today, it is estimated that about 10,000 were made, far more than the number of AA-1375 sets sold. The set was on the market from circa March 1923 to the end of the year, although it began to appear in discount stores towards the end of the year.

Radiola VII (AR-905)

The Radiola VII manufactured by Wireless Specialty was originally planned for the 1923 season, but for some reason it was not introduced with the rest of the line in December 1922, nor did it appear in the first edition of the 1923 RCA catalog. Judging from the July 1923 date on the Radiola VII instruction manual, RCA planned to introduce this receiver in the middle of the year, but according to an RCA memo to wholesale distributors dated September 17, 1923,[37] the radio was to be introduced September 25, 1923 long after the original plan. The first ads in magazines did not appear until November 1923,[38] almost at the end of the planned 1923 season. The receiver was sold complete with the FH speaker, tubes, socket antenna, and batteries for $290. Clearly, there must have been some design or manufacturing problems. One possibility is the delay of the GE horn speaker that was to accompany the set. As stated previously, RCA originally planned to provide the GE UZ-1320 speaker with the Radiola VII, but GE could not deliver this speaker on

An unknown manufacturer made an audio amplifier for the Radiola Special to provide additional amplification to drive a speaker by enclosing two UV-199 tubes and associated circuitry in a small green box that fit on top of the Special in place of the lid.

The Radiola VII was contained in a rectangular wood box with eight controls including two vernier tuning dials.

schedule. Perhaps RCA delayed the release of the Radiola VII until September, waiting for delivery of the UZ-1320, and then decided to release it with the FH speaker instead.

This same RCA memo also refers to another problem which presaged the early demise of the Radiola VII: *"It has been our observation that a certain class of customers often call for a receiver which provides a little more than the ordinary degree of selectivity when the set is to be installed close to one or more broadcasting stations. Such receivers require a little more skill to operate than those in more popular demand... However to sense the market demand for something in the nature of a set which requires a little more practice on the part of a customer than the ordinary type of receiver, the Radio Corporation has produced a very limited quantity of the Radiola VII."*[39] The cynic would interpret this memo to say that RCA produced a set which was very difficult to operate, and fully realizing this only after production began, RCA decided to manufacture only a small number sets and move on to something else. Indeed, there were eight separate controls and switches on the front panel, including two vernier dials, and the operating instructions in the manual were amusing if not bewildering.

The Radiola VII was a new type of receiver for RCA using five UV-199 dry cell tubes with two stages of transformer-coupled RF amplification, a detector stage, and two stages of audio amplification. The two stages of RF amplification could be tuned to four pre-selected ranges of wavelengths by a selection switch on the front of the set which was connected to wires wrapped around the core of the two transformers. Each particular arrangement of these wires—controlled by the switch on the front panel—affected the resonant frequency of the RF transformer circuit, thereby allowing the user to tune the RF amplifier stages. The operating range covered the entire broadcasting band of 220 to 550 meters, and the set could be used with either an exterior antenna or an interior loop antenna.

The two RF transformers appear as black cylinders in the upper center portion of the accompanying photograph of the Radiola VII interior. The panel switch controlling the wires leading to the RF transformer core is directly behind the transformer on the right. The large empty volume appearing in this photograph was originally occupied by six "A" batteries, two "B" batteries, and two "C" batteries which have since been removed.

In yet another memo to distributors dated January 10, 1924, RCA announced unfilled orders for the Radiola VII would be filled with an improved Radiola VII-B in early January.[40] Thus, the Radiola VII was only on the market for about three months. The number of sets manufactured can be estimated from an RCA memo to distributors stating that ten sets were available for each

The two tunable RF transformers unique to the Radiola VII appear as black cylinders on the top of the panel in the center right portion of the photograph.

RCA distributor. According to the 1923 RCA Annual Report, there were 173 RCA distribution locations, so there must have been about 1730 sets manufactured. This figure agrees well with the 1692 sets cited by Douglas.[41]

Radiola VII-A

There is no reference to a Radiola VII-A in any RCA price sheets, catalogs or RCA ads, but such a radio was pictured in the 1923-1924 Pettingell-Andrews catalog.[42] It appears to be nothing more than a Radiola VII installed in a console cabinet, almost certainly by the distributor rather than by Wireless Specialty at the factory. The price was to be provided upon request, and the distributor most likely would have made it to order. Since each distributor was allotted only ten sets, it is unlikely that many of these console sets were actually made up and sold.

AG-1385 Rotating Loop Antenna

The original AG-1380 loop antenna introduced in June of 1922 had a fixed base and therefore was awkward to reorient, requiring the user to raise it off the floor to rotate it. A year later the antenna was modified by adding a rotating mechanism to the base, and given the new designation of AG-1385. It first appeared in the RCA price list dated September 15, 1923 at a price of $24, four dollars more than the AG-1380 that sold for $20. Both loops were on the market for several years, and both were discontinued by April 15, 1924. The rotating base of the AG-1385 is pictured on page 164 of Chapter 4 along with the AG-1380 antenna. Many collectors and historians seem to be unaware of the existence of the AG-1385.

Cleartone Model UD-1430 and UD-1432 Crystal Detectors

In May of 1923, RCA introduced two Cleartone detectors manufactured by Wireless Specialty Apparatus that were less expensive than the DB detector manufactured by Westinghouse. The Cleartone Model UD-1430 detector was enclosed in a protective glass cylinder retailing for $2.30, while the Model UD-1432 shown here was an open type retailing for $1.60. These detectors were all discontinued at the end of the 1923 season, marking the end of the crystal set era for RCA broadcast receivers.

Radiola VII-A

The Radiola VII-A consisted of a Radiola VII receiver which was installed in a console by the distributor with a price provided upon request. (1923-24 Pettingell-Andrews catalog)

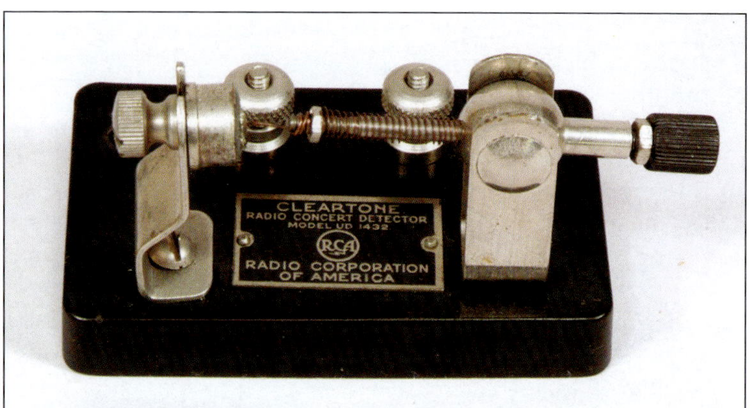

Wireless Specialty manufactured two inexpensive Cleartone crystal detectors for RCA in 1923, one a UD-1430 with a glass tube to protect the detector, and the other a UD-1432 without a glass tube as shown here to the left.

Rebroadcasting–Origins of Network Radio

By 1923, there were over five hundred broadcast radio stations, most serving the larger metropolitan areas. In a very short time, two problems became evident. First, only about half the population in the U.S. lived within reliable receiving range of these stations using the average receiver of the day, while the other half had to buy sophisticated and expensive sets to receive distant stations, if they could receive any stations at all. Second, much of the programming and entertainment of the day was live, which meant that owners of multiple stations–like Westinghouse–had to replicate live programming in each city with the attendant additional cost.

The obvious solution to both problems was to form a network whereby programming could be generated at one or a few locations and distributed through a radio network. This solution became evident to at least two influential people, apparently quite independently. Editor Hugo Gernsback appears to be the first to publicly articulate such a concept in an editorial in the December 1922 issue of *Radio News*, in which he suggested that amateur radio operators rebroadcast programming from distant stations.[43] However, a few months before, Westinghouse Vice President H. P. Davis had articulated among Westinghouse executives the idea of a few centrally located broadcast stations that would serve many low-powered and non-interfering repeating stations.[44]

Westinghouse initially considered two approaches to distributing the signal from a central location—one using telephone lines and one using short wave radio. However, RCA and Westinghouse were having a dispute with AT&T over rights to manufacture and distribute broadcast radio equipment, and so distribution via telephone lines was not an attractive option at that time. Further, AT&T was already performing an historic network experiment with long distance telephone connections in early 1923 using their broadcast station WEAF in New York and John Shepard's station WNAC in Boston.[45] As a result, Westinghouse turned to short wave radio, a technology with which Frank Conrad had been experimenting. Conrad had demonstrated that short waves in the 80-100 meter range could provide a more reliable and economical means of communication between distant points than the longer waves above 360 assigned to the broadcast band. The shorter waves also had the advantage of avoiding interference in the broadcast band such as the chirps produced by the numerous regenerative receivers in operation at that time.

In early 1923, Westinghouse began to experiment with short wave relaying, first using the experimental station KDPM in Cleveland and later using experimental station KFKX in Hastings, Nebraska. Signals were transmitted from KDKA in Pittsburgh and WBZ in East Springfield, Mass. on both the normal broadcast band and at 100 meters. These experiments were so successful that Westinghouse officially opened KFKX on November 21, 1923 as the first radio relay station in the world. After the first program was received from KDKA at 100 meters and re-transmitted in the broadcast band from the Hastings station, Westinghouse received letters from all over the country telling of the wonderful results.

A description of the Hastings station was reported in the March 1924 issue of *Radio News*, along with photographs of the interior and exterior of the station.[46] Of particular interest is a photograph of the short-wave receiver

The Westinghouse rebroadcast receivers consisted of a regenerative receiver and a separate amplifier unit based on Radiola Grand technology.

The rebroadcast receiver consists of one stage of untuned RF amplification and a regenerative detector/amplifier, the output of which feeds an audio transformer connected to the output terminals (left). This signal is fed into the audio amplifier unit which contains two stages of audio amplification—the first stage consisting of a single tube and transformer, and the other unusual stage consisting of four tubes and two large audio transformers in a parallel push-pull configuration capable of driving a broadcast transmitter, telephone lines, or a speaker (right).

developed by Westinghouse to receive signals in the 100 meter band from the KDKA station in Pittsburgh. This receiver consisted of two cabinets, one a regenerative receiver unit with an untuned RF amplifier and a regenerative detector, and the other an amplifier unit consisting of one stage of audio amplification followed by an audio output stage consisting of two push-pull stages wired in parallel. The output stage was capable of driving a transmitter, telephone lines, or a speaker. The panel of this short-wave receiver has a strong resemblance to the Radiola Grand—and in fact it used the same technology and many of the same components—but it operated in the 100 meter band. The story about how these receivers were discovered in 1976 in the attic of a Heinz official is recounted by Alan Douglas.[47]

In the end, RCA and AT&T settled their differences. AT&T agreed to abandon broadcast radio, and the RCA radio group set up radio networks. However, the technology of choice for distributing the programming was AT&T's telephone lines and not short-wave radio. The story of how network radio arose in mid-1920 is recounted in some detail by Gleason Archer.[48]

A 1923 Season Retrospective

The 1922 season had been dominated by sales of Westinghouse apparatus, and as a result, General Electric was way behind on the agreed-upon 60%-40% split in revenues between the two companies at the end of 1922. RCA made a concerted effort in 1923 to meet their commitment by contracting with GE to manufacture more new receivers than Westinghouse, and by preferentially advertising GE apparatus at the expense of Westinghouse equipment. This approach succeeded, by and large, because revenues from the sales of GE equipment dominated revenues from Westinghouse equipment in virtually all competing categories.

At the low end, GE's Radiola I crystal set was the only one marketed by RCA during most of the season. The Westinghouse Radiola Jr. (the Aeriola Jr. had been renamed Radiola Jr. in the fall of 1922) had been removed from the market before the season got underway, and the WSA AR-1375 was discontinued a few months after the season began. In the lower-priced regenerative receiver category, GE's Radiola II outsold the competing Westinghouse RS receiver by a margin of more than two to one. In the higher-priced regenerative receiver category, the Radiola V outsold Westinghouse's RC/RT/AR by more than two to one—neglecting the large number of RC's sold well below cost on closeout. At the high end, the Radiola IV also outsold the Radiola Grand by more than two to one—assuming that about half the Radiola Grand's made were used to replace Aeriola Grand's at no charge.

Endnotes and References for Chapter 5

1 *Radio Merchandising,* Vol. 2, No. 5, February 1923, pp. 36-37.

2 *Saturday Evening Post,* February 3, 1923, pp. 90-91.

3 "Specifications for Broadcast Receivers for 1924," RCA Research Department Report, January 16, 1923 (kindly provided from the private collection of John Terrey). Excerpts of this report were reproduced in: John P. Wolkonowicz with Alan Douglas, John Terrey and Richard G. Wolven, "RCA's Intended Models for 1923-1924 Season 'What Might Have Been,'" *Antique Radio Classified*, Vol. 16, No. 6, June 1999, pp. 4-11.

4 *Radioana*, "Minutes of General Sales Committee Held June 26, 1923," Vol. 1, No. 18, CWC5 2021A, p. 2.

5 *Radio Merchandising*, Vol. 2, No. 10, July 1923, pp. 36-37.

6 *Radioana*, Memo to RCA General Sales Manager E. E. Bucher and B. H. Clark from an RCA salesman dated January 25, 1923 summarizing the results of a three-day show in January 1923 at Bamberger's Department store in New York City.

7 William C. White, *"The Early History of Electronics in the General Electric Company*," (William C. White Papers Collection, Schenectady Museum & Suits-Bueche Planetarium, Schenectady, New York, prepared 1953-1955), Chapter VI, p. VI-9.

8 *Radioana*, Memo to RCA General Sales Manager E. E. Bucher and B. H. Clark from an RCA salesman dated January 25, 1923 summarizing the results of a three-day show in January 1923 at Bamberger's Department store in New York City.

9 *The Wireless Age*, Vol. XI, No. 3, December 1923, p. 15.

10 *Radioana*, Memo to RCA General Sales Manager E. E. Bucher and B. H. Clark from an RCA salesman dated January 25, 1923 summarizing the results of a three-day show in January 1923 at Bamberger's Department store in New York City, p. 2.

11 *The Wireless Age*, Vol. 10, No. 7, April, 1923, p. 37.

12 Memo from W. M. Derrick to A. D. McKenzie and P. H. Boucheron regarding assignment of model designation, July 25, 1922 (kindly provided by Alan Douglas).

13 *The Wireless Age*, Vol. 10, No. 5, February, 1923, p. 1.

14 *RCA Victor Service Notes for 1923-1928*, (RCA Victor Company, Camden New Jersey, no date), Radiola V, Model AR-885, p. 5.

15 *The National Geographic Magazine*, Vol. XLV, No. 10, October, 1923.

16 *A Catalog of Radio Apparatus*, (Pettingell-Andrews Co., Boston, November 1923) p. 8.

17 *Radio Supplies and Equipment*, (Catalog by Erner and Hopkins Company, Columbus, Ohio 1922-1923) p. 7.

18 *Radioana*, Quinton Adams, District Sales Manager, "Memo to Wholesale Distributors RCA Eastern District; Subject: - Instruction Books Radiola VII," RCA Sales Department, October 9, 1923.

19 *Radiola VII Manual*, Wireless Specialty Apparatus Co., July, 1923, p. 4.

20 Floyd Paul, "Radio Horn Speakers that RCA Didn't Make," *The Old Timer's Bulletin*, Vol. 28, No. 3, Nov. 1987, pp. 24-25.

21 The prototype was from the Hassleman Collection auctioned at Estes Auctions, Seville, Ohio, on January 17, 2004; the production unit was held for a long time in a private estate until 2003.

22 Floyd Paul, "Radio Horn Speakers that RCA Didn't Make," *The Old Timer's Bulletin*, Vol. 28, No. 3, Nov. 1987, pp. 24-25.

23 *Radioana*, A. Goldsmith, letter from RCA Research Department to E. E. Bucher, RCA Sales Department, re: General Electric Company Loud Speaker UZ-1344, December 22, 1922.

24 Beginning in 1927, Westinghouse may have produced speakers designed by GE as apart of an integrated production plan.

25 *The Wireless Age*, Vol. 10, No. 3, December 1922, p. 92.

26 *The Wireless Age*, Vol. 10, No. 5, February 1923, p. 1.

27 "Recently Discovered," *Radio Age*, Vol. 18, No. 7, July 1993, p. 7. (This article contains only the first page of a five-page undated memo from the manager of the RCA sales department to wholesale distributors.)

28 *Radio Merchandising*, Vol. 2, No. 10, July 1923, pp. 36-37.

29 Alan Douglas, *Radio Manufacturers of the 1920's, Vol. 3 - RCA to Zenith* (Vestal Press, Ltd., Vestal, New York, 1991) p. 11.

30 *Radioana,* George H. Clark, SRM Class 45, Book No. 13, November 13, 1947, p. 16 (item 15).

31 *Radioana*, E. E. Bucher, Memo to Wholesale Distributors of Radio Corporation of America, Subject: Radiola VII, September 17, 1923.

32 *The Wireless Age*, Vol. 11, No. 2, November 1923, p. 83.

33 Floyd Paul, "RCA Horn Speakers," *The Old Timer's Bulletin*, Vol. 30, No. 3, Nov. 1989, p. 37.

34 *"Specifications for Broadcast Receivers for 1924,"* RCA Research Department Report, January 16, 1923 (Excerpts of this report were reproduced in the *Antique Radio Classified*, Vol. 16, No. 6, June 1999, pp. 4-11).

35 Alan Douglas, *Radio Manufactures of the 1920's*, p. 26.

36 *The New York Times*, November 9, 1923, p.7.

37 *Radioana*, E. E. Bucher, Memo to Wholesale Distributors of Radio Corporation of America, Subject: Radiola VII, September 17, 1923.

38 *The Wireless Age*, Vol. 11, No. 2, November 1923, p. 83.

39 *Radioana*, E. E. Bucher, Memo to Wholesale Distributors of Radio Corporation of America, Subject: Radiola VII-B, January 10, 1924

40 Ibid.

41 Alan Douglas, *Radio Manufacturers of the 1920's*, p. 27.

42 *A Catalog of Radio Apparatus*, (Pettingell-Andrews Co., Boston, November 1923) p. 11.

43 Hugo Gernsback, "Popularizing Radio [-] A Double Barreled Scheme," *Radio News*, Vol. 4, No. 6, December 1922, p. 1049.

44 "Re-Broadcasting, a New Era in Radio," *Radio News*, Vol. 5, No. 9, March 1924, p. 1242.

45 Gleason L. Archer, *History of Radio to 1926*, (The American Historical Society, Inc., New York, 1938) pp. 284-287.

46 *Radio News*, Vol. 5, No. 9, March 1924, p. 1242.

47 Alan Douglas, *Radio Manufacturers of the 1920's*, p. 34; Richard Elskamp and Alan Douglas, "Westinghouse Short-Wave Broadcasting," *The Old Timers Bulletin*, Vol. 18, No. 4, March 1978, pp. 18-19.

48 Gleason L. Archer, *Big Business and Radio*, (The American Historical Society, Inc., New York, 1939) Chapters 12 and 13.

Radiola
Super-Heterodyne
—gets distance with no antenna—no ground wires—no connections. Once mark the stations on the dials, and you can turn to the marked spots and tune in—clear—true—real!
Operates entirely on dry batteries.
With 6 Radiotrons UV-199 and Radiola Loudspeaker. With compartments to hold the batteries. Entirely complete except batteries, $269
Extra Radiola Loop, Model AG-814, $12
This symbol of quality RCA is your protection
The "Super-Het" brings home the fun!
Where's the fun today? At home, wherever there's a "Super-Het!" City home and farm home hear the same good music—laugh at the same good comedy—thrill to the same big games!
At the family dinner table, there's the dinner music of the greatest hotels. For home parties, famous orchestras play their best dance music. So many of the things that folks once went out of the home to get, come into the home now—with a Radiola Super-Heterodyne.
"There's a Radiola for every purse"
RADIO CORPORATION OF AMERICA
Sales Offices: Suite No. 191
233 Broadway, New York
10 So. La Salle St., Chicago, Ill.
28 Geary St., San Francisco, Cal.
Radiolas manufactured and distributed in Canada by the Canadian General Electric Company, Limited, Toronto, and the Westinghouse Company, Limited, Hamilton, Ont.
Radiola
REG. U. S. PAT. OFF.

Chapter 6

Super-Hets are all the Rage

1924-1925

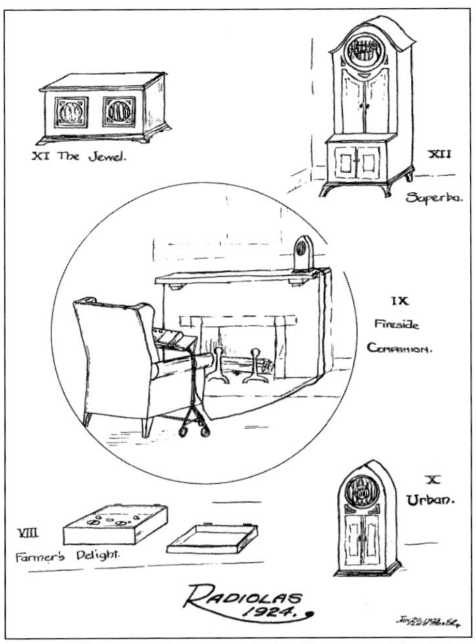

The RCA receivers originally specified for the 1924 season were abruptly cancelled in early 1923 to make way for a new line featuring the improved Armstrong Superheterodyne. (John Terrey Collection)

The 1924 season was most unusual because the entire line of Radiolas planned for 1924 was cancelled only two days after RCA placed a $5 million order for the line early in 1923. RCA had originally planned to introduce five new Radiolas in the 1924 season bearing model numbers Radiola VIII through Radiola XII. These cancelled Radiolas were described in an RCA planning document dated January 16, 1923[1], a summary of which appeared in an *Antique Radio Classified* article in 1999.[2] Sketches of the cancelled Radiolas are pictured to the left, and the general characteristics of each are summarized in Table 6-1. It is notable that all sets were to use either the new dry-cell UV-199 tubes, or UV-201A tubes powered by the lighting circuit.

Of particular interest is the top-of-the-line Radiola XII, or Superba, which according to the planning document was to be similar to a prototype that had been developed by the RCA Research Department during 1922. A photograph of a set believed to be that prototype was found in the *Radioana Collection* and is reproduced here. There is no other reference in the document to the existence of other prototypes.

The cancellation of the 1924 line was prompted by Edwin H. Armstrong's demonstration of his improved superheterodyne receiver to RCA general Manager David Sarnoff and RCA Chairman Owen D. Young. By early 1923, Armstrong and his colleague Harry Houck had built the improved superheterodyne model using the "second harmonic principle," which reduced the number of tubes required, and allowed him to build a set which was complete with a built-in loop antenna, loudspeaker, and batteries. Houck had developed the second-harmonic oscillator in which the first detector and oscillator

Table 6-1. Radiolas originally planned for the 1924 season

Radiola	Cabinet	Tubes	Circuit	Speaker	Cost to RCA[a]
VIII (Farmer's Delight)	Horizontal panel; wood box and lid	2 UV-199	Similar to II or RS	External	$25-$30
IX (Fireside Companion)	Horizontal panel; shallow wood box with cover	3 UV-199	Similar to IV or Grand	External	$35
X (Urban)	Vertical panel with door and grill	3 UV-199	Same as IX	Internal	$45-$50
XI (The Jewel)	Similar to IV	4-6 UV-199	Superior to IV and Grand	Internal	$70-$80
XII (Superba)	Floor model; vertical panel which could fit other custom cabinets	5-8 UV-199 or UV-201A[b]	Superior to XI	Internal	$125-$140

[a] The retail price would be slightly more than twice the cost to RCA

[b] Designs with UV-201A tubes were to be powered by the AC lighting circuit

functions are combined in a single tube (e.g., self-heterodyning), thereby increasing stability, reducing the number of tuning controls to a manageable two, and drastically reducing radiation.

Historian George Clark describes what happened next in a September 10, 1937 document entitled "The Story of the Superheterodyne:"[3]

> *"Mr. Sarnoff was tremendously impressed, and suggested that the set be shown to Mr. O. D. Young, then chairman of the Board of RCA, whose apartment was not far away. So to his apartment they went, Armstrong carrying the set under his arm, playing grand opera as he walked along."*

> *"Mr. Young in his turn was greatly impressed by the possibilities of the new receiver. So he and Mr. Sarnoff decided, on the basis of this one test, to make a radical change in RCA's broadcast receiver plans. Two days before the historic event, Mr. Sarnoff had placed an order for five million dollars worth of receivers, of a simpler type, but at once cancelled these and substituted the superheterodyne type."*

In a presentation to the Institute of Radio Engineers the following year, Edwin Armstrong identified the demonstration model he showed to RCA management.[4] An image of his demonstration set from this presentation is reproduced here. This set measured 18 x 10 x 10 inches, and was completely self-contained including batteries, loop antenna, and a driver for the horn speaker which protruded out from the top of the cabinet. This five-tube demonstration model should not be confused with a widely-publicized six-tube model—which, with an additional amplifier stage for improved performance, became the prototype for the production unit.[5]

A prototype of the Radiola XII Superba pictured in the original 1924 lineup document had been developed and photographed by October 1922. (George H. Clark Collection, Smithsonian Institution)

It is surprising that Sarnoff and Young would make a decision to cancel a multi-million dollar order for the entire 1924 line based on a single observation of Armstrong's superheterodyne prototype in Young's apartment one evening. Nevertheless, by all accounts that appears to be the case. Armstrong's five-tube prototype superheterodyne receiver was immediately sent to the RCA Research Department for test and evaluation circa February 1923 where it was tested against several receivers—including a special standard superheterodyne receiver built by the RCA Research Department and several superheterodyne models built by Westinghouse and GE. RCA Research Department Director Alfred Goldsmith reported the preliminary test results back to David Sarnoff in a letter dated March 29, 1923:[6]

> *"Without going into details, it may be stated broadly that this receiver can be made as selective as the best (that is, the super-heterodyne), is of extreme sensitiveness, thereby enabling long distance reception on small loops, if desired, does not radiate appreciably (thereby avoiding "birdies"), and that it is probably possible to build it in a very easily manipulable form.*

> *In other words, it does everything which the most elaborate present-day receivers do but with a distinct economy in tubes. While it is true all the component elements (with the possible exception of the Houck contribution) were known to our engineers, and while it is also true that we were very close to the device itself, the fact remains that Mr. Armstrong's contribution has been not only extremely timely but very helpful from a technical point of view and that he deserves real credit and recognition therefor [sic]."*

This superheterodyne receiver is the one that Armstrong used to convince David Sarnoff and RCA Chairman Owen Young to cancel the receivers ordered for 1924 and replace them with a line featuring the Super-Heterodyne. (*Proc. IRE*, October 1924, p. 548)

The 1924-1925 Radiola Line

This abrupt change of plans sent the three RCA manufacturers back to the drawing board to design an entirely new line of receivers for 1924. With such a late start, the three manufacturers had the daunting task of providing the new line in time for the 1923 Christmas season. GE, with responsibility for

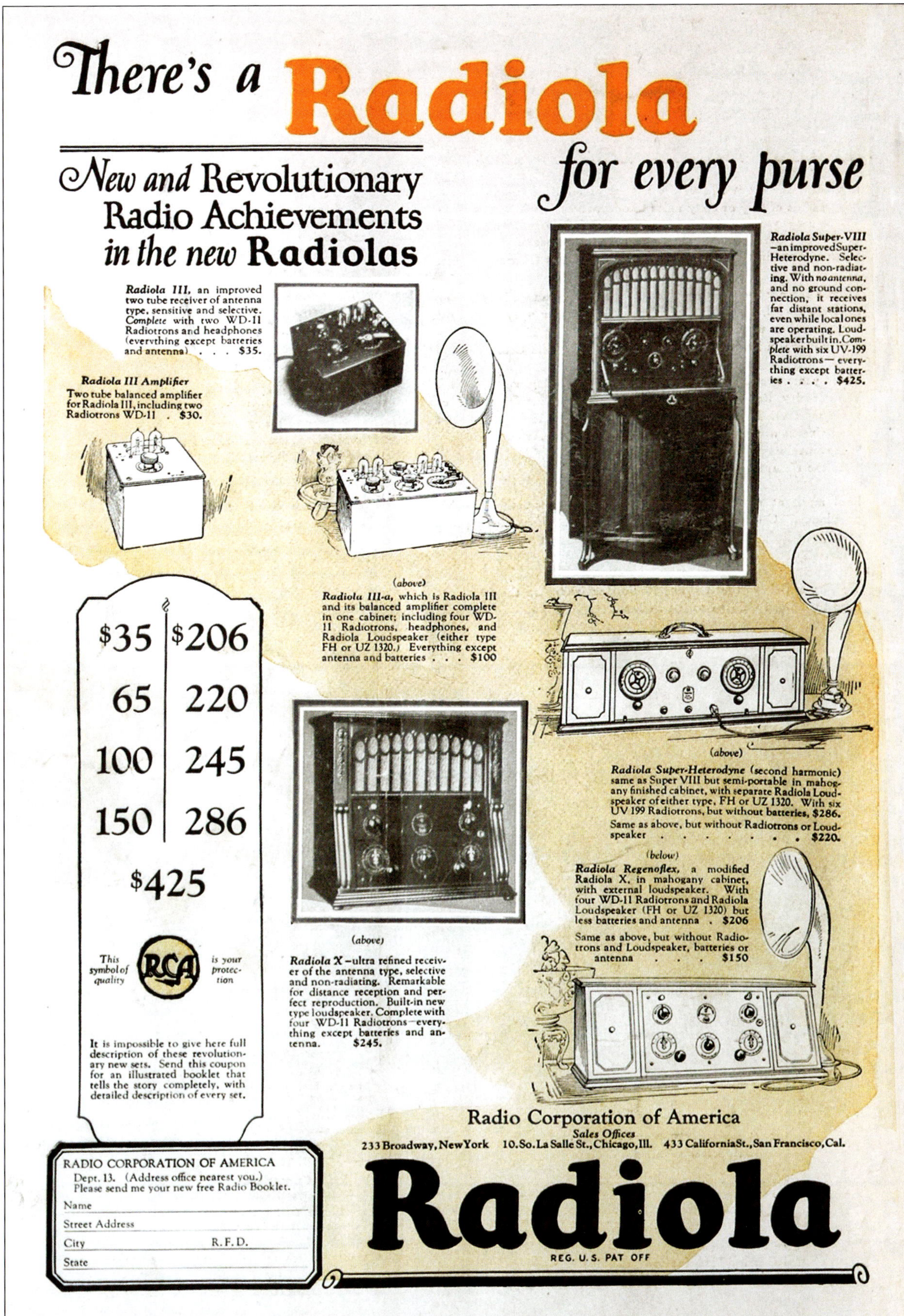

The new Radiola line featuring seven receivers introduced in 1924 appeared in numerous national advertisements beginning in March 1924. (*The Wireless Age*, March 1924, back cover)

A number of different superheterodyne prototypes were developed by RCA/GE during 1923 including the two shown here. (George H. Clark Collection, Smithsonian Institution)

designing and mass producing the radically different superheterodyne, had significant trouble with a noise problem (thermal hiss) in the superheterodyne receiver, and as a result, the 1924 line was not introduced until February 1924, well after the Christmas season. RCA took a significant gamble on the superheterodyne, and undoubtedly lost substantial sales during the Christmas

Table 6-2. Radiola apparatus marketed by RCA in 1924

	Model	Apparatus Type	Cabinet Style	Tubes	Intro Date	Discon'd. By	List Price[a]	Mfg. By
1924-25 Season New Models	Radiola III (AR-805)	Regenerative Receiver	Box	2	2/2/24	Sep. 1926	$35[b]	W
	Balanced Amp (AR-806)	Push-pull Amplifier	Box	2	3/4/24	Sep. 1926	$30	W
	Radiola III-A (AR-802)	Regenerative Receiver	Box	4	3/27/24	Feb. 1927	$65[b]	W
	Regenoflex (AR-817)	Regenerative TRF Receiver	Cabinet	4	4/25/24	Sep. 1925	**$150** $206	W
	Radiola IX (AR-1404)	TRF Receiver	Phono panel	5	Nov. 1923	Feb. 1925	$215	WSA
	Radiola X (AR-816)	Regenerative TRF Receiver	Wood Cabinet	4	3/21/24	Sep. 1925	$245[b]	W
	Radiola VII-B (AR-907)	TRF Receiver	Wood Cabinet	5	Jan. 1924	Feb. 1925	$275	WSA
	R. Super-Heterodyne (AR-812)	Super-heterodyne	Wood Portable	6	3/4/24	Jun. 1926	**$220**	GE
	Radiola Super-VIII (AR-810)	Super-heterodyne	Console	6	3/18/24	Feb. 1927	$425	GE
	AG-814	Antenna	None	N/A	Apr. 1924	Sep. 1925	$12	GE
	UZ-1325	Horn Speaker	None	N/A	Oct. 1924	Feb. 1927	$25	GE
Holdover	FH	Horn Speaker	None	N/A	Sept. 1923	Feb. 1925	$36.50	W
	UZ-1320	Horn Speaker	None	N/A	Nov. 1923	Oct. 1924	$36.50	GE
	AG-1380	Loop Antenna	None	N/A	Jun. 1922	Feb. 1925	$18	WSA

[a] All apparatus included tubes in advertised price except those noted with the price in bold font

[b] Headphones were included in advertised price

The new RCA catalog for 1924 featured the seven new receivers introduced in March 1924 plus Radiolas IX and VII-B introduced a few months earlier. (RCA Catalog 1924)

season because of the delay it caused, but the gamble paid off because the RCA superheterodyne receivers were all the rage in 1924, and became the standard for receivers thereafter.

The new line was introduced to the public in national ads beginning in March of 1924, and was also pictured in the 1924 RCA catalog distributed about the same time. The new line, summarized in Table 6-2, was radically different from the original plan. The least expensive receiver was designated the Radiola III—the III designation still being available because the original III was dropped the previous year. The Radiola Balanced Amplifier was designed for use with the III, and the combination of the Radiola III and the Radiola Balanced Amplifier in a single box became the Radiola III-A. The most expensive set, the new superheterodyne in a console, was assigned the model name Super-VIII, the "Super" alluding to both performance and the superheterodyne principle. A less expensive version of this set was placed in a smaller cabinet without a speaker, but oddly was not referred to as the VIII-A, but rather as the Radiola Super-Heterodyne without any number.

The Radiola IX and VII-B were actually introduced before the rest of the 1924 line, and while they were considered part of the 1924 line—they were not featured in ads such as the one pictured here introducing the new RCA in early 1924. They did however appear in the RCA catalog for 1924 which is also pictured here. The Radiola IX was designed with a thin panel similar to one of the planned sets, but had five tubes instead of the original three, and was designed to be an adjunct to a phonograph. The Radiola VII-B was based on the Radiola IX, but was enclosed in a wood cabinet with a self-contained speaker. A more appropriate model designation might have been Radiola IX-A, but RCA needed a replacement for the ill-fated Radiola VII manufactured by Wireless Specialty and the IX panel was the most logical choice, having been manufactured by Wireless Specialty, as well.

The Radiola X was a four-tube reflexed receiver, nothing like the three-tube receiver originally planned. The Radiola Regenoflex was virtually identical to the Radiola X except for a less expensive cabinet and the absence of an enclosed speaker. In addition to these receivers, the AG-814 antenna and UZ-1325 speaker were introduced in 1924.

What is most interesting and refreshing about the 1924 line is that no prior-year receivers were held over—and for the first time, none of the receivers in the line competed directly with each other. All receivers were uniformly designed for the new broadcast band, which was expanded in May of 1923 from 360-485 meters to 222-545 meters (550-1350 kHz). Name tags were prepared for each of the RCA receiver models uniformly specifying 220-550 meters as the operating range. Further, more attention had been paid to the appearance and styling of the sets, all of which were contained in wood cabinets with varying degrees of attractiveness. Clearly, RCA as a design and marketing enterprise for consumer radios was maturing.

This ad for the Super-Heterodyne appeared in both color and black and white beginning in March 1924. (*Boys Life*, April 1924)

General Electric Offerings

General Electric was assigned the task of designing and mass producing the superheterodyne receiver for the 1924 season. The same superheterodyne receiver was packaged both as a console (AR-810 Super-VIII) and what was described as a portable radio (AR-812 Super-Heterodyne). In addition, GE designed the AG-814 loop antenna specifically for use with the Radiola Super-Heterodyne, and an improved horn speaker designated the UZ-1325. The UZ-1325 replaced both the FH and UZ-1320 horns when it was introduced later in the year.

Radiola Super-Heterodyne (AR-812)

The Radiola Super-Heterodyne was officially released on March 4, 1924 at a list price of $286 complete with Radiotrons and loudspeaker but excluding batteries, or $220 without tubes or other accessories. It was heavily advertised

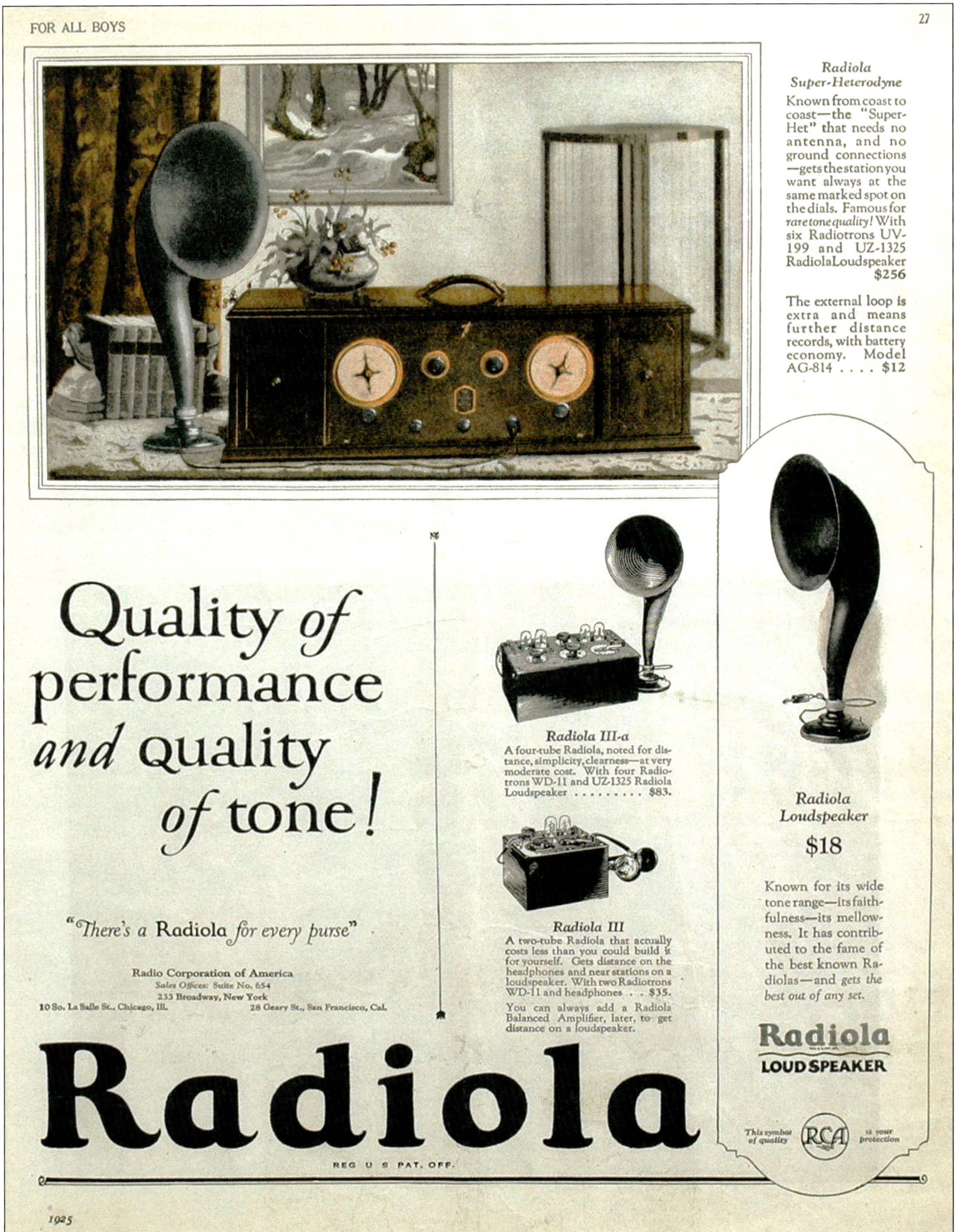
FOR ALL BOYS
27
Radiola Super-Heterodyne
Known from coast to coast—the "Super-Het" that needs no antenna, and no ground connections—gets the station you want always at the same marked spot on the dials. Famous for rare tone quality! With six Radiotrons UV-199 and UZ-1325 Radiola Loudspeaker $256
The external loop is extra and means further distance records, with battery economy. Model AG-814 . . . . $12
Quality of performance and quality of tone!
Radiola III-a
A four-tube Radiola, noted for distance, simplicity, clearness—at very moderate cost. With four Radiotrons WD-11 and UZ-1325 Radiola Loudspeaker . . . . $83.
Radiola III
A two-tube Radiola that actually costs less than you could build it for yourself. Gets distance on the headphones and near stations on a loudspeaker. With two Radiotrons WD-11 and headphones . . $35.
You can always add a Radiola Balanced Amplifier, later, to get distance on a loudspeaker.
Radiola Loudspeaker
$18
Known for its wide tone range—its faithfulness—its mellowness. It has contributed to the fame of the best known Radiolas—and gets the best out of any set.
"There's a Radiola for every purse"
Radio Corporation of America
Sales Offices: Suite No. 654
233 Broadway, New York
10 So. La Salle St., Chicago, Ill.
28 Geary St., San Francisco, Cal.
Radiola
LOUDSPEAKER
This symbol of quality
RCA
is your protection
Radiola
REG U S PAT. OFF.
1925

The Radiola Super-Heterodyne was equipped with a carrying handle, which according to various ads qualified it as a portable or semi-portable (reproduction handle pictured).

in national publications for the first year and half it was on the market using ads in both black and white and color. The set was often pictured with the UZ-1320 speaker and/or the external AG-814 loop antenna which was designed to extend its range even further.

The Radiola Super-Heterodyne was a so-called "portable" receiver enclosed in a wood case with battery compartments on either side and a leather handle on top to carry the forty-pound receiver.[7] A loop antenna was built into the wood case, obviating the need for an external antenna. Oddly, the AR-810 console version of this set was assigned the designation Super-VIII, but the AR-812 version was not assigned any Radiola number.

The front panel contained a minimum number of controls, including two tuning dials, a filament current control, a volume control, and an on-off switch. The instructions specified that the user was to select a wavelength range of interest on the left dial and then tune the right dial in the same general wavelength location to home in on a station. When a station was found, the user could write the station letters at the appropriate position on both paper dials so that it could be readily located for future listening. Many of the sets found on the market today have paper behind the dials with numerous station locations marked by users from the 20's. The instructions also advised that if a station could not be found, the set should be rotated by ninety degrees to reorient the loop antenna. It was said that the set was often used as a direction finder.

The front panel of the set opened downward to reveal the six UV-199 tubes sitting atop a chassis filled with brown sealing wax contained in an outer metal wrapper and seal, presumably to protect the circuitry from prying eyes. This so-called catacomb was the bane of the radio repairman and collector alike because it made it difficult to repair or to trace a circuit. The repairman was instructed as follows: *"If the catacomb fails to pass any of the tests in part 1 or part 2 of these instructions [from the RCA Service manual] it should be removed from the panel and replaced by a new one. Under no circumstances should the lead seals on the cover plate be broken."*[8] RCA replaced defective catacombs for $73.50, but only if the defective catacomb was returned without the seals broken.[9] Oddly, none of these safeguards would have prevented a determined individual from melting the wax and reverse engineering the circuitry.

The sensitivity and selectivity of this set when compared to other commercial sets of the day was nothing short of phenomenal. There were numerous reports of both transoceanic and transcontinental reception, even from the larger cities where a number of local high-powered stations often interfered with all but the strongest signals. As an added benefit, the set did not appreciably interfere with

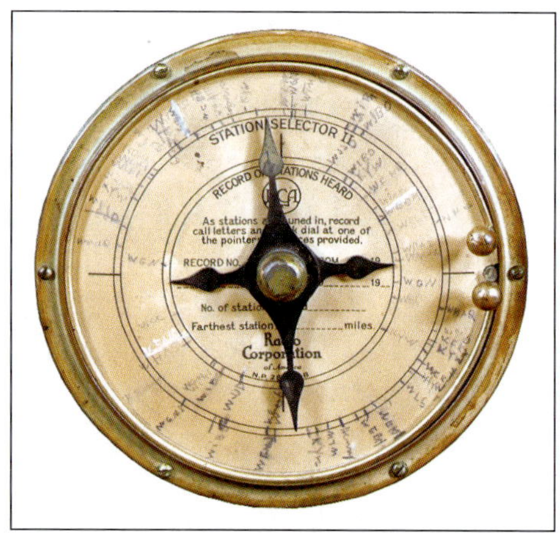

Cardboard disks were supplied with the AR-812 allowing the user to mark the location of each station. These disks now provide a nostalgic record of many old stations—many of which no longer exist—and also may identify where the radio was used, such as this one from Chicago, IL.

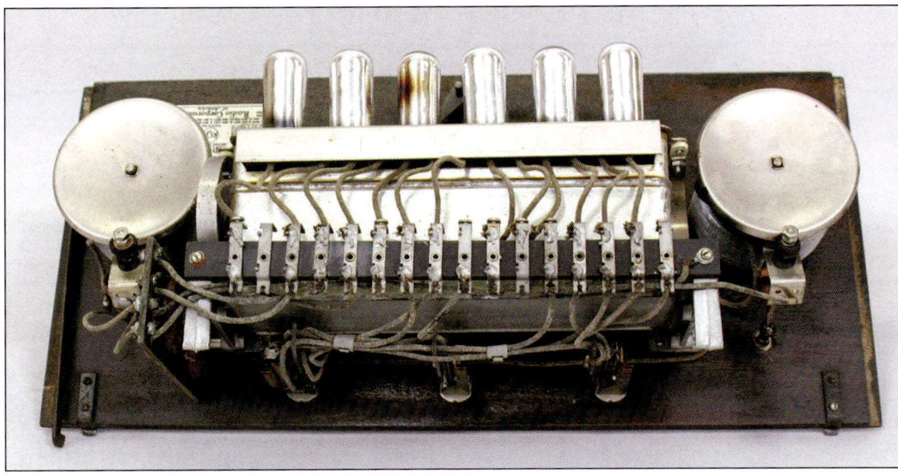

The interior of the AR-812 Super-Heterodyne is dominated by the sealed catacomb used to hide most of the circuit components from prying eyes.

A General Electric employee is featured on the front page of the *General Electric News* soldering the catacomb to its panel during the assembly of the Super-Heterodyne receiver in building 77 at the Schenectady Works in New York.

neighboring sets as did the regenerative receivers of the day. The public obviously took to RCA's superheterodynes in a big way despite the high price tag, because 148,300 sets, including both the Super-Heterodyne and the Super-VIII, were shipped from Schenectady by May 30, 1925.[10]

Each set had a metal name plate attached to the inside of the front panel with the Radiola Super-Heterodyne name, the AR-812 designation, and the operating wavelengths specified as 220-550 meters. Each plate was also stamped with an individual serial number, the lowest believed to be 100,000 and the highest observed being 532,324. There were three "production runs" of the AR-812, each with a different range of serial numbers as shown in Table 6-3. The only model known to be interleaved with the Super-Heterodyne is the Super-VIII, with serial numbers concentrated between 174,000 and 197,000. As detailed in Appendix D, there is a gap in AR-812 serial number samples between 172,943 and 188,948 which is believed to be filled with about 16,000 Super-VIII sets. In addition, there are another estimated 3,000 Super-VIII sets interleaved with the AR-812 serial numbers above the gap—for a total of 19,000 Super-VIII sets in the first production run of about 81,000 AR-812's. Since there are no sets interleaved with AR-812 in the second and third production runs, the number in each run is determined by the end points of the serial number samples. Thus, the total number of Super-Heterodynes for all three production runs is estimated to be 164,000—about 10% higher than the total of 148,300 reported by both Douglas and production designer W. L. Carlson of RCA to one Mr. E. A. Laport in a letter dated August 23, 1961.[11]

The new Radiola Super-Heterodyne was so popular that many owners held Radiola parties using an invitation supplied by RCA such as this one.

Dear ....

You are cordially invited to attend a Radio party we have planned for the evening of ....

We have just installed a new Radiola and expect to have a delightful evening "listening in" to a variety of radio broadcast programs.

Please drop us a line, or call up to say that you will come. Sincerely,

(OVER)

Table 6-3. Serial number evidence for the AR-812 Super-Heterodyne

Production Run	Serial numbers	Estimated No.
1	100,000-173,000 188,000-199,999	73,000 8,000
2	350,000-399,999	50,000
3	500,000-533,000	33,000
	All	164,000

There were at least seven versions of the Super-Heterodyne manual—the first dated February 1924 was assigned the identification number 86963, while the manual dated December 1924 was assigned the identification number 86963F. The last production run occurred circa February 1925, although the set remained on the market for another year. New manuals were often printed as a result of some change in the design, but that was not the case with the AR-812. Most of the changes to the AR-812 manuals dealt with operating instructions. For example, the May 1924 manual was changed to include instructions for connecting the new AG-814 loop antenna, which had been designed specifically for use with this set.

One important change in operating inductions was also accompanied by a visible change on the filament rheostat dial. A small arrow mark was added to the dial, the purpose of which was explained by the following warning which appeared as an addition to the user's manual: *"WITH FRESH BATTERIES, DO NOT TURN THE POINTER PAST THE ARROW MARK. THIS IS EXTEMELY IMPORTANT, as you will otherwise shorten the useful life of the Radiotrons and batteries..."*[12] This change, which was made sometime before September of 1924 during the first production run, was obviously made because many users had damaged the relatively fragile UV-199 tubes by not resetting the filament control when the batteries were replaced. The user, who had to reduce the resistance of the filament rheostat to compensate for the batteries as they drained, often forgot to reset the filament rheostat before installing the new batteries and tuning the set on. The mark on the dial was intended to be a reminder for the user to reset the rheostat. Another change was the replacement of control knobs with inscribed circles on the face used in early sets with knobs having a plain smooth face.

The set without accessories was reduced to $116 in the RCA price list dated September 21, 1925, and was discontinued by the time the June 1, 1926 price list was released. However, this venerable set could still be found in discount catalogs as late as 1933 for the bargain price of $8.75 without accessories.[13]

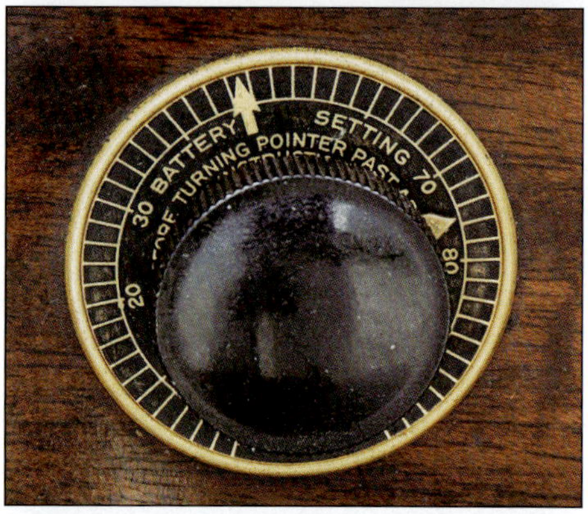

Sometime during the first production run the knobs with round circles (upper figure) were replaced with ones having a plain surface (lower figure), and an arrow was added to the battery dial with a warning to the user not to turn the dial beyond the arrow when using fresh batteries (lower figure).

Radiola Super-Heterodyne (AR-813)

It is reasonably well known that RCA provided a version of the Radiola Super-Heterodyne in a square panel with the model number AR-813 for Brunswick phonograph-radio combinations in 1924, but it is less well known that a version of the AR-813 was manufactured for use in a console style radio-only receiver sold under the RCA name without any reference to Brunswick. The receiver using the RCA AR-813 panel is addressed here, while the Brunswick phonograph-Radiola combinations using the Brunswick AR-813 panel are addressed in Chapter 11.

The set designated Radiola Super-Heterodyne AR-813 had the same name (Super-Heterodyne) as the AR-812, but was totally different in appearance, with significantly different name plates, as can be seen from the comparisons pictured here. The console-type cabinet with a horizontal panel accessible from the top was smaller than the Radiola Super-VIII console. The thick front door, which concealed the horn speaker grill, contained a loop antenna which could

The RCA AR-813 Radiola Super-Heterodyne radio was distinctly different than the Brunswick phonograph-Radiola combinations using the AR-813 panel.

The AR-813 panel used in the Radiola Super-Heterodyne radio was virtually identical to the AR-813 panel used by Brunswick in their phonographs with the notable exception that the Brunswick name plate was replaced by one with a warning about the rheostat setting.

A loop antenna was hidden in the relatively thick front door of the AR-813 Radiola Super-Heterodyne cabinet.

be positioned for optimum reception. The set has a rather unusual horn speaker (for RCA) with an oval shaped opening having dimensions of approximately eight by twelve inches—the same type used by Brunswick in their phonographs. The design of the speaker grill on this set is virtually identical to the one Brunswick used on their Model 30 offered for sale in the last half of 1924.

While the set never appeared in an RCA price list or catalog, it did appear in at least one brochure where it was described as being "*manufactured by the Radio Corporation of America.*" The brochure described the cabinet as "*finished in Adam Brown Mahogany and American Walnut,*" a description used by Brunswick for their cabinets of the same era. In fact, the brochure was published by Brunswick as evidenced by the identification number (Form 616 JPM 9-18-25) which is the same sequence of characters used on virtually all Brunswick brochures of the era.

There are several clues that reveal the genesis of this set. First, there is a wood strip and a metal panel support arm used on the Brunswick AR-813 panel that

The name plate appearing on the RCA AR-813 Super-Heterodyne (top) was distinctly different than one appearing on Brunswick phonographs (bottom), but both were stamped by GE as indicated by the initials N. P.

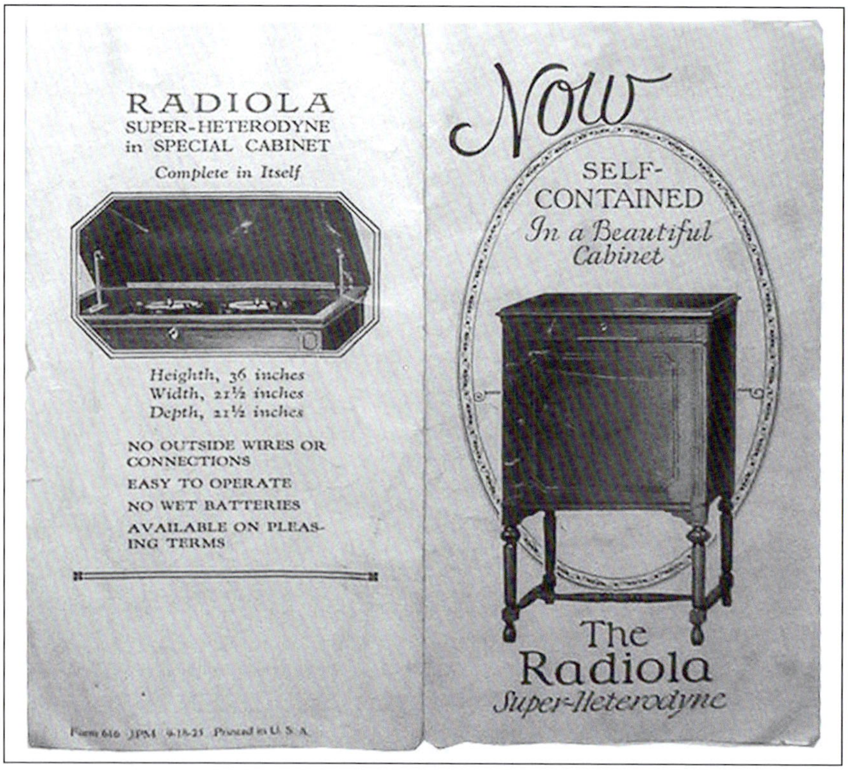

The RCA Radiola AR-813 Super-Heterodyne brochure was actually printed by Brunswick, although the Brunswick name never appeared. (Courtesy of Joe Formanek)

has been clearly removed from the RCA AR-813 Super-Heterodyne panel. Second, the serial numbers for the Brunswick AR-813 panels (which are stamped in the wood panel just behind the row of tubes) are in the range of 300,000 to 325,000, while serial numbers for the few RCA AR-813 panels seen by this author have been consistently above those for the Brunswick panels found in phonographs. Thus, one can conclude that the panels used in the RCA radio were excess inventory from a production run for Brunswick phonographs. It appears that the excess panels were used in the radio-only console as a way of clearing the inventory of panels. What is not known is whether Brunswick or RCA owned the excess inventory, and therefore whether it was Brunswick or RCA that instigated the manufacture of this radio.

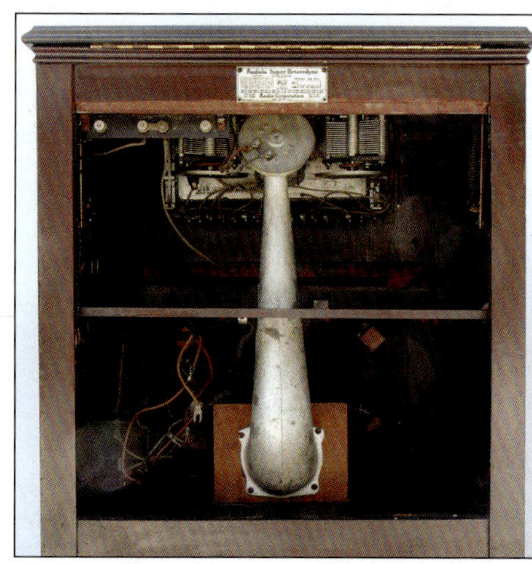

The speaker used in the RCA AR-813 Super-Heterodyne radio was the same as the ones used by Brunswick in their phonographs.

Regardless of whether RCA or Brunswick owned the excess inventory, it appears that the RCA AR-813 Super-Heterodyne radio was not an after market adaptation of the Brunswick panel by some distributor. This conclusion is based on the following four points. First, the brochure clearly states that it was manufactured by the Radio Corporation of America. Second, the name plate affixed to the back of the radio cabinet contains the initials N. P. which clearly identifies it as a GE name plate—GE used the initials N. P. on virtually all metal name plates for radios they manufactured. The existence of the GE name plate affixed to the cabinet, as opposed to the chassis or radio panel, would suggest that GE manufactured this set (but the similarity of the cabinet and speaker to Brunswick hardware would suggest than Brunswick manufactured the set). Third, the Brunswick name plate on the center of the panel was replaced with a warning label regarding the rheostat setting for fresh batteries, a plate that was clearly printed at a factory. No distributor in the after market would have taken it upon itself to add such a warning. Fourth, the antenna door, speaker and receiver panel were integrated so well with the cabinet that it is unlikely the components could have been assembled in the after market using a generic cabinet—it has the look and appearance of a factory model.

A very interesting instruction manual prepared specifically for this set provides some additional insights. The schematic diagram in the instruction manual is almost identical to the one prepared by Brunswick for the AR-813 panel published in Rider's *Perpetual Trouble Shooter's Manual*—but with one noticeable difference. The modified diagram shows a UX-120 output tube (in lieu of the UV-199) using an adaptor (RCA's UR-556) with wiring to separate batteries described as "B-X" and "C-X." Apparently, this adaptation appeared on all the AR-813 Super-Heterodyne receivers, although it would not be apparent to the collector or historian on sets in which the output tube and associated adaptor had been removed. The UX-120 tube and higher plate voltage would have provided improved audio output power and performance as compared to the UV-199 output tube for which the AR-813 panel was originally designed.

Very little seems to be known about this set, including the date it first appeared, the date it was discontinued, the list price at introduction, or the number manufactured. Based on the date of the brochure, it must have been introduced sometime in the fall of 1925. It is not even clear whether RCA distributors or Brunswick distributors—or both—distributed this set. That Brunswick was somehow involved in the distribution is suggested by the fact that Brunswick printed the sales brochures. Perhaps Brunswick acted as an RCA distributor (or perhaps the sole distributor) for this set. Since the AR-813 panel was developed specifically for Brunswick, why the Brunswick name did not appear on this set is also a mystery. Perhaps RCA did not want the public to think that Brunswick was directly competing with them in the radio receiver market. Judging by the number of RCA AR-813 Super-Heterodyne sets appearing in collections and on the auction market, there appear to be a relatively large number of these sets—hundreds and perhaps thousands.

Radiola Super-VIII (AR-810)

The Radiola Super-VIII was officially introduced on March 18, 1924, several weeks later than the Super-Heterodyne portable. It was essentially the same as the semi-portable Super-Heterodyne except that it was contained in an

The finish on many Radiola Super-VIII sets has aged so they appear to have a uniform mahogany color rather than the two-toned appearance they had when new (right).

This Super-VIII console has been refinished in two tones to simulate the original. (Sonny Clutter collection)

attractive console cabinet with a self-contained horn and rotatable loop antenna selling for $425 exclusive of batteries. The loop antenna was similar to the self-standing AG-814 antenna, but its orientation could be controlled by a knob in the lower center of the front panel with a mechanical connection to the antenna using a large geared wheel. It was actually the first floor console set marketed by RCA, the forerunner of many more to come in future years. The Super-VIII was the top-of the-line set, and was heavily advertised.

The Super-VIII was heavily advertised in numerous national magazines. (*National Geographic*, March 1924)

The earlier Super-VIII consoles used a single horn speaker with a U-shaped neck that folded back on it self such that the driver was on top of the horn opening. The loop antenna in the rear could be rotated with a knob mounted on the receiver control panel (left). The later Super-VIII consoles were equipped with a Doublette horn with two drivers and two apertures, possibly the first woofer and tweeter combination used in a radio receiver.

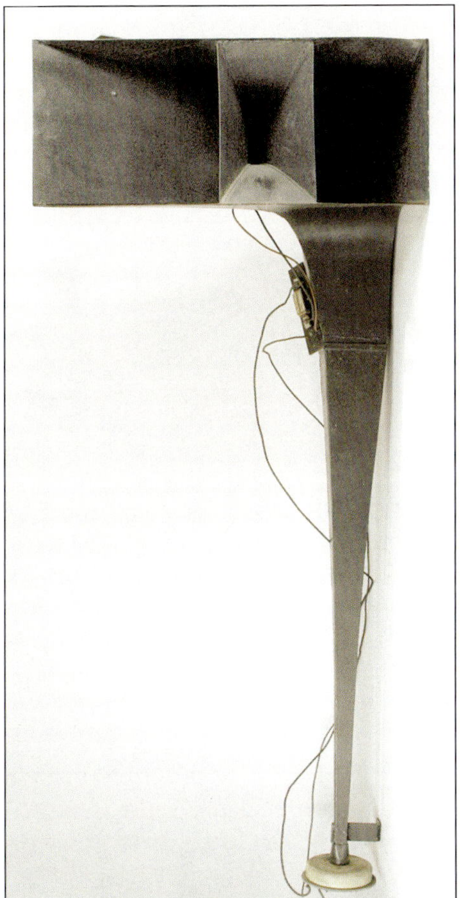

The smaller high-frequency horn is located in the throat of the larger low-frequency horn; the low-frequency driver is located at the bottom of the antenna, the high-frequency driver is located behind the large horn aperture, and the cross-over filter is located on cabling connecting the two drivers.

There were a relatively small number of Super-VIII sets made, most having serial numbers in the two intervals of 10,000-11,000 and 175,000-195,000. Based on the serial number evidence, it is likely that there were less than 20,000 manufactured, but is difficult to ascertain the number with any precision from the serial number evidence because the semi-portable Super-Heterodynes had serial numbers which were interleaved among the Super-VIII's in the 175,000-195,000 range.

There must have been an issue with the performance of the single horn speaker in the Super-VIII because sometime after production began circa February 1924 but before September 1924, GE replaced the single horn with the "famous Doublette Loudspeaker."[14] From observations of Super-VIII's in collections and in the auction market, sets with serial numbers below 183,000 had the original single horn, while receivers with serial numbers above 192,000 had the Doublette, and so the serial number at which the Doublette was introduced is somewhere between the two. Sets with these serial numbers were manufactured circa September 1924, and indeed, an article in the *The Wireless Age* on the Doublette horn states that the Doublette speaker was introduced into the Super-VIII in late summer of 1924.[15] Despite all the evidence that the Doublette was not introduced until September 1924, several Super-VIII's containing Doublettes have surfaced with serial numbers in the mid-10,000's. Either these sets with low serial numbers were made later than sets with the higher numbers, or the chassis in sets with Doublettes were switched with chassis having the lower numbers. Based on the dates from *The Wireless Age* article and other documentation, it is highly doubtful that Doublettes were placed in sets manufactured much before September of 1924.

The Doublette horn speaker consisted of a long tapered metal structure with a large opening for the low frequencies, and a small short high-frequency metal horn with a taper that penetrated the wall of the low frequency horn. Thus, the high frequency horn was actually mounted in the throat of the low frequency horn, with the driver on the back side of the large horn. Each horn had its own driver with a capacitive filter to direct the high and low audio frequencies to the appropriate drivers, thus possibly forming the first woofer and tweeter arrangement in broadcast radio![16] The UV-199 was initially used to drive both the single horn and Doublette, but later the new higher-power UX-120 Radiotron with the UR-556 base adaptor was used in the Super-VIII and the Radiola Super-Heterodyne, among other receivers.

The impetus for replacing the original horn speaker may have been the competition between RCA and Western Electric over their respective superheterodyne receivers—RCA engineers were quoted as saying that the Western Electric superheterodyne 4-B receivers were superior to the top-of-the-line Super-VIII receivers primarily because of their new Rochelle-salt loudspeaker.[17] RCA may have wanted to improve the performance of their top-of-the-line set using the Doublette horn for purposes of prestige and positioning in the marketplace—regardless of cost. It should be noted that the price of the Super-VIII was not increased after the obviously more-expensive Doublette horn was substituted for the single horn.

The Super-VIII was on the market from February 1924 through 1926, six to nine months longer than the Radiola Super-Heterodyne portable version. Even though the cost to RCA had increased with the addition of the Doublette speaker in the fall of 1924, the price was actually reduced from $425 to $340 by September 1, 1925, a sure sign that sales were lagging.

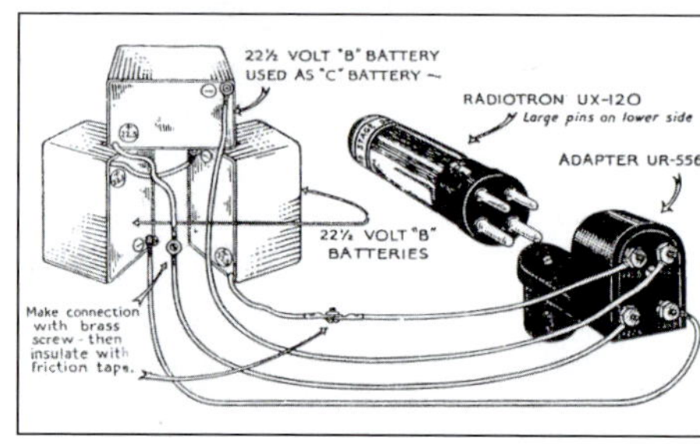

The RCA UR-556 adaptor was designed to allow replacement of the UV-199 audio output tube of the early RCA six-tube superheterodyne receiver with the new UX-120 power output tube. (*RCA Victor Service Notes for 1923–1928*)

UR-556 Adaptor

RCA developed the UX-120 high-power audio output tube for the new line of receivers appearing in the fall of 1925. The performance of the UX-120 was so much better than the UV-199 it replaced, RCA decided to manufacture an adaptor so it could be used both in the after-market and to retrofit sets already in production—most notably the AR-812 and Super-VIII Super-Heterodyne. This adaptor was also used on all AR-813 Super-Heterodynes, which were manufactured beginning sometime in the fall of 1925.

The UR-556 was more than just a mechanical adaptor to match the new UX base to the older UV-199 sockets—it also had two terminals to attach two additional 22½-volt "B" batteries to achieve the 135 volt plate voltage needed to drive the UX-120 tube, and two additional terminals to attach a 22½-volt battery used as a "C" battery for grid biasing. Thus, the modification to accommodate the new UX-120 tube could be made with the UR-556 adaptor without having to rewire the set in any way. The socket is designed such that the UX-120 tube lies on an axis perpendicular to the plate of the rest of the tubes. The adaptor first appeared in the September 21, 1925 price list at $1.50.

AG-814 Loop Antenna

The Model AG-814 was designed primarily for the portable Super-Heterodyne, which according to the instructions provided directional selectivity and more effective pickup than the loop provided with the set. It is similar in appearance to the one used with the Super-VIII with the addition of the sculptured base.

The loop was easily assembled and disassembled by inserting the dowels into the various predrilled holes in the wood structure. Insulated antenna wire was wrapped around the wood structure and connected to terminals 2 and 3 on the panel in the rear. The separate antenna had the advantage that it could be rotated on its base to obtain maximum signal, obviating the need to rotate the entire Super-Heterodyne receiver, a plus in many home settings.

The antenna first appeared on the market with the AR-812 in March of 1924 with a retail price of $12. Like the AR-812, it had been dropped from the line by mid-1926, as evidenced by its absence from the RCA price list dated June 1, 1926.

The AG-814 antenna designed for the AR-812 Super-Heterodyne was based on the loop used in the Super-VIII.

The AG-814 was sold as a kit with the instruction card shown here to assist in assembly and installation.

The name plate, screwed onto the wood base of the antenna, clearly identified it as the AG-814 antenna.

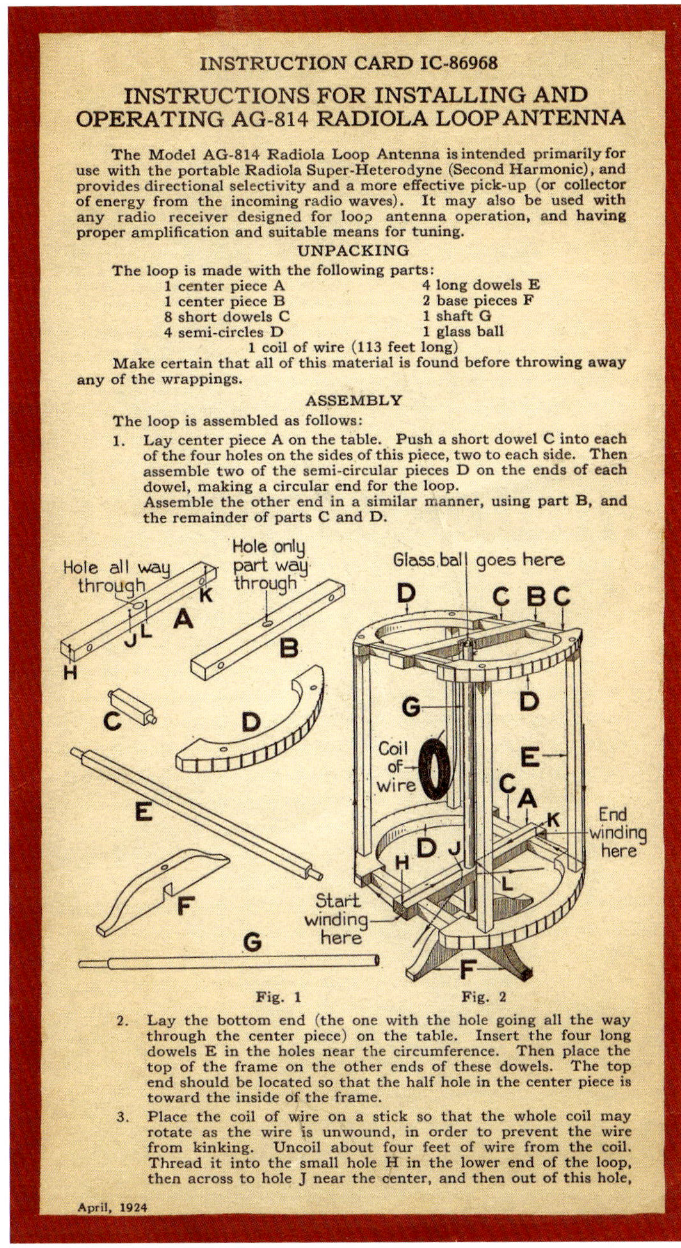

INSTRUCTION CARD IC-86968

INSTRUCTIONS FOR INSTALLING AND OPERATING AG-814 RADIOLA LOOP ANTENNA

The Model AG-814 Radiola Loop Antenna is intended primarily for use with the portable Radiola Super-Heterodyne (Second Harmonic), and provides directional selectivity and a more effective pick-up (or collector of energy from the incoming radio waves). It may also be used with any radio receiver designed for loop antenna operation, and having proper amplification and suitable means for tuning.

UNPACKING

The loop is made with the following parts:

1 center piece A	4 long dowels E
1 center piece B	2 base pieces F
8 short dowels C	1 shaft G
4 semi-circles D	1 glass ball

1 coil of wire (113 feet long)

Make certain that all of this material is found before throwing away any of the wrappings.

ASSEMBLY

The loop is assembled as follows:

1. Lay center piece A on the table. Push a short dowel C into each of the four holes on the sides of this piece, two to each side. Then assemble two of the semi-circular pieces D on the ends of each dowel, making a circular end for the loop.
Assemble the other end in a similar manner, using part B, and the remainder of parts C and D.

Fig. 1 Fig. 2

2. Lay the bottom end (the one with the hole going all the way through the center piece) on the table. Insert the four long dowels E in the holes near the circumference. Then place the top of the frame on the other ends of these dowels. The top end should be located so that the half hole in the center piece is toward the inside of the frame.
3. Place the coil of wire on a stick so that the whole coil may rotate as the wire is unwound, in order to prevent the wire from kinking. Uncoil about four feet of wire from the coil. Thread it into the small hole H in the lower end of the loop, then across to hole J near the center, and then out of this hole,

April, 1924

The UZ-1325 had an appearance very similar to the UZ-1320 horn which it replaced.

UZ-1325 Horn Speaker

The Radiola UZ-1325 horn, an improved version of the UZ-1320, first appeared on the market circa October 1924 for $25. The new horn immediately replaced the UZ-1320 horn, and by the end of the year, it had replaced the Westinghouse FH horn as well. The speaker was heavily advertised with ads first appearing circa November 1924.[18] It was RCA's most popular horn, outselling the UZ-1320 and the FH combined by a wide margin in the two or so years it was on the market.

The differences between the UZ-1325 and the predecessor UZ-1320 horn were documented by Floyd Paul.[19] Perhaps the most obvious difference was the upright lever arm on the UZ-1325 which was used to adjust the spacing between the pole pieces and the metal diaphragm, whereas the UZ-1320 had a knurled thumb wheel used to perform the same function. This lever arm is often mistaken for a volume control. The two can also be distinguished when placed next to each other by the lower profile of the UZ-1325 base, although the size and shape of the bells are virtually identical. The base of the UZ-1325 was made of metal while the base of the UZ-1320 was made from wood. A metal screen was

The UZ-1325 horn was placed on the market at the end of 1924 just in time for the Christmas season. (*The Wireless Age*, December 1924)

located in the base of the UZ-1325 to keep out foreign material, while a screen located in the neck of the bell was used for the same purpose on the UZ-1320.

The UZ-1325 horns were manufactured in significantly larger quantities than either the UZ-1320 or the Westinghouse FH horns, although the exact number is not known. In a survey conducted by Floyd Paul, the relative number of speakers held by collectors was approximately five UZ-1325 horns for each UZ-1320 and FH horn, with slightly more UZ-1320 horns reported than FH horns.[20] In a survey of speakers sold on a popular auction site by this author over a two year period from late 2003 to mid-2005, the ratio of the number of UZ-1325 to UZ-1320 horns sold was also about 5:1, but the ratio of UZ-1325 to FH horns was a whopping 25:1. One logical interpretation of this data is that there were far fewer FH horns made than either the UZ-1320 or UZ-1325, and as a result they are so desirable that most of them are in the hands of collectors! Obviously, neither of these two surveys is based on solid science, and neither is conclusive. Whatever this data may mean, it is clear that the collector will have a much easier time finding either a UZ-1320 or UZ-1325 horn than an FH horn.

The UZ-1325 can be easily distinguished from the UZ-1320 by the vertical bar that is used to adjust the diaphragm position.

The early base on the UZ-1325 was cast metal with the name plate secured by either a screw (top) or knurled pin (middle), while later the base was stamped metal with the name plate secured by a rivet with a hole in the center (bottom).

There were two different versions of the UZ-1325, the earlier being a non-ferrous cast-metal base that accepted a threaded earphone driver, and the later being a sheet-metal base in which the driver screwed into a simple cup housing which in turn screwed into the base. The earlier version is most easily recognized at a glance by either the screws or the solid round pins used to attach the nameplate to the base. The solid round pins had knurled shafts so they could be driven firmly into the base with the tap of a hammer. The name plates on the later version with the sheet metal base were attached with rivets, which are readily identifiable by a hole in the center of each rivet.

The price of the UZ-1325 was reduced from $25 to $18 by February 1925, and remained at $18 until the end of 1926 when it was removed from RCA price sheets. The UZ-1325 was the last horn marketed by RCA, thus marking the end of the horn era in so far as RCA was concerned. From that point on, RCA marketed cone type speakers exclusively.

Westinghouse Offerings

While both GE and Westinghouse were interested in the new superheterodyne receivers, in the end it was decided that Westinghouse would manufacture four of the more conventional regenerative receivers and a two-tube audio amplifier along the lines of the models originally planned for the 1924 season. Their task was relatively easy because they used only two different circuits—the four-tube regenerative circuit used in the Radiola III-A series, and the four-tube regenerative circuit used in both the Regenoflex and Radiola X. At the low end, they manufactured the Radiola III based on the old Radiola RS receiver, a two-tube push-pull Balanced Amplifier, and the Radiola III-A which was a combination of the two housed in one box. For the middle-of-the-line, they designed a new reflexed regenerative receiver which was used in two different types of cabinets—a less expensive cabinet with a horizontal orientation evocative of the GE portable Super-Heterodyne designated the Radiola Regenoflex, and a more expensive cabinet with vertical orientation and a self-contained speaker designated the Radiola X.

Westinghouse assigned "Type" designations to all the receivers they manufactured for RCA in 1924, while RCA assigned all of them "AR" designations. Oddly, the Westinghouse "Type" designations (two letters beginning with an "R") appeared on at least one version of all Westinghouse receivers for 1924 while the RCA "AR" designations did not appear on any of the Westinghouse receivers. Both the RCA and Westinghouse designations are shown in parenthesis in the headings below for each set—the RCA designation is shown first followed by the Westinghouse designation.

Radiola III (AR-805) (RI)

According to George Clark, the Radiola III was officially released on February 2, 1924, one of the earliest release dates for sets for the 1924 season. Ads for the Radiola III began to appear in March 1924 with a price of tag of $35 including two WD-11 tubes and earphones but without batteries. This price was extremely attractive in comparison to the two-tube RS which had been selling for $85 complete, including about $10 worth of batteries.

The Radiola III was similar to the RS with the same tuning coil and an improved audio transformer. Unlike the RS with two antenna binding posts, the Radiola III had four antenna binding posts and a ground link which, according to the earliest instruction manuals, could be arranged in seven different combinations to select various ranges of wavelengths between the overall operating range of 200 and 760 meters. The set had two types of tuning circuits including a "more sensitive" single-tuning circuit similar to the RS and Radiola Senior, and a "more selective" coupled circuit to be used when operating near interfering stations. Shortly after production began, there was a change in the antenna circuitry that, according to the revised instruction manual, reduced the number of combinations to six, and had the effect of

Henry Albert Botkin, a commercial artist commissioned by RCA, created this widely distributed ad featuring the Radiola III. (*National Geographic*, March 1925)

reducing the maximum wave lengths which could be received from 750 meters to 640 meters. The rather complicated arrangement of switching wires for different stations could be somewhat simplified by purchasing a special antenna switch to facilitate the selection process.

The Radiola III was a very popular set with almost 300,000 sets made in four production runs which were sold over a 2½-year period. Each set was stamped with a serial number on the bottom of the wood box in groups corresponding to four different production runs (see Table 6-4). Because many of the Radiola III's were assigned serial numbers that overlap the ranges assigned to Radiola III-A and Balanced Amplifiers, the number of each type can be determined from the difference between the high and low serial numbers only if the numbers for each model are not interleaved. Evidence presented in Appendix D strongly suggests that the serial numbers were not interleaved–implying that serial numbers among the three models were duplicated.

Based on the serial number evidence reported in Appendix D, the total number of Radiola III's is estimated at 293,000, the number of Radiola III-A's at 166,000, and the number of Balanced Amplifiers at 55,000–for a total of 514,000. Had the serial numbers not been interleaved, then the corresponding numbers would be 226,100, 51,300, and 15,600 respectively, for a grand total of 293,000.

The Radiola III is one of the most common Radiolas available from the 1920's with an estimated 293,000 sets manufactured (left). The interior of the Radiola III is dominated by the large variometer which was used for both station selection and control of regeneration (right).

An antenna switch was sold on the secondary market which allowed the user to make the necessary change in the antenna circuit for different wavelengths without having to physically change the antenna wiring.

It is interesting to compare these numbers with those reported by Douglas: "*either 164,624 or 'nearly 500,000' Radiola III-A's were produced*," and "*103,701 Radiola III's were made.*"[21] The 166,000 Radiola III-A's predicted from the serial numbers is very close to the 164,624 sets from RCA records. However, it is readily apparent from the serial number data that the 500,000 option cited by Douglas for the Radiola III-A is not possible because the total number for all three sets could not have exceeded 514,000. Perhaps the 500,000 figure cited by Douglas referred to the total number of all three sets.

It is also apparent from the serial number data that there were many more Radiola III's manufactured than Radiola III-A's—about 293,000—and so the 103,701 figure for the Radiola III reported by Douglas must be incorrect. There are other reasons to believe this is true. First, the rate at which the Radiola III appears at auction and in flea markets is two to three times that of the Radiola III-A—a ratio which is also reflected by the number of serial numbers collected for both. Second, the Radiola III's were manufactured throughout all four production runs, while the Radiola III-A was manufactured throughout only one run and part of a second. Third, the Radiola III was less expensivce to own and operate. With regard to the Balanced Amplifier, there is no independent estimate of the number of these sets manufactured, but the serial number data suggests that the number was approximately 55,000.

All these factors point to a larger number of Radiola III's, and so at best, it

Table 6-4. Combined serial number evidence for Radiola III, Radiola III-A, and Balanced Amplifier

Production Run	Serial numbers	Models	Estimated No.
1	40,000-52,000	Radiola III	11,000
2	200,000-299,999 200,000-299,999 200,000-255,000	Radiola III Radiola III-A Balanced Amp[a]	100,000 100,000 55,000
3	400,000-499,999 400,000-466,000	Radiola III Radiola III-A	100,000 66,000
4	600,000-699,999	Radiola III	82,000
Total	All	All	514,000

[a]A few Balanced Amps have been observed with serial numbers outside the interval 200,00-299,000

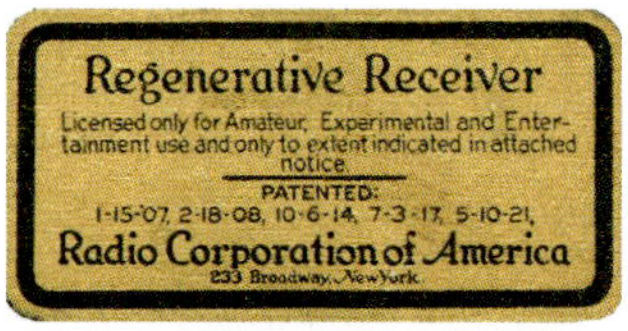

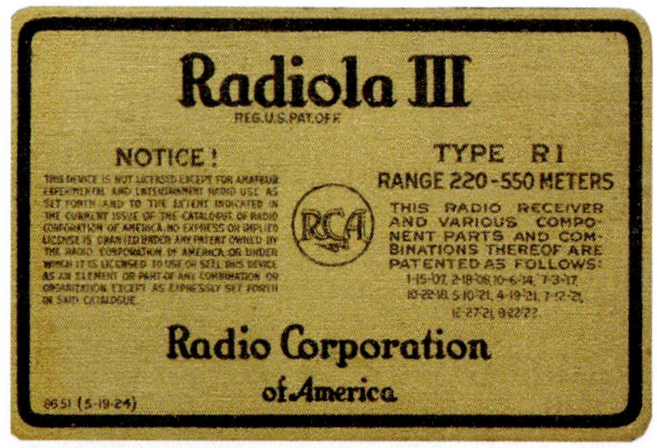

Four obvious changes were made in the Radiola III during its lifetime including (from top to bottom) a change in the decal in which the description Regenerative Receiver was replaced by Radiola III and a production date of (5-19-24) was added, a deletion of the handle from the tuning control, the addition of a fifth capacitor to the antenna circuit, and a deletion of the spring contacts on the variometer.

would appear that the production numbers reported by Douglas for the III and III-A are reversed, and at worst they are incorrect. It is likely that they are incorrect because, even if the Douglas numbers were reversed, the ratio of III's to III-A's at best would be 1.6, which is a factor of two lower than the data suggest. It is interesting to note that the total combined production of both sets of 268,325 reported by Douglas is reasonably close to the total of 260,000 inferred from the serial number data, although the later number also includes about 20,800 Balanced Amplifiers. Douglas does not give any production figures for the Balanced Amplifier.

During the last three production runs, there was one circuit change and a number of minor design changes, some of which appear to be intended to reduce production costs while others appear to be merely cosmetic in nature

Table 6-5. Design changes in the Radiola III during the second through fourth production runs

Design Feature	1st Production run 40,000-52,000	2nd Production run 200,000-299,999	3rdProduction run 400,000-499,999	4th Production run 600,000-699,999
Name on decal	Regenerative Receiver	Radiola III after 237,000-252,000	Radiola III	Radiola III
Decal Size	Small	Large after 237,000-252,000	Large	Large
Letter size on terminal labels	Small only	Small and large	Small and large	Small and large
Antenna circuit	Original design with three capacitors	Fourth capacitor added after 239,000-288,000	Four capacitors	Four capacitors
Bottom of box	Bottom visible from side	Bottom recessed into side panels	Bottom visible from side	Bottom visible from side
Wood type for box	Mahogany	Mahogany	Mahogany	Poplar after 625,000-638,000
Tuning control	Upright knob	Upright knob	Upright knob	Finger hole after 625,000-638,000
Tuning coil connectors	Spring coil	Spring coil	Spring coil	Solder lugs used after 625,000-638,000

(see Table 6-5). Sometimes the design changes coincided with the beginning of a new production run, but most of the time the changes occurred part way through a production run. Estimates of the serial numbers at which the design changes occurred are indicated in the table when known.

The earliest detectable modification to the Radiola III was an increase in the letter size of the words "antenna" and "phones" that were used to label these respective terminals. Both small and large letters can be found in the second through fourth production runs, but it is also possible that some of the panels in the first production run were made with large letters, as well. There is no logical explanation for this variation in size—perhaps there may have been two different suppliers.

The most visible change during the second production run was undoubtedly the new and much larger decal affixed to the side of the set in which the name "Radiola III" was substituted for "Regenerative Receiver." Presumably, this change was prompted by the increasingly bad reputation that regenerative receivers were rightfully earning because of the squealing they caused in neighboring receivers. While the change did nothing to reduce the squealing, it did reduce visibility into the type of receiver the consumer was purchasing. The new decal also carried the Westinghouse "Type RI" designation and a production date of (5-19-24), which may have been the approximate date that the new decal was added. It should be noted that many collectors and dealers mistakenly use the numeral "1" rather than letter "I" when describing this set.

The one circuit change made during the second production run was the addition of a capacitor between the RC grid leak device and the antenna, which had the effect of decoupling the detector grid from the antenna at low frequencies. According to the manual, this circuit change reduced the range of operating wavelengths from 200-750 meters to 200-640 meters, indicating a loss of sensitivity at the longer wavelengths. It is interesting to note that the new decal referred to in the previous paragraph listed the operating wavelengths as 220-550 meters, which was certainly inconsistent with the instructions which accompanied this set. It is quite transparent that the range appearing on the decal was selected to correspond (approximately) to the 222-

545 meter band adopted by the Commerce Department in May of 1923 for broadcast radio, and not the actual range, which would have been dependent on the antenna design among other factors.

There were other circuit changes suggested by the RCA Sales Department such as the ones to accommodate the UV-199 and UX-120 tubes, but these were intended for the after-market and were not implemented at the factory. The suggested wiring changes for the III, the III-A and the Balanced Amplifier can be found in the RCA Victor Service Notes for 1923-1928.[22] Wiring instructions were also prepared and distributed to consumers showing them how to wire batteries for use with both the UV-201A and UX-201A tubes equipped with appropriate adaptors.

There were few, if any further changes until partway through the last production run when several changes were made, presumably to reduce the manufacturing costs. First, the vertical knob on the tuning dial was removed, and a hole was provided in its place for the finger tip. Second, the mahogany in the box was replaced with a cheaper wood—most likely poplar—and the finish appeared more like flat black paint than varnish. Finally, the terminal connections to the tuning coil were changed by eliminating the spring coils and replacing them with more conventional solder lugs.

The Radiola III, III-A, and Balanced Amplifier were the first receivers marketed by RCA for which a concerted effort was made to suppress all evidence as to which manufacturer made the set, so that even the most discerning customer could not identify the manufacturer. There is no direct evidence of the manufacturer on the sets themselves, nor in advertising, instruction manuals, repair manuals, or catalogs that are generally available. As a result, many collectors and historians are confused to this day as to whether GE or Westinghouse made the Radiola III, III-A, and Balanced Amplifier. It has been claimed at various times that GE was the manufacturer,[23] that Westinghouse was the manufacturer,[24] and that both were the manufacturers.[25]

The fact that these sets were designed to use Westinghouse WD-11 tubes rather than GE UV-199 tubes has been suggested as proof that they were designed by Westinghouse, and while that may be true, it is no guarantee that they were actually *manufactured* by Westinghouse. There were other models—such as the Radiola 26—which were designed by GE using their UV-199 tubes, but which were actually manufactured by Westinghouse. William C. White writing of his experiences at General Electric states, *"As was to be expected, a real competitive spirit continued and each company [GE and Westinghouse] did its best to initiate tube and receiver designs that appealed to the RCA and would result in forcing the other company to manufacture its winning competitor's design."*[26] RCA's practice of transferring one company's design to the other company for production was, of course, intended to maintain the 60-40% sales-volume ratio for GE and Westinghouse respectively.

There are several pieces of evidence, which when taken together, make a compelling case that Westinghouse alone—and not GE—made all three sets. First, there is testimony from Westinghouse Patent Counsel Le Roi J. Williams, who stated under oath at the Federal Trade Commission hearings in 1927 that Westinghouse made a number of receivers for RCA including the Radiola III-A (there was no testimony from either Westinghouse or GE regarding the Radiola III or Balanced Amplifier).[27] Second, all three of these sets have the telltale "Type" designation on the name tags affixed to these sets, a designation that was a hallmark of all receivers and amplifiers manufactured by Westinghouse for RCA. The Radiola III decal is labeled "Type RI," the Balanced Amplifier is labeled "Type AP," and the Radiola III-A is labeled "Type RL." It is difficult to imagine that GE would have manufactured a set and affixed a Westinghouse label.

Third, the *General Electric News* published by the Schenectady Works of GE chronicled the manufacture of virtually all of their broadcast receivers during this period—particularly the ones with large production runs such as the Radiola

The radio manuals published by Westinghouse (top) and GE (bottom) support the contention that Radiolas III, III-A and Balanced Amplifier were manufactured by Westinghouse and not GE.

Super-Heterodyne AR-812 which was featured on the cover of the January 23, 1925 issue—but there is no mention at all of the Radiola III or III-A receivers in any of these monthly publications.[28] It is difficult to believe that sets with a combined production of perhaps 500,000 sets would not have been mentioned.

Fourth, the instruction manuals for the Radiola III, Radiola III-A, and Balanced Amplifier are identical in style to those of the Radiola X and Regenoflex—two sets known to be made by Westinghouse, whereas the manuals for the Super-Heterodyne and Super-VIII made by GE are distinctly different. The five manuals all have the same numbering system characteristic of previous Westinghouse manuals, while the two GE manuals have a different numbering system characteristic of previous GE manuals and instruction sheets.

Last but not least, General Electric engineer L. P. Richmond documented the approach used to bias grids on virtually every Westinghouse and GE receiver made for RCA through 1926, and in doing so he documented which company manufactured each receiver. He clearly identified Westinghouse as the manufacturer of the Radiola III and III-A.[29]

Radiola III was on the market until September 1926, at which time the remaining sets were reduced for clearance at Gimbel Brothers in New York City at the price of just $10 including Brandes headphones and two WD-11 tubes.[30] The Radiola III was one of the most popular sets on the market up to that time, and regardless of whether the number of sets was 293,000 or 226,000, it would have been *the* most popular—or at least the one with the most sets manufactured—far more than the Radiola RC and the portable Super-Heterodyne.

Radiola Balanced Amplifier (AR-806) (AP)

The Balanced Amplifier, specifically designed for use with the Radiola III, was released on March 4, 1924 shortly after the Radiola III was released. The Balanced Amplifier, as its name suggests, was a balanced push-pull arrangement, which had certain advantages for driving a speaker—not the least of which was increased power output and reduced distortion. The push-pull circuit was first used by Westinghouse in the Radiola Grand to drive a self-contained horn speaker.

The panel and case of the Balanced Amplifier matched those of the Radiola III, and the terminals of both sets were arranged so the output of the III could

The Radiola Balanced Amplifier was designed to match the Radiola III with binding posts that line up with those on the Radiola III (left). The interior of the Balanced Amplifier is dominated by the two audio transformers used in the push-pull output circuit (right).

The Balanced Amplifier was designed to connect with the Radiola III using two short connector wires.

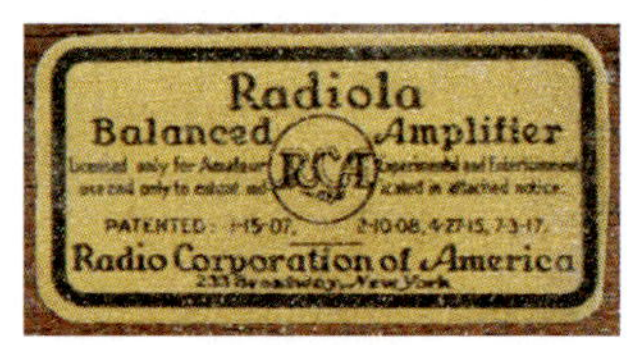

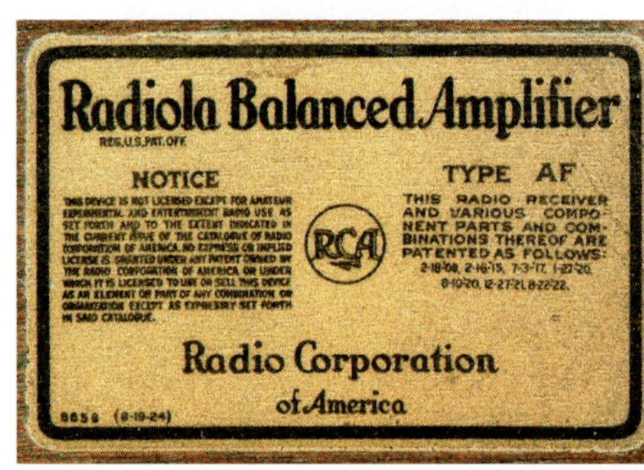

The decal of the Balanced Amplifier was changed to a larger decal, and the Type AF designation and manufacturing date of (8-19-24) were added to the decal at the same time (circa August 1924).

be fed directly into the input of the amplifier with two short interconnecting wires. In this configuration, the sets were electrically equivalent to the Radiola III-A, which was housed in a single box with a general appearance and dimensions approximating those of the pair side-by-side. It is interesting that RCA priced the Balanced Amplifier at $30 so that the price of the combination was exactly equal to the price of the electrically equivalent Radiola III-A at $65. Such a pricing policy did not encourage the average consumer to step up to the III-A at the outset, knowing full well the amplifier could be purchased later without a cost penalty.

Most of the Balanced Amplifiers have serial numbers in the interval 200,000-260,000 although a few have been observed with numbers in the low 300,000s and also some in the 600,000s. There are apparently no data from RCA regarding the number of sets manufactured, but from the available serial numbers, the number is estimated to be about 55,000. Like the Radiola III and III-A, there was a change in the decal, and in this case the Westinghouse designation Type AP was added, as well as the production date of (8-19-24). The RCA designation AR-806 did not appear on either decal or anywhere else on the Balanced Amplifier, for that matter. There were few if any other modifications to the Balanced Amplifier.

Radiola III-A (AR-802) (RL)

The Radiola III-A was electrically equivalent to the combination of the Radiola III and the Balanced Amplifier units. It was officially released on March 27,

The exterior of Radiola III-A appeared as if the Radiola III and Balanced Amplifier had been joined together at the middle.

The interior of the Radiola III-A appeared as if the Radiola III and Balanced Amplifier were joined at the middle.

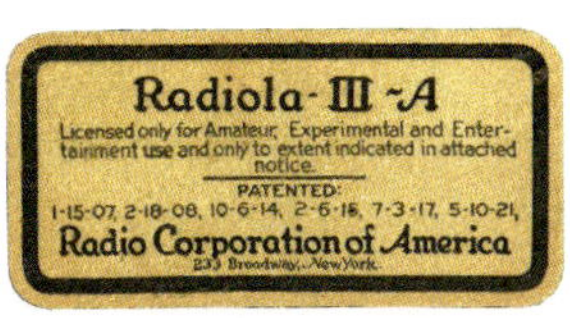

The decal of the Radiola III-A was enlarged, and the Type RL designation and manufacturing date of (8-19-24) were added to the decal at the same time (circa August 1924).

1924 at a suggested retail price of $65 with tubes shortly after the official release of the Radiola III on February 2, 1924. Unlike the early Radiola III sets with the generic designation "Regenerative Receiver," the early III-A receivers had a decal with the model designation Radiola III-A. This decal was soon replaced with one having the same model name, but containing additional information including the Westinghouse Type RL designation, a production date of (8-19-24), and the new operating wavelengths of 220-550 meters. While RCA had designated this receiver the AR-802, it did not appear anywhere on the receiver, or in the instruction manuals for that matter. For the most part, design changes to the Radiola III-A mimicked those made to the Radiola III—with the notable exception of the changes made to the Radiola III during the last production run (serial number beginning with a "6"), for which there was no corresponding Radiola III-A production run.

Douglas reported that there were either 500,000 or 164,624 Radiola III-A's manufactured,[31] and indeed the latter number is supported by the available serial number data. The 500,000 number is far too large—approximately equal to the estimated number of all three sets combined. As detailed in the paragraphs on the Radiola III, it is estimated from the serial number data that 166,000 Radiola III-A sets were manufactured in two production runs, the first run with serial numbers in the interval 200,000-299,999 and the second run with serial numbers in the interval 400,000-499,999. Unlike the Radiola III, there were no Radiola III-A sets manufactured with serial numbers in either the range of 40,000-52,000 or 600,000-690,000.

The RCA price lists suggest the Radiola III-A was on the market slightly longer than the Radiola III and Balanced Amplifier. The Radiola III and Balanced Amplifier had been removed from the RCA distributor price list dated June 1, 1926, while the III-A was not removed until the price list dated February 1, 1927.

Radiola III-A De Luxe

While the Radiola III and III-A were advertised and sold in a number of after-market cabinets in combination with a small horn speaker, RCA introduced the Radiola III-A De Luxe with a cabinet and enclosed speaker specifically designed for the Radiola III-A at a list price of $67.50 with Radiotrons, or $57.50 less accessories. The announcement of the De Luxe was made by RCA in the October 15, 1925 issue of *The Town Crier*, the magazine published by RCA for their distributors and dealers. It was described as follows: "*A beautiful cabinet worthy of any living room, great loud speaker volume, and so easy to control that any child can operate it, Radiola III-A De Luxe offers the greatest value on the market today.*"

This set consisted of a cabinet with a small horn speaker specifically designed for the Radiola III-A, but it was also used with the Radiola III and Balanced Amplifier. The Radiola III-A De Luxe was designed to use one UX-199 detector tube and three UX-120 amplifier tubes in conjunction with an adaptor such as the Naald 421-X. Unlike other after-market cabinets used with the Radiola III-A, this one was specifically sanctioned by RCA and was tested at the RCA Research Department in New York.

The Radiola III-A De Luxe, an after-market integration of the Radiola III-A with a speaker cabinet, underwent testing and evaluation at the RCA Research Department. (George H. Clark Collection, Smithsonian Institution)

The Radiola III-A De Luxe was first announced to RCA distributors and dealers at the price of $57.50 less accessories in the October 15, 1925 issue of *The Town Crier*, the house organ for RCA's distribution system.

The Westinghouse three-tube Radiola 18 prototype, while not distributed by RCA in the U.S., almost certainly formed the basis for the Canadian Westinghouse Model 53 and the Canadian GE Model 93 receivers distributed in Canada. (George H. Clark Collection, Smithsonian Institution)

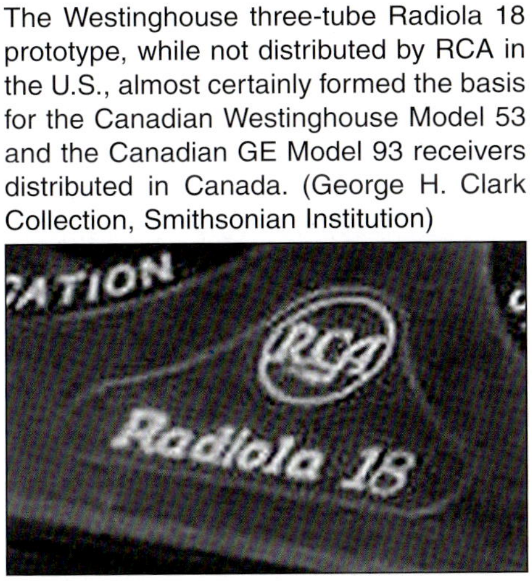

Radiola 18 Prototype

A photograph of a three-tube radio bearing the model name Radiola 18 with an appearance similar to the Radiola III was found in the George Clark's *Radioana Collection*. This set has a Westinghouse name plate with a serial number stamped on the plate which appears to be 120. While this radio was not manufactured or distributed by RCA, it is of interest for several reasons. First, this radio is virtually identical in appearance to the Canadian Westinghouse Model 53 and the Canadian GE Model 93, which were manufactured and marketed by the two respective companies in Canada (see Appendix H). The existence of this prototype would explain how the two Canadian companies—which were competitors in Canada—came to manufacture sets which were virtually identical in appearance, even though there was no similar set marketed by RCA in the U.S. to use as a copy. It almost certainly means that the two Canadian companies were authorized by RCA to copy the prototype set which had been designed for them by American Westinghouse. (It was common practice for RCA to authorize both companies to manufacture and distribute Radiolas that RCA marketed in the U.S.)

Second, it means that RCA contemplated the manufacture of a set nearly identical to the Radiola III-A but with a single audio output tube in lieu of two tubes in a push-pull configuration used in the Radiola III-A—similar to the circuits found in the two Canadian sets. Because of the Radiola 18 designation, it is more likely that this design was considered for a follow-on to the Radiola III-A in 1925 rather than in lieu of the Radiola III-A for the 1924 season lineup. It is not known why Canadian Westinghouse and Canadian GE decided to market their respective versions of this three-tube set *in addition* to the Radiola III-A while RCA marketed only the Radiola III-A. As an aside, the Radiola 18 that RCA introduced several seasons later bore no resemblance to this prototype.

Regenoflex (AR-817) (RJ)

The Regenoflex name was derived from the combination of regeneration and reflex, two key circuit features used in this receiver. It was housed in a table top cabinet with the panel in the center and two battery cabinets on either side, giving it a look similar to the GE AR-812 Super-Heterodyne. The front panel came in three different styles including a black bakelite-type material, a marbled black and brown bakelite-type material, and a wood grained panel.

The Regenoflex was manufactured with three different panel types including a black bakelite panel like the one shown here, a black and brown marbled panel, and a wood-grained panel similar to the one which appears on the Radiola X in the next section.

According to George Clark, the Regenoflex receiver was officially released on April 25, 1924, and although it appeared in national magazine ads as early as March 1924 with the rest of 1924 Radiola line, it was not featured by itself in national ads until circa May of 1924.[32] It was a middle-of-the-line receiver priced at $206 complete with Radiotrons and a horn speaker less batteries, or $150 less Radiotrons, speaker, and batteries. The set was designated AR-817 by RCA, but that designation never appeared on the set or in the instructions. However, the Westinghouse designation RJ did appear on some, but not all of the name plates affixed to the rear of the cabinet. The nameplates with the Type RJ designation had the production date of (8-19-24).

It was suggested by RCA historians that the Regenoflex was "a natural evolution" of the Radiola Grand, a four-tube regenerative receiver with a detector and two stages of audio amplification—the last stage utilizing two tubes in a push-pull configuration. The Regenoflex used four WD-11 tubes as did the Radiola Grand, but the tubes were used in a different configuration to correct some of the shortcomings of previous regenerative receivers. The first stage was a tunable RF amplifier—as opposed to a detector—to increase sensitivity and to isolate the antenna from the regenerative detector, thereby reducing re-radiation to eliminate "birdies," which were most annoying to neighboring listeners. The second tube became the regenerative detector, the output of which was reflexed

The Regenoflex came with three different panel types, one of which was this wood-grained panel.

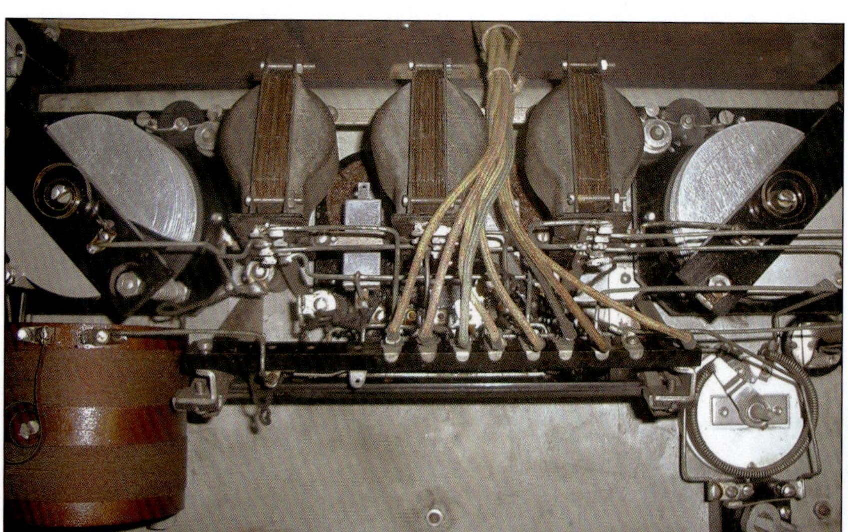

The original Regenoflex was designed with three transformers, two of which were used in the push-pull output circuit. (Stephen Brown collection)

The Regenoflex ad placed by RCA just before Christmas 1924 was one of the early color ads by commercial artist J. F. Kernan. (*The Saturday Evening Post*, December 6, 1924)

The Regenoflex was manufactured with two types of name plates, one with the Westinghouse Type RJ designation and one without any designation.

back through the first stage to provide one stage of audio amplification before the signal was fed into the final audio amplifier stage using two tubes in a push-pull configuration to drive the speaker.

The set was on the market for only a few months when it became apparent that something was amiss. While there appears to be no record of exactly what was amiss, there were major circuit changes made in the fall of 1924 in which the reflex circuit was eliminated, and the push-pull stage was rewired to form two conventional audio stages in an attempt to make up for the amplification lost by removing the reflex circuit. Further, a relatively minor change was made in the tuning coil. These sets can be easily distinguished from the earlier sets by the missing audio transformer. A revised operating manual was issued with these sets containing wording that clearly reflected the changes.[33]

Apparently the first set of changes was not quite sufficient, because additional modifications were made even later to both the tuning and the filament current control circuit. The tuning circuit was changed from a two-circuit to a single-circuit tuner configuration, which would have had the effect of increasing sensitivity at the expense of selectivity. Perhaps the first changes had diminished sensitivity to a level noticeable to the consumer. The variometer—which had provided the volume control function—was replaced with single coil for the antenna circuit and a rheostat to control the filament

The interior of this Regenoflex has only two transformers, signifying that it was from a later production run in which the two-tube push-pull output circuit using two transformers was replaced with a single-ended design using only one tube and one output transformer.

current to the RF amplifier. The rheostat was used to change the gain of the RF amplifier, thereby controlling the volume.

What is most interesting about this design change is the way it was implemented. The antenna coil was raised up from the panel to provide space underneath for the rheostat, which was positioned such that the shaft protruded through the front panel at the same position as the old variometer shaft. Thus, there was no need to change the location or marking for the volume control in any way. In fact, there is no way to distinguish among the three designs from the exterior. However, from the interior, sets with the second change can be distinguished from the earlier designs by the absence of a secondary coil within the antenna coil, and the presence of a rheostat directly under the antenna coil. Both design changes were reflected in wording of the Radiola Regenoflex instruction manuals, and the three resulting circuit designs were documented in the RCA Victor Service Notes.[34]

With all these relatively drastic circuit changes, it is odd that RCA did not also change the model designation in some fashion—perhaps by adding a dash and a letter—to distinguish among the various sets. There is no evidence on the receivers themselves, nor in manuals, ads, or any RCA documents found in the *Radioana Collection* that the Regenoflex name was changed in any way.

The Regenoflex was on the market for about a year before the price was reduced in the February 2, 1925 price sheet by $71 for the complete set and $60 for the set without Radiotrons or speaker, a sure sign that sales were flagging. By September 1, 1925 the set had been dropped from the RCA price sheets.

The variometer used as a volume control in a two-tuner circuit on the original Regenoflex (above) was replaced by a fixed coil in a single-tuner circuit mounted over a rheostat in later sets (below); the rheostat was used as a volume control by varying the filament current in the RF amplifier.

Radiola X (AR-816) (RK)

The Radiola X was electrically equivalent to the Regenoflex with the exception of a high-quality horn speaker that was integrated into the attractive wood cabinet. According to George Clark, the set was officially released on March 21, 1924, a full month before the Regenoflex. The set was advertised for $245 complete with Radiotrons and headphones—even though it had a self-contained speaker—but without batteries. RCA designated this set the AR-816, but like all the receivers Westinghouse made in 1924, none contained this designation anywhere on the set or in the instructions. A majority of the Radiola X sets have name plates with no model designation other than Radiola X, but several sets have been observed with a name plate containing the Westinghouse Type RK designation

The circuit changes made in the Regenoflex were also made in the Radiola X, presumably at the same time and in the same order. In contrast to the Regenoflex, there is some evidence that RCA intended to change the designation of the X by adding an A as an alteration designation. In a list of release dates

192 THE SATURDAY EVENING POST May 24, 1924

RADIOLA X—ultra refined receiver of the antenna type, selective and non-radiating. Complete with four WD-11 Radiotrons—everything except batteries and antenna. $245.

Operates on dry batteries.

New Joys in Radio

3,000,000 Homes listening in, when Presidents speak—when famous artists sing—when the colleges broadcast their lectures, and the theatres broadcast their programs. And with the new Radiolas, with their simplicity, and their tremendous advance in clear reception, hundreds of thousands more homes will soon be tuning in.

Of the new receivers, one of the most important is the beautifully cabineted Radiola X. For music clearer and truer. Selectivity so sharp that no near station can interfere with the programs from far away. Simplicity so perfect that any beginner can get the distant stations — quickly — easily!

Radiola X is the result of new discoveries, developed by world-famed radio engineers. A special new loudspeaker built into the cabinet is one of its outstanding achievements. It is new in essential parts that make for greater performance — and in little details that mean convenience and simplicity. Most important is the fact that improvement in reception, which usually means greater complexity, here means greater simplicity than ever!

Tune in! Turn the knob and pick your program out of hundreds—all clear—loud—and *real,* with a Radiola X.

RADIOLA REGENOFLEX, a modified Radiola X, in mahogany cabinet, with external loudspeaker. With four WD-11 Radiotrons and Radiola Loudspeaker but less batteries and antenna. . $206.

Same as above, but without Radiotrons and Loudspeaker . $150.

Operates on dry batteries.

"There's a **Radiola** *for every purse"*

Radio Corporation of America

Sales Offices:

233 Broadway, New York | 10 So. La Salle St., Chicago, Ill. | 433 California St., San Francisco, Cal.

There are many Radiolas at many prices. Send for the free booklet that describes them all.

RADIO CORPORATION OF AMERICA
Dept. 175 (Address office nearest you.)
Please send me your free Radio Booklet.

RCA — *This symbol of quality is your protection.*

Name

Street Address

City R. F. D.

State

Radiola

REG. U. S. PAT. OFF.

The Radiola X was often advertised as "Bringing *New* Joys in Radio." (*The Saturday Evening Post*, May 24, 1924)

The Radiola X with a wood-grained front panel was designed with an upright cabinet with a self-contained speaker and battery compartment.

The Radiola X has been observed with two different types of name plates, one with the Type RK designation and (9-19-24) as a date of manufacture, and another without a Westinghouse Type or date of manufacture.

prepared by George Clark dated January 19, 1939, there is an entry for a Radiola X-A with a release date of September, 1924. It is likely that this entry corresponded to the Radiola X with the reflex circuitry removed. However, since there is no other reference to an X-A on the receiver itself or any associated documentation, it is likely that RCA decided not to signal these circuit changes to the public by renaming either the Radiola X or the Regenoflex.

The Radiola X chassis is virtually identical to the Regenoflex chassis, both having three different versions (version three characterized by the rheostat is shown).

Like the Regenoflex, the Radiola X was on the market for about a year before the price was reduced in the February 2, 1925 price sheet to $150 complete from the price of $245, a sure sign that sales of the X were flagging as well. By September 1, 1925 this set, too, had been dropped from the RCA price sheets.

Wireless Specialty Offerings

Wireless Specialty was assigned the task of manufacturing two receivers for the 1924 season, both introduced several months before the official introduction of the new 1924-25 line in March of 1924. The first was the Radiola IX, designed as an adjunct to phonographs, while the second was the Radiola VII-B, a replacement for the ill-fated Radiola VII which had been introduced as a part of the 1923 season. Wireless Specialty's task was relatively simple because the Radiola VII-B was a simple adaptation of the Radiola IX panel to a table-top set in a cabinet with an enclosed speaker having an appearance evocative of the Radiola Grand.

The Radiola IX panel was designed to fit into the lid or panel of a phonograph.

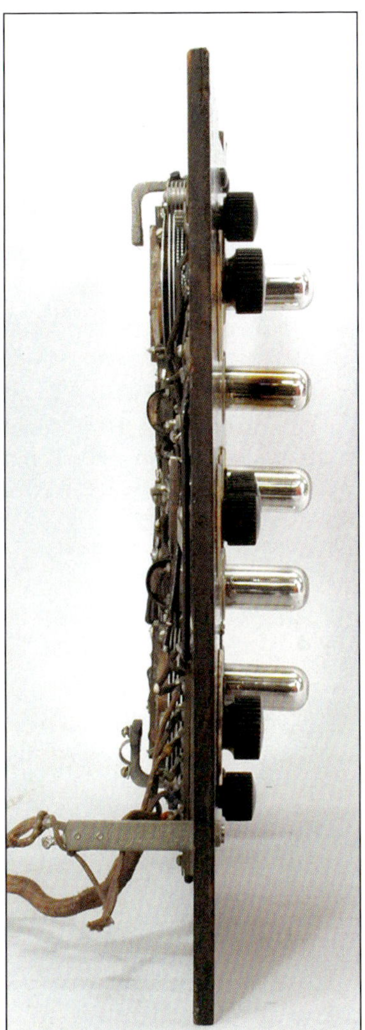

With a chassis depth of less than 3¾ inches, the Radiola IX panel easily fit into the lid of many phonographs of the day.

Radiola IX (AR-1404)

According to a dealer memo from RCA Sales Manager E. E. Bucher, the Radiola IX was developed in response to inquires from many phonograph and musical dealers for a radio set which could be installed in a phonograph.[35] RCA contracted with Wireless Specialty to design and manufacture a receiver constructed in a flat panel form that would fit into a phonograph lid compartment with as little as a 3¾ inch clearance, a top panel, or a drawer in the cabinet. A separate box was provided for the batteries, since the cabinet was not originally designed for batteries.

The Radiola IX had a highly selective two-circuit tuner with five UV-199 tubes which were used for two stages of RF amplification, one stage of detection, and two stages of audio amplification. Special radio frequency transformers equipped with electrical switches were used to obtain a relatively flat amplifier response across the broadcast band. Like other Radiolas of this period, it was designed to operate on the recently expanded broadcast band from 220-550 meters. It could receive local stations with a short antenna made by a wire draped over a picture frame or lamp, but an external antenna was recommended for long-distance reception.

The components selected for the Radiola IX had a low profile to minimize the panel depth.

A battery box was provided with the Radiola IX since the phonograph cabinets generally were not designed for battery space. (Stew Oliver collection)

The Radiola IX chassis with the Sonora name plate was first used in the Sonoradio phonograph advertised in national magazines such as the *The Saturday Evening Post* circa October 1924. (*The Saturday Evening Post*, October 25, 1924)

The Radiola IX was often advertised with the panel featured in a phonograph lid. (*The Ladies Home Journal*, March 1924)

The Radiola IX was initially produced in three versions: 1) one to fit the Victrola Model VV-100, 2) one to fit the Colombia Grafonola Model G2, and 3) a universal model that could be trimmed down to fit almost any phonograph. Some of the models came with a frame that can be seen in the *RCA Victor Service Data Book*.[36] RCA also offered custom-made panels for any set in quantities of 25 or more for the same price. Obviously, at least one other phonograph manufacturer took RCA up on their offer of custom panels because a number of Radiola IX panels appeared in Sonora phonographs with Sonora nameplates on the panel. One of the first Sonora phonograph combinations to use the Radiola IX was a Sonoradio model, which is shown on page 240 in an early ad with the Radiola IX.

The set was announced to the trade in November of 1923 and promised for delivery to dealers toward the end of December 1923, but the sets probably did not reach the public until early January 1924.[37] The earliest manuals for this set bore the date December 1923. The retail price of $215 included the receiver panel, six UV-199 Radiotrons, a phonograph loudspeaker attachment, a battery box with interconnect cables, and a special eyelet to accommodate the wiring. Unlike the Radiola VII-B, the Radiola IX was nationally advertised as early as March 1924, about the same time that ads appeared for the rest of the 1924 line.[38] The ads illustrated the flat-paneled Radiola IX in both the lid of a phonograph and horizontally positioned as a drawer panel.

The Radiola IX was on the market for about a year until it was dropped from the RCA price list dated February 2, 1925. There is no information on the number manufactured, but based on the number appearing on the resale market today and the price it fetches, it is estimated that only several thousand were made.

Radiola VII-B (AR-907)

The Radiola VII-B was designed to replace the ill-fated Radiola VII, which was removed from the market at about the same time the VII-B was distributed to dealers in early January 1924.[39] Advertisements for the set were placed in local newspapers such as the one appearing in the April 12, 1924 edition of the *New York Sun*, but it was apparently never advertised in national magazines. The set was housed in an attractive mahogany cabinet evocative of the Radiola Grand

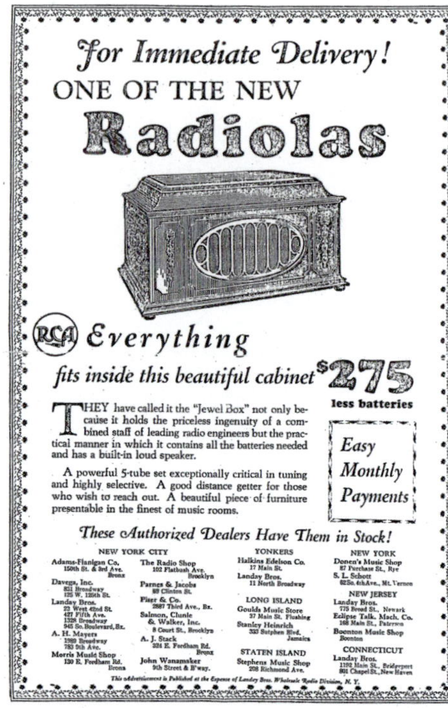

The Radiola VII-B was advertised in local publications such as the *New York Sun*.

The exterior of the Radiola VII-B was undoubtedly inspired by the Aeriola Grand.

The control panel of the Radiola VII-B was evocative of the Radiola Grand panel with the exception of a few more controls.

RCA placed a decal with the Model VIIB and AR 907 designation in center of the box lid.

Radiola
REG. U.S. PAT. OFF.

VII-B

RCA

This symbol of quality is your protection

MANUFACTURED BY
WIRELESS SPECIALTY APPARATUS CO.
for
Radio Corporation of America

JANUARY, 1924

COPYRIGHT 1924 BY WIRELESS SPECIALTY APPARATUS CO.

The instruction manual used the designation Radiola VII-B with a dash, whereas the designation on the receiver itself was Radiola VIIB without the dash.

or perhaps a table phonograph. The self-contained speaker faced forward to produce sound through an open grill structure faced with gold gauze. The set had a phone jack in the rear for listening to distinct stations that had insufficient signal strength to drive the speaker.

The Radiola VII-B design was based on the Radiola IX, and in fact it used the Radiola IX panel. The VII-B was favorably priced at $275 versus the $290 price for a comparably equipped VII. RCA assigned this receiver the designation AR-907 which appeared on the decal—unlike Westinghouse receivers, which for some reason did not have their respective AR designations assigned by RCA on their name tags. The decal on the lid and name plate on the underside of the panel characterized this set with the designation Radiola VIIB (without a dash between the Roman numerals and alteration letter), although the instructions and RCA catalog designated the set as the Radiola VII-B with a dash.

This set was on the market for about a year at the original price without any reduction. However, it was abruptly dropped from the RCA price list dated February 2, 1925, although all of the GE and Westinghouse sets for the 1924 season remained on the market. Clearly, the Radiola VII-B was not considered to be much of an asset or a source of significant revenue. Undoubtedly, this set would have been more popular if it had been nationally advertised like the rest of 1924 line, but like most other Wireless Specialty receivers offered by RCA, it languished from lack of exposure to the public. According to Douglas, 2,569 Radiola VII-B receivers were manufactured,[40] which is about what one would expect from the number of sets appearing on the current resale market available to the public and the prices they fetch (a few each year at $3000-$5000).

Early Superheterodynes

The superheterodyne receivers introduced by RCA in 1924 were neither the first nor the best, albeit the two RCA Radiola superheterodyne models were the first commercially successful receivers with combined sales of more than 148,000 sets in a period of just over a year. While it is beyond the scope of this book to detail the early history of the superheterodyne, certain receivers designed by the Experimenters Information Service and Western Electric—both of which predate the RCA superheterodyne—are noteworthy.

Superheterodynes Designed by Experimenters Information Service

Charles R. Leutz, a former engineer with American Marconi and Liberty Electric Division of Mallory, and his partner, Claude Golden formed Experimenters Information Service (E.I.S.) with the objective of providing blueprints and technical support to radio enthusiasts and experimenters. They placed their first ad in the February 1921 issue of the *QST* magazine and continued to advertise in the *QST*, *The Wireless Age*, and *Radio News* magazines for a number of years thereafter.[41] At first, they offered technical advice, then

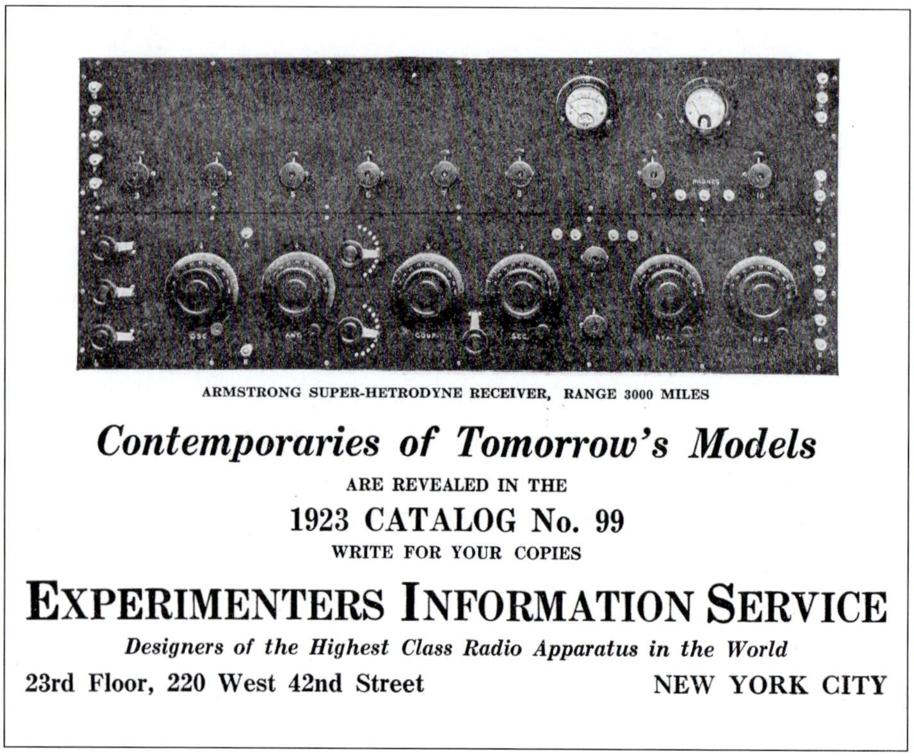

The E.I.S. Model "L" was the first superheterodyne receiver offered to the American public circa November 1922, albeit in kit form. (*Radio News*, November 1922, p. 993)

Leutz's E. I. S. Model "L" may have been the first superheterodyne offered to the American public, but it was also one of the most complicated and difficult to operate as can be inferred from the imposing control panel on the model shown here—a later version of the model shown in the original ads. (Stew Oliver collection)

The Leutz Model "C," shown here in a deluxe display cabinet, was easier to operate than the original Model "L", but it was still formidable and still required assembly.

blueprints for sets designed by Leutz, and later "knock-down" sets complete with all the parts and instructions for building receivers, much like the kits offered by Heathkit many years later. E.I.S. did not dare to sell completed sets for fear of patent infringement lawsuits by the RCA radio group.

E.I.S. offered a series of superheterodyne designs beginning with the Model "L," which was first advertised in the November 1922 issue of *Radio News*.[42] The Model "L" designation did not appear in these ads, but Leutz in his book entitled *Modern Radio Reception* clearly identified the sets appearing in the *Radio News* ads in late 1922 and early 1923 as the Model "L."[43] This set was one of the most complex superheterodyne designs ever offered to the public, requiring substantial skill to operate. Leutz soon developed a simpler and more popular design, Model "C," for which ads were place in *Radio News* beginning with the July 1923 issue.[44] These superheterodyne models, although not pre-assembled, were offered to the public at a time when no other manufacturer was offering superheterodynes, and even before they were the topic of articles in radio magazines intended for the public. Thus, in a very real sense, Leutz, through E.I.S., was the first to offer superheterodyne sets in the United States, an accomplishment in which he obviously took great pride. In his book entitled *Modern Radio Reception*, written in a distinctly self-aggrandizing tone, he claimed that he was solely responsible for popularizing the superheterodyne:[45]

> *"The facts presented in this book proves [sic] conclusively that Charles R. Leutz was solely responsible for the successful introduction of the Super-Heterodyne System to the Broadcast [sic] listeners. His designs have always been far in advance of the copying competitors and new-born 'experts' and this book gives full details of the rapid progress made to date, and gives some idea of what to expect in the future."*

Leutz not only pointed out that he introduced his line of superheterodynes to the public 14 months before RCA introduced the Radiola superheterodyne, but also that his designs were vastly superior and could be purchased at a fraction of the cost. While these assertions were true, the listener still had to assemble his own set, and then spend a month or more learning how to operate it. With regard to the operation of the *improved* Model "C," Leutz observed:[46]

> *"It must be remembered that the operator cannot consider himself to be fully proficient with less that several weeks' experience. It is, however, possible to secure very good results after three or four weeks' practice. However, a thoroughly experienced operator can bring in signals to a high audibility that an inexperienced operator would not hear at all."*

Even Leutz admitted that the operation of the earliest Model "L" was even more daunting:[47]

"The Model L Super-Heterodyne was designed principally for the experienced operator and the proper control of its component parts call for considerable skill to obtain maximum results. The adjustment of the low frequency amplifier is the most difficult operation and once that is correct the other operations are just as simple as the Model C Receiver."

It was said that RCA Engineer Elmer E. Bucher believed RCA broadcast radios should have no more than two tuning controls,[48] and so he would have eschewed the Leutz Model "L" superheterodyne. Nevertheless, shortly after introducing the Radiola superheterodyne in February 1924, RCA sued E.I.S. for patent infringement, and as a result, E.I.S. was enjoined from selling knock-down superheterodyne kits. E.I.S. was reduced to selling blueprints and parts for superheterodyne receivers, but not the kits. E.I.S. with its limited resources and relatively complicated knock-down kits was never a serious threat to RCA, so clearly RCA intended to make an example of E.I.S. to discourage other larger companies from manufacturing superheterodyne receivers.

Superheterodynes Designed by Western Electric

From the beginning of the broadcast era, AT&T and their Western Electric subsidiary took an interest in designing and marketing radio transmission equipment for broadcast stations–including the receivers used by broadcast stations to monitor transmissions. They became intimately familiar with superheterodyne receivers as a result of their participation in duplex telephony experiments using the GE superheterodyne receiver described earlier.[49] The first superheterodyne receiver designed by Western Electric specifically to monitor broadcast radio reception was the 4-A, which had a design similar to an instrument developed earlier to measure and characterize radio transmission. They favored the superheterodyne receiver because of its high selectivity, a feature which was very useful in monitoring transmissions in cities such as New York, where a number of stations were forced to occupy a relatively narrow band of wave lengths.

Western Electric, noticing the substantial increase in broadcast radio receiver sales at RCA, decided to make a move into the market, as well. The agreements between AT&T/Western Electric and RCA were made before the advent of broadcast radio, and so the agreements were not clear about the respective rights of RCA and Western Electric in the broadcast receiver market. Several significant events occurring in 1922 signaled Western Electric's intent to compete with RCA in broadcast receivers, thus initiating a protracted period of hostilities which would ultimately be resolved by binding arbitration between the two industrial giants.

First, in the summer of 1922 Western Electric opened the broadcast station WEAF in New York which directly competed with the Westinghouse station WJZ. By all accounts, this station was superior to all other stations including WJZ in terms of signal quality and clarity, and in programming content–much to the chagrin of RCA and Westinghouse. Second, in the fall of 1922, AT&T refused to lease landlines to Westinghouse for the purpose of relaying a broadcast of the World Series from the ballpark to the transmitter at WJZ.[50] Third, and perhaps most ominous, Western Electric began to sell receiving equipment to the public in the form of amplifiers and loudspeakers–although they had not yet begun to sell tuners or complete receivers to the public.

Towards the end of 1922, Western Electric decided to reveal the details of their new 4-A superheterodyne receiver to RCA management. According to Lloyd Espenschied, an engineer at AT&T from 1910 to 1954, *"A copy of an engineering report on its development was given to Dr. Alfred N. Goldsmith of the Radio Corporation of America in October, 1922, and a model of the receiver itself was given to him a few months later for test purposes."* By all accounts, this receiver was far better than any other receiver available at that time.[51]

News of the new 4-A superheterodyne sent shock waves through the senior management of RCA. According to Archer, *"Knowledge of this device literally gave Elmer E. Bucher and others responsible for RCA sales the jitters."*[52] This knowledge almost certainly played a role in the RCA decision in early 1923 to scrap the original line of regenerative receivers and develop a superheterodyne receiver during 1923 for the 1924 season.

While RCA was developing their new superheterodyne receiver in 1923, Western Electric was developing the model 4-B, a more compact follow-on receiver to the relatively large and bulky 4-A. At the time RCA began to market the Super-VIII circa March 1924, RCA serendipitously discovered that Western Electric had installed one of their new 4-B superheterodyne receivers in the White House just before an RCA Super-VIII was to be installed there, and that Western Electric was actively marketing their receiver to the public.[53] The following letter from one F. P. Guthrie of the RCA Washington office to David Sarnoff provided the tip off:[54]

The Western Electric Model 4-B superheterodyne was used in the White House demonstration much to the chagrin of RCA management.

The Western Electric Superheterodyne receiver using six peanut tubes was a splendid piece of engineering, although it did require an amplifier with three additional vacuum tubes to make a complete radio that provided the same functionality as the RCA six-tube Radiola Super-Heterodyne.

April 30, 1924

To: Mr. David Sarnoff

From: F. P. Guthrie

Subject: New Western Electric Superheterodyne—Type 4-B

Dear Mr. Sarnoff:

This afternoon I went to the Navy Department to see the new Western Electric set which has been submitted as being "much better" than our Superheterodyne.

The set which they have bears the type number 4-B, serial number twenty-six, and seems to be a six tube Superheterodyne set, using the Western Electric "peanut" tubes. The output of this set goes to an ordinary Western Electric three tube power amplifier, the output of which in turn passes to one of their new parchment-cone Rochelle-salt loud speakers.

The set is operated by storage batteries, and while it seems to give good results, it is not at all ornamental, and seems to use nine tubes to do the work that six tubes in our set do.

The tone of the new loud speaker seems to resemble very closely the tone of the loud speaker in our Radiola X. I called Commander Langworthy's attention to the fact that it would be quite a simple matter to attach the Western Electric speaker to our Superheterodyne, so as to get a comparative test of the sets themselves rather than the loud speakers only.

I will keep closely in touch with this situation and advise you of any further developments.

Yours very truly,

(Signed F. P. Guthrie)

RCA Management learned that the Western Electric 4-B Superheterodyne was believed to be much better than the RCA Super-Heterodyne from RCA employee F. P. Guthrie as a result of a visit to the Navy Department.

While the Western Electric superheterodyne was superior to the RCA receiver, in RCA's defense, one could not expect that a receiver designed for the mass retail market could compete with a unit manufactured for the industrial applications in which neither styling nor cost were major design constraints. Nevertheless, as Sarnoff pointed out in a letter dated May 19, 1924 to RCA President Harbord, *"The mere fact that the Western Electric Company is now actively putting superheterodyne sets on the market will 'upset the apple cart,' and require each of us to make some explanations to the trade and the public pending the decision of the arbitrator."*[55]

In the end, RCA and AT&T/Western Electric finally came to terms–albeit not until mid-1926–at which time AT&T and Western Electric agreed to abandon broadcast radio altogether, and terminated the cross-licensing agreements with RCA as well. In the intervening years, Western Electric had produced two additional superheterodyne receiver models, the 4-C and 4-D, both of which were very nearly the same as the 4-B in appearance and function. However, Western Electric did not actively advertise and market these receivers during the protracted arbitration period, so that sales of RCA superheterodynes were never seriously threatened by the Western Electric receivers.

Endnotes and References for Chapter 6

1 "Specifications for Broadcast Receivers for 1924," RCA Research Department Report, January 16, 1923 (kindly provided from the private collection of John Terrey).

2 John P. Wolkonowicz with Alan Douglas, John Terrey and Richard G. Wovlen, "RCA's Intended Models for 1923-1924 Season 'What Might Have Been,'" *Antique Radio Classified*, Vol. 16, No. 6, June 1999, pp. 4-11.

3 *Radioana,* George W. Clark, "The Story of the Superheterodyne," September 10, 1937 (dictated to George W. Clark by Edwin H. Armstrong).

4 Edwin H. Armstrong, "The Super-Heterodyne—Its Origin, Development, and Some Recent Improvements," *Proc. IRE*, Vol. 12, No. 5, October 1924, pp. 539-552.

5 Edwin Armstrong, "The Story of the Super-Heterodyne," *Radio Broadcast*, Vol. 5, No. 3, July 1924, p. 199.

6 *Radioana*, Letter from Alfred M. Goldsmith to David Sarnoff, Subject: Report No. 1 on Armstrong Reflex Superheterodyne Receiver, March 29th, 1923.

7 The AR-812 was initially referred to as a portable, but later it was referred to as a semi-portable.

8 *RCA Victor Service Notes for 1923 - 1928,* (Service Division RCA victor Company, Inc. Camden N.J., N.D.), Chapter on Radiola 26, 24, Super-Heterodyne and Super-VIII, p. 7.

9 *RCA Replacement Parts Catalog* A (Radio Corporation of America, New York, May 1, 1927) p. 15.

10 Letter from RCA production designer W. L. Carlson to one E. A. Laport with subject: FIRST SUPER HETERODYNE RECEIVERS, *Old Timers Bulletin*, Vol. 12 No. 1, June 1971, p. 21.

11 Ibid.

12 These words appear in revisions of both the Radiola Super-Heterodyne and Radiola Super-VIII manuals dated September 1924 and thereafter.

13 *Radio and Shortwave Treatise*, Catalog No. 25 (Radio Trading Company, New York, 1932) p. 25.

14 The Radiola Super-VIII Instruction Manual dated September 1924 contains a schematic clearly showing the Doublette horn.

15 John P. Minton, "Recent Advancement in Loud Speaker Reproduction," *The Wireless Age*, Vol. XII, No. 10, July 1925, pp. 17-18, 51-52.

16 Ibid.

17 Gleason L. Archer, *Big Business and Radio*, (The American Historical Company, Inc., New York, 1939) p. 143.

18 *The Wireless Age*, Vol. XII, No. 2, November, 1924, p. 15.

19 Floyd A. Paul and Ray Thompson, "RCA Radio Horn Speakers," *The Old Timer's Bulletin*, Vol. 30, No. 3, November 1989, p. 37.

20 Floyd A. Paul, *Horn Speaker Encyclopedia*, (Floyd A. Paul, Glendale, CA 1989) p. 26.

21 Alan Douglas, *Radio Manufacturers of the 1920's, Vol. 3 - RCA to Zenith* (Vestal Press, Ltd., Vestal, New York, 1991) p. 34.

22 *RCA Victor Service Notes for 1923 - 1928,* (Service Division RCA Victor Company, Inc. Camden N.J., N.D.) Chapter on Radiola Balanced Amplifier, p. 4-11.

23 "Radiola III," *The Old Timer's Bulletin*, Vol. 12, No. 2, September 1971.

24 Alan Douglas, "Radio Corporation of America Part VIII," *Radio Age*, Vol. 9, No. 5, May 1983, p. 3.

25 Robert Murray, "Broadcast Receiver Manufacture by General Electric and Westinghouse in the First Decade of RCA," *The AWA Review*, Vol. 17, 2004, p. 115-117.

26 William C. White, *The Early History of Electronics in the General Electric Company*, (William C. White Papers Collection, Schenectady Museum & Suits-Bueche Planetarium, Schenectady, New York, prepared 1953-1955) Chapter V, p. V-8.

27 *Federal Trade Commission vs. General Electric Company, et. al., Radio Case, Brief of Facts*, (Sidney C. Ormaby Company, 217 Broadway, New York, 1928) Vol. 7, p. 3515 (testimony given by Westinghouse patent counsel Le Roi J. Williams on November 16, 1927).

28 *General Electric News Schenectady Works*, Vol. 9, No. 2, January 23, 1925, pp. 2, 15.

29 *Radioana*, L. P. Richmond, Notes on Grid Circuits, CWC45-666A, November 23, 1926, page II.

30 *The New York Times*, September 12, 1926, p. 22.

31 Alan Douglas, *Radio Manufacturers of the 1920's, Vol. 3 - RCA to Zenith,* p. 34.

32 *The Wireless Age*, Vol. XI, No. 11, August 1924, back cover.

33 *Radiola Regenoflex Instruction Manual*, RCA Corporation, I. B. 5311-A, N.D.

34 *RCA Victor Service Notes for 1923 - 1928,* (Service Division, RCA Victor Company, Inc. Camden N.J., N.D.) chapter on Radiola Regenoflex.

35 *Radioana*, E. E. Bucher, Memo to Wholesale Distributors, "Subject: Radiola IX (Radio Phonograph Adjunct)," CWC45 6194, November 24, 1923.

36 *RCA Victor Service Data Volume I 1923-1937* (RCA Victor Division of Radio Corporation of America, Camden, NJ, 1st Edition, 1944) p. 10-A.

37 Ibid.

38 *Ladies Home Journal*, March 1924, p. 220.

39 *Radioana*, E. E. Bucher, Memo to Wholesale Distributors, "Subject:-Radiola VII-B," CWC 45 6204, January 10, 1924.

40 Alan Douglas, *Radio Manufacturers of the 1920's, Vol. 3 - RCA to Zenith*, p. 27.

41 *QST*, Vol. IV, No. 7, p. 95.

42 *Radio News*, Vol. 4, No. 5, November 1922, p. 993.

43 Charles R. Leutz, *Modern Radio Reception* (Experimenters Information Service, Inc., New York City, 1924) p. 307. Leutz was inconsistent in his use of parentheses around the model letters (e.g., Model "L" versus Model L) to describe the sets in this book, but the advertisements and most of the instructions used the form with parenthesis and so the form with parentheses is used here.

44 *Radio News*, Vol. 5, No. 1, July 1923, p. 65.

45 Charles R. Leutz, *Modern Radio Reception,* Note page.

46 Ibid., p. 41.

47 Ibid., p. 94. In the particular paragraph quoted here, Leutz chose to use the model form without parentheses.

48 Gleason L. Archer, *Big Business and Radio*, (The American Historical Company, Inc., New York, 1939) p. 92.

49 "Radio Connects Ship and Shore Telephone," *The Wireless Age,* Vol. 10, No. 3, December 1922, pp. 41-43.

50 WJZ was able to lease lines from Western Union to broadcast the 1922 World Series.

51 Lloyd Espenschied, "Discussion of 'A History of Some Foundations of Modern Radio-Electronic Technology,'" Proc. IRE, Vol. 47, No. 7, July 1959, pp. 1253-1258.

52 Gleason L. Archer, *Big Business and Radio*, p. 92.

53 Ibid., pp. 142-143.

54 Ibid., pp. 145-146.

55 Ibid.

This remarkable Radiola operates directly from your Lighting Socket

Single finger tuning. Full orchestra volume. And lighting socket operation.

THESE are new things—but not too new to be thoroughly tried, for the latest developments of radio have had a twelve months' *road test* in RCA Radiolas. And now, where others are taking their first trial steps, RCA Radiolas have had time to be tested for a year—perfected.

In radio, it is *reality* that counts. Tune in a Radiola 28 with a Loudspeaker 104 placed across the room. They are playing a symphony—a famous guest conductor wielding a frenzied baton. Turn up the volume—thirty violins are playing together. Hear the high sweetness of the piccolo—the deep melody of the cellos—the zoom of the big bass drums.

Straight through the concert program, every number comes in with true spirit, capturing the brilliance of the Scheherazade—or the vivid symbolism of a Strauss symphony.

You need never fear that a sudden climax will come with a crash. That day is past, for the Loudspeaker 104 has a reserve of power to handle full orchestral volume!

Unless you have heard the new Radiolas, you can have no conception of what radio is today. When you hear them, you will turn in your old set for a Radiola that you will *keep*.

RCA Loudspeaker 104, power speaker—used with Radiola 28 it operates the set on the house current without batteries. Complete, $275

Radiola 28 is an eight tube super-heterodyne. With loop attached (adaptable for antenna if you prefer it). For operation with dry batteries or *no* batteries. $260

A. C. Package, for adapting Radiola 28 for use with Loudspeaker 104 without batteries, $35

Antenna Coupler, for adapting Radiola 28 for use with antenna, $4.25

Buy with confidence where you see this sign.

Authorized Dealer RCA

RCA Radiola

MADE · BY · THE · MAKERS · OF · THE · RADIOTRON

RCA

RADIO · CORPORATION · OF · AMERICA · NEW YORK · CHICAGO · SAN FRANCISCO

RCA heavily advertised the fact that the new line of Radiolas could be operated directly from the lighting socket. (*Country Life*, November 1926, p. 96)

Chapter 7

Lighting-Socket Power Eliminates the Battery

1925-1927

The 1925 season was unique for RCA in several respects. First, the introduction of sets powered by the lighting circuit would dramatically change the performance characteristics of radios in the future by eliminating the design constraints imposed by the power drain on the relatively expensive dry cell batteries, as well as eliminating the fuss and muss associated with tending to rechargeable storage batteries. While the full potential of the socket-power technology would not be realized until the advent of tubes with filaments powered by AC current which would eliminate the need for expensive power converters, the public immediately embraced these sets despite the high price tags. While AC sets would dominate the RCA agenda in the years to come, battery sets remained a staple in the RCA product line because, while 90% of the urban population had AC power by the early 1930's, only about 10% of the rural population did.

Second, the development of the cone speaker, particularly the Rice-Kellogg electrodynamic speaker, would also dramatically change the future of radio by substantially improving sound fidelity and shrinking speaker size so that it could be contained in a relatively small cabinet. Again, the full potential of this new speaker technology would come over time as AC tubes and the alnico permanent magnet were developed and commercialized. It is interesting to note that it was the introduction of the electrodynamic speaker and the associated power supplies required to drive these speakers that accelerated the adaptation of socket power for the remainder of the radio circuits.

Third, RCA made a strategic alliance with the Victor Talking Machine Company in 1925 similar to the one made with the Brunswick-Balke-Collender Company in 1924 whereby RCA provided Radiola receivers for integration into Victor phonograph lines. While the sales of these Radiolas for phonographs initially represented a relatively small portion of the total revenues for RCA, the strategic alliances formed with Victor in particular would prove to be critical to RCA at the end of the decade when the government initiated antitrust proceedings against RCA. RCA would acquire Victor Talking Machine Co. and shortly thereafter form the RCA Victor Co. to successfully compete with both Westinghouse and GE in the 1930's and beyond after they divested themselves of RCA in the process of settling the antitrust action.

The 1925-1927 Radiola Line

The new model line for 1925-1927 consisted of six new receivers (seven if one includes the Radiola 28 A-C Operated model), three speakers, and a power amplifier and power supply operating from the AC lighting circuit as shown in Table 7-1. These models were introduced in two phases, the first consisting of two portable receivers designated Radiola 24 and Radiola 26 which were introduced in late spring and promoted for outdoor activities in the summer. The two receivers were nationally advertised together in bucolic settings such as campsites, canoes, and roadside picnic areas as early as June of 1925.[1] After

102 **THE SATURDAY EVENING POST** *January 15, 1927*

The high sweetness of the violins carries the theme; the deep drums beat the rhythm; a great symphony holds a million audiences in its spell.

RCA has made radio *not only* greater—*but* simpler

The new RADIOLAS *that are such remarkable* MUSICAL *achievements are at the same time the* simplest *of musical instruments*

EVERY RCA Radiola shows this trend toward simplicity—combined always with new achievements in performance.

Radiola 20—twenty times as selective as the ordinary antenna set. With performance and tone quality that challenge competitors at any price! With Radiotrons, $115

Consider the eight tube super-heterodyne, Radiola 28. It is the culmination of twenty years of radio development—a remarkable instrument of music and of radio. Yet it stands as a simple, charming piece of furniture, and captures the magic of distant music with the turn of a single control!

In the RCA Radiolas of today, single control has been tried and proved . . . . it is

Ads for the 1925-27 line appeared in national magazines from October 1925 to late summer of 1927 when the new line was introduced. (*Saturday Evening Post*, January 15, 1927, pp. 102-103)

no experiment. With these same Radiolas, operation from the lighting socket has had many months of testing and perfecting. They have shown the remarkable musical possibilities of radio when hooked up with the power of the lighting circuit.

And practically every new feature that is heralded in radio today was first developed by RCA engineers—or by their associates in the General Electric and Westinghouse laboratories. In RCA Radiolas, modern radio has had the "road test" of experience and there is nothing to compare with them in tone or in performance.

They have made the old type radio obsolete, and have brought a new conception of radio—and of music—to thousands of homes. Whatever price you plan to pay for radio, be sure to hear the RCA Radiolas. From the least expensive Radiola to the most remarkable "electrical" model, each represents the most advanced radio of today—and the most thoroughly tried and proved.

Radiola 30—combines in one luxurious cabinet all that's newest in radio. An eight tube super-heterodyne, single controlled. An enclosed loop—no antenna. An RCA power speaker. And no batteries. Just plug it in and tune in! Complete, $575

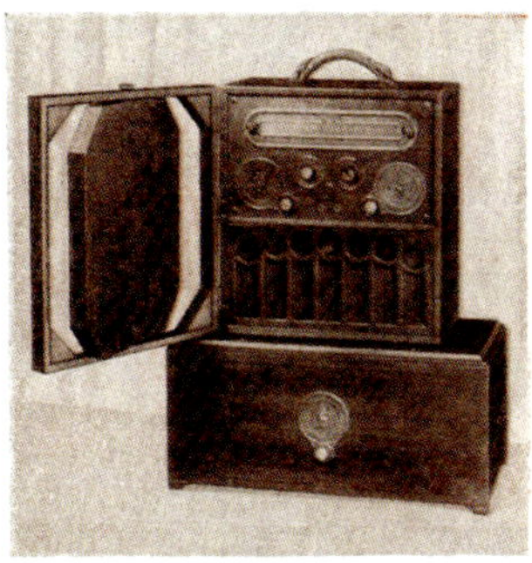

Radiola 26—six tube super-heterodyne. As a portable it has a loop in the cover, loudspeaker inside, place in the back for small batteries. At home it has a fine walnut cabinet that holds larger batteries and has a coupler for an antenna. Cased in richly grained walnut. With 6 Radiotrons, $225

RCA Loudspeaker 104—the speaker that set a new musical standard for radio. Clear at a whisper—and clear at the volume of a brass band! Operates Radiola 25 or 28 (adapted) on 50-60 cycle, 110 volt A. C. lighting circuit. Used with other makes of sets, it eliminates just the "B" batteries. Complete, $275

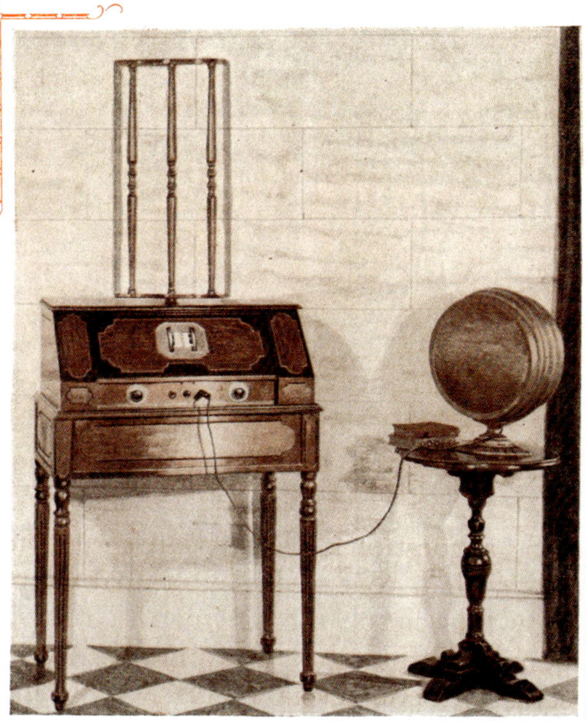

Radiola 28—eight tube super-heterodyne with loop. Tuned with a single finger! With eight Radiotrons, $260
An inexpensive antenna coupler adapts it for an outdoor antenna, if you prefer it. $4.25

RCA Loudspeaker 100—Clear through the whole tone range—and clear at any volume you give it. $35

Radiola 25 or 28 can be adapted from a battery set to a "lighting socket" set for use with RCA Loudspeaker 104. A. C. Package adapter. $35

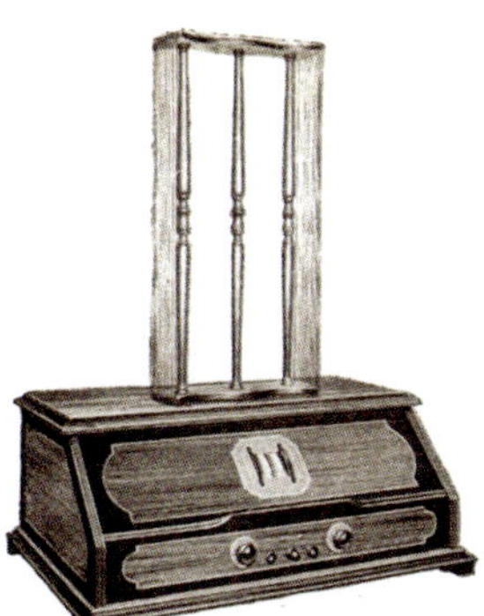

Radiola 25—six tube super-heterodyne with all the far-famed super-heterodyne quality of tone and performance—and the "catacombed" sealing of the vital parts. Single controlled. With 6 Radiotrons, $165

RCA Loudspeaker 102—is the Loudspeaker 100 with the addition of a power amplifier (RCA Uni-Rectron) that puts great volume at your command, with tone that is real—true—undistorted. Complete, $140

(Operates on 50-60 cycle, 110 volt A. C. lighting circuit.)

Buy with confidence Authorized Dealer RCA *where you see this sign.*

RADIO CORPORATION OF AMERICA

RCA Radiola

MADE · BY · THE · MAKERS · OF · THE · RADIOTRON

NEW YORK · CHICAGO SAN FRANCISCO

the Radiola 24 was dropped at the end of 1925, the Radiola 26 was advertised either by itself or with the rest of the line introduced later in 1925.

The remainder of the 1925-1927 line, which included four new receivers designated Radiolas 20, 25, 28, and 30, was introduced to the public at the 1925 Radio World's Fair held September 14-19 at the New York City Armory.[2] The

Table 7-1. Radio receiver apparatus offered by RCA in the 1925-27 season including hold-over models (shaded portion)

	Model	Apparatus Type	Cabinet Style	Tubes[a]	Power	Intro. Date	Discon'd. By	List Price[b]	Mfg. By
1925-27 Season New Models	Radiola 20 (AR-918)	Regenerative (with TRF)	Cabinet or Floor	5	Battery	11/6/25	Sep. 1928	$115	W GE
	Radiola 24 (AR-804)	Super-heterodyne	Portable (Leather)	6	Battery	5/12/25	Dec. 1925	$195	GE
	Radiola 25 (AR-919)	Super-heterodyne	Wood Cabinet	6	Battery[c]	9/10/25	Jul. 1927	$165	W GE
	Radiola 26 (none)	Super-heterodyne	Portable (Wood)	6	Battery	6/23/25	Jul. 1927	$225	W
	Radiola 28 (AR-920)	Super-heterodyne	Floor	8	Battery[d]	Aug. 25	Sep. 1928	$260	W GE
	Radiola 28 A-C Operated (AR-969)	Super-heterodyne	Floor	8	AC Only[e]	9/29/25	Sep. 1928	See text	W GE
	Radiola 30 (AR-921)	Super-heterodyne	Console	8 (2)	AC	Nov. 25	Feb. 1927	$575	W GE
	Uni-Rectron (AP-935)	Power Amplifier	Metal Box	3	AC	Sep. 25	Dec. 1927	$105	GE
	Duo-Rectron (AP-937)	"B" Battery Eliminator	Metal Box	3	AC	Sep. 25	Dec. 1927	$65	GE
	100 (UZ-915)	PM Speaker	None	N/A	N/A	Sep. 25	Jul. 1927	$35	GE
	102 (UZ-913)	PM Speaker	None	N/A	AC	Sep. 25	Jul. 1927	$140	GE
	104 (UZ-914)	Dynamic Speaker	Floor	1 (4)	AC	Sep. 25	Sep. 1928	$245	GE
Holdover Models	Radiola III (AR-805)	Regenerative	Box	2	Battery	2/2/24	Sep. 1926	$24.50	W
	Radiola III-A (AR-806)	Regenerative	Box	4	Battery	3/27/24	Feb. 1927	$49.50	W
	Balanced Amplifier (AR-806)	Push-pull Amplifier	Box	2	Battery	3/4/24	Sep. 1926	$23.00	W
	Radiola Super-Het (AR-812)	Super-heterodyne	Cabinet (Portable)	6	Battery	3/4/24	Jun. 1926	**$116**	GE
	Radiola Super-VIII (AR-810)	Super-heterodyne	Console	6	Battery	3/18/24	Feb. 1927	$340	GE
	UZ-1325	Horn Speaker	None	N/A	N/A	Oct 24	Feb. 1927	$18	GE

[a] Additional rectifier and ballast tubes are shown in parentheses

[b] Prices of all apparatus appearing in the September 1, 1925 price list included tubes except for Radiola Super-Heterodyne

[c] Operated on AC power with Model 104 Radiola loudspeaker using AC Package UP-971 @$35

[d] Operated on AC power with Model 104 Radiola loudspeaker using AC Package UP-972 @$35

[e] Radiola 28 (A-C Operated) was an AC-only set in which the UP-972 AC Package was installed at the factory rather than by the dealer

full line also appeared in RCA distributor and retail dealer price lists dated September 1, 1925,[3] but according to George Clark, only Radiolas 25 and 28 were "officially released" by RCA in September, while the official release of Radiola 20 was delayed until November. However, none of the receivers was advertised until November 1925 when the full line appeared in *Radio Merchandising*, published for the trade,[4] and it was not until December 17, 1925—almost too late for Christmas—that the first ad for the new line appeared in the *Saturday Evening Post*, the major national magazine used by RCA for advertising Radiolas in that era. George Clark reported the release date of February 15, 1925 for the Radiola 30, but this is clearly in error. He may have meant February 15, 1926, but even that date is doubtful because the Radiola 30 appeared in ads beginning in December 1925.

The RCA catalog published in 1925 carried all the Radiolas for the new season including the Radiola 26, but for some reason the Radiola 24 was not included. (RCA catalog for 1925)

The new receivers in the 1925-1927 line were essentially of three types: a new tuned-radio-frequency or TRF circuit, the six-tube superheterodyne circuit using basically the same catacomb used in the 1924 line but fitted for UX-based tubes, and a new superheterodyne circuit using an eight-tube superheterodyne catacomb, also with UX-based tube sockets. The Radiola 20 used the new TRF circuit, Radiolas 24, 25, and 26 used the six-tube superheterodyne catacomb, and Radiolas 28 and 30 used the new eight-tube superheterodyne catacomb. The eight-tube superheterodyne differed from the six-tube superheterodyne in that the two new tubes were used to create a separate oscillator stage and an additional IF amplifier stage to avoid reflexing the IF signal back through the first RF stage. This design had the benefit of eliminating noise problems that plagued the six-tube superheterodyne, and it also eliminated the problem of tuning at shorter wavelengths inherent with the second-harmonic design.

The RCA catalog for the 1925-1927 line published in 1926 contained all of the apparatus listed in Table 7-1 with the notable exceptions of the portable Radiola Super-Heterodyne and Radiola 24. The Radiola 24 had been dropped by the end of 1925, and while the portable Super-Heterodyne introduced in 1924 was still on the market, RCA obviously opted to market the portable Radiola 26 in lieu of the Radiola Super-Heterodyne and Radiola 24 portables. The catalog was supplemented with a set of slick brochures for each of the receivers—except for Radiolas 24 and 26 for which a single brochure was prepared covering both sets but featuring only the Radiola 24 on the cover.

To supplement its catalog, RCA developed a set of attractive brochures for dealers to help market their 1925-27 line.

The Expanded Broadcast Band

Beginning in November of 1924, fifteen additional frequencies from 1360 to 1500 KHz (200-220 meters) were authorized for Class A broadcast stations, which were outside of the previous band of 220-550 meters (550-1360 KHz). Despite the creation of this new band, RCA felt sufficiently comfortable to introduce three new sets in 1925 (Radiolas 24, 25, and 26) based on the six-tube superheterodyne catacomb from the previous year which could not receive the new band. This decision was probably based on the fact that there was no immediate rush for license applications in this new band for several reasons—not the least of which was that many existing broadcast receivers could not receive stations transmitting in the new band.

However, RCA realized it would not be long before these bands would be populated by broadcast radio stations, and so they also developed two new sets which did cover this band, including a low-end TRF receiver, the Radiola 20 and a high-end eight-tube superheterodyne, the Radiola 28 with a separate oscillator that could easily tune to the higher frequencies. The six-tube second-harmonic superheterodyne would eventually become obsolete, but not until mid-1927 after a respectable 1½-year run.

Standardization of Receiver Manufacturing

The year 1925 would mark the end of the era of direct competition between GE and Westinghouse whereby each company vied to have RCA accept their respective apparatus for sale each year. While RCA might pick the design of one company or the other, RCA arranged for the two to share their design drawings so that both would manufacture essentially identical sets in the ratio of 60-40% respectively, thus assuring that their revenue agreement was satisfied independently of consumer demand for each model. With this change, it became very difficult for the consumer at that time—as well as the historian and collector now—to determine by inspection which company made a particular set.

Consolidation of manufacturing was reflected in the standardization of owner's manuals beginning with the Radiola 20. Note the difference in the styles of the manuals for Radiola 24 and 26 which were printed before consolidation.

The standardization in manufacturing was reflected by standardization of the owner's manuals, which—with the notable exception of Radiolas 24 and 26—had virtually identical covers and interiors with a common formatting for the 1925-27 season and beyond. The manuals for Radiolas 24 and 26, which were published before standardization, not only had covers which were different in detail, but also had interiors with a different identification numbering scheme and different formatting. Before the 1925-27 season, these differences could easily be used by an informed observer to determine which company manufactured which set, but not so with the manuals published after standardization.

Reliable documentation clearly shows that both GE and Westinghouse manufactured Radiolas 20, 25, 28, and 30.[5] However, there were many variants of these sets, and most variants were assigned different AR designation numbers depending on whether they were intended for battery power, AC power, a Victor Victrola, or a Brunswick phonograph. Evidence suggests that GE and Westinghouse did not manufacture each variant, particularly those for Brunswick and Victrola applications where small production runs were often the case. The evidence is also clear that GE designed and manufactured the 100, 102, and 104 speakers and associated power supplies, and while this author has not found any direct evidence that Westinghouse also manufactured these items for the 1925 season, it is certainly possible. Evidence relating to subsequent model years shows that both companies manufactured a full range of apparatus for RCA including receivers, power supplies, speakers, and other accessories.[6]

The Radiola 24 and Radiola 26 receivers were exceptions in that GE alone manufactured the Radiola 24 while Westinghouse alone manufactured the Radiola 26. These two receivers were distinct from the rest of the 1925 line because they were developed, ordered, manufactured and introduced to the market significantly ahead of the remainder of the 1925 line. It is likely that the joint manufacturing agreements had not been put in place at the time RCA ordered these sets.

Since it is not possible to associate most of the apparatus with a particular manufacturer from 1925 onward, the apparatus presented here are arranged according to the category of apparatus rather than by manufacturer—the approach used in previous chapters. The remaining part of this chapter is divided into the following sections: receivers, AC power amplifiers and power supplies, and loudspeakers.

Receivers for the 1925-1927 Season

The Radiola receivers are presented in the approximate chronological order they were introduced by RCA beginning with the Radiola 24 and 26. The remaining sets, which were offered later in the year, are presented in ascending numerical order—which also happens to be the order of increasing price and quality.

Radiola 24 (AR-804)

According to George Clark, the Radiola 24 was introduced on May 12, 1925, a date which correlates well with the date of April 1925 appearing on the earliest instruction manuals and the first national ads which appeared in June 1925.[7] Oddly, it was always advertised in conjunction with the Radiola 26, but unlike the Radiola 26 it was never advertised alone. This set which was assigned the preliminarily Model No. 100 had been under development by GE for almost a year, as evidenced by the date of October 1, 1924 on a photograph of what appears to be a mature prototype. The production units had name plates with individual serial numbers and the RCA model number AR-804, but no date of manufacture. Receivers have been observed with serial numbers between 540,375 and 548,068, indicating that approximately 8,000 were manufactured.

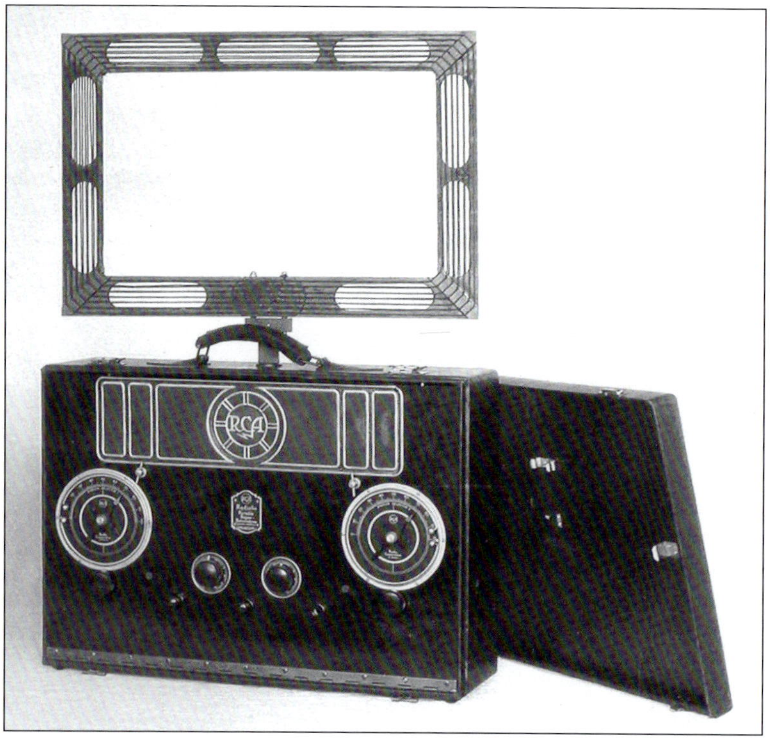

A prototype of the Radiola 24 clearly shows the leather handle without a buckle that was also used on the production unit. (George H. Clark Collection, Smithsonian Institution)

The genuine leather used on the Radiola 24 has deteriorated on many of the brief-case style cabinets, revealing an orange color characteristic of the leather underneath the black dye.

The Radiola 24 was basically the 6-tube Radiola Super-Heterodyne in a smaller package than the AR-812 portable. It was enclosed in a genuine black cowhide leather case that according to RCA advertising could not be distinguished from an ordinary traveling bag. Unfortunately for today's collectors, the leather on the cases of many sets have deteriorated badly, and the leather handles have also deteriorated and are often missing. Some sets on the collector market today have been equipped with incorrect replacement handles with buckles such as the one that was used with the AR-812 Super-Heterodyne or Radiola 26. The original Radiola 24 had a handle without buckles which is very difficult to replace without damaging the set in some fashion.

The antenna stored in the lid of the case was easily mounted on the top of the case with a two-pronged plug (left). The interior of the Radiola 24 reveals the presence of a rather large folded horn in addition to the catacomb used in the larger Super-Heterodyne portable the year before (right).

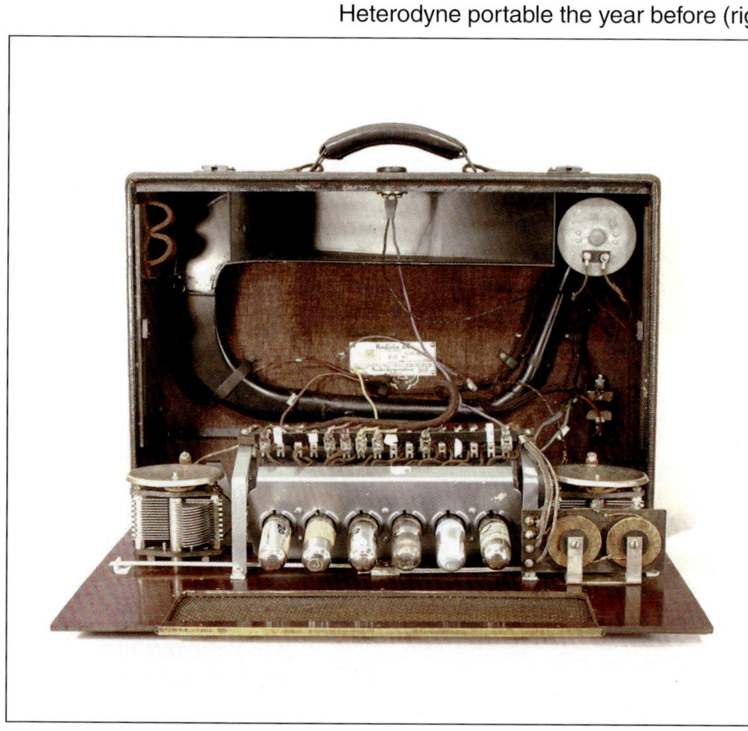

The name plates were stamped with individual serial numbers in the range of 540,000 to 549,000.

For normal operation, the cover to which the loop antenna was attached for storage was opened to expose the front panel. The antenna was removed from the lid and inserted into a socket on the top where it could be freely rotated. The two tuning dials were equipped with cardboard disks similar to those on the Radiola Super-Heterodyne upon which stations could be entered. The receiver was also equipped with a new horn speaker designed specifically for the set. The entire set was self contained, and with batteries weighed about 40 pounds. The set was equipped with a four-way plug such that the lighter-weight portable batteries could be cut out in favor of standard size batteries for home use. Unlike the Radiola 26, the Radiola 24 did not have a separate battery box for home use.

The Radiola 24 was the last receiver manufactured by GE alone before RCA instituted the joint manufacturing arrangements whereby virtually all receivers were made by both manufacturers from the same drawings. The Radiola 24 was a portable receiver with the standard six-tube superheterodyne catacomb used in the Radiola Super-Heterodyne the year before. The Radiola 24 was introduced in the late spring of 1925 in time for use out of doors during the summer. This receiver was heavily advertised in different rustic settings along with the very similar Radiola 26 portable superheterodyne manufactured by Westinghouse using the same superheterodyne catacomb. The Radiola 24 was offered at $195 while the Radiola 26 was offered at $225.

It is difficult to understand why RCA would decide to offer two very similar sets at the same time with a similar price point. Portable radios had become very popular by 1925,[8] and so perhaps RCA wanted to have two entries into the market, or perhaps RCA did not want to choose between the two models presented by GE and Westinghouse, and decided to let the market make the choice. Alternatively, since the sets were not introduced at the same time, perhaps there was a production problem and so both were introduced as a hedge. For whatever reason RCA chose to offer both models, it is clear that the Radiola 24 was the poor stepchild from the outset. The Radiola 24—unlike the Radiola 26—did not appear in any RCA catalog, in any RCA dealer catalog, or on any RCA dealer price list. While it did appear in national ads, it appeared only with the Radiola 26 and not by itself.

The major feature distinguishing the two receivers was captured in this statement appearing in an RCA advertising brochure:

> *"Either you want a receiver that is exclusively portable and that can be laid aside for one in a more luxurious cabinet when you are at home, or else you want a receiver that can be transported about, yet handsome enough to match the fittings of the living room. Radiola 24 supplies the former need. Radiola 26 the latter."*

It is unlikely that many people wanted to spend $195 for a portable receiver like the Radiola 24 only to be put away at home in favor of yet another equally expensive receiver, when one could purchase a single receiver like the Radiola 26 at $225 which was suitable for both portable and home use. Indeed, the Radiola 26 outsold the Radiola 24 by a wide margin, and the Radiola 24 disappeared from the market by the end of 1925. According to Douglas, 7,079 Radiola 24 sets were manufactured versus 19,507 Radiola 26 sets.[9]

Ads for the Radiola 24—which invariably referenced the Radiola 26, as well—had bucolic themes ranging from picnics to camping trips. (*National Geographic*, July 1925)

Radiola 26 (no AR model number)

The Radiola 26 was introduced on June 23, 1925—less than a month after the Radiola 24—at a suggested list price of $225. It was developed by Westinghouse at the same time GE was developing the Radiola 24. A number of photographs of developmental models, such as the one shown on page 262, were taken during the second half of 1924 and the first part of 1925. The Radiola 26 was heavily advertised—first in conjunction with the Radiola 24, but later by itself after the Radiola 24 was discontinued.

Name plates were attached to production units with individual serial numbers and a production date of (4-22-25). Receivers have been observed with serial numbers between 200,000 and 219,937, indicating a total of about 20,000 were manufactured. This number agrees well with Douglas, who reports 19,507 were manufactured.[10] It is interesting to note that, unlike all the other receivers manufactured by Westinghouse for RCA in the 1924-25 era, the Radiola 26 was not assigned an AR number by RCA. It also appears that Westinghouse did not assign the receiver its usual two-letter "Type" designation either.

The Radiola was very similar to the Radiola 24 with a few notable differences. Most noticeable was the handsome walnut cabinet with a handle on

The Radiola 26 name plate bears witness to the fact that no RCA AR number or Westinghouse Type letters were assigned to this receiver.

The American Magazine 91

There is no loneliness where there is a Radiola.....

Radiola 26—at home

Sealed against summer heat and moisture

In the mountains, where lake mists rise at night . . . on the coast, where sea-laden breezes blow inland . . . in the city, where open windows let the dust fly in . . . you can take a Radiola Super-Heterodyne, knowing that neither weather nor dust can harm it.

Every Radiola Super-Heterodyne is permanently sealed against air, dust and moisture. The delicate parts that hold the secret of its accurate selectivity and its fine tone quality are fitted into a stout little metal box, dubbed a "catacomb"—then a weather-proof, fast-hardening compound is poured in—and the cover is permanently *sealed* on.

The catacomb—sealed

Super-Heterodynes have withstood severe tests. A Radiola 26 has safely pioneered with daring explorers into the Sierra Madre mountains of old Mexico—carried by pack-mule—often following an unmade trail high above the clouds.

When an RCA dealer tells you that a Radiola is a *permanent* investment, you can prove it for yourself by listening to the quality of its performance. And you know that in the Super-Heterodyne that quality is proof against dust, temperature, and moisture—sealed in to outlive the years.

For summer camping trips, Radiola 26 is a *portable* Super-Heterodyne with loop and loudspeaker built in—and battery space inside. Yet, see its catacomb—with the full, exact, sealed-in quality of the "Super-Het." For fine city or country homes, it has a bigger battery box to stand upon—but for picnic and camping days, just lift off the portable, and take it along to the fields. It is made to be carried. And it is made to give Super-Heterodyne performance. Complete with 6 Radiotrons and extra battery box, with antenna coupler $225.

RCA

RCA-Radiola

MADE · BY · THE · MAKERS · OF · RADIOTRONS

Authorized Dealer RCA

RADIO · CORPORATION · OF · AMERICA · NEW YORK · CHICAGO · SAN FRANCISCO

The Radiola 26 was heavily advertised by itself after the Radiola 24 was discontinued. (*The American Magazine*, 1926)

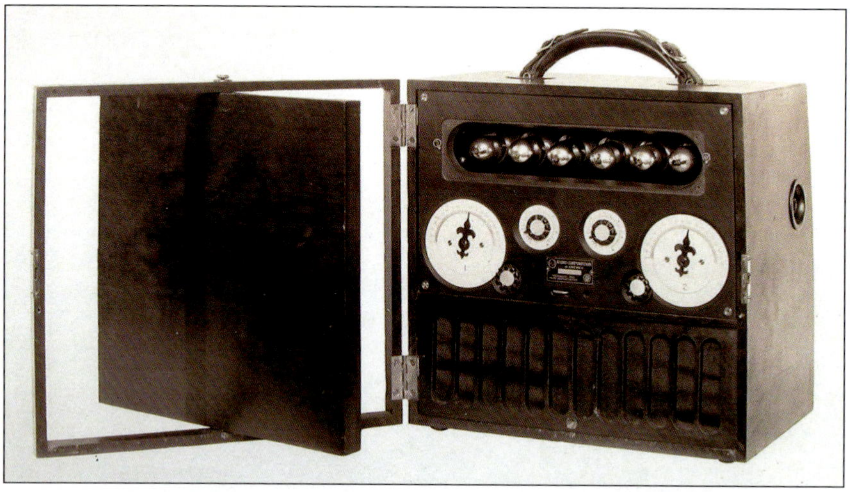

The Radiola 26 prototype clearly shows the leather handle with the buckle-type strap that was used on the production model. (George H. Clark Collection, Smithsonian Institution)

The tuning dials of the Radiola 26 production units did not have provisions for paper disks to mark the location of stations.

The speaker on the Radiola 26 was folded back on itself which made it much smaller than the one used on the Radiola 24, resulting in a more compact cabinet.

the top. The loop was integrated into the front door of the cabinet, where it could be rotated when open. The front door could also be removed and attached to the back for reception and tuning while it was being carried, or for use with an external antenna in conjunction with the battery box (see next paragraph). The set used the same standard six-tube superheterodyne catacomb as the Radiola 24, but unlike the Radiola 24, the tuning dials did not have cardboard disks for marking stations. A "Radiola Calibration" card was provided with each set which was ruled with horizontal and vertical columns with headings for "Wavelength," "Frequency," "Call-Letters" and "Settings of Station Selectors." The instructions advised the owner to write down the pertinent information whenever a station is heard so the setting can be duplicated whenever desired.

The Radiola 26 used a unique horn speaker in which the narrow portion of the neck was folded back through the flared opening to reduce the overall dimensions. As a result, the Radiola 26 cabinet was smaller than the Radiola 24 case. The popularity of the Radiola 26 portable as compared to the Radiola 24 portable may well have been due to the smaller cabinet which can be traced to the speaker design.

This receiver was also designed for home use in conjunction with an equally attractive walnut battery box which had two distinct functions. The set could be placed on the top of the box where it made electrical connections to the batteries

For use in the home, the Radiola 26 was designed to sit on top of a battery box with an exposed mating plug which lined up with a mating plug on the bottom of the Radiola 26—a design which presaged the modern laptop computer docking station.

The loop antenna when hung from the rear of the cabinet lined up with a coil in the rear of the Radiola 26 battery box, which provided for optimal RF coupling so that the set could be used with a long-wire antenna when attached to a terminal on the rear of the battery box.

via a protruding plug, not unlike the docking station for a laptop computer. It was also equipped with a tuning circuit for use with a long-wire antenna. What is most interesting about this tuning circuit consisting of a coil and adjustable capacitor is that it made no direct contact to the Radiola 26 circuitry. Instead, the loop antenna was removed from its position on the front lid and attached to the rear of the set with thumb screws. The loop was then in position so that it could be inductively coupled to the coil directly beneath in the battery box. The set was tuned in the usual way and then the tuning knob on the front of the battery box was adjusted for maximum volume. Presumably, the Radiola 26 with the two-circuit tuner had better selectivity that the Radiola 24, which did not have such a tuning circuit. It is possible that the increased selectivity in the home was also a major factor in making this set more popular than the Radiola 24.

The Radiola 26 was on the market until mid-1927 despite the fact that like all RCA six-tube superheterodyne sets, it could not tune to the broadcast band from 1350-1500 KHz. The portable Radiola 26 was absent from the RCA price list dated July 15, 1927, and it was the last portable RCA offered until 1932. The portable boom ended in 1926 almost as fast as it began in 1924, ending portable offerings from virtually all manufacturers for a number of years.[11]

Radiola 20 (AR-918)

The Radiola 20 was officially released by RCA on November 6, 1925, although instruction manuals were dated as early as September 1925, and ads began to appear in the *Saturday Evening Post* with the rest of the new line in late October 1925. It was offered at $102.50 without Radiotrons, $115 with Radiotrons, and at prices between $133 and $350 for a choice of speakers including the Model UZ-1325, Model 100, Model 102 and Model 104. The Radiola 20, a tuned-radio-frequency or TRF receiver, was priced at the low end of the RCA line, whereas all other models offered by RCA in the 1925-27 season were of the superheterodyne type which commanded higher prices because of their superior performance characteristics.

The receiver was housed in an attractive wood cabinet with a lid for access to the interior and a sloped panel on which the tuning controls were mounted—a new configuration for RCA which was also used with the Radiola 25 and 28

April, 1927 27

The Home Run A. C. manicures the club quarters to honor the star outfielder of the Blue Sox

Yes! RCA has sold more sets than any other maker

Wherever you go you see them. George got a Radiola 20 for Christmas. So did Chub—he's got his on battery eliminators. And Tom's still a bit cocky because they got a 28, with the Loudspeaker 104. He's always turning it up loud as a brass band, to impress the crowd.

You see Radiolas everywhere, because there are more of them sold. And there are more of them sold because they're better sets and better values. Put that up to Dad. He'll see it.

When the family begins thinking about getting a new set, here are some facts you will want to have handy, about one of the best buys they can make today. It's the set that thousands are changing to now—Radiola 20.

The Radiola 20—operated either with batteries—or from the lighting socket—is an outstanding example of popularity based on performance—and value. It is not the lowest priced set you can buy. But it has improvements you will not find in sets at twice the price.

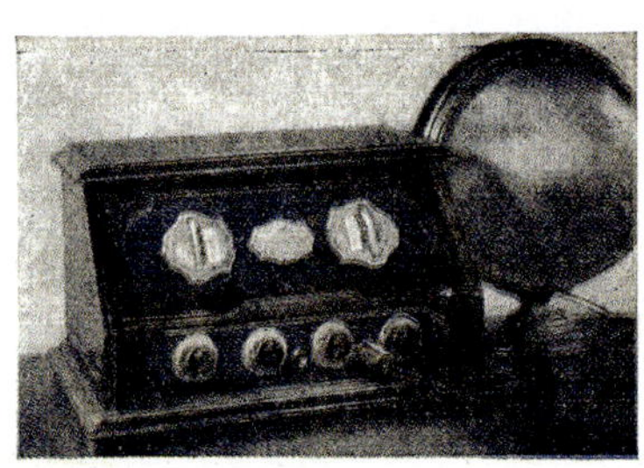

Radiola 20, less equipment *$78*
RCA Loudspeaker 100 . . . . *$35*

Even in far costlier sets you will not find higher quality of musical reproduction . . . a big statement . . . but one your own ears can prove. Thousands have made the test!

In distance reach, its five tubes have the sensitivity of seven tubes in larger sets. This is an important new improvement, for it makes possible great economy of operation.

In picking out only the station you want, it is twenty times as selective as the ordinary antenna set! Another big improvement.

And its power tube in the last stage keeps the tone clear and true, when you turn up the volume.

Ask for a demonstration at any RCA Authorized Dealer.

Authorized Dealer RCA

RADIO CORPORATION OF AMERICA

RCA-Radiola

MADE · BY · THE · MAKERS · OF · THE · RADIOTRON

NEW YORK · CHICAGO
SAN FRANCISCO

RCA commissioned commercial artist Harry Beckhoff to produce this unusual cartoon ad for the Radiola 20. (*The American Boy*, April 1927)

The Radiola 20 was designed with a two-tone rectangular cabinet and front panel with large thumb-wheel tuning dials—design themes which were carried through to Radiolas 25 and 28, as well. The battery and tube compartment was accessed through the hinged top.

The first RF amplifier tube on the Radiola 20 was set well apart from the other tubes, presumably to avoid some type of unwanted interaction.

superheterodynes. It was designed with four UX-199 Radiotrons and one UX-120 Radiotron which were configured with two stages of balanced RF amplification, a regenerative detector, and two stages of audio amplification. The UX-120, a new dry-cell power amplifier tube that superseded the UV-199 for the audio output stage, was used to produce higher speaker volumes with less distortion. The new UX-199 was essentially the same as the UV-199 with the exception of a different pin arrangement in the base. A volume control was provided on the front panel which functioned by controlling the filament current of the first RF amplifier tube. There were two jacks on the front panel to accommodate a volt meter for the purpose of assisting the operator in keeping the filament voltage at or below 3.3 volts to maximize performance and filament life of the relatively fragile UX-199 tubes.

The Radiola 20 used the latest design features to achieve higher sensitivity and selectivity than previous designs without increasing the number of tubes. RCA advertised the selectivity to be a factor of 20 higher than equivalent receivers.[12] Undoubtedly, this was due in part to a combination of a two-circuit antenna coil and two stages of tuned radio-frequency amplification preceding the detector. The station tuning was simplified by ganging together the three tuning capacitors in a straight line configuration mounted on a new vertical drum control dial. The variable condensers were linear with frequency which permitted the stations to be received at uniform intervals on the selector drum, and so the scale was calibrated for frequency rather than wavelength. A provision was made for recording the call letters of the stations directly on the dial with a soft pencil, and many of the early stations can still be seen on receivers found today. Vernier tuning controls were provided for use in fine tuning of stations when receiving signals over more than ordinary distances.

Radiola 20 sets used in cabinets with legs can be easily be distinguished from table-top sets by the name plate affixed to the lid with the words "Floor Model."

The drum on the right side of the sloping panel controlled regeneration, which was relatively independent of frequency over the operating bandwidth because the inter-electrode capacity was compensated by small neutralizing capacitors. As a result, once the regeneration was set for a given signal strength, it did not have to be readjusted when tuning to other stations of a similar signal strength. The set was designed to operate over the new broadcast band from 550 Khz to 1,500 KHz (approximately 550-200 meters). An external antenna of 75-150 feet was recommended, with connections using pin jacks on the rear of the chassis marked "long," "med," and "short" referring to long, medium, and short antennas.

At the beginning of 1927, the price of the Radiola 20 with Radiotrons was decreased to $89.50. Towards the end of 1927, the set was also sold as a floor model at $105 in an attempt to stimulate sales. The floor model which appeared in an RCA ad in *Radio Retailing* looked very much like the Radiola 28 floor model.[13] In fact, the Radiola 28 cabinet was used for the Radiola 20 floor model, modified by inserting two wood strips on either side to fill the space left by the narrower Radiola 20 control panel.[14] Sets designed for these cabinets can easily be distinguished by the name plate affixed to the top with the words "Radiola 20 Floor Model," and by the battery wiring diagram on the inside lid entitled "Radiola 20 Floor Model" which shows the batteries in a rectangular configuration in a cabinet. The Radiola 20 was on the market until the summer of 1928, most likely at the end of the radio season in September—it was no longer listed in the October 23, 1928 RCA price list. According to Douglas, a total of 135,121 were manufactured.[15]

Radiola 25 (AR-919)

According to George Clark, the Radiola 25 was officially released by RCA on September 10, 1925. The set was offered for $165 with tubes but without batteries or speaker, and from $200 to $410 depending on the choice of speakers from among the UZ-1325 horn or Models 100, 102, or 104 cone speakers. It used the same standard six-tube second-harmonic superheterodyne catacomb introduced in 1924, but with UX tube sockets for five UX-199 Radiotrons and one UX-120 power amplifier for the audio output stage. The set

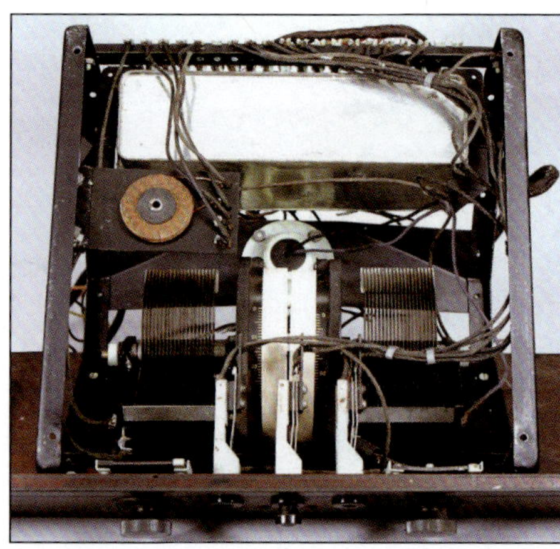

The thumbwheel dials and attached tuning condensers dominated the bottom of the Radiola 25 chassis.

The Radiola 25 had an appearance similar to the Radiola 20 except that the two thumb wheel tuning controls were set together, loosely joined by a friction clutch to allow coarse tuning with a single finger.

The UV-887 mounted on the right front side of the cabinet base was removed in later sets.

also contained a special protective tube, the UV-877, which was described in a sales pamphlet as *"an electrical brain, if you will—which protects the set against accidental short circuits."* It was actually a ballast tube that would prevent the plate current from an accidental short circuit through the Radiotrons by lighting up and limiting the voltage.

This feature must not have been very successful, because it was removed from the set by December of 1925, not long after the set was introduced. The Instruction Manual dated December 1925 had a change notice inserted with the following statement: *"The design of the Radiola 25 Super-Heterodyne with which this booklet is shipped, has been changed to eliminate the use of the Radiotron Model UV-877 (Protective Tube), and the socket in which it was inserted."* Thus, the modified version of the Radiola 25 can be easily identified by the absence of the UV-887 tube and socket, which was attached to the base at the front right side of the receiver cabinet.

This set was one of the first A-C operated radios sold by RCA. It was actually designed to operate in three separate ways including 1) complete dry cell operation, 2) combination dry cell and 110 volt AC lighting-circuit operation, and 3) complete 110 volt AC lighting-circuit operation. For combination dry-cell operation, the plate voltage ("B" battery) was supplied by the AP-937 Duo-Rectron described later in this chapter. For complete AC operation, the "A," "B," and "C" battery voltages were provided from the Model 104 speaker in conjunction with the UP-971 AC package, which could be purchased for an additional $35.

The major parts of the UP-971 AC package consisted of a Condenser Unit, a cable harness, a resistance strip, a volume control, and a UP-591 resistor unit that was mounted on a tube base which was used to replace the UX-874 ballast tube in the Model 104 speaker. Model 104 speakers used for AC operation with either the Radiola 25 or Radiola 28 can be easily identified by the presence of the UP-591 resistor mounted on a UX tube base in lieu of the UX-874 ballast tube. The installation required significant rewiring which required the use of a soldering iron, and so the instructions that came with the kit recommended that arrangements be made with the RCA dealer from whom the package was purchased to convert the set.

An odd-looking UP-591 resistor unit mounted on a UX tube base replaced the UX-874 ballast tube in the Model 104 power amplifier chassis when the Radiola 25 was converted to AC operation using the UP-971 AC package.

The UP-971 AC package was used to convert the Radiola 25 to AC operation in conjunction with the Model 104 speaker. (RCA Catalog 1925)

The style of the cabinet was similar to the Radiola 20 with a sloped panel and two tuning controls similar to the Radiola Super-Heterodyne from 1924. However, the two station selectors on this set were mounted on two drums with a friction clutch such that the drums could be turned either simultaneously or independently. The drums had graduated scales upon which the various station locations could be written, preferably in pencil. The two were rotated together to locate a particular station, and then were adjusted independently when fine tuning was required. A single counterweight was used in earlier sets to keep the drum from rotating after it was set, but one was not sufficient for that purpose, so in later sets a counter weight was added to both drums.

A loop antenna fit into the top of the lid which could be rotated to maximize signal strength and minimize interference from stations located in directions away from the desired station. There were jacks for head sets and a separate volume control for speaker operation on the lower front panel. There were also two jacks in the lower front panel so the operator could monitor filament voltage with a small volt meter, sold separately. Since the cabinet contained both the batteries and the antenna connection, the entire set was totally self-contained.

Unfortunately the standard six-tube superheterodyne was designed when the broadcast band was 550-1,350 KHz, while in 1925 the broadcast band was extended to 1,500 KHz, so the set could not receive stations at the high end of the frequency band. Undoubtedly this caused complaints, particularly as that portion of the band became occupied by more and more stations. This shortcoming was probably one of the reasons the Radiola 25 was discontinued in mid-1927 well ahead of the Radiola 20 and the eight-tube superheterodyne receivers. This set was absent from the RCA price list dated July 15, 1927. Douglas reported that 94,815 sets were made, significantly fewer than the lower priced Radiola 20, but slightly more that the higher priced Radiola 28.

Radiola 28 (AR-920)

According to George Clark the Radiola 28 was released in August of 1925, although the entire line was not officially introduced until a month later at the 1925 Radio World's Fair held September 14-19, 1925. The Radiola 28 was not introduced to the trade until November of 1925 when an ad appeared in *Radio Merchandising* introducing the receiver with the rest of the line.[16] The set was first offered with eight Radiotrons and no speaker at $260, with the UZ-1325 horn at $278, the Model 100 speaker at $295, the Model 102 speaker at $400, and the Model 104 speaker at $505. The Radiola 28 was RCA's first eight-tube superheterodyne, with one of the additional tubes being used to create a separate oscillator stage and the other to create a separate IF stage, thereby eliminating the need to reflex the signal through the first RF stage. The receiver used seven UX-199 tubes and one UX-120 power amplifier in the audio output stage. The set also used the same UV-877 protective tube used in the Radiola 25, but it too was removed a few months after the set appeared on the market.[17]

The Radiola 28 had an appearance very similar to the Radiola 25 with the same cabinet style including a sloping panel with two tuning drums, filament and volume controls along the lower front, and a rotatable antenna protruding from the top. The controls and operation were very similar to the Radiola 25 with the notable exception that the Radiola 28 covered the full broadcast band from 550 to 1500 KHz. Like the Radiola 25, the Radiola 28 also came with a rotatable antenna which could be attached through a hole in cabinet lid.

The receiver came with a small table cabinet which contained either the batteries, or the AC package when configured for light-circuit operation. The table had an appearance similar to the base cabinet of the model 104 speaker with which it was designed to be used. Because the table held the batteries, the Radiola 28 cabinet itself was smaller than the Radiola 25 cabinet, which had to be large enough to contain all the batteries. The Radiola 28 cabinet was hinged

The tuning dial clutch was nothing more than a bracket attached to one dial holding a friction pad which rubbed against the other dial.

The antenna which was inserted into a hole in the top of the Radiola 25 could be rotated for directional reception.

A volt meter fitting into jacks on the front panels of both Radiolas 25 and 28 could be purchased separately to monitor the filament voltage for more accurate rheostat adjustment.

The Radiola 28 came with a desk table which contained either the batteries or the UP-972 AC package for AC operation with the Model 104 speaker.

The eight-tube Radiola 28 interior was significantly different than the six-tube Radiola 25 with its three tuning capacitors and one long coil.

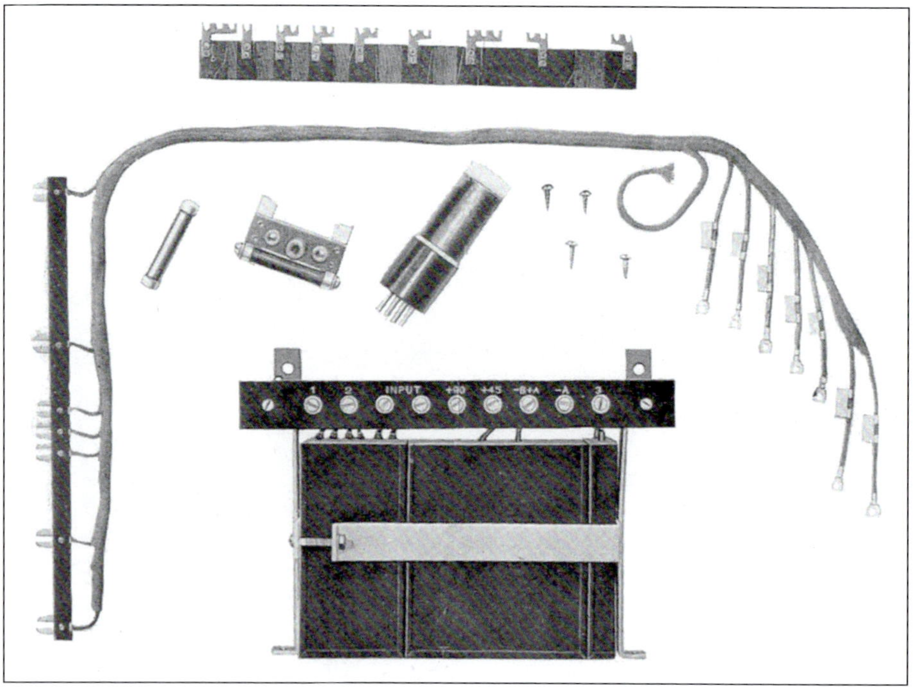

The UP-972 AC package was used to convert the Radiola 28 to AC operation in conjunction with the Model 104 speaker. (George H. Clark Collection, Smithsonian Institution)

to the back of the table so it could be raised to gain access to the battery cabinet. Many of the sets available to the collector today have been separated from the table, but the tell-tell screw holes for the screws that held the hinges are invariably found at the back of the base.

Like the Radiola 25, the Radiola 28 was designed for three modes of operation including battery only, combination battery and AC, and AC only. The same Duo-Rectron used with the Radiola 25 could be used to supply "B" battery voltage for combination operation, but a different AC package, the UP-972, was required for full AC operation in conjunction with the Model 104 speaker.

The Radiola 28 was on the market well into 1928, the only set from the 1925-27 season to remain on the market into 1928. The set appeared in the RCA distributor price list dated February 1, 1928, but was absent from the list dated October 3, 1928. According to Douglas, 84,798 receivers were manufactured.[18]

Like the Radiola 25, the Radiola 28 had a hole in the cabinet lid for a rotatable loop antenna.

Radiola 28 (A-C Operated) (AR-969)

The Radiola 28 (A-C Operated) was a factory built AC model designed specifically to be used with the Model 104 speaker. It came with a separate instruction manual and was assigned the RCA model number AR-969—distinct from the AR-920 model number assigned to the set designed for battery operation. According to George Clark, the AC version was introduced on September 29, 1925, just a month later than the battery version. The AC version of the Radiola 28 was apparently never advertised, and it never appeared in any price list known to this author. Consequently, the list price of the AC version is not known. It may have been set at the price of the battery version ($260), or it may have been set at $295 to equal the price of the battery set plus the price of the required UP-972 AC package ($35) which was included in the AC version.

There are a number of differences between the AC and battery version, although the escutcheon plates identify both sets as a Radiola 28 without distinguishing between the two. The differences are noted in the *RCA Victor Service Notes* and in the Instruction Manual entitled *Radiola 28 Super-Heterodyne (A-C Operated)*, with the most noticeable differences being the presence of the filament voltmeter pin jacks and second-stage audio phone jacks on the lower front panel of the battery version which are absent from the AC version.[19] Another major difference is that the AC version did not use the audio output tube UX-120. Instead, it was wired to use the UX-210 audio

The RCA Victor Service Data book, which characterizes the AR-969 as the Radiola 28 DC, is clearly in error; the name plate from a battery version of the Radiola 28 has the number AR-920, while a name plate from an A-C Operated version has the number AR-969.

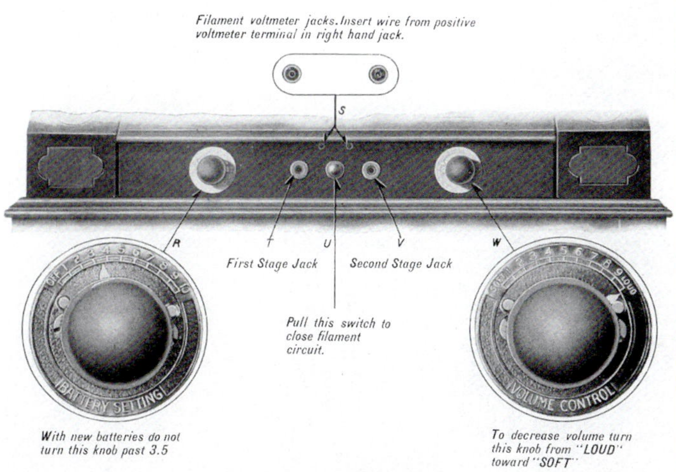

Fig. 10—Vertical Control Panel

1—Insert Loud Speaker plug in jack *V*.
2—Pull out filament switch *U*.
3—Turn "VOLUME CONTROL" knob *W* to "LOUD".
4—Turn "BATTERY SETTING" knob *R* to 3.5.

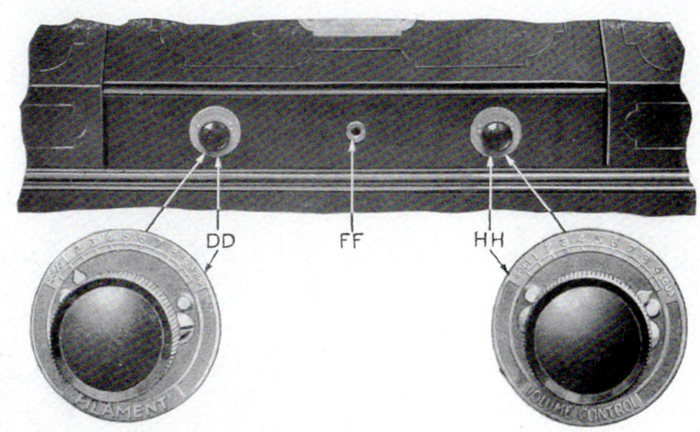

Fig. 5. Radiotron Control Panel of Radiola 28 (A-C Operated)

DD "FILAMENT" Knob
FF Telephone Jack
HH "VOLUME CONTROL" Knob

The most noticeable difference between the Radiola 28 (left) and Radiola 28 A-C Operated (right) version is the arrangement of jacks and switches on the front panel of each. (RCA Radiola 28 and Radiola 28 A-C Operated Instructions)

output tube that came with the Model 104 speaker, leaving the last socket on the catacomb unused. Also, the AC version came with the UP-972 AC package already installed in the battery compartment in the table, and so it was not necessary to purchase it separately.

Radiola 30 (AR-921)

The Radiola 30 was the first AC-only radio offered by RCA.[20] According to George Clark, the Radiola 30 was released February 15, 1925, but this date is clearly in error. Clark may have meant 1926, but that date would have been too late, because the set not only appeared in the September 1, 1925 RCA distributor price list, but was featured in national ads in December 1925, a sure sign in previous years that the receiver was in the stores.[21] However, the date of February 1926 is found on the *first* version of the instruction manual, which would support George Clark's date of February 15—assuming his 1925 entry for the year was a typographical error. In any event, the Radiola 30 was the last set introduced in the 1925-27 season, significantly later than the other sets.

RCA engaged Everett Shinn to create an elaborate color ad for the Radiola 30 featuring "*Week-end evenings at the Wainrights'*." (Unknown source, 1926)

The Radiola 30 consisted of the Radiola 28 chassis, Model 100 speaker, and loop antenna in a very elaborate cabinet with inlays. (Joe Formanek collection)

This receiver consisted of four assemblies: the Radiola 28 A-C Operated, the standard Model 100 cone speaker, the Model No. AP-947 Rectifier-Power-Amplifier Unit which was similar but not identical to the one found in the Model 104 speaker, and the loop antenna. It was described in RCA brochures as a receiver *"primarily intended for those who desire the ensemble in one cabinet, and who desire not only all which the art is capable of in the way of radio reproduction, but who also wish their home to be graced by a fine piece of furniture."*

That the Radiola 30 used the A-C Operated version of the Radiola 28 is evident from the front panel with only a single phone jack. (Dan Middleton collection)

The Radiola 30 was divided into four compartments containing the power amplifier, the Model 100 speaker, the Radiola 28 chassis and the rotatable antenna. (Dan Middleton collection)

That the A-C Operated version was used in the Radiola 30 is evident from the front panel which has only one phone jack.

The Radiola 30 was introduced at the price of $575 complete. It is interesting to note that RCA heavily advertised the combination of a Radiola 28 at $260 with the Model 104 speaker at $245 and the AC package at $35 for a total of $540. Consequently, one could have purchased the same receiver with the superior Model 104 speaker for $35 less, although it would have been in two cabinets.

As shown in the accompanying figure, the Radiola 30 is divided into four compartments containing the receiver, the speaker, the AC power supply, and the antenna. The receiver is located at the lower center of the set behind a narrow desk-fall which can be lowered to access controls and tune stations. The receiver can then be closed up for listening. The right hand compartment (facing the rear of set) contains the loop antenna which can be rotated by means of a thumb wheel which is exposed when the compartment door is opened. The upper center compartment contains the Model 100 speaker which is accessible only from the rear. The left hand compartment (facing the rear) contains the AP-947 Rectifier-Power-Amplifier Unit which converts the AC power to the DC power needed for plate voltages, filament current and biasing voltages.

The Radiola 30 was on the market until the end of 1926, but it was absent from the RCA dealer price list dated February 1, 1927, although Radiolas 20, 25, 26, and 28 were still on the list. It was also dropped from RCA advertising beginning in March, leaving the RCA without any console until the replacement Radiola 30A was introduced in August 1927. According to Douglas, 10,835 were manufactured, which was a respectable number for a set priced at $575—although not nearly as large as the numbers for the other receivers in the lineup.[22]

AC Power Amplifiers and Power Supplies for the 1925-1927 Season

The introduction of the new UX-210 audio power amplifier tube and the increase in the number of tubes in the superheterodyne receiver from six to eight increased the power drain on the "B" battery to an excessive level. In response to the increased "B" battery current requirements for their sets, RCA marketed two separate devices: a stand-alone Uni-Rectron power amplifier with a UX-210 audio output tube deriving power from the AC lighting circuit, and a Duo-Rectron "B" battery eliminator to provide plate current to all the tubes including the high-power audio output tube found in sets sold by RCA in 1925.

The common theme in both devices was the rectifier tube—marketed by RCA as Rectron tubes—which converted the AC current from the lighting

The Uni-Rectron and Duo-Rectron were marketed as a power amplifier and "B" battery eliminator respectively for all battery radios—not just for Radiolas. (*Popular Radio*, December 1925)

circuits to the DC current needed for the plate circuit. The power amplifier was designated the AP-935 Uni-Rectron, while the battery eliminator AP-937 was designated the Duo-Rectron. Presumably the "Uni" referred to the fact that the Rectron used in the Uni-Rectron had a single plate used as a half-wave rectifier, while the "Duo" referred to the Rectron used in the Duo-Rectron with two plates used as a full-wave rectifier.

Uni-Rectron Power Amplifier (AP-935)

The Uni-Rectron was designed to connect to *"the first audio stage of any receiver, and to any loudspeaker—to supply a stage of power amplification for tremendous volume increase and marked improvement in tone."* It was introduced in September 1925 at a price of $105, and appeared in a December 1925 ad along with the Duo-Rectron.[23]

The Uni-Rectron power amplifier consisted of an input and output audio transformer, a UX-210 audio power amplifier Radiotron, a power transformer,

The Uni-Rectron was a power amplifier operating from the lighting circuit which could be used with any speaker to provide greatly amplified output (left). The Uni-Rectron used a UX-210 as a power amplifier and a UX-216-B as the "Rectron" half-wave rectifier (right).

a UX-216-B Rectron rectifier tube and associated capacitors, and resistors and chokes to convert the AC current to DC. The output from the radio which normally went to headphones could be patched into the input of the Uni-Rectron for amplification. A speaker of choice could be connected to the output transformer terminals of the Uni-Rectron to obtain a loud signal without unduly loading the "B" battery. The entire unit was enclosed in a metal case which was described in RCA sales literature as being *"small and compact and presenting a good appearance."*

The Uni-Rectron was actively marketed by RCA until late 1927—it was absent from the December 17, 1927 price list. There appears to be no information on the number of these power amplifiers manufactured, but they are relatively rare on the auction market.

The Duo-Rectron, with an appearance similar to the Uni-Rectron, was a "B" battery eliminator operating from the lighting circuit which could be used with most radios to eliminate the need for the "B" battery (left). The Duo-Rectron used the Rectron UX-213 as a full-wave rectifier and the UX-874 "glow tube" to maintain a constant voltage under varying load conditions (right).

Duo-Rectron "B" Battery Eliminator (AP-937)

The Duo-Rectron appeared on the market at the same time as the Uni-Rectron and was featured in a December 1925 ad in *Popular Radio*.[24] This device was designed to provide "B" battery voltages of 22½, 45, 90, and 135 volts for receivers of any size up to 10 tubes. It used a UX-213 full-wave rectifier tube and a new UX-874 voltage regulator tube which was designed to produce a constant voltage for all loads up to those drawing a maximum of 50 milliamperes. There is no information available on how many of these units were manufactured, but these units are also relatively rare. Like the Uni-Rectron, the Duo-Rectron was on the market until December of 1927.

Loudspeakers for the 1925-1927 Season

Model 100 Loudspeaker (UZ-915)

The Model 100 loudspeaker was introduced in the fall of 1925 with the rest of the 1925 line. It appeared in the RCA distributor price list dated September 1, 1925 at a price of $35, but it did not appear in national ads until circa November 1925.[25] It was also featured in a two-page ad appearing in *Radio Retailing*.[26] The Model 100 was the first cone speaker offered by RCA with a

The Model 100 speaker was heavily advertised with full page ads making it one of the most popular cone speakers of the time. (*Saturday Evening Post*, October 30, 1926, p. 96)

The Model 100 speaker was 12 inches in diameter with a dull bronze finish that had a distinct art-deco appearance.

Model 100 speakers without serial numbers had filters that were rectangular shaped which were located at the base of the speaker.

permanent magnet. It was of the "floating cone type" in which the cone was suspended at the outer edge and attached to the speaker mechanism at the center. The Model 100 cone speaker was truly a great improvement over the horn speaker, rendering them obsolete over night.

The speaker was 12 inches in diameter with a dull bronze finish that had a distinct art-deco appearance. It had a name plate affixed to the top of the circular housing, some of which had serial numbers while others did not. It is not known why some had serial numbers and other did not–it is possible that one version was manufactured by GE and the other by Westinghouse. There were three distinctly different types of name plates as shown in the accompanying figure. The two with serial numbers had different licensing agreements–one referred to specific usages including talking machines, while the other referred to licensing on the box. The one referring to talking machines had higher serial numbers than the other, and so the plate was obviously changed to license the speaker for use with phonographs in addition to broadcast reception.

Certain interior components, most noticeably the filter, were slightly different as well. The filter was described in the sales literature as an "electric tone clarifier" which helped *"materially to subdue those occasional atmospheric and other noises of remote source which are no faults of the receiving apparatus."* The filters came in two shapes and were placed in two different locations. Speakers without serial numbers had filters that were rectangular in shape

Model 100 speakers with serial numbers had filters which were cylindrical in shape and attached to the speaker mechanism (filter was unattached from the frame in this photograph for added clarity).

The Model 100 came with three distinctly different types of name plates, two types with a serial number and one without. One type with a serial number referred to a license for use with talking machines among other uses (top), and the other type referred to licensing information on the carton (middle).

and located at the base of the speaker, while speakers with serial numbers had filters which were cylindrical in shape and attached to the speaker mechanism. While the two filters were physically different, they were electrically identical.

The Model 100 was on the market until mid-1927 when it was replaced by the improved Model 100-A speaker—it was absent from the July 15, 1927 RCA price list. The number of Model 100 speakers manufactured is not known, but judging by the large number of speakers now available at auction, it must have been one of the best selling speakers of the 1920's. According to available serial number data, well over 150,000 were manufactured, but many speakers were manufactured without serial numbers, so the actual number could be twice that amount.

Model 102 Loudspeaker (UZ-913)

Loudspeaker Model 102 was introduced at the same time as Model 100, and the two often appeared together in ads and sales literature.[27] While the Model 102 was generally referred to as a loudspeaker, it also included a power amplifier unit consisting of the AP-935 Uni-Rectron operated from the AC lighting circuit. The speaker was essentially the Model 100 cone assembly and housing but without the filter. The Uni-Rectron provided the necessary grid, plate and filament voltages for operation of the UX-210 power amplifier tube which was also part of the Uni-Rectron. Because the power amplifier tube and the loudspeaker drew current from the AC lighting circuit, the Model 102 reduced the load on the batteries from the rest of the receiver to reasonable levels.

The Model 102 could be used with any receiver, but was especially recommended for the Radiola Super-Heterodyne, Radiola 24 and Radiola 26, all of which were supplied with low-power UV-199 tubes. The audio output from these or any other receiver could be connected to the input of the Uni-Rectron via the two binding posts on the side of the unit.

A nameplate was affixed to the top of the metal speaker housing with the designation "Radiola Loud Speaker Model 102." The few that have been observed by this author do not have individual serial numbers. The speakers with the Model 102 designation must be relatively rare because they are rarely seen in flea markets or at auction. This speaker was also discontinued

The Model 102 designation on the name plate is the only way to distinguish it from the Model 100 without inspecting the interior for the missing filter.

The Model 102 loudspeaker was actually the Uni-Rectron power amplifier and a loudspeaker that had the same visual appearance as the Model 100, although it was electrically different.

The Model 102 was almost identical to the Model 100 but did not have the filter.

in July 1927 when the Model 100-A was introduced. However, the Uni-Rectron remained on the market until the end of the year for use with the 100-A and other similar types of speakers.

Model 104 Loudspeaker (UZ-914)

The Model 104 loudspeaker was actually a revolutionary new type of electrodynamic loudspeaker with a model AP-951 Rectifier-Power-Amplifier Unit designed for use with the Radiola 25 and Radiola 28 for all AC operation. Both units were housed in an attractive hardwood cabinet with four legs matching the look of the Radiola 28. The Model 104 was introduced at the same time as the companion Radiolas 25 and 28 in September of 1925 at a price of $245. While this may have seemed like a very expensive loudspeaker at the time, it must be remembered that it also housed a power supply that eliminated the need for batteries to operate any part of the loudspeaker/receiver combination. The introductory price must not have produced acceptable profit margins because it was not long before the price was raised to $275.

The heart of the Model 104 was the Rice-Kellogg loudspeaker that had been under development in the GE laboratories since 1922-1923. Chester W. Rice and Edward W. Kellogg identified three key requirements for an improved loudspeaker,[28] and developed a design that achieved their requirements. The result was an electrodynamic loudspeaker with a stiff, light paper cone with a diameter of approximately six inches that was attached at the center to the loudspeaker mechanism and to a flat baffle plate that formed one surface of a rectangular box enclosing the loudspeaker. The electromagnet—referred to as the "pot magnet" because of it appearance—was wound with over 3½ miles of

The new Rice-Kellogg electrodynamic cone loudspeaker dramatically improved sound and at the same time accelerated the use of the AC lighting circuit to power radios. (Chris Hicks collection)

The Model 104 loudspeaker enclosure had an appearance very similar to the Radiola 28, both having a table base with matching trim. (Sonny Clutter collection)

The interior of the Model 104 loudspeaker contained the Model AP-951 power amplifier unit and the new Rice-Kellogg electrodynamic loudspeaker. (Sonny Clutter collection)

The three different Rice-Kellogg electrodynamic loudspeaker prototypes shown here were photographed in the GE laboratory in 1924. (Schenectady Museum & Suits-Bueche Planetarium, Schenectady, NY)

wire capable of pulling ten pounds. An important part of the solution was the development of high-power vacuum tubes to drive both the electromagnet and the loudspeaker used in the loudspeaker mechanism. The power requirements for both were beyond the capabilities of existing "B" batteries, and therefore it was necessary to develop a new power supply operating from the AC circuit.

GE and RCA decided to commercialize the new loudspeaker for the 1925-26 season, but the economics proved to be formidable. The combination of the loudspeaker, rectifier-power-amplifier unit and receiver totaled $445 for the least expensive Radiola 25, so the market was quite limited. Ultimately, the introduction of the AC heater tube in 1927 and the development of the alnico magnet circa 1935 would bring the price down.[29]

Despite the steep price, the Model 104 was on the market until well into 1928. It was absent from the RCA distributor price list dated October 3, 1928 and was most likely discontinued at the beginning of the new season in September of 1928. It is not known how many were made, but they are relatively rare, although they do appear on the market occasionally.

Several prototypes of the Rice-Kellogg electrodynamic loudspeaker such as this exist in museums and private collections. (Schenectady Museum & Suits-Bueche Planetarium, Schenectady, NY)

Epilog

According to the RCA Annual Report for 1926, RCA made the decision in early 1926 to continue the sale of receivers introduced in 1925 rather than developing a new line of receivers for the 1926-27 season. There were two major reasons for this decision. First and most obvious was the glut of unsold receivers that RCA and their distributors had on hand from the 1924 and 1925 seasons. This apparatus had to be sold before introducing a new line of receivers.

Second and less obvious was the fact that RCA needed time to develop new AC-filament tubes which would reduce the cost of expensive power supplies needed for high-performance using lighting-circuit power–thereby reducing prices sufficiently to create a mass market. The alternative was to develop a new line of battery-operated receivers with inferior performance characteristics. RCA opted to skip the 1926-27 season to develop a better, less expensive set for the 1927-28 model year.

Endnotes and References for Chapter 7

1 *National Geographic*, Vol. XLVII, No. 6, June 1925.

2 *Radio News*, Vol. 7, No. 3, August 1925, p. 231; *Radio News*, Vol. 7, No. 3, December 1925, p. 764.

3 The Radiola 24 did not appear in the September 1, 1926 price list, or in any other RCA price list discovered by the author.

4 *Radio Merchandising*, Vol. 5, No. 1, November 1925, pp. 52-53.

5 Westinghouse patent counsel Le Roi J. Williams testified at the FTC hearing that Westinghouse manufactured Radiolas 20, 25, 26, 28, and 30: *Federal Trade Commission vs. General Electric Company, et. al., Radio Case, Brief of Facts*, (Sidney C. Ormaby Company, 217 Broadway, New York, 1928) Vol. 7, p. 3515 (testimony given on November 16, 1927); *General Electric News* magazine described the manufacture of Radiolas 20, 28, 30, and Loudspeaker 100: *General Electric News, Schenectady Works*, Vol. 10, No. 5, February 19, 1926, pp. 20-21.

6 *Radioana*, J. W. Rafferty, letter dated April 7, 1928 to Mr. F. E. Eldridge, Manager, Radio Commercial Department, Westinghouse Electric, CWC5 23634.

7 *National Geographic*, Vol. XLVII, No. 6, June 1925, p. 4.

8 The portable radio had become so popular by 1925 that editor Hugo Gernsback of *Radio News* magazine published the first annual review of portable radios: Hugo Gernsback, "First Annual Portable Radio Set Directory," *Radio News*, Vol. 7, No. 3, August 1925, pp. 161-163.

9 Alan Douglas, *Radio Manufacturers of the 1920's*, Vol. 3 - RCA to Zenith (Vestal Press, Ltd., Vestal, New York, 1991) pp. 36-37.

10 Ibid.

11 Michael Brian Schiffer, *The Portable Radio in American Life* (The University of Arizona Press, Tucson & London, 1991) pp. 82-83.

12 *Saturday Evening Post*, Vol. 199, No. 29, January 15, 1927, page 102.

13 *Radio Retailing*, Vol. 4, No. 10, October 1927, p. 183.

14 The versatile Radiola 28 cabinet was also used by RCA for other radios such as the Radiola 16—see, for example: *The Town Crier*, Vol. VI, No. 4, January 1928, p.3.

15 Alan Douglas, *Radio Manufacturer's of the 1920's*, p. 40.

16 *Radio Merchandising*, Vol. 5, No. 1, November 1925, pp. 52-53.

17 The UV-877 tube appeared in earlier instruction manuals, but had been eliminated by the time Edition A with a date of October 1925 was published.

18 Alan Douglas, *Radio Manufacturer's of the 1920's*, p. 41.

19 *Radiola 28 Super-Heterodyne (A-C Operated)*, Instruction 86999, (Radio Corporation of America, New York, N.D); Edition B.*RCA Victor Service Notes for 1923 - 1928,* (Service Division RCA Victor Company, Inc. Camden N.J., N.D.), Chapter on RCA Radiola 25 and 28 A. C., pp. 14;

20 The Radiola 28 A-C Operated set was actually introduced before the Radiola 30, but there was also a battery version of the Radiola 28, and so strictly speaking the Radiola 28 was not an AC-only set.

21 *National Geographic*, Vol. XLVII, No. 12, December 1925.

22 Alan Douglas, *Radio Manufacturer's of the 1920's*, p. 39.

23 *Popular Radio*, Vol. VIII, No. 6, December 1925, inside cover.

24 Ibid.

25 *Popular Radio*, Vol. 5, No. 1, November 1925, pp. 52-53.

26 *Radio Retailing*, Vol. 3, No. 3, March 1926, pp. 278-279.

27 Ibid.

28 The three requirements were to remove the natural resonant period of metallic diaphragms or other devices that resonate, to move the air with a relatively large surface area, and to minimize distortion by designing an amplifier of adequate power and fidelity to drive the loudspeaker.

29 According to the November 4, 1935 issue of *Time* magazine (page 40), GE Schenectady had been experimenting with alnico, and a GE spokesman was quoted as saying: "Alnico is being groomed to displace small electromagnets in motors, transformers, and loudspeakers, lowering the cost and simplifying the construction."

Radiola 17 ads proclaimed that socket-power operation had been perfected by employing the wonderful new alternating current Radiotrons. (*Saturday Evening Post*, December 10, 1927, p. 69)

Chapter 8

AC Tubes Perfect Socket-Power Operation

1927-1928

The introduction of the UX-226 and UY-227 tubes with filaments operated directly from the AC lighting circuit dramatically reduced the size and cost of AC-operated sets.

By early 1926, it became clear to RCA that vacuum tubes with filaments heated directly from the AC lighting circuit would be necessary to reduce the cost of broadcast receivers. RCA turned to General Electric and Westinghouse to develop such a tube, and since Westinghouse had been working on AC tubes since 1922, they were able to provide samples of two tubes with AC filaments in a very short time. Vacuum tube historian Gerald Tyne describes the colorful story of the development and extravagant claims made for these tubes, which were eventually assigned the designations UX-226 and UY-227.[1]

RCA announced the development of these two new tubes to their distributors in a memo dated May 19, 1927, which stated that RCA would be introducing these tubes to the trade on or about July 1st, 1927.[2] The UX-226 was a 1½-volt AC filament-type tube drawing 1.05 amps, and was intended for use as a radio frequency and audio frequency amplifier. The UY-227 was a 2½-volt AC heater-type tube drawing 1.75 amps, and was intended as a detector tube in sets using the UX-226 as an amplifier. Because of the extra heater element (cathode), the tube had a special five-prong base with the new "Y" designation. This memo would also presage the design of a new type of AC receiver for the 1927-28 season:

> *"Application of 'A.C.' Tubes*
>
> *The initial demand for the new amplifier and detector Radiotrons, UX-226 and UY-227, will arise from amateur and experimenters. These tubes will eventually appear in sets designed to accommodate them, at which time a growing "repeat" demand may be anticipated.*
>
> *It is likely that newly designed sets will employ Radiotron UX-226 in all r.f. sockets and in the first a.f. stage, UY-227 in the detector stage, and one of the Radiotron power amplifiers in the last audio stage, ie., UX-112, UX-171 or UX-210. The requisite low-voltage supply for the filament may be obtained from an independent step-down transformer or from low-voltage windings on the usual 'B' or plate supply transformer."*

This description fit exactly the new Radiola 17 designed for the 1927-28 season.

The 1927-1928 Radiola Line

The new RCA line for the 1927-28 season was announced to RCA dealers in the June 15, 1927 issue of RCA's trade magazine, *The Town Crier*,[3] and to the public in national magazines circa September 1927.[4] The models shown in these ads also appeared in the early edition of the RCA catalog for model year 1927-28 published in 1927. In addition to these models, the Radiola 18, 50, and 51 receivers and the Model 105 electrodynamic loudspeaker were introduced in the spring of 1928, midway through the model year. The Radiola 18 and Model 105 loudspeaker were included in a revised version of the RCA catalog published circa April 1928, but the Radiolas 50 and 51 were introduced too late to appear in this catalog. All the models either introduced during the 1927-28 season or held over from previous seasons are listed in Table 8-1.

There were basically three types of receivers in the new line: 1) Radiola 16, a battery-operated TRF receiver, 2) Radiolas 17 and 18, new AC-operated TRF receivers, and 3) Radiolas 30A and 32, both of which were AC- and DC-operated superheterodyne consoles based on the standard eight-tube catacomb from the previous season. Radiolas 16, 17, and 18 were all six-tube TRF receivers similar to the Radiola 20 except for the addition of a third stage of RF amplification. The Radiola 17 was an AC version of the Radiola 16 using AC-

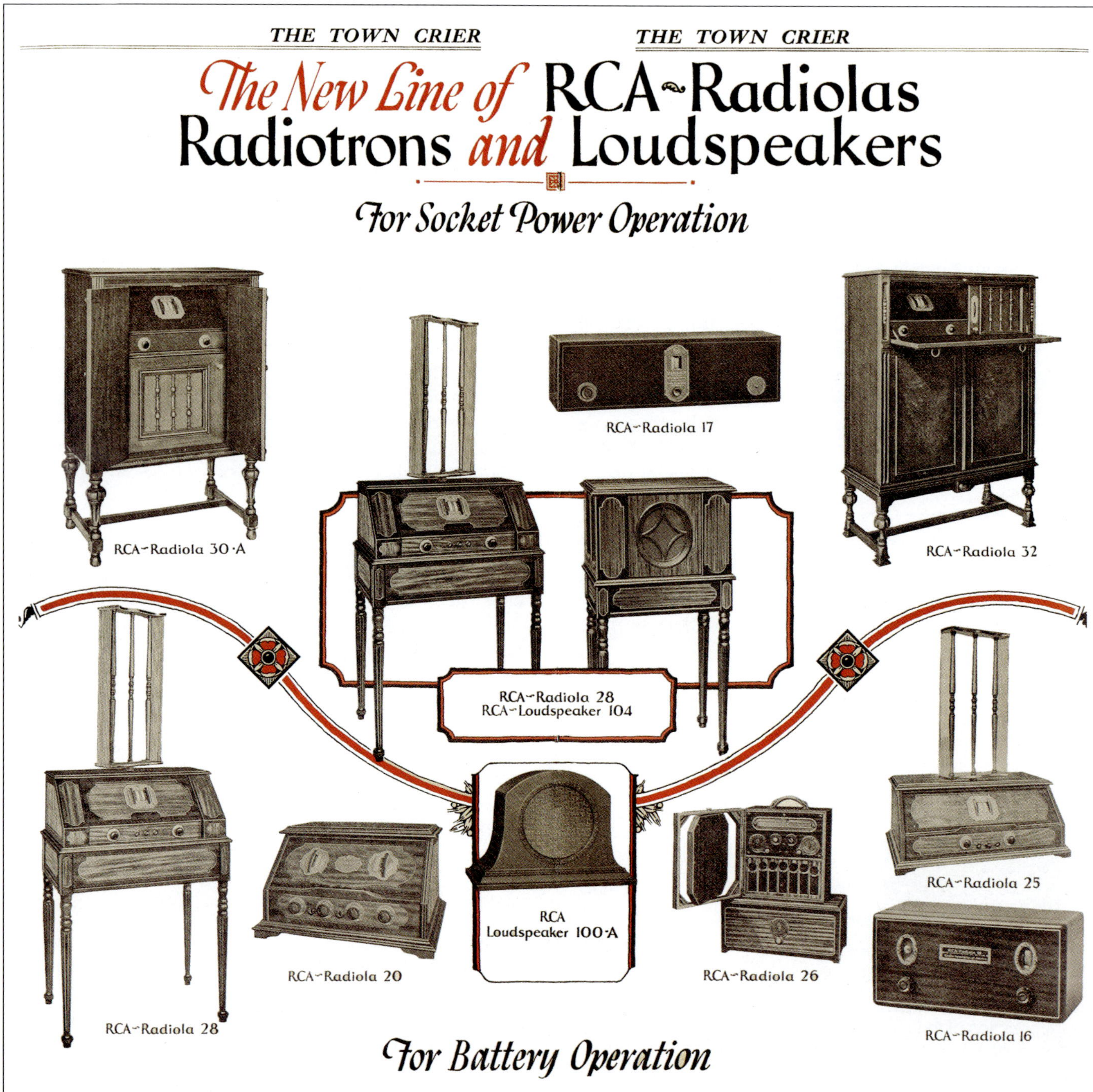

The new RCA line for the 1927-28 season was announced to RCA dealers in the June 15, 1927 issue of RCA's trade magazine, *The Town Crier*.

RCA published two catalogs for the model year 1927-28 with a slightly different lineup of sets, one in late 1927 (left) and one in early 1928 (right).

RCA provided marketing brochures such as the ones shown here for both new and hold-over apparatus offered at the start of the 1927-28 season.

The operating manuals for console models introduced in the 1927-28 season were dressed up with an olive-taupe outer cover, while those for the table-top models had no outer cover.

Table 8-1. RCA apparatus for the 1927-28 model year.

	Model	Apparatus Type	Cabinet Style	Tubes[a]	Power	Intro. Date	Discon'd. By	List Price[b]
1927-28 Season New Models	Radiola 16 (AR-924)	TRF Receiver	Cabinet	6	Battery	8/8/27	Apr. 1929	$82.75
	Radiola 17 AR-927)	TRF Receiver	Cabinet	6 (1)	AC	9/15/27	Apr. 1928	$157.50
	Radiola 18 (AR-936)	TRF Receiver	Cabinet	6 (1)	AC	4/19/28	Jan. 1930	**$115**
	Radiola 30A (AR-926)[c]	Super-heterodyne	Console	8 (3)	AC	8/7/27	Apr. 1929	$495
	Radiola 30A DC (AR 912)	Super-heterodyne	Console	11	DC	9/14/27	Sep. 1928	$495
	Radiola 32 (AR-925)	Super-heterodyne	Console	8 (3)	AC	7/27/27	Sep. 1928	$895
	Radiola 32 DC (AR-928)	Super-heterodyne	Console	11	DC	9/15/27	Sep. 1928	$895
	Radiola 50 (AR-910)	TRF Receiver	Console	6 (1)	AC	3/14/28	Jun. 1928	$285
	Radiola 51 (AR-904)	TRF Receiver	Console	6 (1)	AC	6/13/28	Sep. 1929	**$195**
	100-A (UZ-1076)	PM Loudspeaker	Table	N/A	N/A	5/25/27	1930 +	$35
	104 DC (unknown)	Dynamic Loudspeaker	Floor	4	DC	Jun. 1927	Sep. 1928	$310
	105 (UZ-1082)	Dynamic Loudspeaker	Floor	1 (2)	AC	3/8/28	Nov. 28	$350
	AP-1080	"B" Eliminator	None	0	AC	3/5/28	Apr. 1929	$48
Holdover Models	Radiola 20 (AR-918)	TRF Receiver	Cabinet or Floor	5	Battery	11/6/25	Sep. 1928	$89.50
	Radiola 28 (AR-920)	Super-heterodyne	Floor	8	Battery[d]	9/29/25	Sep. 1928	$260
	Radiola 28 A-C Operated (AR-969)	Super-heterodyne	Floor	8	AC only[e]	9/29/25	Sep. 1928	See text
	104 (UZ-914)	Dynamic Loudspeaker	Floor	1 (4)	AC	Sep. 1925	Sep. 1928	$245

[a] Additional rectifier and ballast tubes are shown in parentheses
[b] All apparatus included tubes in advertised price except those noted with the price in bold font
[c] Model AR-926 is the 60-cycle AC version; AR906 is the 25-cycle AC version
[d] Operates on AC power with Model 104 Radiola loudspeaker using "AC" Package UP-972 @$35
[e] Radiola 28 (A-C Operated) was an AC-only set with the UP-972 AC Package built in at the factory

filament tubes, and the 18 was an early replacement for the ill-fated Radiola 17 that suffered from problems with excessive feedback resulting in unwanted and annoying oscillations. The Radiola 30A was similar to the Radiola 30—both incorporated the same eight-tube superheterodyne catacomb used in the Radiola 28 and the Model 100-A loudspeaker. The Radiola 32 also used the same eight-tube catacomb but incorporated the Model 104 loudspeaker technology in lieu of the Model 100 loudspeaker. Radiolas 51 and 52 introduced later in the season were simply console versions of Radiolas 17 and 18 respectively with the addition of the Model 100-A permanent magnetic loudspeaker.

TRF receiver using the new UX-226 and UY-227 AC-filament tubes, which significantly reduced the cost of connecting a receiver to the lighting circuit. Otherwise, the receiver line was very similar to the line from the previous year with only very modest improvements. Because there were no new superheterodyne designs in the 1927-28 lineup, one might conclude that RCA had given up on the superheterodyne technology. However, this was not the case as evidenced by the line of receivers introduced the following year based on a new AC superheterodyne design. RCA simply did not have the time to develop a new superheterodyne using AC current tubes in time for the 1927-28 season.

In addition to the four new receivers, RCA also introduced three new loudspeakers including an improved Model 100-A permanent-magnet loudspeaker which replaced a similar Model 100 loudspeaker, a DC version of the Model 104 electrodynamic loudspeaker which operated from DC lighting circuits, and a deluxe Model 105 electrodynamic loudspeaker which also operated from the AC lighting circuit. The DC version of the Model 104 loudspeaker was incorporated into DC versions of the Radiola 30A and 32, but it was also sold as a stand-alone item for use with the Radiola 25 and Radiola 28 for operation from the DC power line. The Model 105 loudspeaker contained a new rectifier-power-amplifier unit which was designed to supply plate and grid voltages to a variety of battery-operated receivers with which it could be used.

Finally, it should be noted that much of the apparatus introduced the previous year was held over for the new season. The apparatus available in the new season—about half old and half new—was captured in a collage of sales brochures advertised to dealers in the October 1927 version of RCA dealer's magazine, *The Town Crier*.

The Radiola 16 cabinet represented a departure from large boxy sets to a more streamlined look.

RCA emphasized the small size of the Radiola 16—achieved by using external batteries which were never pictured with the radio in their ads and brochures. (RCA sales brochure for Radiola 16)

Receivers for the 1927-1928 Season

Radiola 16 (AR-924)

According to George Clark, the Radiola 16 was officially released on August 8, 1927, a date which correlates well with the date of the first ads for the entire line which appeared circa September 1927. The first ads featuring the Radiola 16 itself began to appear circa October 1927 in magazines directed at the rural community without power lines such as *Country Gentlemen*, *Capper's Farmer*, and *Standard Farm Paper Unit*. The introductory price of $82.50 including Radiotrons was $32.50—lower than the introductory price of the similar but somewhat inferior Radiola 20 at $115 in late 1925. Even though the price of the Radiola 20 had been reduced to $89.50 at the time the Radiola 16 was introduced, it is difficult to see why anyone would have bought it after the Radiola 16 became available at $82.75.

Ask the dealer to demonstrate this fine Radiola 16 in your own home

ALTHOUGH low in price, the compact Radiola 16 is designed and built with the same scientific care as Radiolas that cost several times as much.

It has back of it the combined resources of General Electric, Westinghouse and the Radio Corporation of America.

It is built for fine performance and long life. It will give thousands of hours of enjoyment for all the family.

Ask the R C A dealer in your town to demonstrate it for you in your own home. Tune it in to one of the famous music hours, or pick up the daily market reports. Take note of its high sensitivity and selectivity, and particularly of its superb tone quality. In combination with the RCA Loudspeaker 100A, it will give you a new conception of the perfection of modern radio.

If your home is wired for central station electric service, the RCA dealer will be glad to demonstrate the Radiola 18, which takes all its power from the house current.

RCA sets may be purchased from RCA Radiola Dealers on the RCA Time Payment Plan.

RADIOLA 16

RCA RADIOLA 16—Very compact, sturdy, battery operated, 6-tube receiver. Single dial control. Perfected RCA tuned-radio-frequency circuit. Mahogany finished cabinet. $82.75 (with Radiotrons)

RADIOLA 18

RCA RADIOLA 18—For direct operation from A. C. house current (110 volt, 60 cycle). Employs special A. C. tubes and rectifier. Electrically lighted dial. Finest set of its kind. $115 (less Radiotrons)

The best reproducer to get the full tone qualities of the "16" or the "18" is the RCA LOUDSPEAKER 100A—$29.

Buy with confidence Radiola Dealer RCA *where you see this sign.*

RADIO CORPORATION OF AMERICA

RCA Radiola

MADE · BY · THE · MAKERS · OF · THE · RADIOTRON

NEW YORK · CHICAGO SAN FRANCISCO

Ads for the battery-operated Radiola 16 were often set in country stores, farm kitchens and other scenes evocative of regions where rural electrification had not yet occurred. (By David Hendrickson in *The Country Gentlemen*, September 1928, p. 83)

The Radiola 16 chassis was dominated by a large 3-stage tuning condenser mounted in a base evocative of a bathtub—whence the name bathtub condenser.

Ads for the Radiola 16 in a floor model configuration with AC tubes and an AC power pack were circulated to authorized RCA dealers later in the 1927-28 season. (mailer from Penn Electrical Engineering Company, Scranton, Pa., circa 1928)

The receiver was housed in a compact and attractive mahogany case with a lid that was hinged in the rear to give access to the chassis and tubes. The tuning and volume controls as well as the on-off switch were all mounted on the front of the cabinet. This new cabinet style would be used on other receivers in the new RCA line including the Radiolas 17 and 18. Attachments to the receiver for batteries, antenna and ground wires, and loudspeaker leads were made through connections on the rear of the set, giving a neat and symmetrical appearance to the cabinet. RCA recommended that an external antenna be used for best reception, and also recommend the new Model 100-A loudspeaker for the most faithful reproduction of sound. In addition, RCA literature pointed out that the set could be operated from the AC line with an appropriate battery eliminator such as the Duo-Rectron.

The Radiola 16 was a six-tube TRF receiver using five UX-201-A tubes and one high-power UX-112-A audio output tube. It was one of the first RCA "uni-control" receivers designed to operate with only two controls—a tuning control to select stations and a volume control. Clearly gone from the front panel were the usual vernier tuning controls for fine tuning, rheostats to adjust filament current, and the regeneration control to adjust feedback. A single tuning control was achieved by using a three-gang tuning condenser, and an aperiodic antenna circuit to eliminate the need for a vernier control for fine tuning. Regeneration was eliminated in favor of three RF amplifier stages. The ganged "bath-tub" tuning condenser common to all three new sets—Radiolas 16, 17, and 18—dominated the chassis.

The Radiola 16 was on the market for a respectable 20 months. Toward the end of its run, ads appeared for the Radiola 16 in both a Radiola 28 desk cabinet and third-party cabinets complete with a battery eliminator and speaker. It was retired in circa April, 1929, absent from the RCA distributor price list dated April 1, 1929. According to Douglas, a total of 82,320 were made.[5]

Radiola 17 (AR-927)

George Clark listed September 15, 1927 as the date the Radiola 17 was released to the market, and indeed, it appeared in ads grouped with the entire new Radiola line circa September 1927. It was featured in national ads appearing in the *Saturday Evening Post*, *Literary Digest*, *Liberty*, and *Colliers* circa December 1927 with the caption "*Socket-power operation perfected by RCA, Westinghouse and GE*"—obviously referring to the use of the new AC tubes.[6] The receiver was advertised at $157.50 including Radiotrons.

The general appearance of the Radiola 17 was similar to that of the 16, although the cabinet was noticeably longer to accommodate the additional AC power supply chassis. The Radiola 17 receiving circuit was almost identical to that of the Radiola 16 except that it used four AC-filament UX-226 amplifier

The Radiola 17 had the general appearance of the Radiola 16, but with an extended cabinet length needed to contain the new AC power supply.

The Radiola 17 chassis was divided into two parts—one for the radio circuitry and one for the AC power supply (top). Many of these Radiola 17 chassis show some type of repair work, such as the obvious replacement of the AC cord appearing in the set pictured here (bottom).

tubes, one AC-filament UY-227 detector tube, and one UX-171-A power amplifier tube which produced over ½ watt of audio power. A small electric bulb was used in the escutcheon plate which served to illuminate the tuning dial and provide an indication that the set was turned on. This was the first receiver marketed by RCA that used AC-filament tubes, and as a result, the price was low relative to the AC receivers from the previous model year.

The chassis for the receiver circuit was similar in appearance to that of the Radiola 16, except for the placement of the audio output transformer, which was moved to the top of the chassis in the Radiola 17. The receiver used the same uni-control design and bath-tub tuning condenser as the Radiola 16, in which the feedback control was eliminated in favor of a single tuning knob located in the center of the cabinet panel. The power supply chassis was more than half the length of the receiver-circuit chassis, accounting for the increased cabinet length. The chassis was dominated by a power transformer, chokes and the UX-280 full-wave rectifier tube. Antenna and ground wires protruded from the rear of the set.

Two versions of this set were made, one for 60-cycle AC operation and one for 25-cycle AC operation.[7] It is likely that different model numbers were assigned to the two versions because it would have been necessary to distinguish between the two. However, there does not appear to be any record of the model number for the 25-cycle version, or in fact, any reference to a 25-cycle version in the RCA Victor repair manuals or price lists.

It was not long after the Radiola 17 was introduced before reports of "oscillating trouble" came back to RCA from across the country. RCA basically whitewashed this problem with the following message addressed to RCA dealers in the December of 1927 issue of *The Town Crier*:

> *"Trouble in this direction [unwanted oscillations] can generally be attributed to improper installation. If the suggestions on antenna length, ground systems, proper adjustment of line switch, etc., contained in RCA 'Radiola 17 Service Notes,' are observed little or no trouble from uncontrolled oscillations will be met."* [8]

RCA quietly responded to these complaints by adding an 800 ohm resistor across the primary of the second RF transformer during production to correct the problem,[9] but that apparently resulted in additional complaints about a noticeable lack of sensitivity.[10] The basic problem appears to be that the designers tried to squeeze too much amplification out of the three RF amplifier stages by using grid resistors that were not of sufficient size to prevent oscillations resulting from feedback through inter-electrode capacity in the RF amplifier tubes under all conditions faced by the receiver in the field. By adding additional resistance, they were able to suppress the oscillations, but at the expense of gain and receiver sensitivity. The Radiola 17 was fundamentally flawed in that it could not provide sufficient sensitivity under certain conditions, while at the same time remaining stable under other conditions. The principal factors affecting sensitivity and stability were individual tube characteristics and antenna design—hence the admonition that the oscillation problem could be attributed to improper antenna installation.

Apparently, RCA had made the decision many months before to redesign the Radiola 17, rename it Radiola 18, and withdraw the Radiola 17 from the market—which is exactly what they did at the end of April 1928. In the May issue of *Radio Retailing*, RCA declared victory by introducing the new Radiola 18 as follows: "*ANNOUNCING the new and improved alternating current Radiola RCA RADIOLA 18. A finer instrument than the sensational "17."*" In fine print the ad went on to say: *"An even finer broadcasting receiver than the wonderful '17,' of which more than $27,000,000 worth have been sold since October."*[11] Thus, RCA was able to withdraw yet another defective receiver from the market without ever admitting it by using the time-honored method of replacing it with a new and improved model—and at a lower price!

At $157.50 a set, the $27,000,000 in Radiola 17 revenues translated into about 171,000 sets sold in seven short months, which agrees well with Douglas who reported a total of 179,917 manufactured. Clearly, the AC-powered receiver was very popular because the Radiola 17 outsold the almost identical battery-operated Radiola 16 by more than 2-to-1, even though the 17 cost almost twice as much and was on the market for less than half the time! The Radiola 17 proved to RCA that AC-powered receivers were the wave of the future.

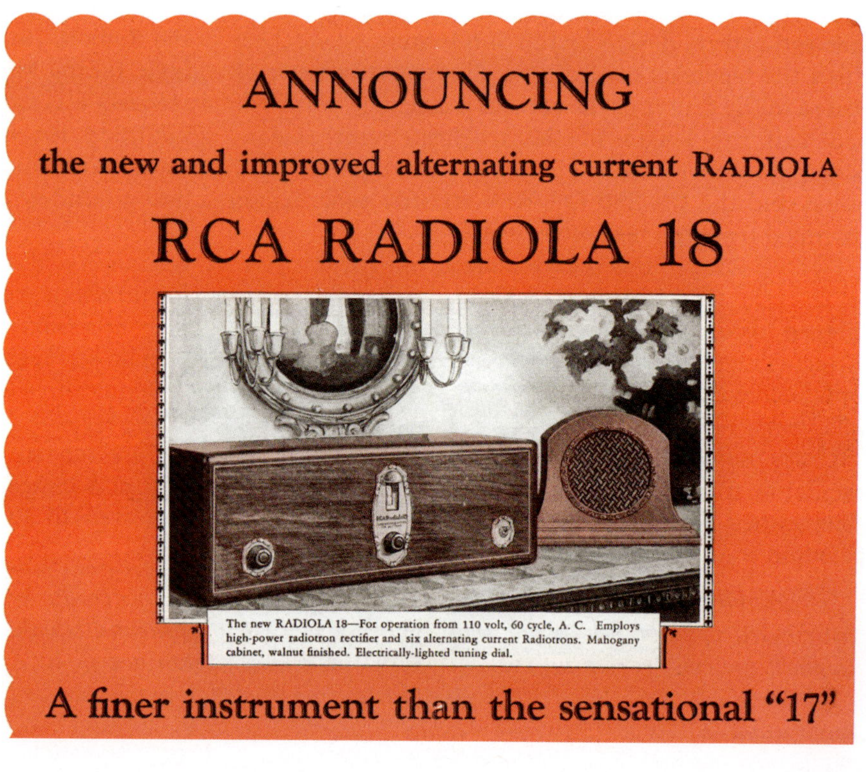

In this portion of an ad appearing in the May 1928 edition of *Radio Retailing*, RCA announced to the trade that the new and improved Radiola 18 was a finer instrument than the Radiola 17.

The external appearance of the Radiola 18 cabinet is almost identical to the Radiola 17, differing only in the detail of the escutcheon plates.

Radiola 18 (AR-936)

The Radiola 18 was introduced on April 19, 1928 as a replacement for the ill-fated Radiola 17, which initially suffered from oscillations in the RF amplifiers of many sets, and later from lack of sensitivity resulting from the addition of a resistor across the primary of the second RF transformer during production to correct the oscillation problem. The Radiola 18 was introduced at the reduced price of $115 less tubes, and sold for $137.50—including $22.50 for tubes—as compared to the introductory price of $157.50 including tubes for the Radiola 17. A new sales brochure was also prepared for the Radiola 18 which replaced a very similar brochure for the Radiola 17.

The Radiola 18 chassis is very similar in appearance to the Radiola 17.

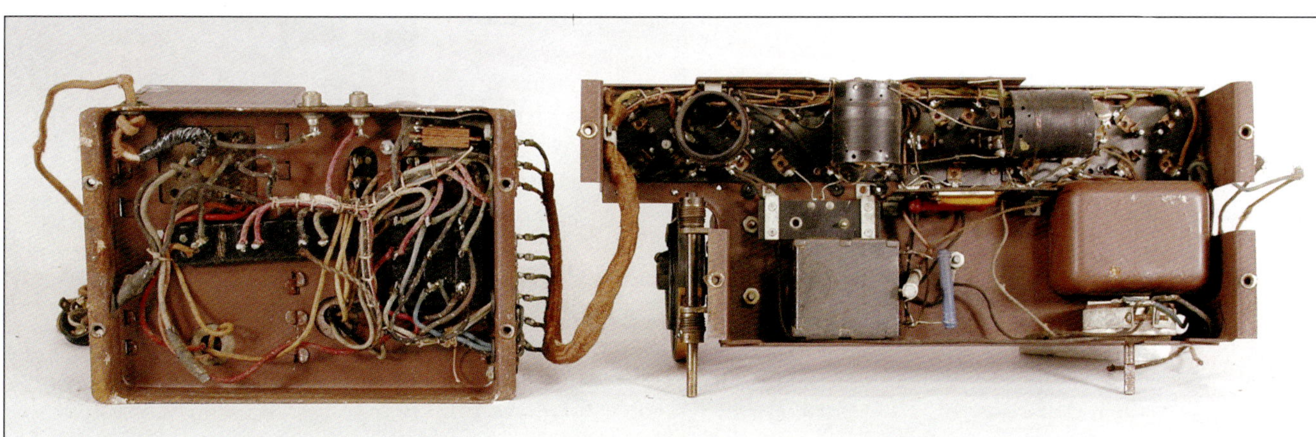

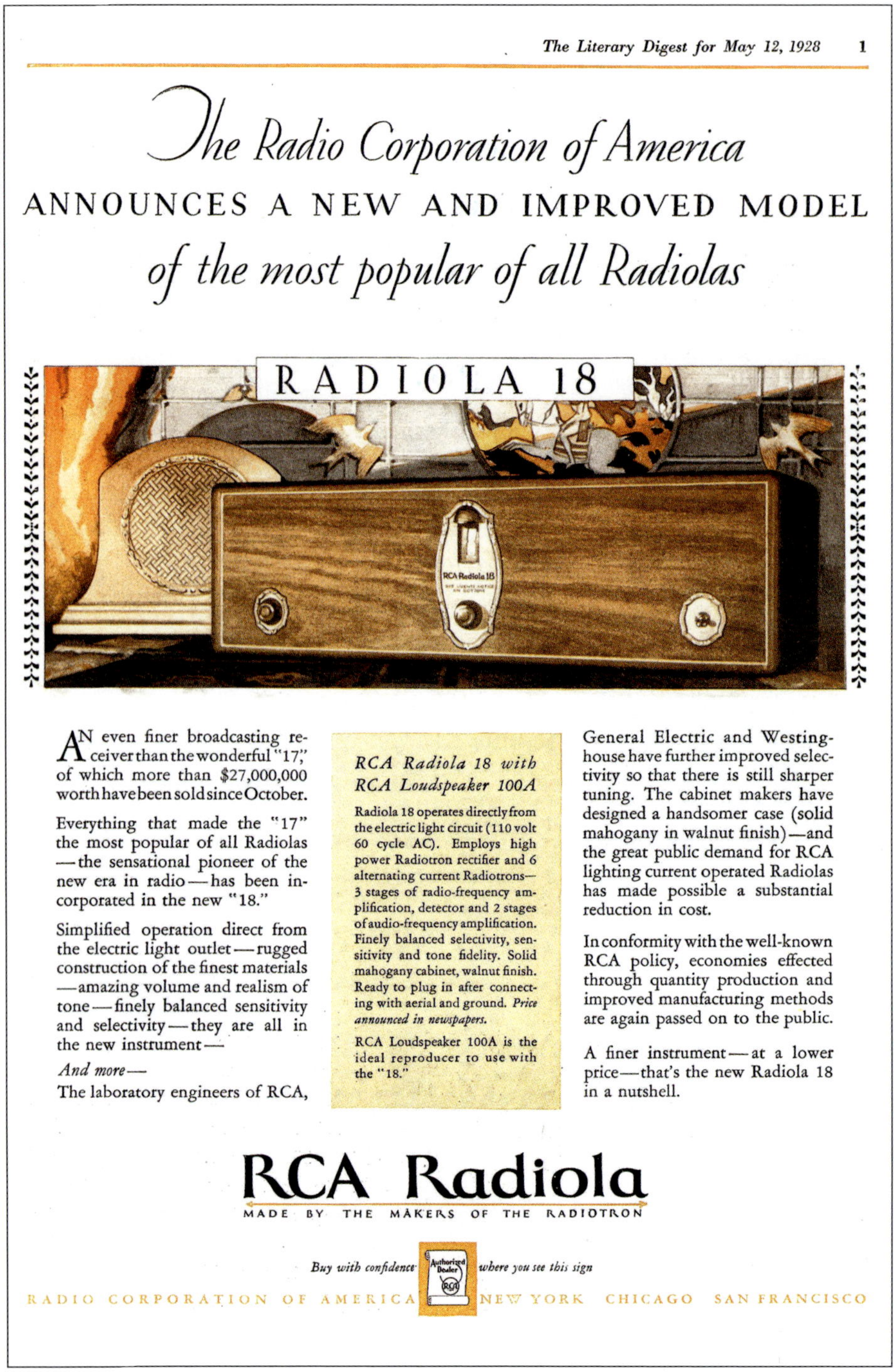

The Literary Digest for May 12, 1928 1

The Radio Corporation of America

ANNOUNCES A NEW AND IMPROVED MODEL

of the most popular of all Radiolas

RADIOLA 18

AN even finer broadcasting receiver than the wonderful "17," of which more than $27,000,000 worth have been sold since October.

Everything that made the "17" the most popular of all Radiolas—the sensational pioneer of the new era in radio—has been incorporated in the new "18."

Simplified operation direct from the electric light outlet—rugged construction of the finest materials—amazing volume and realism of tone—finely balanced sensitivity and selectivity—they are all in the new instrument—

And more—

The laboratory engineers of RCA, General Electric and Westinghouse have further improved selectivity so that there is still sharper tuning. The cabinet makers have designed a handsomer case (solid mahogany in walnut finish)—and the great public demand for RCA lighting current operated Radiolas has made possible a substantial reduction in cost.

In conformity with the well-known RCA policy, economies effected through quantity production and improved manufacturing methods are again passed on to the public.

A finer instrument—at a lower price—that's the new Radiola 18 in a nutshell.

RCA Radiola 18 with RCA Loudspeaker 100A

Radiola 18 operates directly from the electric light circuit (110 volt 60 cycle AC). Employs high power Radiotron rectifier and 6 alternating current Radiotrons—3 stages of radio-frequency amplification, detector and 2 stages of audio-frequency amplification. Finely balanced selectivity, sensitivity and tone fidelity. Solid mahogany cabinet, walnut finish. Ready to plug in after connecting with aerial and ground. *Price announced in newspapers.*

RCA Loudspeaker 100A is the ideal reproducer to use with the "18."

RCA Radiola

MADE BY THE MAKERS OF THE RADIOTRON

Buy with confidence where you see this sign Authorized Dealer RCA

RADIO CORPORATION OF AMERICA NEW YORK CHICAGO SAN FRANCISCO

The Radiola 18 was heavily advertised as a replacement for the Radiola 17. (*The Literary Digest*, May 12, 1928, p.1)

The escutcheon plate on the dial of the Radiola 18 has a curved outline evocative of an "8" (left),while the escutcheon plate on the dial of the Radiola 17 has straight lines evocative of a "7"(right).

The new compensating condenser in the feedback circuit of the Radiola 18 was accessed through a hole in the base of the bathtub condenser which appears in the center of this photograph.

The adjustable compensating condenser was mounted on a bakelite panel directly below the bathtub condenser which appears in the center of this photograph.

The exterior of the Radiola 18 so closely resembles the Radiola 17 that it is difficult to distinguish between the two at a glance. Both cabinets are virtually identical in size and finish, and both have the same controls in the same location. The most distinguishable exterior feature is the rounded escutcheon plate on the Radiola 18 as compared to the escutcheon plate on the Radiola 17 which has distinctly squared corners.

The chassis, circuits and tubes for the two are also very similar, with the most notable difference being what RCA termed "a new method" to stabilize the tuned RF amplifier stages. The Radiola 17 used grid resistors to control instabilities, while the Radiola 18 used a neutralizing circuit consisting of two coils mounted within the second and third stage RF transformers in conjunction with an adjustable "compensating condenser" in the first RF

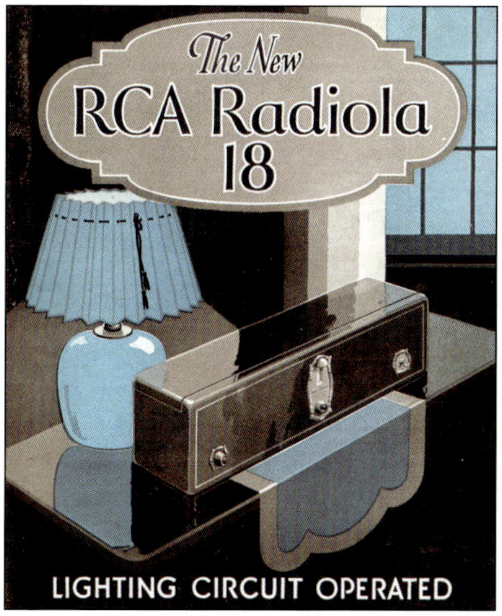

A new brochure was prepared for the Radiola 18 which replaced a very similar brochure for the Radiola 17.

The Radiola 18 is inspected and packed for shipping at the Westinghouse Plant in East Pittsburgh in 1929. (Schenectady Museum)

stage to adjust feedback from the plate to the grid, thereby allowing one to optimize receiver sensitivity for each set with its own particular tube and antenna characteristics.

The feedback adjustment was intended to be a one-time affair performed by the dealer at the time of installation, and so a brown paper seal was placed over the hole to discourage adjustment by the owner. In fact, this control was not even mentioned in the owner's manual for the Radiola 18. The compensating condenser was accessed from the top of the receiver with the lid open by inserting a screw driver through a hole at the bottom of the tuning condenser assembly. Many dealers reported back to RCA that the adjustment was one of the most desirable features of the Radiola 18.[12] Obviously, RCA did not want to regress from its uni-control design by placing the feedback control on the front panel, something they had worked hard to eliminate in the design of the new Radiola line in order to simplify tuning.

The Radiola 18 remained on the market after the new 1928-1929 line was announced in September of 1928, but the price was then dropped to $115 with tubes included. It remained on the market until circa October 1929 when it was replaced by the virtually identical Radiola 33. The set was dropped from the RCA lineup in advertising beginning in September 1929, but it was still listed in the October 1, 1929 RCA price list. It was absent from RCA price lists by January 30, 1930.

The set was manufactured in four different versions: 110 volts AC, 60 cycles;

110 volts AC, 25 cycles; 200-250 volts AC, 50-60 cycles; and 110 volts DC. The DC version, which is addressed in the next chapter, was introduced in the 1928-29 model year. According to Douglas, more than 250,000 were manufactured,[13] a number which presumably includes all four types.

Radiola 30A (AR-926/AR-906)

The Radiola 30A was officially released on August 7, 1927, and was featured in ads beginning in October 1927. It was introduced at a price of $495 complete, which was $80 less than the $575 price tag of the nearly identical Radiola 30 of the previous year. One significant difference between the 30 and 30A was that the 30A did not have an internal loop antenna, so it required an external antenna, either indoors or outdoors. As a result the 30A cabinet was much narrower than the cabinet for the 30.

The Radiola 30A was an attractive, but relatively unremarkable console set that combined the Radiola 28 superheterodyne and the new Model 100-A cone loudspeaker. The set used the standard eight-tube catacomb with seven UX-199 tubes and a UX-171 power amplifier tube located on the power-rectifier-amplifier unit. The power unit also contained two UX-281 tubes in a full-wave rectifier configuration, and a UV-876 ballast tube. The only significant differences between the 30 and the 30A were the use of the improved Model 100-A loudspeaker, the new tubes in the power-rectifier-amplifier unit, and the lack of a loop antenna. RCA sold both a 60-cycle and 25-cycle version of this set, assigning the 25-cycle version model number AR-906 and the 60-cycle version model number AR-926.

The Radiola 30A was on the market until circa April 1929. It was absent from the RCA distributor price list dated April 1, 1929. According to Douglas, 17,363 were manufactured.[14]

The Radiola 30-A was constructed of select dark walnut veneers with a finish described in RCA sales literature as "exquisite" and proportions described as "graceful" (left). The Radiola 30-A featured the Radiola 28 AC-only control panel adapted to a console (right). (Joe Formanek collection)

The Radiola 30-A was heavily advertised in ads such as this one which featured a large image of the attractive console. (*National Geographic*, rear cover, circa 1928)

A rear view of the Radiola 30-A reveals two compartments—an upper compartment for the Radiola 28 chassis, and a lower compartment for the power amplifier and a reproducer housing (covered by wiring instructions) in which the Model 100-A cone loudspeaker was mounted. (Joe Formanek collection)

Radiola 30A DC (AR-912)

The DC version of the Radiola 30A was released on Sept. 14, 1927 at a price of $495. The DC version of the Radiola 30A was different from the AC version in several interesting ways. First, the DC version used the Model 104 electrodynamic loudspeaker rather than the Model 100-A loudspeaker used in the AC version. Second, the audio output stage used four UX-171-A tubes connected in a parallel push-pull configuration to obtain the same output power as a single UX-171 tube used in the AC version. Third, an external "C" battery was used to bias the UX-171-A tube in the DC version whereas the bias in the AC version was supplied by the power-rectifier-amplifier unit.

Connecting to the DC power line also required a special accommodation for the different line voltages encountered in different locations. The set was designed to operate on voltages between 105 and 125 volts, but tube performance and filament life was sensitive to line voltage and so the set was designed with a four-position switch with the following four ranges: 105-110,

110-115, 115-120, and 120-125. The instructions called for the installer or owner to accurately measure the line voltage and select the proper switch position for each installation. The set was also sensitive to the orientation of the plug in the wall socket. It would only work in one position, and that was to be determined by trial and error.

RCA soon found that providing a full line of receivers for the power line actually required three different designs for each model to cover 25-cycle and 50-60-cycle AC lines as well as DC lines—versus a single design for battery operation. Undoubtedly, this complicated the development, marketing, sales and distribution, and after-sale support for each model.

The DC version was on the market from September 1927 to circa September 1928—it was absent from the RCA distributor price list dated October 3, 1928. Presumably, the manufacturing figure of 17,363 cited by Douglas for the 30A includes both the AC and DC versions.

Radiola 32 (AR-925)

The Radiola 32 was officially released on July 27, 1927 according to George Clark, which was after the first announcement in *The Town Crier*, but about a month before the first ads appeared in national magazines for the new line circa September 1927. The Radiola 32 was simply the Radiola 28 eight-tube superheterodyne and the Model 104 loudspeaker released the year before, but placed in a single cabinet along with the rotating loop antenna. The most curious part of the Radiola 32 was the price tag of $895 which was $325 above the $570 price of the identical equipment when purchased separately as a Radiola 28 and Model 104 loudspeaker. In essence, there was a $325 premium for the single cabinet, which RCA described as being *"of exquisite artistry in a beautifully grained, dark brown walnut [which] has just enough decorative carving and fluting to give it that quiet dignity of appearance so desired by the discriminating."*

The Radiola 32 consisted of four sections including the Radiola 28 receiver, the Model 104 electrodynamic loudspeaker, the Model AP-832-A power amplifier

The Radiola 32 was described in RCA literature as being *"of exquisite artistry in a beautifully grained, dark brown walnut [which] has just enough decorative carving and fluting to give it that quiet dignity of appearance so desired by the discriminating"* (left). As with the Radiola 30-A, the control panel of the Radiola 32 was the Radiola 28 A-C Operated panel adapted to a console (right). (Joe Formanek collection)

The Radiola 32 was divided into three main compartments—a small compartment in the upper right containing the receiver circuitry, a larger compartment directly below containing a rotating loop antenna, and the largest compartment to the left containing the power amplifier and Model 104 loudspeaker. (Joe Formanek collection)

unit, and the rotatable loop antenna. The set was equipped with seven UX-199 tubes, one UX-210 power output tube, two UX-281 rectifier tubes, and a UV-886 ballast tube used to maintain a relatively constant plate voltage. The set featured a small electric lamp to illuminate the tuning dial and give notice that the power was applied, and the set was automatically switched off when the desk fall was closed.

The set was designed to operate over the broadcast frequency band of 550-1,500 KHz (approximately 545 to 200 meters). There was almost surely a 25-cycle and 60-cycle version of this set, as there were for most other AC sets made by RCA in that time frame. The set was on the market for about a year, and was withdrawn from the market sometime between June and September of 1928—most likely in September when the new line was introduced. According to Douglas, there were 3,041 Radiola 32 consoles manufactured—as compared to 17,363 for the Radiola 30A console which had a price tag of only $495.[15]

Radiola 32 DC (AR-928)

The DC version of the Radiola 32 was released on Sept. 15, 1927 at the price of $895, the same as the AC version. What is most interesting is that the DC version of the Radiola 30A for $495 had the Model 104 loudspeaker, and therefore was virtually identical to the DC version of the Radiola 32 that sold for $895. The main differences between the two were the imposing cabinet of the Radiola 32 and the loop antenna which was not included with the Radiola 30A.

The Radiola 32 DC was described in *The Town Crier* as being identical to the AC version except for the power supply apparatus,[16] but that is slightly misleading because the audio output stage was actually part of the power supply unit, which, in fact, was significantly different. As in the DC version of the Model 30A, the DC version of the Radiola 32 had four UX-171-A tubes connected in a parallel push-pull configuration to obtain the same output power as the single UX-210 tube used in the AC version.

The DC version was released on September 15, 1927, several weeks after the AC version, and it was discontinued circa September 1928 along with the AC

The Radiola 32 was often featured in ads with the Radiola 30-A. (*Radio Retailing* April 1928, p. 63)

version. There is no information available on how many of theses sets were manufactured for DC versus AC lighting circuits—presumably, the 3,041 sets quoted by Douglas include all sets regardless of the type of line voltage.

Radiola 50 (AR-910)

Radiola 50, a console version of Radiola 17 with the Model 100-A loudspeaker, was introduced on March 14, 1928 at a price of $285 including tubes. This ill-fated set was introduced just as the Radiola 17 was withdrawn from the market and replaced by the Radiola 18. Within a very short period of time RCA stopped producing the Radiola 50, and began to produce the Radiola 51—which was nothing more than the Radiola 50 console with the new Radiola 18 chassis and the Model 100-A loudspeaker mounted in a slightly different position.

The custom-built console cabinet consisted of beautifully grained and matched walnut with doors opening outward to reveal the controls and the loudspeaker opening concealed by a grill cloth trimmed with pillared openings. The loudspeaker was suspended from a baffle-board with a heavy flat pad to prevent cabinet vibration and other unwanted resonance effects. A large hinged door on the rear afforded access to the tubes and chassis. A voltage switch was accessible in the rear to accommodate line voltage variations outside of the specified range of 105-125 volts.

The Radiola 50 consisted of a Radiola 17 chassis mounted in an attractive walnut console cabinet with a Model 100-A loudspeaker (left). The operating panel of the Radiola 50 was similar to the Radiola 17 with the notable exception of the volume control and on-off switch which were co-located and trimmed with a single escutcheon plate mimicking the one used on the tuning control (right). (George H. Clark Collection, Smithsonian Institution)

The Radiola 50 was on the market for such a short period of time that there were few if any national ads, and this author has not found it in any RCA distributor price lists. It did not appear in any RCA catalogs, and apparently no marketing brochures featuring this set were circulated. However, it did appear in a paragraph in *Radio Retailing* for April, 1928 describing new apparatus from various manufacturers.[17] It may also have been advertised in local newspapers for a short while. According to Douglas only 3220 were made.

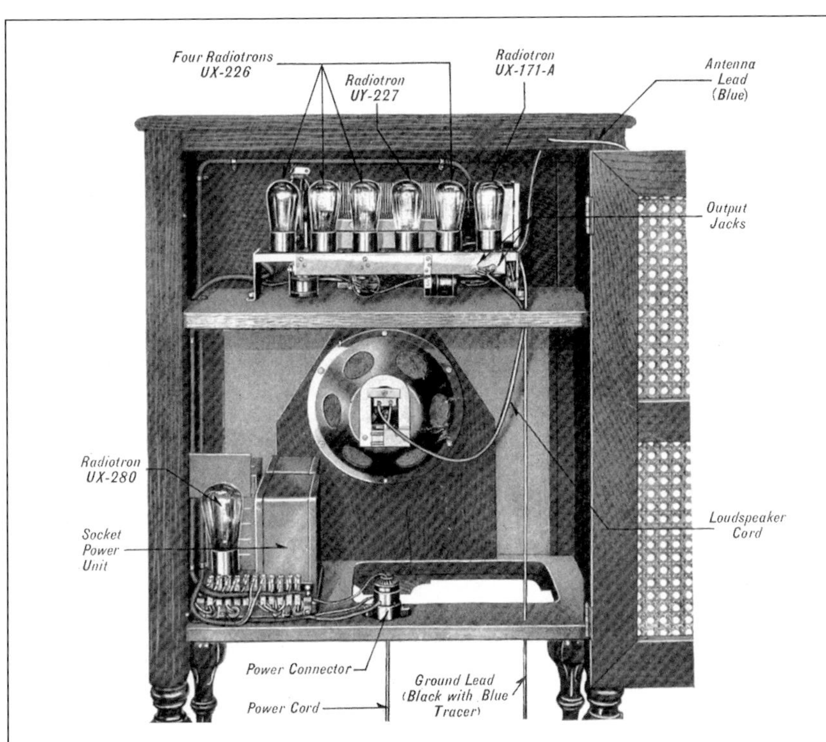

The Radiola 50 contained three major components consisting of the Radiola 17 receiver, the Radiola 17 power amplifier chassis, and the Radiola 100-A loudspeaker. (RCA Radiola 50 operating manual)

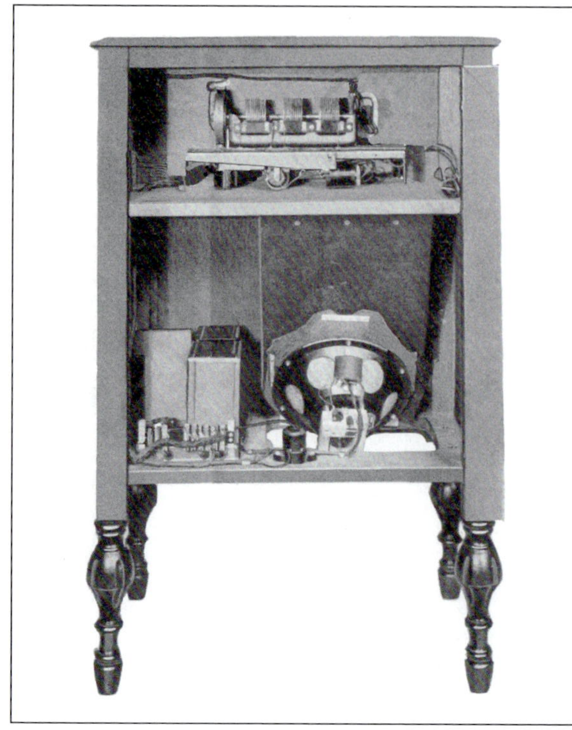

The operating panel of the Radiola 51 was similar to the Radiola 18 with the notable exception of the volume control and on-off switch which were co-located and trimmed with a single escutcheon plate mimicking the one used on the tuning control (left). (George H. Clark Collection, Smithsonian Institution) The interior of the Radiola 51 was similar to that of the Radiola 50 with the notable exception of the slanted baffle for the 100-A loudspeaker—a modification which presumably provided better audio response. (*The Town Crier*, July/August, 1928, pp. 8-9)

Radiola 51 AC Model (AR-904)

The Radiola 51 AC model—a combination of the Radiola 18 and the Model 100-A loudspeaker incorporated into a console—was released on June 13, 1928, replacing the console Radiola 50 which had just come on the market in March 1928 using the ill-fated Radiola 17 chassis. The set was advertised for $195 without Radiotrons in the September 1928 issue of *Radio Retailing* and the September 22, 1928 issue of the *Saturday Evening Post*, but it appeared at $195 with seven Radiotrons in the October 3, 1928 RCA price list. This model was the least expensive console of the 1928-1929 model line. The custom-built console cabinet was identical to the one used in the Radiola 50, and the chassis was similar but not identical to the Radiola 18 chassis.

The AC model came in a 25 cycle version and a 50-60 cycle version for the same price, the only real difference being in the design of the power transformer. Both types of sets were on the market for more than a year until the new models came out in the fall of 1929. These sets were absent from the dealer price lists dated October 1, 1929.

Loudspeakers for the 1927-1928 Season

Model 100-A Loudspeaker (UZ-1076)

The Model 100-A loudspeaker was released on May, 25, 1927, well ahead of the receivers introduced in the 1927 season, and indeed it appeared in the July 15, 1927 RCA price list at $35, replacing the Model 100 loudspeaker. While it was advertised from the outset with the rest of the Radiola line, it rarely received top billing in national publications. One of the rare ads highlighting this loudspeaker appeared in the February 1928 edition of *Radio Retailing* in which it was described as "*The reproducer that is the standard of comparison in the radio industry.*"

RCA prepared a pleasing brochure extolling the virtues of the Model 100-A loudspeaker. It was advertised as having *"more faithful reproduction over the entire musical scale, greater sensitivity and the ability to handle substantial volume of music and speech with remarkable clarity..."* This brochure featured a pictorial diagram of the loudspeaker highlighting the key features as follows:

The external view of Radiola 51 as it appeared in the 1927-28 RCA catalog was identical to that of the Radiola 50, which never appeared in a Radiola catalog.

- An 8" cone which is corrugated and treated to prevent paper rattles
- A felt enclosure located behind the cone to dampen resonances
- A large permanent magnet to give greater sensitivity and volume
- A step-up lever to provide efficient low frequencies
- An extra-heavy armature to enable great volume without saturation
- A heavy loudspeaker case to act as a baffle to preserve deep full tones
- An electrical filter to remove distortion from high-frequency harmonics

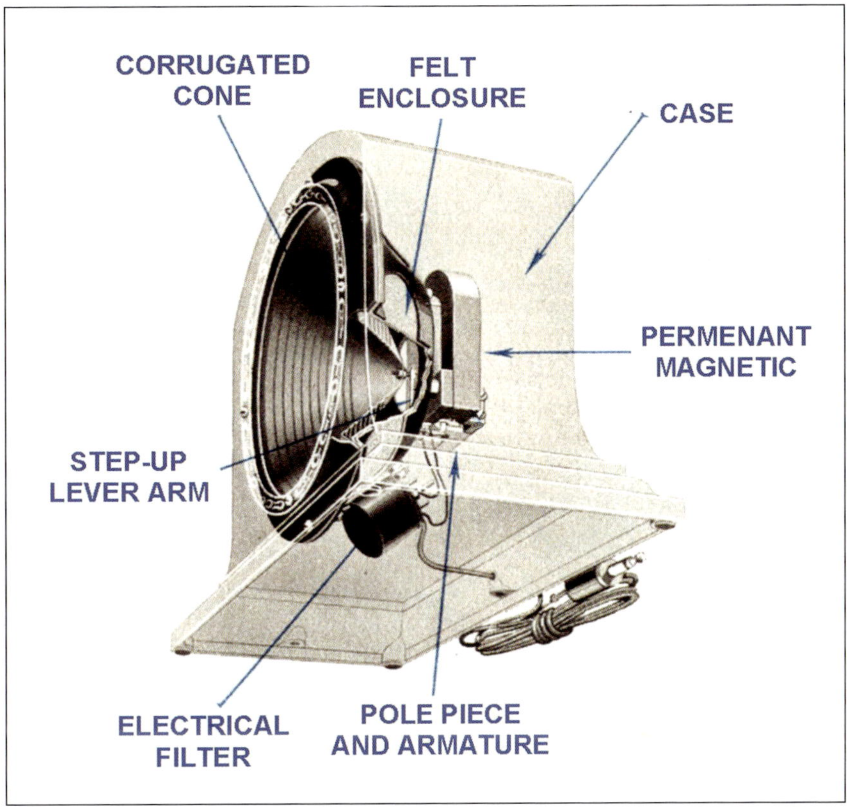

This diagram of the Model 100-A loudspeaker taken from an RCA sales brochure shows the key features of this new cone loudspeaker.

The Model 100-A loudspeaker was cast from a non-ferrous metal and coated with a dull bronze finish.

The main features of the Model 100-A reproducer were the large permanent magnet mounted on the rear, an electrical filter mounted on the frame to remove distortion from high-frequency harmonics, and a felt layer positioned behind the cone to dampen resonances.

The Model 100-A loudspeaker came with two different styles of name plate decal, one with a second space for additional numbers—the significance of which is not known.

The heavy loudspeaker case referred to in the brochure is cast from a non-ferrous metal and finished on the exterior with a dull bronze finish. It has an ornamental grille on the front and rear, with four screws supporting both the grille frame and the loudspeaker housing. Certain of the features identified in the brochure are evident in the accompanying photograph of the Model 100-A reproducer. For example, the electrical filter is attached to the loudspeaker frame appearing at the front of the photograph, and the felt backing can be seen situated between the cone and rear housing. The loudspeaker appears to have a relatively modern appearance, even by current standards.

It is not known exactly how many of these loudspeaker were made, but in a recent sampling of loudspeakers appearing on the auction market, there were more Model 100-A loudspeakers sold than any other RCA model—60% more than the ubiquitous RCA Model 100 loudspeaker—suggesting that it was the most popular loudspeaker offered by RCA in the 1920's. The loudspeaker was assigned serial numbers which appear on a paper decal affixed to the base of the loudspeaker. Most of the serial numbers appearing on the bottom of the loudspeaker observed by this author have been rather uniformly distributed in the range of 100,000 to 1,000,000, although some loudspeakers have been observed with numbers in the 2,000,000 range. It is difficult to believe that were as many as one million of these loudspeakers manufactured—much less two million—so one cannot draw any conclusions from the relatively small sample size of serial numbers observed. Many of the serial numbers are followed by two letters, but the meaning of these letters appears to be unknown.

Two different types of decals have been observed, one with a second space for an additional sequence of numbers. The significance of the difference in decals and the additional numbers in the second space is unknown. It is interesting to note that all the decals have the designation N. P. 37601, the N. P. being characteristic of the name plate designation used by GE on all their radio apparatus. Based on the presence of the N. P. designation, it is possible, and even probable, that GE manufactured all Model 100-A loudspeakers for RCA.

This loudspeaker—used both as an external loudspeaker for table-top radios and integrated into radio consoles and phonographs marketed by RCA, Brunswick and Victor—was truly exceptional for its time. It was on the market until the early 30's, well after the RCA merger with Victor which later resulted in the creation of the RCA Victor Company. The price was reduced from the introductory price of $35 to $29 by the time the October 3, 1928 price list was published, and it was sold at that price into the early 1930's, even after RCA merged with Victor.

One of the first national ads featuring the Model 100-A loudspeaker appeared in the February 1928 issue of *Radio Retailing*, a magazine aimed at the radio trade.

Model 104 DC Loudspeaker (Model No. unknown)

According to George Clark, the DC version of the model 104 loudspeaker was introduced in June of 1927. It appeared in the July 15, 1927 price list for $310, which was $35 more than the AC version. The external appearance of the DC version is identical to the AC version. The DC lighting circuit version was designed to operate with the battery-operated version of either the Radiola 25 or 28. The DC version of the 104 was essentially the same as the one used in the Radiola 30A, and is briefly described in the paragraph on the Radiola 30A. The DC version of the 104 did not appear in the October 3, 1928 RCA dealer price list, and was discontinued in September 1928 along with the AC version at the beginning of the new model year.

Model 105 Loudspeaker (UZ-1082)

According to George Clark, the Model 105 electrodynamic loudspeaker was introduced on March 8, 1928, about half way through the 1927-28 model year. The Model 105, an improvement on the Model 104, was described as being able to handle three times the load of the 104 with greater power depth and tonal beauty.[18] Like the Model 104, the Model 105 had both a reproducer unit and a rectifier-power-amplifier unit which connected to the AC lighting circuit. However, unlike the 104 which was designed for use with Radiolas 25 and 28, the Model 105 was designed for use with any receiver, capable of supplying "B" voltages up to 90 volts and suitable "C" voltages for battery sets.

The loudspeaker was enclosed in an attractive floor-model cabinet with a rectifier-power-amplifier unit utilizing the new "super-power" audio amplifier Radiotron UX-250 and two Radiotron UX-281 rectifier tubes. The reproducer used a new type of moisture-proofed cone with corrugations to eliminate "paper rattles." It was said that the electromagnet used three and one-half miles of wire and was capable of a magnetic pull of ten pounds. There was also a "high-low" line switch to compensate for voltage variations on the lighting circuit.

The 105, described in the RCA catalog as a "De Luxe Loudspeaker" was priced at $350 versus the $275 price tag for the 104 at the time the 105 was introduced. RCA must have realized that the $350 price tag might be a problem

The Model 105 loudspeaker was housed in a console cabinet described in RCA sales literature as "a gracefully designed floor model of beautiful walnut" (left). (*The Town Crier*, April 1928, p. 8) The Model 105 loudspeaker consisted of a large rectifier-power-amplifier unit using one UX-250 power amplifier and two UX-281 rectifier tubes, and a 6-inch electro-dynamic reproducer based on Model 104 loudspeaker technology (right).

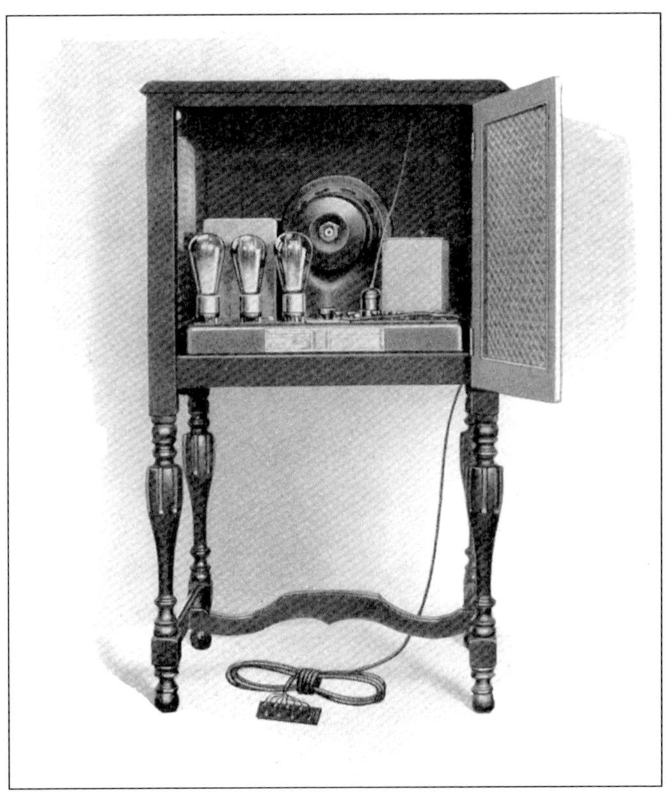

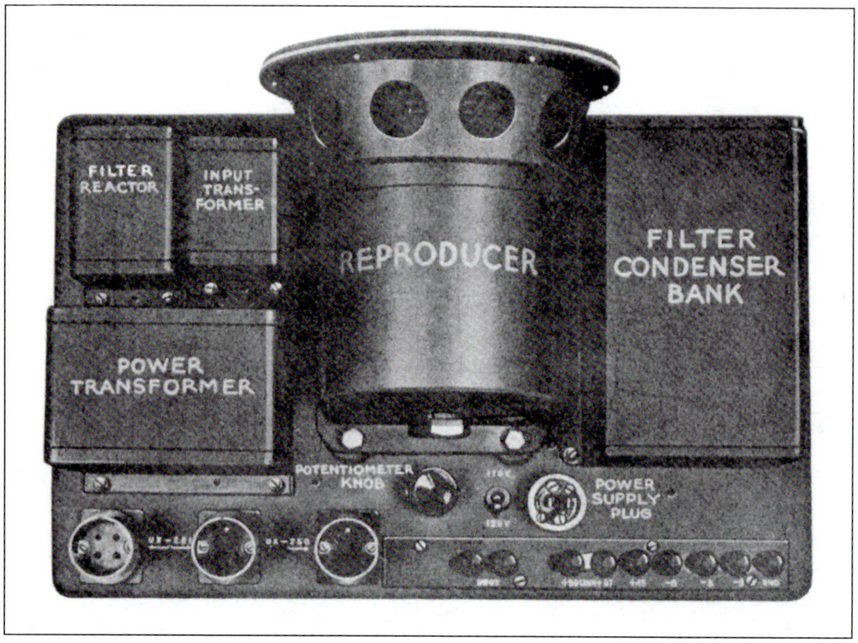

The rectifier-power-amplifier unit for the Radiola 105 loudspeaker was dominated by filters, transformers, tubes and an electrodynamic reproducer. (*RCA Victor Service Notes* for 1923-1928)

because in their marketing literature they provided dealers with metaphors of luxury watches and automobiles to use with their prospective customers:

> *"A five dollar watch will keep good time, but no one would think of comparing it with a watch in which the movement alone cost hundreds of dollars,"*

An even better metaphor was the automobile:

> *"A thousand dollar automobile will give excellent service, but it cannot be compared with the riding and driving qualities of a $10,000 car. In the same way RCA loudspeaker 105 represents the very finest form of a radio loud-speaking instrument because it embodies the best there is in technical design, materials and workmanship."*[19]

Clearly, the 105 was the top-of-the-line loudspeaker of the day, although it did not appear in any of the known RCA retail price lists, a sure sign that it was not intended to be a big revenue producer. Indeed, it must not have sold well because it was discontinued a short time after its introduction in March 1928. The 105 appeared in Radiola ads throughout the spring and summer of 1928, but only in conjunction with rest of the Radiola line—it was rarely of ever featured alone. It was removed from these group ads after October 1928.[20] It is not known how many Model 105 loudspeakers were produced, but the number must have been very low.

Power Supply for the 1927-1928 Season

"B" Eliminator (AP-1080)

RCA released a new type of "B" battery eliminator on March 5, 1928 at $48, well below the price of the Duo-Rectron "B" battery eliminator introduced in September of 1925 at $65. The unit used dry rectifiers in place of vacuum tubes and was sealed in permanent steel containers. The device was extremely compact, designed to fit neatly inside a radio cabinet, or to be used as a stand-alone exterior to the receiver similar to the Duo-Rectron. The eliminator could be used with any radio having up to eight tubes including a power amplifier tube such as the 171 or 171A. The maximum voltage was limited to 135 volts. The AP-1080 was on the market until the spring of 1929—it was absent from the April 1, 1929 RCA price list.

The new "B" Eliminator, which replaced the larger and heavier Duo-Rectron, used dry rectifiers in place of vacuum tubes, and was sealed in permanent steel containers to make the unit sufficiently compact to fit neatly inside a radio cabinet.

Endnotes and References for Chapter 8

1 Gerald F. J. Tyne, *Saga of the Vacuum Tube*, (Prompt Publications, Berkeley Heights, N. J., 1977) pp. 319-321.

2 *Radioana*, Memo from E. E. Bucher, General Sales Manager of RCA to all RCA distributors titled "Subject: - New Types of Radiotrons," May 19, 1927.

3 *The Town Crier*, Vol. V, Nos. 1 and 2, June 15, 1927.

4 *The Literary Digest*, September 10, 1927, p. 55.

5 Alan Douglas, *Radio Manufacturer's of the 1920's, Vol. 3 - RCA to Zenith* (Vestal Press, Ltd., Vestal, New York, 1991) p. 44.

6 "The Advertisements for December," *The Town Crier*, Vol. VI, No. 3, December 1927, p. 12.

7 "Operating Precautions for Socket Power Radiolas," *The Town Crier*, Vol. VI, No. 4, January 1928, p. 7.

8 "Hints on the Proper Installation of Radiola 17," *The Town Crier*, Vol. VI, No. 3, December 1927, p. 7.

9 "Radiola 17—1200 Ohm Resistor," *The Town Crier*, Vol. VI, No. 7, April 1928, p. 12.

10 "Increasing the Sensitivity of Radiola 17," *The Town Crier*, Vol. VI, No. 7, April 1928, p. 12.

11 *Radiola Retailing*, Vol. 7, No. 5, pp. 66-67.

12 "An Important Feature of Radiola 18 Installation," *The Town Crier*, Vol. VI, No. 10, July/August, 1928, p. 5.

13 Alan Douglas, *Radio Manufacturer's of the 1920's*, (Vestal Press, Ltd., Vestal, New York, 1991) p. 44.

14 Alan Douglas, *Radio Manufacturer's of the 1920's*, p. 43.

15 Ibid.

16 *The Town Crier*, Vol. 5, Nos. 1 & 2, June 15, 1927, p. 10.

17 "What's New in Radio and Where to Buy It," *Radio Retailing*, Vol. 7, No. 4, page 65.

18 "Super-Power Speaker," *Radio Retailing*, Vol. 7, No. 4, April 1928, p. 66.

19 "Answers to Questions you may be asked about RCA De Luxe Loudspeaker 105," *The Town Crier*, Vol. VI, No. 7, April 1928, p. 10-11.

20 While the Radiola 105 loudspeaker was dropped from the October 3, 1928 RCA price list, it appeared (for the last time) in the October 20, 1928 edition of the *Saturday Evening Post.*

Radio Retailing, December, 1928

The Radiola line for the 1928-29 season—for the first time ever—contained more consoles than table models, including three new consoles but only one new table model. (*Radio Retailing*, November 1928, p. 51)

Chapter 9

Radio Consoles Dominate the Scene

1928-1929

RCA introduced their first console model, the Radiola Super-VIII, in 1924. Over the following four years, the Company introduced one or two new consoles each model year, but the sales were not brisk because of the relatively high prices in comparison to table models with comparable sound performance. However, by mid-1928 the prices had dropped, and the concept of radio as a furniture piece captured the public's imagination. For the first time ever, the RCA line in 1928-29 contained more consoles than table models—even when considering the table Radiola 33, which was not introduced until 1929 and therefore did not appear in the original lineup first advertised nationally circa September 1928.[1]

The craze in radio consoles was well documented in the contemporary press for both the trade and the public. Articles began to appear in trade magazines such as *Radio Retailing,* extolling the virtues of radio furniture and explaining to dealers how best to market these expensive pieces of furniture. One article pointed out that *"with elimination of separate power-supply equipment, furniture becomes the radio merchant's largest profit-getting accessory."*[2] The article went on to say that "*according to authorities*" the talking points about radio furniture fall into seven sub-divisions: wood, construction, workmanship, design, period authenticity, finish, and differences of various historical periods as applied to design. Salesmen were urged to become familiar with these talking points and articulate them to potential customers.

The potential customer had two basic choices when it came to purchasing consoles: 1) factory-made consoles such as those sold by RCA, and 2) custom-made consoles that were assembled by a distributor or retail store using a combination of cabinets, power supplies, speakers and/or receivers available from a variety of manufacturers. In the latter case, the consoles were pre-designed and stocked by the distributor or retailer, or they were custom made to meet the needs of the individual customer.

The receivers used in the custom consoles were often table models manufactured by reputable companies such as RCA, and often with their tacit approval. By 1927, receivers were generally of a reasonably standard size and shape such that the cabinet manufacturers could make panels which could be cut down or otherwise adjusted to accommodate a wide range of receivers on the market. As a result, there were many consoles sold with RCA receivers in unfamiliar cabinets that do not appear in any RCA catalog or price list.

There were certain advantages to buying a custom-made console. First, of course, the customer could get exactly what he wanted, and generally at a lower price. Second, since the cabinets were styled to accept a wide range of components, the individual components could be replaced when they became obsolete, thus obviating the need to replace the entire set including the very expensive cabinet. On the other hand, there were also advantages to buying a factory-made console. The components were designed and tested as a unit to assure proper operation, and the manufacturer, if reputable, would stand behind the product. Also, it was generally easier and less expensive to have the set serviced and to locate replacement parts.

RCA and Victor Talking Machine Companies Merge

Merger negotiations between RCA and Victor—well underway at the beginning of the 1928-29 radio season—culminated in a merger of the two companies effective March 1929. While the merger was widely announced to the industry and the public, there was minimal impact on the 1928-29 radio season, not only because the RCA radio line had been designed and was ready for release at the time of the merger, but also because the operations of the two companies remained quite separate during the entire calendar year 1929. In fact, even the ads placed by RCA for the 1928-29 season apparatus during the first half of 1929 continued to use the Radio Corporation of America name.

It would not be until later in 1929 that the merger would become more apparent to the public through a name change which appeared in national ads, sales brochures and operating manuals. However, major changes would not occur until 1930 when the manufacturing of all RCA radio apparatus was moved to the Victor plant in Camden, New Jersey a story which is recounted in the *Afterwords*.

The 1928-1929 Radiola Line

The 1928-29 Radiola line was first announced to dealers and the trade in September of 1928,[3] although additional models were added to the line in December of 1928 and March of 1929. The Radiola line for 1928-29 is shown in Table 9-1, where it can be seen that there were more console receivers in the RCA lineup than table-top cabinet-style receivers—a first to be sure. The console had finally come of age. Also, for the first time there were more AC models than battery models, and in fact there was only one battery set in the lineup—the Radiola 16 introduced the previous year. Finally, for the first time there were as many new superheterodyne receivers introduced in the 1928-29 line as there were new TRF receivers.

Loudspeaker technology was making rapid advances, and as a result a number of new speaker models were introduced in rapid succession. The Model 100 table speaker had just been replaced by the Model 100-A the year before, and two new table speakers, Models 100-B and 103, were introduced in 1928. The Model 105 console speaker introduced in March of 1928 would be replaced by an even newer and far less-expensive Model 106 console in December 1928 as well.

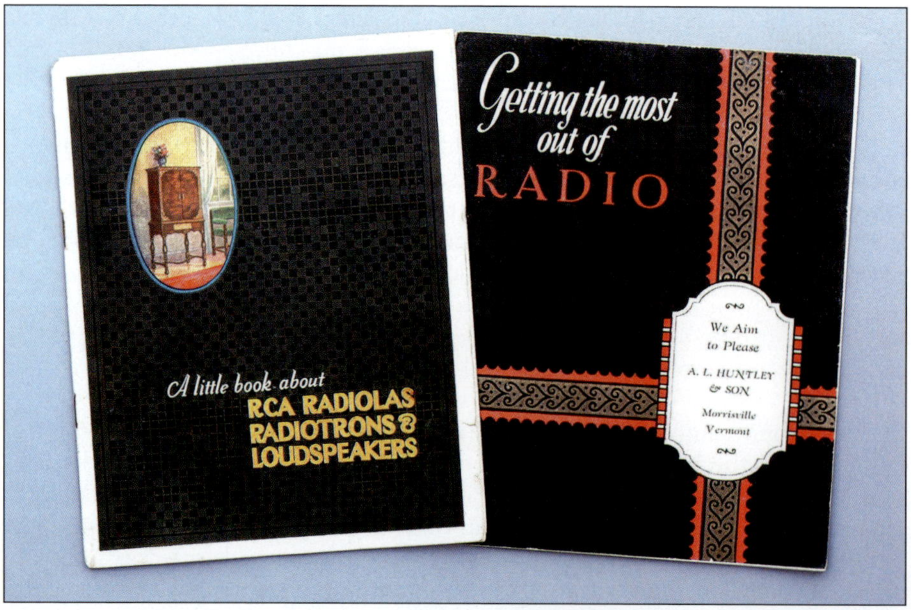

RCA produced two identical catalogs for the 1928 season with different covers—one entitled *A little book about Radiolas, Radiotrons and Loudspeakers* for general distribution (left), and another entitled *Getting the Most out of Radio* with the name of individual dealers printed on the front (right).

As in previous years, RCA provided its dealers with a full range of marketing material during the 1928-29 season such as the brochures shown here.

The instruction manuals for the three new consoles introduced in the 1928-29 season were produced with the heavy olive-taupe covers similar to ones from the previous season, while the ones for the two table models were produced with the white paper covers.

As in previous years, RCA supported the dealers with a full line of marketing materials such as catalogs, brochures, mailers, clocks, blotters, and tapestries, all of which were advertised to the dealer in almost every issue of *The Town Crier*. Examples of the brochures provided to dealers for distribution in the 1928-29 model year are shown in the collage above.

Receivers for the 1928-1929 Season

Of the five new receivers introduced during 1928, Radiolas 33 and 41 were based on the standard six-tube TRF design using the AC-filament UX-226 and UY-227 tubes for all but the audio output stage, while Radiolas 60, 62 and 64 were based on a new eight-tube superheterodyne circuit designed specifically for AC lighting circuits. The Radiola 41 was a modified version of the Radiola 17 with a new electrodynamic speaker, and the Radiola 33 was a less expensive replacement for the Radiola 18. In December 1928, DC versions of the Radiolas 18 and 51 were quietly introduced in limited regions having DC lighting circuits.[4]

The new Radiola 60 was an eight-tube fundamental superheterodyne circuit using seven UY-227 tubes for all stages but the audio output, which used the UX-171-A, and a UX-280 tube as a full-wave rectifier in the power supply.[5] The rectifier was not generally included in the tube count in the late 20's, a holdover from tube counting practices for battery sets. The Radiola 60 circuit was incorporated into two consoles, the Radiola 62 and the Radiola 64. The Radiola 64 was a deluxe version with greater power output and other luxury features.

The only battery set in the lineup was the Radiola 16, a holdover from the previous year. The Radiola 16 must have continued to sell well, because the

Table 9-1. Radio receiver apparatus offered by RCA in the 1925-27 season including hold-over models (shaded portion)

	Model	Apparatus Type	Cabinet Style	Tubes[a]	Power	Intro Date	Discon'd. By	List Price[b]
1928-29 Season New Models	R. 18 DC (AR-891)	TRF Receiver	Cabinet	6	DC	12/28/28	Jan. 1930	$125
	Radiola 33 (AR-784)	TRF Receiver	Cabinet & legs	6 (1)	AC	3/26/29	Sep. 1930	$86.50
	R. 33 DC (Unknown)	TRF Receiver	Cabinet & legs	7	DC	7/29/29	Sep. 1930	$79.75
	Radiola 41 (AR-782)	TRF Receiver	Console	6 (1)	AC	12/15/28	Sep. 1929	$238.50
	R. 41 DC (AR-871)	TRF Receiver	Console	10	DC	3/1/29	Sep. 1929	$255
	R. 51 DC (AR-903)	TRF Receiver	Console	6	DC	12/28/28	Sep. 1929	$205
	Radiola 60 (AR-954)	Super-heterodyne	Cabinet	8 (1)	AC	8/13/28	Sep. 1930	$210
	Radiola 62 (AR-982)	Super-heterodyne	Console	8 (1)	AC	9/14/28	Jan. 1930	$410
	Radiola 64 (AR-894)	Super-heterodyne	Console	9 (2)	AC	10/23/28	Sep. 1930	$608.50
	100-B (UZ-783)	PM Speaker	Tabletop	0	N/A	3/26/29	Sep. 1930	$22
	103 (UZ-749)	PM Speaker	Tabletop	0	N/A	8/17/28	Sep. 1930	$37.50
	106 (UZ-642)	Dynamic Speaker	Floor	0	AC	11/26/28	Sep. 1930	$88
	106 DC (Unknown)	Dynamic Speaker	Floor	0	DC	4/12/29	Sep. 1930	$65
Holdovers	Radiola 16 (AR-924)	TRF Receiver	Cabinet	6	Battery	8/8/27	Apr. 1929	$82.75
	Radiola 18 (AR-936)	TRF Receiver	Cabinet	6 (1)	AC	4/19/28	Jan. 1930	$115
	Radiola 30A (AR-926)	Super-heterodyne	Console	8 (3)	AC	8/7/27	Apr. 1929	$285
	100-A (UZ-1076)	PM Speaker	Tabletop	0	N/A	5/25/27	Sep. 1930	$29
	105 (UZ-1082)	Dynamic Speaker	Floor	1 (2)	AC	3/8/28	Nov. 1928	$350

[a]Additional rectifier and ballast tubes are shown in parentheses
[b]Prices include Radiotrons

price had not been reduced from the introductory price of $82.75 the year before. In stark contrast, the holdover Radiola 30A superheterodyne was reduced to a modest $285 from the introductory price of $495, a reduction which was necessary to clear out the old stock in the face of competition from the new Model 62 superheterodyne priced at $410.

Radiola 18 DC Model (AR-891)

The DC version of the Radiola 18 was released without fanfare on December 28, 1928 in regions served by DC power lines. The DC version was similar to the AC version in appearance and performance, but electrically was significantly different. There were a number of minor wiring differences in the receiver portion, including different placements of capacitors and the addition of capacitors in the antenna circuit. The DC version used five UX-112-A tubes in lieu of the four UX-226 and one UY-227 tubes used in the AC version. Both used a UX-171-A tube in the audio output stage.

The power supply portion of the set was completely different. The transformer and rectifier tube were absent from the DC version. Oddly, while this receiver was connected to the DC lighting circuit, it still required a separate 18 volt "C" battery for grid biasing. Despite the fact that the set had no transformer or rectifier tube, the set sold for $125 versus the $115 price tag for the AC version. The DC version was on the market for about a year—it was absent form the RCA dealer price sheets dated January 30, 1930.

Radiola 33 Model (AR-784)

The Radiola 33 was introduced on March 26, 1929, the last receiver to be introduced in the 1928-29 season. Ads such as the art deco rendering by commercial artist William Rienecke shown below began to appear in April 1929. Advance copies of full-page color ads such as this were routinely inserted in *The Town Crier* beginning in 1929, presumably to give RCA dealers a preview of ads that would be appearing in future issues of national magazines.[6]

Ads for the Radiola 33 with the Model 100-B loudspeaker such as this art deco rendering by commercial artist William Rienecke began to appear circa April 1929. (Unknown magazine)

The companion Model 100-B speaker (described on page 329) with a similar exterior was designed specifically for the Radiola 33, and the two were often shown together in ads and other marketing material.

RCA undertook a national newspaper and magazine campaign, predicting that fully half of the families in America would see and read the advertising material in April 1929 alone.[7] RCA prepared the usual sales materials to promote the 33, but the keynote of the sales program was the "Big Blue Book" used by salesmen to emphasize that the Radiola 33 and its companion 100-B loudspeaker created its own art and added to the beauty of its surroundings—which were amply illustrated in the book by photographic reproductions of the 33 in various home settings.[8] The Big Blue Book was indeed big by any standard for radio pamphlets with dimensions slightly larger than 14 by 20 inches!

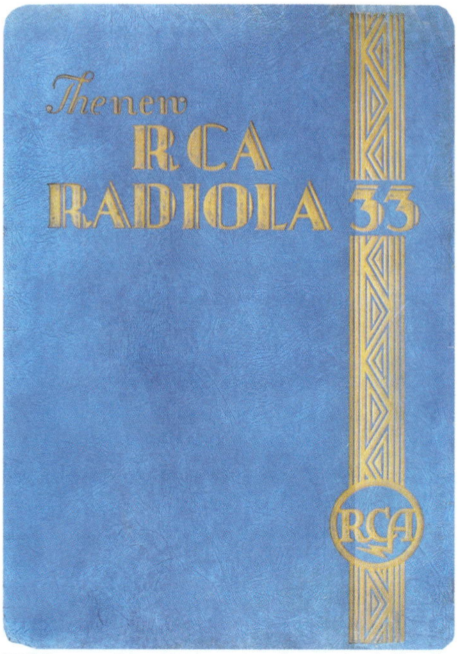

RCA produced an uncharacteristically large booklet with dimensions slightly larger than 14 by 20 inches to extol the virtues of the Radiola 33—not surprisingly referred to as "The Big Blue Book."

The Radiola 33 was made as inexpensively as possible, beginning with the metal case attached to a wood base which emulated the metal cases of many other manufacturers including those of Atwater Kent. The introductory price of the Radiola 33 was advertised at $77.50 without tubes, $86.50 with tubes but without legs, and $94.50 with legs—prices on par with other similar sets. In order to make the metal cabinet more appealing to potential customers, RCA suggested that dealers use the following monolog to describe the set to potential customers:

> *"The inherent strength of metallic construction has been combined with the warmth and visual appeal of wood through the skillful use of panels which reproduce, in rich brown tones, the grain of rare African mahogany. The Champleve enamel border of triangles and acute angles takes its design from decorations used by the Indian tribes of Central and Southwestern America. So you see how skillfully the old American art has been blended with the present trend in furniture design."*[9]

The metal cabinet was not the only cost-saving feature on the Radiola 33. Unnecessary parts were eliminated such as terminal strips on the power supply and the AC line voltage adjustment switch, and a new type of knob requiring less assembly time was used. Other minor design changes were made such as moving the RF compensating and detector-trimming capacitors to the rear of the cabinet where they could be more easily accessed. The AC version came in models for both 25-40 and 50-60 cycle lighting circuits.

The Radiola 33 used the standard six-tube TRF circuit which was almost identical to the one used in the Radiola 18. It is difficult to understand why RCA would introduce another set virtually identical to the popular Radiola

The Radiola 33 was sold with or without legs which readily attached to wood base. The wood base in turn supported a less expensive metal cabinet described in RCA sale literature as one with a *"Champleve enamel border of triangles and acute angles* [which] *takes its design from decorations used by the Indian tribes of Central and Southwestern America."*

The Radiola 33 chassis shown here was physically and electrically similar to the Radiola 18 chassis pictured in Chapter 8.

18—particularly when the new screen-grid receivers were to be introduced only a few short months later. The only reasonable explanation seems to be the price pressure that RCA faced in late 1928 from receivers offered by other popular manufacturers such as Atwater Kent's very similar Model 40 TRF priced at $77 without tubes versus $95 for the Radiola 18 similarly equipped.

Judging by the number of sets manufactured during the year or so that the Radiola 33 was on the market, both the set and the RCA advertising campaign were successful. Sales were undoubtedly spurred by the price reduction to $54 ($68.50 with tubes) in June 1929 when the first receivers for the new model year were announced. According to Douglas, 236,961 sets were manufactured.[10] The set was discontinued at the beginning of the new model year circa September 1930.

Radiola 33 DC (Model Number Unknown)

RCA introduced two versions of the Radiola 33 for DC power lines in 1929, one for 110 volts and one for 220 volts. According to George Clark, the official release date for the Radiola 33 DC was July 29, 1929. While both versions appeared on export price lists beginning with the September 1, 1929 list, only the 110-volt version appeared on the domestic price lists, the first entry beginning on the October 1, 1929 list. The suggested retail price for the 110-volt DC version was $79.75, but the AC version had been reduced to $68.75 by this time.

The DC sets had the same appearance as the AC version, but the tubes and circuitry were somewhat different—as were all RCA sets made for DC power lines. While the AC version used UX-226 and UY-227 AC tubes for the first five stages, both DC versions used the UX-112-A dry cell tubes. All three sets used the UX-171-A tube in the audio output section, and while the 110-volt version used two tubes in a push-pull configuration—as was customary in RCA's DC sets—the 220-volt version only used one tube, as did the AC version. Obviously, a plate voltage of 110 volts was not sufficient to drive a single UX-171-A, but the 220-volt line provided the necessary plate voltage. The AC version used a UX-280 full wave rectifier which could provide the necessary plate voltage to the audio output tube, while of course the DC versions required no rectifier tube. The DC version also used a line voltage regulating switch to compensate for variations in line voltages in different areas. This set was discontinued at the beginning of the new model year circa September 1930.

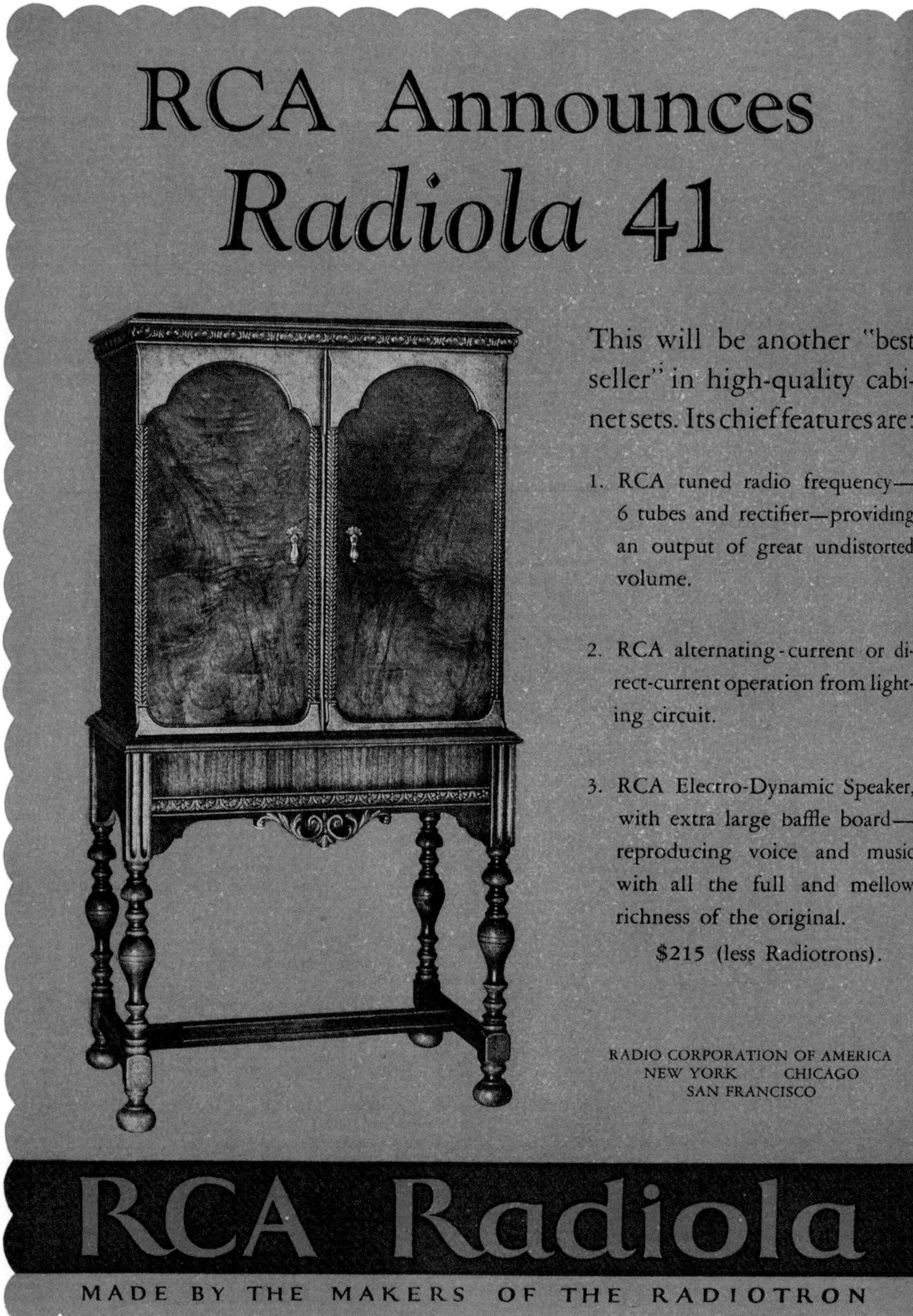

The Radiola 41 was introduced to the trade in November 1928 with this full page ad in *Radio Retailing*. (*Radio Retailing*, November 1928, p. 50)

Radiola 41 (AR-782)

According to George Clark, the Radiola 41 was introduced on December 15, 1928, somewhat later than the introductory dates of the superheterodyne line in late August to early September 1928. However, it was introduced to the radio industry a month before the Clark release date in an ad appearing in the November 1928 issue of *Radio Retailing.*[11] The set was designed around the *"famous time-tested RCA tuned radio frequency circuit,"* which was similar, but not identical, to the Radiola 17. The TRF receiver was installed in an attractive walnut finished console with cabinet doors which were hinged so that, when opened, they could be flush with the sides of the cabinet. The set was equipped with the new Model 106 electrodynamic speaker, which is described later in this chapter.

This TRF receiver used four UX-226 tubes and one UY-227 tube, but unlike the Radiola 17, it used the UX-210 tube for the audio output. Incredibly, grid resistors were once again used in the tuned RF stages to prevent unwanted oscillations, the same design approach that led to uncontrollable oscillations in the Radiola 17. Not surprisingly, the RCA Victor repair manual once again had a section devoted to uncontrolled oscillations in which technicians were advised:

> *"Where everything tests O.K. and the Radiola still oscillates, the following remedy should be applied: Connect an 800-ohm fixed resistance in series with the plate supply to all Radiotron UX-226. This will reduce the plate voltage to these tubes and should prevent any oscillation. If, however, the oscillation continues, do not put the resistor in the plate supply to the UX-226 Radiotrons, but connect it across the primary of the second R. F. transformer. This will effectively prevent any case of oscillation."*

This fix is essentially the same one recommended for the Radiola 17 that led to complaints regarding lack of sensitivity.

The Radiola 41 was described as a combination of the well-known RCA tuned radio frequency circuit and the new dynamic speaker encased in a handsome walnut-finished cabinet (left). (*The Town Crier*, November 1928, pp. 8-9) The doors of the Radiola 41 cabinet—artistically decorated with carved moldings—were hinged so that they could be turned back flat against the cabinet. (*The Town Crier*, November 1928, pp. 8-9)

The Radiola 41 was introduced on December 15, 1928 at a price of $215 without Radiotrons or $238.50 with Radiotrons. The AC version came in both 25 cycle and 50-60 cycle versions, and both were on the market for less than a year, discontinued when the new model sets came out in the fall of 1929. The set was absent from the distributor price list dated October 1, 1929.

The speaker and chassis of the Radiola 41 with four UX-226 Radiotrons, one UY-227 Radiotron, and one UX-210 power output Radiotron were located in the upper portion of the cabinet, while the power supply with one UX-280 rectifier tube was located on the bottom shelf. (*The Town Crier*, November 1928, pp. 8-9)

Radiola 41 DC Model (AR-871)

The DC version of Radiola 41 was introduced on March 1, 1929, several months after the AC version. It had the same appearance as the AC version, but the circuitry was somewhat different, as was the circuitry for all the other six-tube TRF receivers operating from the DC lighting circuit. The 41 DC used six UX-112-A tubes—three as RF amplifiers, one as a detector, and two as audio amplifiers. However, it also had an additional power amplifier stage consisting of four UX-171-A tubes in the now-familiar parallel push-pull configuration to drive the Model 106 electrodynamic reproducer—and so, the Radiola 41 had the distinction of having six tubes devoted to audio amplification! The Radiola 41 was on the market until the end of the model year in the fall of 1929, and like the AC version, the set was absent from the distributor price list dated October 1, 1929.

Radiola 51 DC Model (AR-903)

The DC version of the Radiola 51 was released on December 28, 1928 with little fanfare at a suggested retail price of $205. The DC version of the 51 was identical to the DC version of the Radiola 18 addressed above, except that it was placed in a console—which had an appearance identical to the AC version of the Radiola 51. The sale of these sets was confined to limited regions that were serviced by DC lighting circuits, and so no formal announcement to the public was made. RCA dealers were instructed that a limited number of DC sets could be obtained from their distributors. It was discontinued by September of 1929.

Radiola 60 (AR-954)

The Radiola 60 was a table-top version of the new AC superheterodyne line which was officially introduced on August 13, 1928 at a price of $175 without Radiotrons and $210 with nine Radiotrons. It was first featured in an ad placed in the September 1928 issue of *Radio Retailing* with the new Model 103 loudspeaker, but the most spectacular image was featured in an ad created by Saul Tepper in the March, 9, 1929 issue of the *Saturday Evening Post*. Saul Tepper produced illustrations for many major magazines of the time, as well as advertisements for major corporations including General Motors, Coca-Cola, Mobil, Texaco, and GE in addition to RCA.

The Radiola 60 had a very attractive cabinet with a two-toned walnut finish, and three decorative cast-metal escutcheon plates—one each for the tuning control, volume control, and on-off switch.

A most spectacular ad for the Radiola 60 superheterodyne and Model 103 loudspeaker by commercial artist Saul Tepper appeared in the March 19, 1929 issue of the *Saturday Evening Post*.

The new Radiola 60 superheterodyne receiver chassis with certain modifications was used in all of the 60-series console sets.

This Radiola 60 chassis, like many other AC sets of this era, shows some sign of repairs.

The Radiola 60 power supply using a single UX-280 rectifier tube was also used in the Radiola 62 console.

The set was housed in an attractive two-toned walnut cabinet with a bronze escutcheon plate in the center surrounding the selector dial and a small concealed bulb to illuminate the tuning dial. The only controls were an on-off switch and a volume control, making the Radiola 60 the first superheterodyne with just two controls. The molded feet painted in a bronze tone were designed to match the shell design on the escutcheon plate. These feet are often broken or missing on sets found today.

The Radiola 60 consisted of an eight-tube circuit using seven AC type UY-227 tubes and a type UX-171-A audio output tube, plus a type UX-280 rectifier tube used in the power supply situated in a separate chassis. The eight tubes were used in the same way as in the Radiola 28 superheterodyne: two stages of tuned RF amplification, a detector, two stages of IF amplification, a second power detector, and one stage of power audio amplification. The set was operated by a single tuning control which was attached to the same three-gang bathtub condenser used in the Radiola 18. The use of the UY-227 tube with its indirect heaters (i.e., cathodes) allowed the use of AC-powered loudspeakers such as the 104 and 105 without producing objectionable hum.

The Radiola 60 chassis with slight modification was used in the Radiola 62 and 64 consoles as well. The entire line of superheterodynes was advertised as having "inclusive selectivity" which was the ability to separate broadcast stations in the most congested areas without excluding any of the musical frequencies that were necessary for perfect reception. It was also advertised to be four times as sensitive as ordinary TRF sets. Indeed, The Radiola 60 was an amazing set at that time, and according to Douglas, 135,336 were made.[12] The set was on the market until well into 1930, although the price had been reduced to $152.75 including tubes by October 1, 1929.

Radiola 62 (AR- 982)

The Radiola 62 was officially released on September 14, 1928, about a month later than the Radiola 60, for $375 without Radiotrons and $410 with nine Radiotrons including one rectifier tube. It was featured along with a Radiola 60

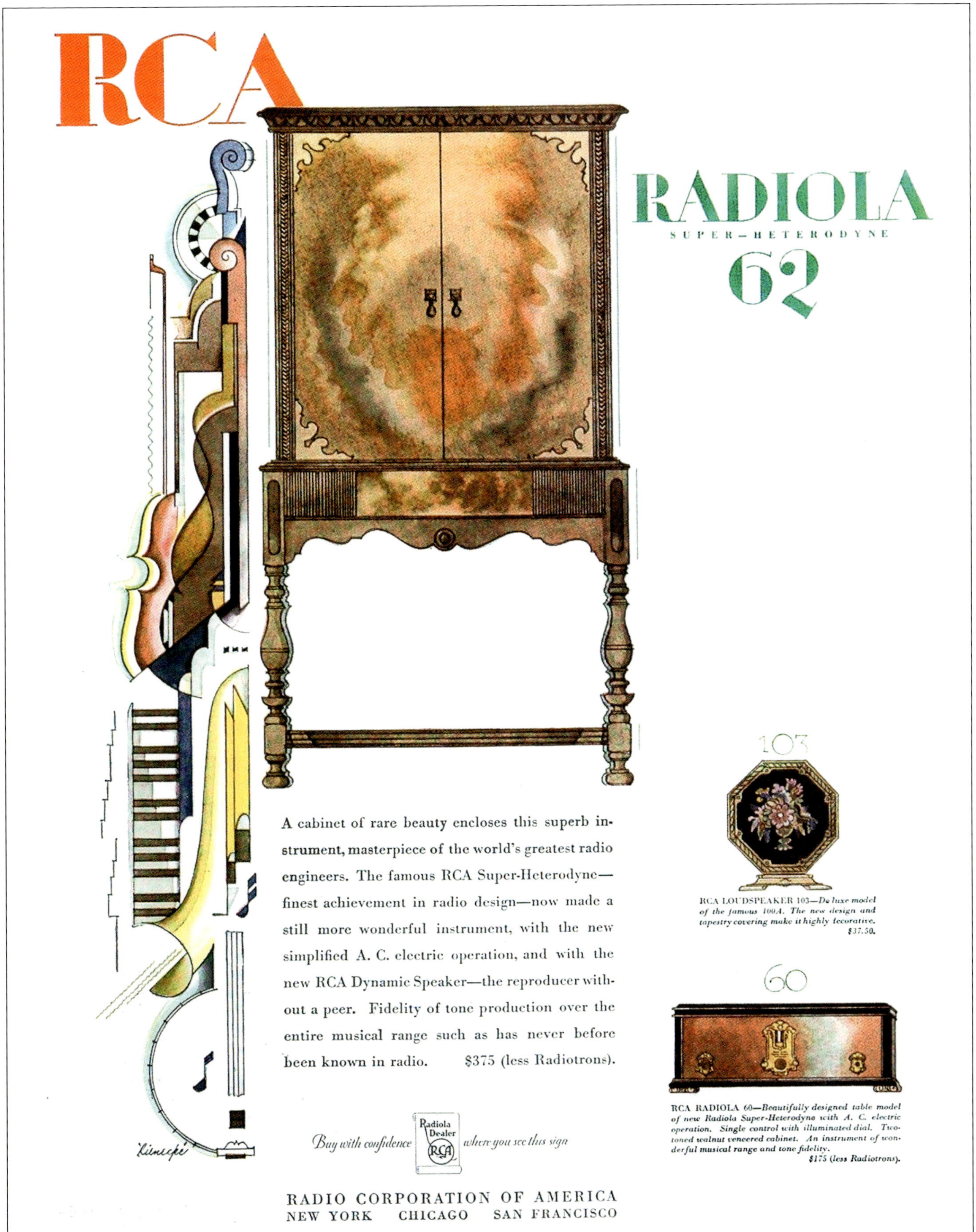

The Radiola 62 was featured in colorful art deco ads by commercial artist William Rienecke. (Unknown magazine circa 1928)

The Radiola 62 was described as a "*beautiful walnut console cabinet... artistically decorated with maple inlays and carved moldings in accordance with the best furniture practice* (left)." (RCA Catalog, 1928) The cabinet doors of the Radiola 62 opened flat against the sides to reveal two control escutcheons in the upper portion of the front panel and a speaker opening covered by a grille cloth in the lower center of the panel (right). (*The Town Crier*, January 1929, pp. 8-9)

in a two page advertisement in the September 22, 1928 issue of the *Saturday Evening Post*. This set also appeared in a series of colorful ads by commercial artist William Rienecke.

This receiver was a combination of the Radiola 60 and the new Model 106 8-inch electrodynamic speaker enclosed in what was described as *"a beautiful walnut console cabinet."* Advertising brochures focused as much on the furniture aspect of the radio as the performance: *"The beautiful walnut console cabinet is artistically decorated with maple inlays and carved moldings in accordance with the best furniture practice. It is so designed that its appearance is equally pleasing when the doors exposing the control panel and loudspeaker grill are open or closed."*

The cabinet doors opened out to the sides, revealing the front panel with a tuning control on the right side and the volume control and on-off switch co-located on the left side–both with a cast metal escutcheon plate. The plate for the tuning control was identical to the one on the Radiola 60, and a new matching escutcheon plate was designed for the volume control and on-off switch. The speaker opening covered with a grille cloth for the electrodynamic speaker was located towards the bottom of the front panel.

The interior of the set was divided into two sections, an upper compartment for the receiver circuit, and a larger compartment at the bottom for the power supply, speaker and associated field supply unit for the field coil in a new type of dynamic speaker. The chassis for the receiver and power supply were essentially identical to the Radiola 60, and the 8-inch dynamic reproducer was identical to the one used in the new Model 106 loudspeaker, but a new field supply unit with disc rectifiers was used to supply current for the speaker coil.

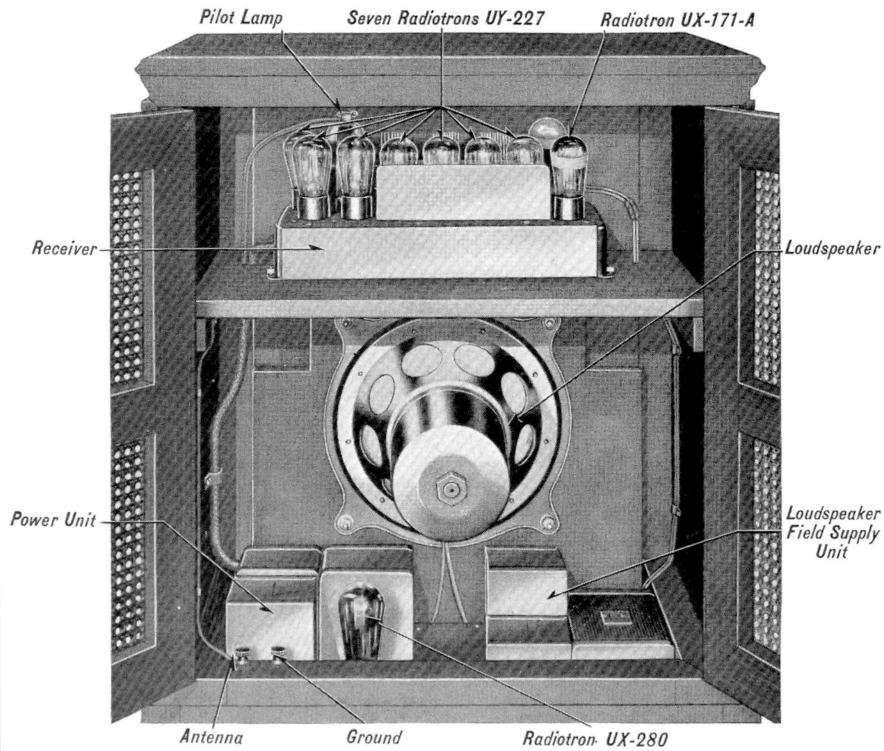

The Radiola 62 was divided into two major compartments—a smaller compartment at the top for the receiver chassis, and a larger compartment at the bottom for the power supply, the 8-inch dynamic reproducer used in the new Model 106 loudspeaker, and a new field supply unit for the dynamic speaker coil using disc rectifiers. (*RCA Radiola 62 Instructions*, IB-62)

There is little to see of the field supply unit from the exterior because the three main components—the disc rectifiers, the condenser bank, and the coupling circuits are contained in three separate metal cans. This field supply unit was located on the right side when viewing the set from the rear.

The Radiola 62 was on the market until circa January 1930, not as long as the electrically equivalent but less expensive Radiola 60. It was absent from the RCA dealer price list dated January 30, 1930. According to Douglas 19,972 were made.[13]

Radiola 64 (AR-894)

The Radiola 64 was officially released on October 23, 1928 at a price of $550 without Radiotrons and $608.50 with Radiotrons. The set, which was also based on the Radiola 60 receiver circuit, had an appearance very similar to that of the Radiola 62 priced at $375 without Radiotrons. It had several new features described below which presumably justified the additional $175 the owner had to pay for the Radiola 64 as compared to the Radiola 62. The set was heavily advertised using colorful art deco ads by commercial artist William Rienecke similar to the ones he produced for the Radiola 62.

This top-of-the-line receiver was housed in a large walnut cabinet with doors that swung around to the side to reveal the controls and tuning meter in the center of the panel. It was described as the crowning achievement in radio receiving set design, having no equal. The layout of this set on the front panel was distinctly different than the Radiola 62. The speaker opening was located at the top of the cabinet above the tuning and volume controls and on-off switch. The escutcheon plates for the controls were evocative of—but yet distinctly different than—those used in the Radiola 62. A new tuning meter described below was centered between and just above the controls.

The Radiola 64 was similar to the Radiola 62 except for a new higher-power UX-250 amplifier tube and the addition of several innovations including the following: a tuning meter allowing the user to obtain the position of maximum

RCA commissioned commercial artist William Rienecke to produce a series of art deco ads for the Radiola 64 similar to the ones he produced for the Radiola 62. (*Country Life*, circa 1928)

signal, an automatic volume control to counter the effects of fading and keep the volume at a constant level for both local and distant stations, and a new sensitivity control to minimize static and other electrical disturbances originating at points more distant than the station being received. Marketing literature of the day admitted that the extra output power produced by the new UX-250—almost five watts or three times that of the next most-powerful tube on the market—was not needed for greater volume, but rather to *"provide superb quality of reproduction at any volume."*

Like Radiolas 60 and 62, the Radiola 64 used seven UX-227 Radiotrons in the receiving circuit, but an additional UX-227 tube was required for the automatic volume control circuit. The new UX-227 for the tuning control was placed in the position occupied by the UX-171-A output tube on the Radiola 60 chassis, and the new UX-250 power output tube was moved to a position on the new power-amplifier chassis. Thus, it was possible to use the same Radiola 60 chassis for the Radiola 64 with some minor wiring variations discernable only to a keen observer. Also located on the new power-amplifier chassis were two UX-281 Radiotrons used in a full-wave rectifier configuration, bringing the total tube count to eleven. The interior layout was reversed from that of the Radiola 62 with the speaker mounted on a baffle at the top of the cabinet. The chassis, with an appearance almost identical to the Radiola 60 chassis, was mounted on a shelf just below the speaker, and the rather large and heavy power-amplifier circuitry sat on the base of the cabinet at the bottom.

The Radiola 64 was made in two versions, one for 50-60 cycle AC operation and one for 25 cycle operation. It was on the market until late 1930—absent from the RCA price sheet dated November 14, 1930. At the price of $608, it is not surprising that only 7,406 were made.[14]

The Radiola 64 constructed of walnut was said to be distinguished by *"a studied attention to detail and design that is usually found only in the finest custom built furniture* (left)." (RCA Catalog, 1928) The panel of the Radiola 64 can be distinguished from the very similar panel of the Radiola 62 most easily by the position of the controls, which were located below the speaker on the Radiola 64 (right). (*The Town Crier*, December 1928, pp. 8-9)

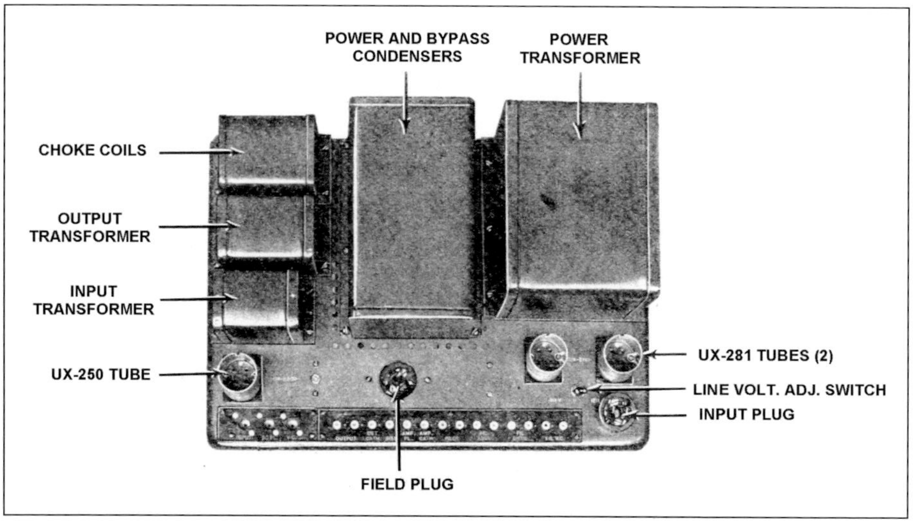

While the Radiola 64 is based on the Radiola 60 circuitry, the power-amplifier was quite different with one UX-250 power output tube and two UX-281 rectifier tubes. (Adapted from the *RCA Replacement Parts Catalog D*, April 1929, p. 28)

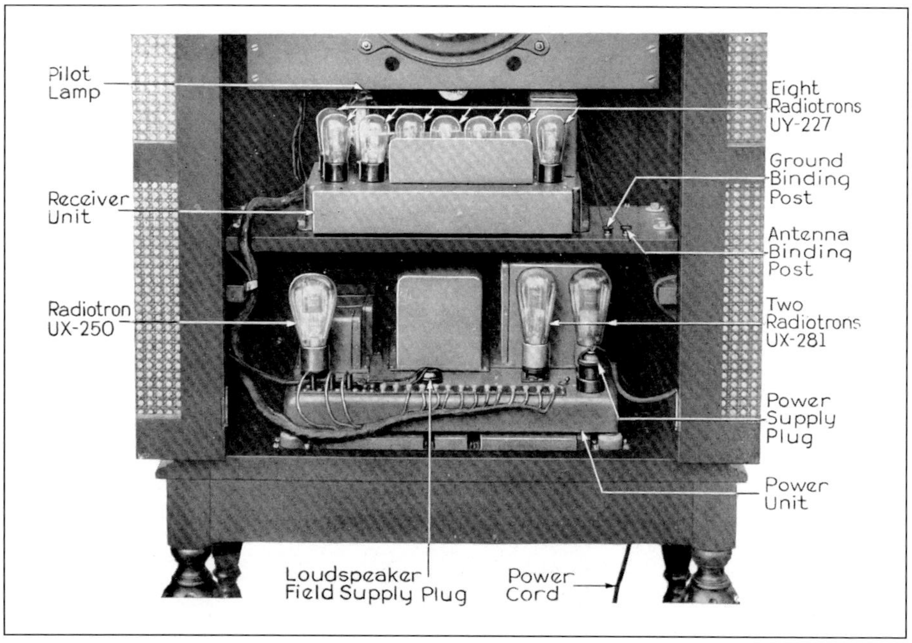

An interior view of the Radiola 64 clearly shows the Radiola 60 chassis with eight UY-277 Radiotrons—in contrast to the seven UY-227 tubes and one UX-171-A power output tube used in the Radiola 60. (*RCA Radiola 64 Instructions*, IB-64)

Loudspeakers for the 1928-1929 Season

RCA had been marketing two types of cone speakers, a permanent-magnet type to provide the required magnetic field, and an electrodynamic type with a self-contained rectifier-power-amplifier unit designed to supply power for both the speaker coil and the rest of the receiver. Both types were sold either as a stand-alone unit for use with virtually any radio, or integrated together with a receiver in a common cabinet—generally a console, which provided sufficient space for the receiver, the power unit and the reproducer.

RCA had been making improvements in both types of speakers, and this year would be no different. RCA introduced the Model 100-B and Model 103 loudspeakers, both improvements over the Model 100-A permanent-magnet speaker introduced the previous year. RCA also introduced the Model 106 electrodynamic speaker, but this speaker was markedly different than its two predecessors, Models 104 and 105. The Model 106 loudspeaker was a self-standing, electrodynamic speaker with a small power supply designed to drive only the speaker coil but not any of the receiver tubes. Absent the large power supply and power amplifier tube, RCA would market this new electrodynamic speaker for substantially less than the Model 105 electrodynamic loudspeaker introduced the previous model year.

The Model 100-B was clearly designed for the Radiola 33 receiver, generally appearing in ads sitting atop the Radiola 33.

The appearance, operating principles, and construction details of the Model 100-B speaker were virtually identical to that of the Radiola 100-A.

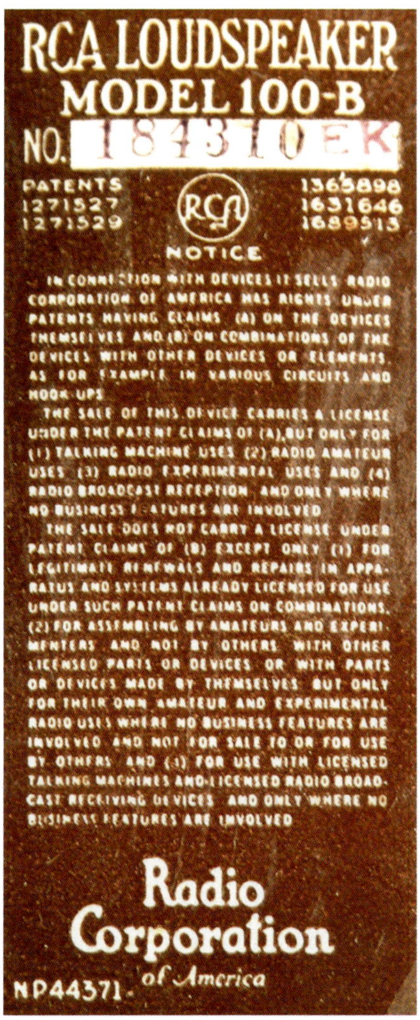

The Model 100-B had a name plate decal affixed to the bottom of the speaker frame visible through the open bottom of the cabinet.

Model 100-B Loudspeaker (UZ-783)

The Model 100-B permanent-magnetic table-top loudspeaker was introduced simultaneously with the Radiola 33 on March 26, 1929 at a price of $22—versus the $29 price of the almost identical Radiola 100-A. The speaker was designed specifically to match the Radiola 33, and was generally photographed sitting atop the Radiola 33. It rarely if ever appeared in an ad by itself. However, the RCA sales literature was quick to point out that it harmonized with all receivers and any home setting.

This speaker was a simplified version of the Model 100-A designed to provide a low-cost option to compete with other manufacturers. The extra heavy pole pieces and armature gave greater volume without saturation, and an improved filter was designed to remove high-frequency harmonics. The appearance, operating principles, and construction details of the reproducer were virtually identical to that of the Radiola 100-A.

It is not known how many were made, but judging by the number on the auction market today, there were relatively few made in comparison to Models 100, 100-A, and 106. There is a paper decal on the bottom of the speaker frame visible from the underside of the speaker stamped with individual serial numbers—most between 100,000 and 200,000. However, an insufficient number of these decals have been observed by this author to make any estimate of the number of speakers from the serial numbers. The price was reduced to $17.50 when the new line for the 1929-30 season was introduced in the fall, and the speaker was on the market well into 1930.

Model 103 Loudspeaker (UZ-749)

The Model 103 permanent-magnet speaker was officially released on August 17, 1928 and was initially pictured in ads with the new Model 60 superheterodyne, which was released a few days prior to the speaker. In later ads it was pictured with other receivers such as the Radiola 44, but almost never by itself. The new speaker sold for $37.50, versus $29 for the Model 100-A which it replaced. It

The Model 103 loudspeaker had a molded frame resembling hand carved oak and a grille cover consisting of a flowery tapestry.

The Model 103 reproducer was enclosed with a silk fabric that is now difficult to remove and replace without causing damage.

The Model 103 permanent magnetic reproducer was essentially the same as that used in the Model 100-A.

The Model 103 loudspeaker had a name plate decal with individual serial numbers affixed to the bottom of the molded stand.

was touted as having *"better reproducing qualities and an artistic appearance which entirely removes it from any semblance to a mechanical device."* Indeed, it was a very artistic device with a molded frame to resemble hand-carved oak, and a tapestry cloth grille sporting a flower arrangement.

This reproducer was very similar to that of the Model 100-A with the same basic appearance and construction details. As with other RCA speakers, it was furnished with a filter to reduce undesirable harmonics and noise, thus improving tonal quality. The speaker mechanism and filter extending out the rear of the frame were concealed by a brown silk cover tied over the frame to give it an artistic appearance and yet—according to the repair manuals—allow for easy access. However, removing and replacing the silk cover is not all that easy, particularly for the modern collector who will often find the silk cover to be very fragile and difficult to replace.

The tapestry speaker must have been very popular because a relatively large number were manufactured as evidenced by the large number that appear on the auction and flea markets. Paper decals attached to the bottom of the frame were assigned individual serial numbers—most appearing to be between 100,000 and 350,000. The speaker was on the market well into 1930, although the price had dropped to $18 by the end of 1929.

Model 106 Loudspeaker (UZ-642)

The Model 106 loudspeaker, an improved electrodynamic speaker mounted in a console, was released on November 26, 1928, somewhat later than the introduction of the superheterodyne receivers in September 1928. With an art deco appearance evocative of the Model 103 loudspeaker, it was described with flowery language in sales brochures: *"The cabinet is rich in its simplicity. Its*

The Model 106 loudspeaker was described as one with a "*beautiful walnut finish and panels of Velour de Genes embroidered toile [which] enables it to harmonize with any home furnishings.*"

The Model 106 loudspeaker cabinet was mostly empty except for the reproducer and power supply components located at the center of the front panel. Any AC radio could be plugged into the socket in the rear and controlled with an on-off toggle switch mounted on the side of the cabinet.

The main components of the Model 106 loudspeaker were an 8-inch electro-magnetic reproducer, two dry rectifier stacks on each side of the reproducer, and a filter capacitor enclosed in a metal can.

beautiful walnut finish and panels of Velour de Genes embroidered toile enables it to harmonize with any home furnishings. Its design is the combined handiwork of artists and engineers." The Model 106 appeared in ads beginning circa November 1928 along with the existing 1928-29 Radiola line, but it was rarely if ever featured alone.

The reproducer mechanism itself had an eight-inch corrugated cone similar to the ones used in Radiolas 62 and 64, but with a new type of field coil. Unlike previous electrodynamic speakers distributed by RCA, the Model 106 did not have the usual rectifier-power-amplifier unit used with the Models 104 and 105 speakers. Instead, it had a small power unit utilizing disk rectifiers in a full-wave configuration and two filter capacitors in place of vacuum tubes to provide current to produce the magnetic field in the coil of the reproducer. Because the speaker did not use a transformer, it could be used on both 25 cycle and 50-60 cycle AC power lines. The speaker was also equipped with an AC receptacle for the radio receiver so that both units could be controlled with a single line switch on the side of the 106 cabinet.

Because of the dry rectifiers and the fact that there was no power supply for any part of the receiver, the Model 106 was favorably priced at $88 versus the $350 price tag for the Model 105 electrodynamic speaker from the previous model year. The price began to drop almost immediately after its introduction—reduced to $64 by April 1, 1929 and to $49.50 by October 1, 1929. The speaker remained on the market well into 1930. It is not known how many of these speakers were manufactured, but it was clearly far fewer that the lower priced table-top speakers manufactured by RCA.

Model 106 DC Loudspeaker (Unknown Model No.)

A DC version of the Model 106 dynamic loudspeaker was released on April 12, 1929 at $65, differing from the AC version by the absence of the disk rectifiers and filter condensers. Both sets were on the market well into 1930, although the price had dropped to a mere $49.50 by October 1, 1929.

The Model 106 loudspeaker had a name plate decal with individual serial numbers affixed to the bottom wood rail at the rear of the cabinet.

Endnotes and References for Chapter 9

1 *Radio Retailing*, Vol. 8, No. 3, September 1928, pp. 78-79.

2 *Radio Retailing*, Vol. 7, No. 5, May, 1928, pp. 54-56.

3 *Radio Retailing*, Vol. 8, No. 3, September 1928, pp. 78-79; " New RCA Line is Launched with Big Fall Advertising," *The Town Crier*, Vol. VI, No. 12, September 1928, pp. 6-8.

4 "Direct Current Models of Radiolas 18 and 51," *The Town Crier*, Vol. VII, No. 3, p.6.

5 The separate oscillator allowed the use of the fundamental frequency rather than the second harmonic for hetrodyning.

6 *The Town Crier*, Vol. VII, No. 7, April 1929.

7 "The National Announcement of Radiola 33," *The Town Crier*, Vol. VII, No. 7, pp. 6-7.

8 The "Big Blue Book"—as RCA described their sales brochure entitled *The New RCA Radiola 33*—is described in : "How to Make the Best Use of the New Radiola 33 Sales Help," *The Town Crier*, Vol. VII, No. 7, pp. 2-5.

9 "Selling the New Radiola 33, *The Town Crier*, Vol. VII, No. 7, April 1929, pp. 8-9.

10 Alan Douglas, *Radio Manufacturers of the 1920's, Vol. 3 RCA to Zenith*, (The Vestal Press, Ltd. Vestal, New York, 1991) p. 46.

11 *Radio Retailing*, Vol. 8, No. 5, November 1928, p. 50.

12 Alan Douglas, Radio Manufacturers of the 1920's, Vol. 3 RCA to Zenith

13 Ibid., p. 45.

14 Ibid., p. 47.

THE SATURDAY EVENING POST November 16, 1929

Screen-Grid RADIOLAS

RCA RADIOLA 46—Console cabinet model of the new Screen-Grid Radiola 44, with RCA Electro-Dynamic Speaker $179 (less Radiotrons).

RCA RADIOLA 47—Combination. The newest RCA "all-electric" development in Screen-Grid radio combined with the phonograph. Radio and Phonograph utilize the same improved built-in Electro-Dynamic reproducer. "Music from the air or record." $275 (less Radiotrons).

THE SENSATION OF THE RADIO YEAR

RCA RADIOLA 44—Radio receiver utilizing Screen-Grid Radiotrons—high amplification and great selectivity. Alternating current operation from house circuit. Table cabinet of walnut veneer. $110 (less Radiotrons).

RCA LOUDSPEAKER 103—For use with Radiola 44, $22.50.

In addition to Screen-Grid Radiolas there are eight other models of Radiolas and Radiola Phonograph combinations ranging in price from $54.00 to $690.00. Any of these instruments may be purchased through RCA Radiola Dealers on the convenient RCA Time Payment Plan.

RCA has set a new standard of radio enjoyment in the new Screen-Grid Radiolas! Designed and built by the same RCA engineers who developed the Screen-Grid Radiotrons and Screen-Grid Circuit, they are the sensation of the radio year.

Never before have radio instruments of so few tubes offered such marvelous, well-rounded tonal beauty —such astounding volume without distortion—such balanced reproduction of both high and low notes.

Radiolas 44 and 46 utilize only five tubes—yet give you the performance of sets employing a far greater number. Three of these tubes are the amazing Screen-Grid Radiotrons—an RCA achievement. Included also is a new power amplifying Radiotron capable of tremendous volume without distortion.

In these great Screen-Grid Radiolas you get the freedom from distorting noises and electrical hum without the costly sacrifice of loss in fidelity and tone range . . . without that weakening of power and dulling of low and high notes which owners of ordinary radio sets must suffer when hum is reduced in defiance of electrical research and experience.

Radiola 44 (table model) is in a compact, two-tone walnut veneer cabinet of charming and graceful design. The console model, Radiola 46, makes use of the finest of all reproducers, the famous RCA Electro-Dynamic Speaker, an integral part of the assembly. And in Radiola 47 there is offered the outstanding combination of Screen-Grid Radiola and Electro-Dynamic Speaker, and newest type of phonograph for the electrical reproduction of records! "Music from the air or record."

Visit your RCA dealer today. See and hear these marvelous Screen-Grid Radiolas—designed and built by the creators of the Screen-Grid Radiotron and the Screen-Grid Circuit.

RCA RADIOLA

MADE BY THE MAKERS OF THE RADIOTRON

BUY WITH CONFIDENCE WHERE YOU SEE THIS SIGN

RADIOLA DIVISION

RADIO-VICTOR CORPORATION OF AMERICA

Screen-grid Radiolas were the sensation of the 1929-30 model year. (*Saturday Evening Post*, November 16, 1929, p. 104)

Chapter 10

Screen-Grid Radiolas are a Sensation

1929-1930

The UX-222 screen-grid tube was designed for use in battery operated radios, while the UY-224 screen-grid tube was designed for use in AC-operated radios.

One of the biggest problems facing the designer of TRF receivers of this era was the inability of the triode to provide a satisfactory degree of amplification at RF frequencies without breaking into oscillation. The limiting factor for a well-designed set was generally the inter-electrode capacity between the grid and plate in the RF amplifier vacuum tubes. This problem was addressed either by inserting a lossy element such as a resistor in series with the grid of the RF amplifier tube, or by feeding back an out-of-phase voltage from the plate to the grid to neutralize the effects of grid-to-plate capacity.

The most well-known method of neutralizing the effects grid-to-plate capacity was the neutrodyne method developed by Professor L. A. Hazeltine while consulting for the Navy. His solution was so effective that he was able to create an entire company, the Hazeltine Service Corporation, in 1924 for the purpose of licensing this and other of his patented inventions to other radio manufacturers. Some of the most well-known competitors of RCA such as FADA, Crosley, Freed-Eisemann, Stromberg-Carlson, and AMRAD used the neutrodyne license to manufacture TRF receivers. It was said that by 1927 there had been over 10,000,000 broadcast receivers sold with the neutrodyne circuit.

While many companies elected to use the neutrodyne circuit—and in doing so competed successfully with RCA—RCA did not deign to use this method in their TRF receivers. Instead, they experimented with other techniques, most notably using lossy grid resistors in the Radiola 17, and then a new method of neutralization using the distributed capacity in nested coils in conjunction with an adjustable capacitor to neutralize feedback in the Radiola 18. Neither method was entirely satisfactory, as RCA had come to learn first hand with uncontrollable oscillations which occurred in a limited, but still embarrassingly significant number of Radiola 17 receivers—and to a lesser extent, in Radiola 18 receivers as well.

This problem of instabilities in TRF receivers for broadcast radio was finally solved with the introduction of the screen-grid tube in 1929. RCA must have taken some satisfaction in seeing the value of the Hazeltine neutrodyne patent disappear virtually overnight as other manufacturers rushed to produce sets with screen-grid tubes which were either marketed by RCA or covered by RCA-controlled patents.

RCA Introduces the Screen-Grid Tetrode

Development of the screen-grid tube in the U.S. began in 1924 when Dr. A. W. Hull of GE was investigating the problem of thermal noise in the early portable Super-Heterodyne.[1] The screen grid was a metallic mesh that surrounded the plate to electrically isolate the plate from the grid, yet allow electrons accelerating towards the plate by the screen-grid voltage to actually pass through the mesh and thereby reach the plate. The screen reduced the inter-electrode capacity between the plate and grid to a point where there was virtually no feedback or regeneration.

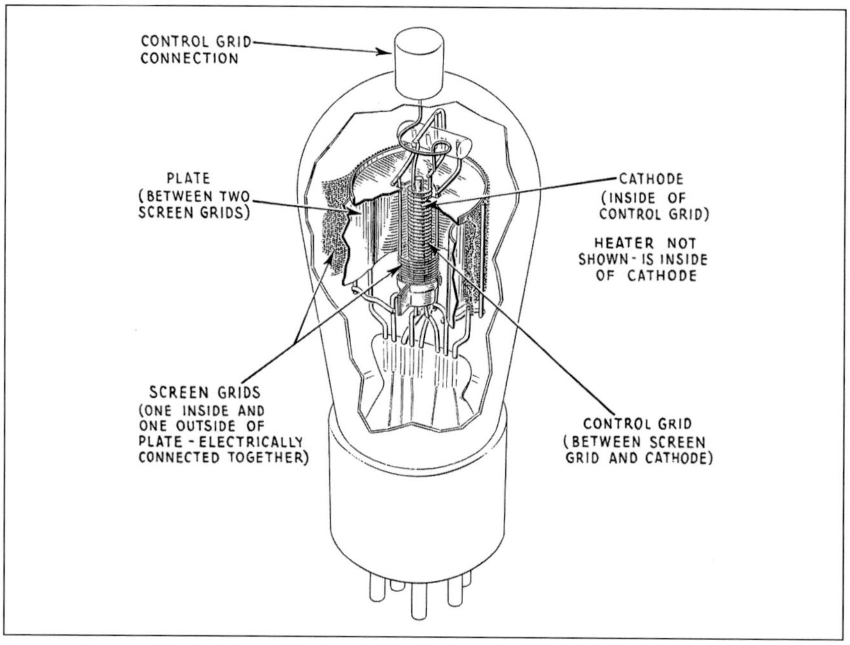

The screen-grid tube actually had two screen grids connected together, one on the outside of the plate and the other inside. (*The Town Crier*, circa 1929)

The screen-grid tube was perfected over the next two years, and RCA was given the first samples in July 1926.[2] The designation UX-222 was assigned in November 1926, and GE began a small but regular production of the tube in February of 1927. The tube was a dry-cell type tube of the UX-199 class with four pins in a UX-type base. The new screen-grid was connected to one of the four pins on the type UX base, so the connection for the control grid was moved to a new grid cap introduced through the glass at the top of the tube. RCA announced the UX-222 to the public in October 1927, listed it in the *RCA Radiotron Retail Dealers Catalogue and Price List* dated December 17, 1927, and featured it in Radiotron ads placed in national magazines beginning in January 1928.

The new screen-grid tube generated interest among radio enthusiasts as evidenced by the number of articles and receiver plans that were published in radio magazines of the day. However, it apparently did not generate a great deal of interest among the radio manufacturers because there was no rush to produce receivers using the new dry-cell version of the tube. By this time, the industry was focused on developing AC-powered broadcast receivers using AC filament tubes, and apparently none of the manufacturers wanted to commit to manufacturing a screen-grid radio until a reliable source of AC screen-grid tubes became available.

GE was working on an AC version of the screen-grid tube at this time as well. According to William White, a set incorporating the new AC screen-grid tube was submitted to RCA for test and evaluation in July 1928—presumably for sale in the 1929-30 season.[3] During the first quarter of 1929, speculation was rife about when the new AC screen-grid tube would be become available, and whether it would be used in any sets for the 1929-1930 season.[4] Sometime in March 1929, RCA announced the new AC screen-grid tube along with its electrical characteristics, but did not give it a designation.[5] According to White, RCA announced the new tube designation UY-224 on April 20, 1929.

In June, RCA introduced two new screen-grid receivers at the radio trade show held in Chicago—the Radiola 44 and 46. Surprisingly, 98 out of 215 new sets introduced at the trade show, or 45 percent of the total number, used screen grids—this according to a survey published by *Radio Broadcast*.[6] Most of these were AC sets because, according to the same survey, a total of only nine new battery sets were introduced. Clearly, a number of radio manufacturers had been quietly designing screen-grid receivers, presumably using the battery-operated version of the screen-grid tube, and had coordinated the timing of the release of their sets with the availability of AC screen-grid tubes from RCA

It appears that the C. R. Leutz Company was the first to advertise a screen-grid radio, which was described as a set with A/C Shield Grid Tubes. (*Radio News*, April 1929, p. 971)

and/or other tube manufacturers. It is interesting to note that the C. R. Leutz Company placed ads in national magazines as early as April 1929 claiming to be the first to offer an AC screen-grid console for sale.[7]

The Screen Grid: A Panacea or a Fad?

"When the screen-grid tube was first introduced, it was heralded by many as a panacea for all the ills associated with r. f. amplifiers. That such is not the case is becoming more and more evident."[8] These were the words of the editor of the venerable *Radio Broadcast* magazine written about a year after the introduction of the battery version of the screen-grid tube. The same debate was raging among radio manufacturers a year later, one group believing it was a panacea and the other group believing it was a fad.

An article in *Radio Broadcast* in August 1929 put a positive spin on the screen-grid tube, pointing out that the screen-grid tube with its greater inherent amplification could be used to design receivers with greater sensitivity, greater selectivity, and improved fidelity without an increase in cost.[9] These features could be translated into receivers with greater range, shorter antennas, less interference, improved fidelity and decreased hum.

The key to achieving these features was the greater inherent gain possible in an RF stage using a screen-grid tube. The practical limit to RF amplification for an amplifier stage with a triode was about a factor of 10 per stage compared to a factor of 30-35 per stage for the screen-grid stage. As a result, the total amplification in a five-tube screen-grid receiver could be as much as 2,300,000 compared to 570,000 for a six-tube triode receiver. Another unexpected benefit of the screen-grid tube was the increased power that could be delivered to the detector with the increased gain from several RF amplifier stages, which made it possible to eliminate one stage of audio amplification. The power output from the detector was sufficient to drive the audio output tube directly without an intermediate stage of amplification, thus reducing both distortion and hum.

The general conclusion was that a well designed screen-grid receiver in 1929 could give the customer a better radio than he could buy in 1928 and at a price no greater than he had paid in 1928. The customer could receive stations further away with better sound quality using a shorter antenna. However, the superheterodyne was still superior to the TRF, albeit at a higher price. With the introduction of the pentode tube a few short years later, the price of the superheterodyne would drop further, and the pentode with a screen grid would become the standard for superheterodyne receivers of the future.

The practical limit to RF amplification for an amplifier stage with a triode was about a factor of 10 per stage compared to a factor of 30-35 per stage for the screen-grid stage. As a result, the total amplification in a five-tube screen-grid receiver could be as much as 2,300,000 compared to 570,000 for a six-tube triode receiver. (*Radio Broadcast*, August 1929, p. 212)

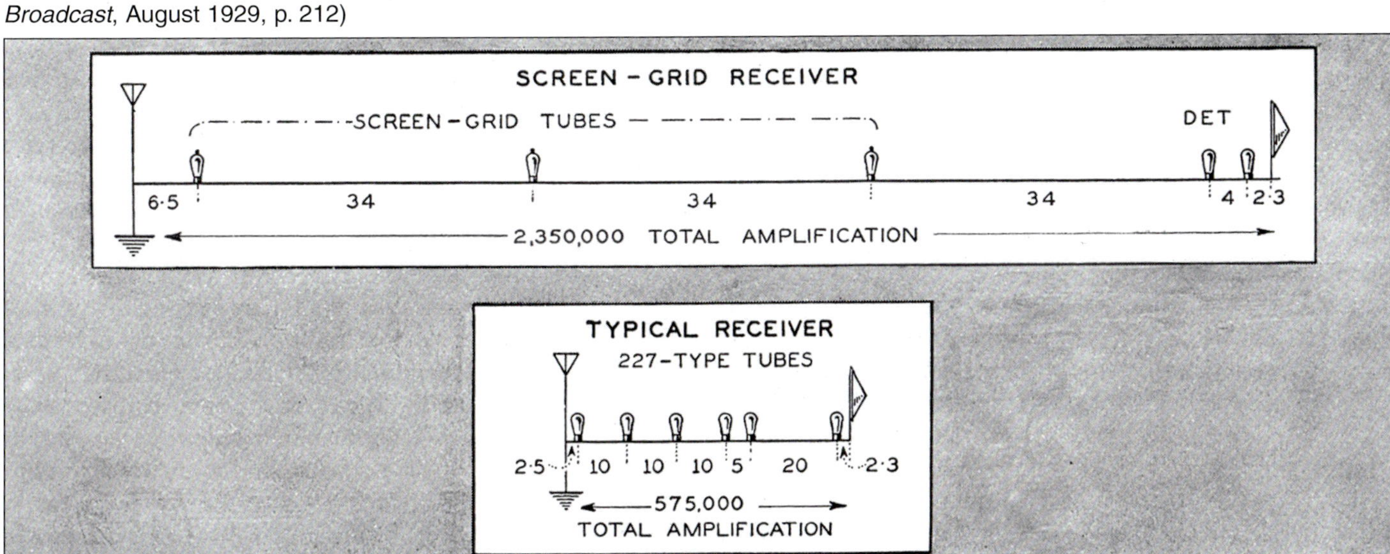

RCA introduced five new screen-grid radios in the model year 1929-30 as compared to only two new superheterodyne models. (*Radio Retailing*, Vol. 10, No. 3 September 1929, pp. 80-81)

The 1929-1930 Radiola Line

The new sets in the Radiola line for 1929-30 can be conveniently divided into three categories: the Radiola 20 series of battery-operated screen-grid TRF sets, the Radiola 40 series of AC-operated screen-grid TRF sets and the Radiola 60 series of AC-operated superheterodynes. The 40 series was introduced first beginning at the end of May 1929, followed by the 60 series introduced beginning in mid-July 1929, and finally the new 20 series beginning in mid-October 1929. All of these new sets—as well as two holdover sets—are listed in Table 10-1 and pictured in the accompanying ad placed by RCA in *Radio Retailing* magazine for the trade. These sets also appeared in the new gold-covered RCA catalog for the 1929-30 season. Oddly, the Radiola 18

Table 10-1. Radiola apparatus for the 1929-30 season including holdover sets in the shaded portion of table

	Model	Apparatus Type	Cabinet Style	Tubes[a]	Power	Intro Date	Discon'd. By	List Price[b]
1929-1930 Season New Models	Radiola 21 (AR-1258)	Screen-Grid TRF + Regen	Cabinet	5	Battery	10/19/29	After 1930	$85.25
1929-1930 Season New Models	Radiola 22 (AR-1265)	Screen-Grid TRF +Regen	Console	5	Battery	11/7/29	After 1930	$150.75
1929-1930 Season New Models	Radiola 44 (AR-594)	Screen-Grid TRF Rcvr.	Cabinet	4 (1)	AC	5/29/29	Sep. 1930	$128.50
1929-1930 Season New Models	Radiola 46 (AR-596)	Screen-Grid TRF Rcvr.	Console	4 (1)	AC	5/29/29	Sep. 1930	$197.50
1929-1930 Season New Models	R. 46 DC (AR-597)	Screen-Grid TRF Rcvr.	Console	6	DC	11/6/29	Sep. 1930	$212.25[c]
1929-1930 Season New Models	Radiola 47 (AR-1147)	Screen-Grid TRF Rcvr.	Console Phono	4 (1)	AC	10/7/29	Sep. 1930	$293.50
1929-1930 Season New Models	Radiola 66 (AR-598)	Super-heterodyne	Console	7 (1)	AC	7/18/29	Sep. 1930	$246.50
1929-1930 Season New Models	Radiola 67 (AR-1168)	Super-heterodyne	Console Phono	8 (2)	AC	9/28/29	Sep. 1930	$733.00
Holdover Models	Radiola 18 (AR-936)	TRF Receiver	Cabinet	6 (1)	AC	4/19/28	Jan. 1930	$109.75
Holdover Models	R. 18 DC (AR-891)	TRF Receiver	Cabinet	6	DC	12/28/28	Sep. 1930	$123.50
Holdover Models	Radiola 33 (AR-784)	TRF Receiver	Cabinet & legs	6 (1)	AC	3/26/29	Sep. 1930	$68.75
Holdover Models	Radiola 60 (AR-954)	Super-heterodyne	Cabinet	8 (1)	AC	8/13/28	Sep. 1930	$152.75
Holdover Models	Radiola 64 (AR-894)	Super-heterodyne	Console	9 (2)	AC	10/23/28	Sep. 1930	$595.50
Holdover Models	100-A (UZ-1076)	PM Speaker	Tabletop	0	N/A	5/25/27	Sep. 1930	$29
Holdover Models	100-B (UZ-783)	PM Speaker	Tabletop	0	N/A	3/26/29	Sep. 1930	$17.50
Holdover Models	103 (UZ-749)	PM Speaker	Tabletop	0	N/A	8/17/28	Sep. 1930	$22.50
Holdover Models	106 (UZ-642)	Dynamic Speaker	Floor	0	AC	11/26/28	Sep. 1930	$49.50
Holdover Models	106 DC (Unknown)	Dynamic Speaker	Floor	0	DC	4/12/29	Sep. 1930	$49.50

[a]Additional rectifier and ballast tubes are shown in parentheses
[b]Prices taken from RCA Distributor's Price List dated October 1, 1929 and include Radiotrons
[c]Price estimated from RCA Export Price List dated Jan 30, 1930

remained on dealer price lists until 1930 but did not appear in ads or RCA catalogs with the rest of the 1929-30 line.

RCA and their suppliers at Westinghouse and GE were clearly busy designing and manufacturing screen-grid TRF receivers for the 1929-30 season because they introduced three new AC-operated models, Radiolas 44, 46, and 47, and two new battery-operated models, Radiolas 21 and 22. While Radiola 44 was a table model, Radiola 46 a console, and Radiola 47 a combination radio-phonograph console, all three used the same basic four-tube circuit with a fifth tube used as a rectifier in the power supply. Similarly, while Radiola 21 was a table model and Radiola 22 was a console, both used the same five-tube circuit requiring no additional tube for the power supply. The AC sets used UY-224 screen-grid tubes for amplification while the battery sets used UX-222 screen-grid tubes. All five TRF receivers for this model year used two stages of screen-grid RF amplification, but the AC receivers required only one stage of audio amplification in addition to a power detector stage, while the battery-operated sets with their low-power detector tubes required an additional stage of audio amplification.

While RCA clearly focused on screen-grid TRF receivers for the 1929-30 season, they certainly did not ignore the superheterodyne. RCA developed a new seven-tube superheterodyne chassis which was used in two new console models, Radiolas 66 and 67. The Radiola 66 was a stand-alone radio while the Radiola 67 was a combination radio and phonograph. The new seven-tube chassis differed from the eight-tube chassis from the previous year in two ways. First, the untuned stage was eliminated such that the antenna circuit was coupled directly into a tuned-RF stage, and secondly, the new UX-245 audio output tube with a maximum output power of 1.6 watts replaced the UX-171A with only half that power rating.

While there were more new TRF's than new superheterodynes, there were actually more superheterodynes offered than any other single category of receiver—four AC superheterodynes versus three AC screen-grid TRF's, two dated AC TRF's, and two battery screen-grid TRF's. For the AC sets, prices ranged from a low of $68.75 for the Radiola 33 TRF to a high of $733 for the new Radiola 67 superheterodyne with a phonograph. It is interesting to note that the Radiola 33 TRF with a total of seven tubes was priced at $68.75 while the new screen-grid TRF with a total of only five tubes was priced at $128.50—almost twice as much!

This model year was the first time in many years that RCA did not introduce any new stand-alone loudspeakers. The one electrodynamic and three permanent-magnet speakers offered were held over from previous years. A DC version of the Model 106 was introduced, but this did not represent a new speaker technology.

Impact of RCA-Victor Merger on the 1929-30 Season

The RCA-Victor merger effective in March of 1929 had a minimal impact on the design of receivers produced by GE and Westinghouse for RCA during the 1929-30 season. In April 1929, RCA created Radio-victor Corporation of America, a subsidiary to market radio receivers and associated apparatus including vacuum tubes. Appending the non-capitalized "victor" to Radio was obviously done so that the new Radio-victor Corporation of America could retain the RCA initials and ubiquitous monogram. The public hardly took note of the merger—the most obvious impact being the slight change in name to Radio-victor Corporation appearing on advertising copy, operating manuals, and radio decals affixed to the sets beginning circa May 1929.

Radio apparatus for the 1929-30 season released prior to July 1929 (Radiolas 44, 46, 66) had instructions and decals with the Radio Corporation of America name, while apparatus released after July (Radiolas 21, 22, 47, and 67) had the Radio-victor Corporation of America name. Since the name

The new RCA catalog for model year 1929-30 carried all the sets offered for sale with the notable exception of the Radiola 18, which appeared on RCA price lists into 1930.

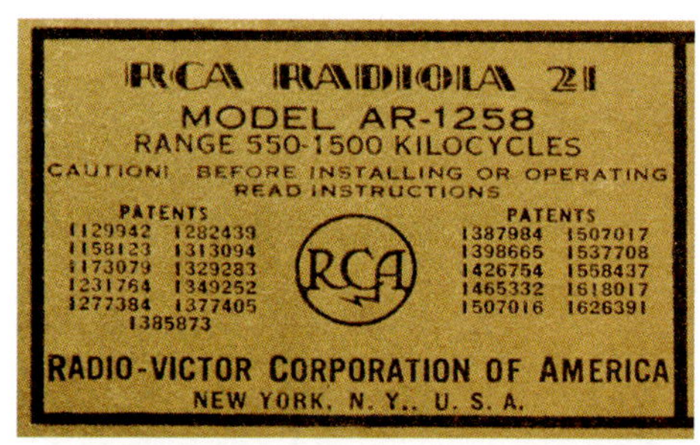

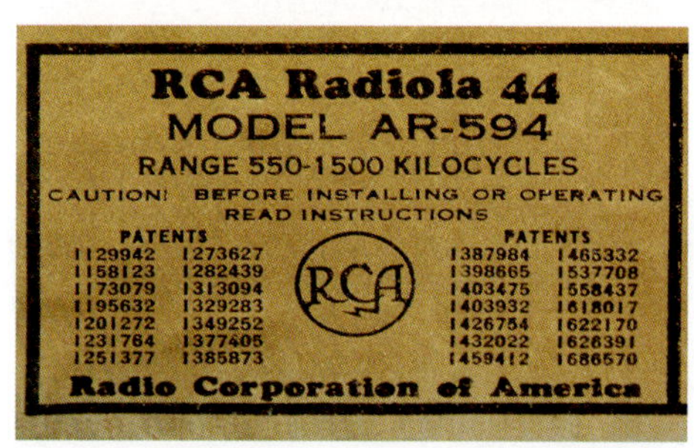

Decals on Radiolas 44 and 66 carried the Radio Corporation of America name and the old-style font, while decals on the Radiola 21 and the rest of the new line carried the Radio-victor Corporation of America name and the new-style font.

As in previous years, instructions accompanying consoles had a stiff olive-taupe cover, while those for table models had a white paper cover. Instructions for Radiolas 44 and 46 (as well as Radiola 66) were printed by Radio Corporation of America with covers similar to those from previous years, while instructions for Radiolas 47 and 21 (as well as Radiolas 67 and 22) were printed by Radio-victor Corporation of America with a new cover design including a new font with "split" letters.

RCA prepared brochures for all of the new sets introduced in model year 1929-30 including a special brochure for the radio-phonograph combinations Radiolas 47 and 67.

change took place in April before the release of any sets, the name tags and instructions for Radiolas 44, 46, and 66 must have been prepared before the name change became effective. In sets released later, not only did the name change appear, but also the instruction manuals had a new look in terms of both the layout of the front cover and the lettering font. Apparently the marketing brochures for all sets were printed after the name change in April because all advertising brochures for the seven Radiola receivers introduced in the 1929-30 observed by this author were printed with the Radio-victor Corporation of America name.

Radiola Receivers for 1929-30

Radiola 44 (AR-594)

The Radiola 44, one of the first screen-grid table top receivers offered by RCA, was released on May 29, 1929 shortly before the Chicago trade show in June 1929. The set was first advertised in July of 1929, often pictured with the Model 103 speaker and the Radiola 46.[10] Another ad featuring the Radiola 44 by itself is shown at the front of this chapter. The set with relatively simple lines in a two-tone walnut veneer was priced at $128.50 with tubes and $110 without tubes. By comparison, the six-tube Radiola 33 offered by RCA at the same time was priced at only $68.75 including tubes.

The tuning and volume controls were arranged on concentric shafts such that they could both be adjusted with one hand. The illuminated tuning dial was configured such that the numbers on the drum were projected—greatly magnified—onto the translucent screen. This unique tuning dial, which was used on all Radiola models for the year, would not function without a pilot lamp, so sets using this dial were outfitted with a dummy screw socket mounted on the cabinet wall which held a spare bulb. The on-off switch was located on the end of the cabinet to the right when facing the set, while a local-distant switch for enhanced tuning was placed symmetrically at the opposite end.

The hinged lid of the Radiola 44 could be opened for easy access to the tubes and the chassis. The set used three UY-224 screen-grid tubes—two as tuned-RF

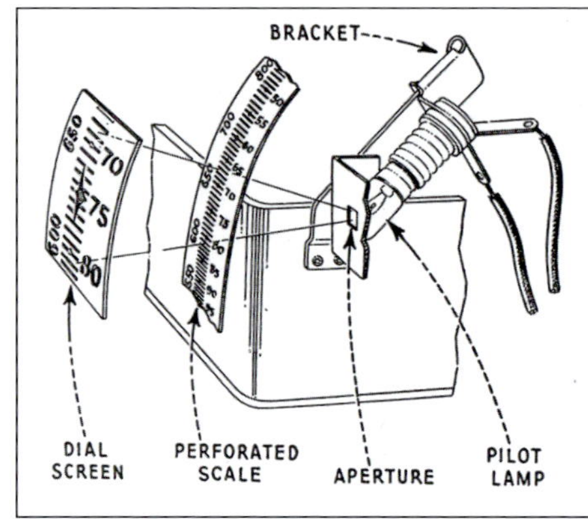

A novel stencil-cut station selector dial requiring back lighting with a pilot lamp was incorporated into all new Radiola sets for the 1929-30 season. (*RCA Victor Service Notes for 1929-1930*, Chapter entitled RCA Radiola 44)

The Radiola 44 tuning controls were arranged concentrically such that there appeared to be only one control.

The 1st and 2nd RF amplifier screen-grid tubes used in the Radiola 44 receiver chassis were individually shielded, and in addition, the 2nd RF amplifier and detector screen-grid tubes were shielded by two larger brown metal cases—both removed and positioned in the rear of this photo. The separate power supply unit is at the right.

amplifiers and one as a power detector—and a UX-245 as a power amplifier. The two screen-grid amplifiers were enclosed in individual pressed copper shields to avoid feedback and resulting oscillations; no other method of neutralization was needed to maintain stability of the screen-grid circuits. The output of the power detector was sufficient to drive the audio output tube without the need for an intermediate audio amplifier. The audio output was fed directly to the speaker through a condenser without a transformer. While the 44 did not come with a speaker, its power supply was designed to supply the current needed to drive the magnetic coil on an electrodynamic speaker.

The Radiola 44 was on the market until September of 1930 when the new season models were introduced. While there are no production figures available for this receiver, Douglas reported that 97,633 electrically equivalent Radiola 46 consoles were manufactured.[11] Since the Radiola 44 priced at $128.50 was significantly less expensive than the Radiola 46 priced at $197.50, it is likely that there were at least as many of the Radiola 44 table model sets sold. Two other AC versions were also available, a 25-40 cycle, 105-125 volt version and a 200-250 volt, 50-60 cycle version.

The Radiola 46 cabinet described in RCA sales literature as having an *"English influence,"* was made of matched walnut veneers and maple overlays with an embroidered silken panel to cover the speaker opening in the center of the front panel.

Radiola 46 (AR-596)

The Radiola 46 combined the Radiola 44 chassis with an electrodynamic reproducer similar to the one used in the Model 106 console. The set was introduced along with the Radiola 44 on May 29, 1929, and was displayed at the Chicago radio show. Ads in national publications began to appear circa July 1929. The retail price was $197.50 with tubes and $179 without tubes—about $70 more that the Radiola 44—which unlike the 46, did not have a speaker. The cabinet, described in RCA sales literature as having an "*English influence*," was made of matched walnut veneers and maple overlays with an embroidered silken panel to cover the speaker opening in the center of the front panel.

The Radiola 46 had the same features as the Radiola 44: the local-distant switch to obtain high-quality performance for both local and distant stations, a translucent screen upon which station markers were projected, and a concentric shaft control for both tuning and volume. According to Douglas, 97,633 sets were manufactured, which presumably includes not only the 105-125 volt version for 50-60 cycles, but also the following versions: a 110 volt AC version for 25-40 cycles, a 200-250 volt AC version for 50-60 cycles, and the DC version described below. The set was on the market until September of 1930 when the new sets for the 1930-31 season appeared.

Ads for the Radiola 46 such as this one by commercial artist Walter Seaton appeared in several different national magazines. (source unknown)

The receiver chassis of the Radiola 46 with the screen-grid tube shields in place is located to the right. (Radiola 46 Operating Instructions)

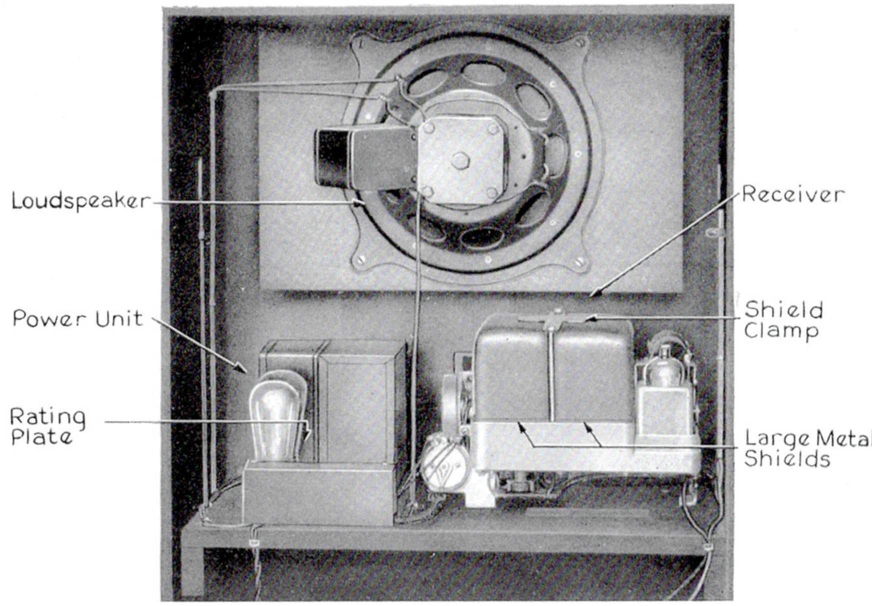

Radiola 46 cabinets were assembled in building 77 of the GE plant at Schenectady, NY in October of 1929. (Schenectady Museum & Suits-Bueche Planetarium, Schenectady, NY)

Radiola 46 DC Model (AR-597)

The DC version of the Radiola 46 was introduced on November 6, 1929, more than four months after the AC version. The appearance of the two sets is virtually identical but the circuitry is substantially different. The first three stages were similar to the AC version—two RF amplifier stages using the UX-222 screen-grid tubes followed by a type UX-112-A detector—but the similarity ended there. The DC version had an extra audio amplifier stage using a second type UX-112-A tube followed by a push-pull audio output stage using two type UX-171A tubes coupled to the Model 106 speaker through an audio transformer. The extra stage of audio amplification was needed because of the smaller gains achievable in the RF amplifier and detectors stages. This set was discontinued at the same time the AC version was discontinued in the fall of 1930.

Radiola 47 (AR-1147)

The Radiola 47 combined the screen-grid Radiola 44 chassis and speaker into a cabinet somewhat larger than that of the Radiola 46 to accommodate an electric phonograph "*of the latest design.*" The set was released on October 7, 1929 at a retail price of $275 without tubes or $293.50 with tubes. Full-color ads such as the one shown here by commercial artist Katharine Sturges began to appear in national magazines shortly after its release.

The Radiola 47 and Radiola 67 were the first two combination Radiola-phonographs marketed by RCA, although RCA had been supplying Radiola

The Radiola 47 cabinet was described in RCA marketing brochures with flowery language such as: "*The simple lines, handsome proportions, and rich color ensembles of the cabinets lends them a subtle charm which enables them to harmonize with almost any interior as well as add a touch of elegance distinctively their own.*" (Radiola 47 Instructions courtesy of Sonny Clutter)

The phonograph panel, accessible by lifting the lid, contained the turntable, the radio-phonograph switch and phonograph volume control. (Radiola 47 Instructions courtesy of Sonny Clutter)

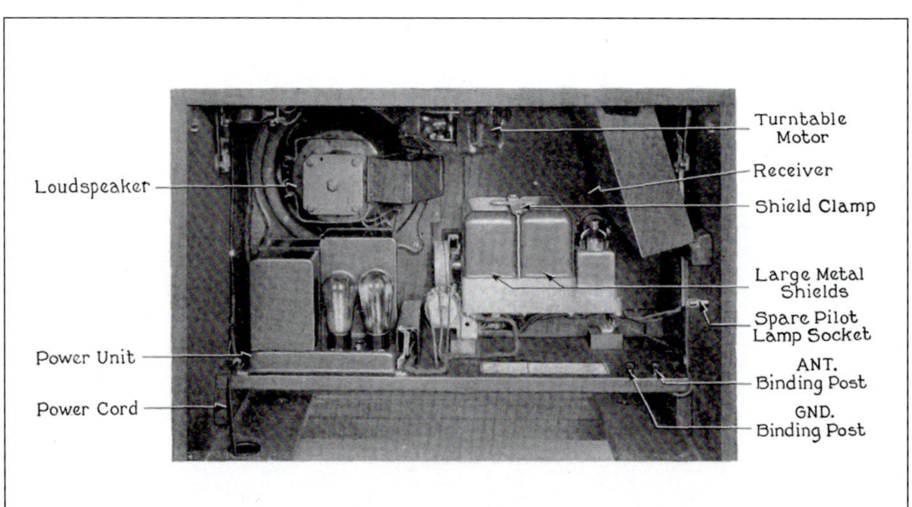

All components of the Radiola 47 were placed in a single compartment. (Radiola 47 Instructions courtesy of Sonny Clutter)

THE HOUSE BEAUTIFUL

Two superb musical instruments

-IN A SINGLE CABINET- *TO GIVE YOU* *"Music from the air or records"*

RCA RADIOLA 47—*Screen-Grid Radiola, Phonograph and Electro-Dynamic Reproducer in a single cabinet for electrical operation.* $275*

RCA RADIOLA 44—*Screen-Grid Radiola, table type.* $110*

RCA RADIOLA 46—*Screen-Grid Radiola and Electro-Dynamic reproducer in cabinet.* $179*

*Radiotron equipment not included in these prices.

When the Radiotrons of the new Radiola 47 are glowing with their incandescent light, you command "music from the air or records" with the amazing realism that only the Radiola can give.

At a touch of the switch the incomparable RCA Electro-Dynamic reproducer within the beautiful cabinet fills your living room with the music of a distant orchestra—with all the beauty and volume of the original—or the wonderfully faithful reproduction of a famous Victor record.

And the new Screen-Grid Radiotrons in this superb instrument—with the special RCA Screen-Grid circuit—give a reserve of power such as has never before been possible in the electrical reproduction of sound.

RADIOLA DIVISION
RADIO-VICTOR CORPORATION OF AMERICA
NEW YORK-CHICAGO-ATLANTA-DALLAS-SAN FRANCISCO

Radiola Dealer RCA — Buy with confidence where you see this sign

RCA RADIOLA

MADE BY THE MAKERS OF THE RADIOTRON

Art deco ads by Katharine Sturges for the Radiola 47 began to appear in national ads soon after its introduction in October 1929. (*The House Beautiful*, November 1929)

receivers and power-amplifier units to both Victor and Brunswick for use in their combination sets since the mid-1920s. The reason that RCA did not offer a phonograph-Radiola combination earlier was likely because of restrictions in the agreements RCA made with Brunswick and Victor when they first began to supply Radiola equipment to the two companies in 1924 and 1925 respectively. The agreements with Brunswick ended in 1929 when RCA merged with Victor, and so RCA was free to manufacture their own sets. It is interesting that RCA would choose to compete directly with phonograph-Radiola combinations offered by Victor, a company RCA had just acquired in March 1929. However, it is not entirely surprising because the Radio-victor origination and the Victor organization in Camden, NJ operated independently for at least a year after the merger. The two organizations continued to market radio-phonograph combinations independently until operations were combined at the Camden facility in 1930 and the two began to operate under the RCA Victor Company, Inc. name as a subsidiary of RCA.

The electric pickup used on the Radiola 47 had an appearance similar to the ones RCA supplied to Brunswick and Victor, but with the RCA logo at the top. (Bob Dobush collection)

The Radiola 47 was enclosed in a cabinet made of walnut veneers. Access to the phonograph was obtained through a hinged lid on the top which had a spring mechanism to allow the lid to open to any position. There was a record well in the top to store a small number of records. The induction motor on the turntable was activated with a radio-to-record switch, and it had an automatic stop which was activated at the end of a recording. The phonograph was equipped with a "*tone arm and electric pick-up of radically new design*." The needle caused movement in a bar which passed through the poles of two coils surrounded by a permanent magnet. The output of the coils was coupled through a transformer to the input of the detector tube. A separate volume control consisting of a variable resistor placed across the output of the coils was used to control record volume with a minimum of distortion.

The features of the radio portion of the Radiola 47 were the same as the Radiola 46. It was made in two AC versions for 110 volts, including the regular 50-60 cycle version and a 25-40 cycle version; there was no DC version. This set was on the market until September of 1930 when the new line for 1930-31 was introduced. According to Douglas, 15,120 sets were manufactured.[12]

Radiola 66 (AR-598)

The Radiola 66 was a new seven-tube superheterodyne design placed in a console style cabinet. The set was officially released on August 18, 1929, and began to appear in national ads in September along with the rest of the new 1929-30 line, priced at $225 without tubes and $246.50 with tubes. It was described in RCA marketing literature as follows: "*The unostentatious beauty of the cabinet is achieved by the use of the finest two-toned walnut veneers, combined with burled maple overlays. The richly embroidered panel in front of the loudspeaker is distinctly decorative.*" The tuning, volume control, and on-off switch appeared on the front panel, as well as the same type of projection dial with kilocycle markings that was used on the TRF receivers. The local-distant switch similar to one appearing on the TRF receivers was located on the right side of the cabinet.

The Radiola 66 using the finest two-toned walnut veneers combined with burled maple overlays and a richly embroidered panel in front of the loudspeaker was described as having "*unostentatious beauty*."

With regard to performance, the RCA marketing literature went on to say that "authorities" acknowledge "*the Super-Heterodyne circuit to be superior in radio's essential requirements—sensitivity, selectivity, and fidelity to the original performance*," a claim which was certainly true. The new seven-tube superheterodyne circuit differed from the previous eight-tube circuit first used in the Radiola 60 in the following ways: the first untuned RF-amplifier stage was eliminated, the UX-171A audio output tube was replaced by the higher-power type UX-245 tube, the IF transformer center frequency was moved from 180 to 175 KHz and sharply tuned to improve overall selectivity and reduce harmonic distortion, a local-distance switch was added to the antenna circuit to permit better control of volume without distortion, and the volume control

Attractive ads for the Radiola 66 by commercial artist Lurelle V.A. Guild appeared in several national magazines for a number of months after its introduction in August 1929. (*Saturday Evening Post*, March 8, 1930, p. 110)

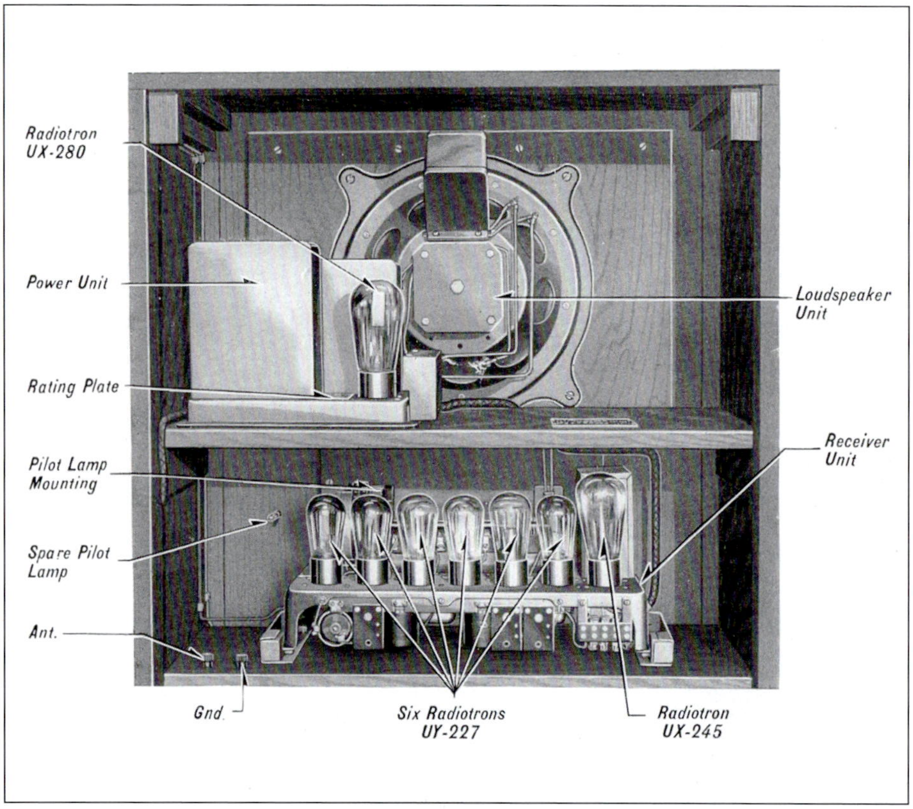

The receiver chassis of the Radiola 66 was mounted to the bottom of the cabinet while the power supply was mounted on a shelf which formed the speaker enclosure. (Radiola 66 Instructions)

The new 7-tube superheterodyne chassis, first used in the Radiola 66 console, was never used in a table-top receiver.

itself was implemented by controlling grid voltages on only the first two amplifier stages as opposed to all the RF and IF amplifier stages. There were a number of other relatively minor changes in this set that are documented in the RCA Victor Service Notes.[13] Unlike previous sets, the new superheterodyne chassis used in the Radiola 66 console was apparently never used in a table-top version.

The Radiola 66 was on the market until September of 1930 when it was replaced by an even newer superheterodyne circuit embodied in the Radiola 80 and 82 using screen-grid tubes and a push-pull audio output stage. The number of Radiola 66 receivers manufactured is unknown.

Radiola 67 (AR-1168)

The Radiola 67, a superheterodyne Radiola-phonograph combination, was released on September 28, 1929 at a list price of $690 without tubes and $733 with tubes about a week before the release of the TRF Radiola 47 Radiola-phonograph combination. The Radiola 47 and 67 were introduced to the trade in the October-November 1929 issue of *The Town Crier*—a copy of which is shown here. *The Town Crier* with humble beginnings as a four-page flyer in 1923 grew to a very sophisticated magazine of about 16 pages by 1929, containing much valuable information for both dealers and repairmen. Typically, one or two sets were featured in each issue, with information on upcoming ad national campaigns sponsored by RCA as well as advertising paraphernalia available to the dealer at a nominal cost.

In the October-November issue, *The Town Crier* promised dealers that the new Radiola-phonograph 47 and 67 combinations would be a new opportunity in the fast growing market of "combination instruments." A new ad campaign was announced in which 25,000,000 prospects were to be told about the new Radiola 47 and 67 through repeated competitive appeals appearing every month in national magazines. Indeed, lavish ads by various commercial artists such as the one shown here by Katharine Sturges soon followed.

The Town Crier, such as this issue showing the new Radiola 67, was used to introduce Radiola-phonograph combinations to RCA dealers and advise them of future advertising campaigns sponsored by RCA as well as advertising paraphernalia available for purchase at a nominal cost.

Colorful ads for Radiola 67 were created by Katharine Sturges, who also created ads with a similar genre for Radiola 47. (*The House Beautiful*, December 1929)

The Radiola 67 receiver and phonograph were enclosed in a cabinet consisting of walnut, rosewood, satin wood, mahogany veneer and maple panels—a design inspired by the Louis XVI period (left). The upper cabinet doors of the Radiola 67 folded back against the sides to reveal the phonograph and record shelf on the right, and the speaker grille on the left. Two smaller doors opened on the bottom to reveal the radio control knobs, tuning dial, tuning meter, on-off switch, and local-long-distance switch (right).

The receiver and phonograph were enclosed in a cabinet consisting of walnut, rosewood, satin wood, mahogany veneer and maple panels—a design inspired by the Louis XVI period. This set used the Radiola 66 superheterodyne receiver circuit but with the deluxe features found on the Radiola 64—an automatic volume control requiring an extra UY-227 tube, and a visual tuning meter activated with a new "silent tuning switch" by pushing in on the tuning dial. A new tone control feature was introduced to allow the user to accentuate the bass tones for both the radio and phonograph.

While the circuit used in the Radiola 67 was similar to the one in the 66, there were some differences. A type UX-250 tube was used in the output stage replacing the UX-245 used in the Radiola 66. The Radiola 67 used two UX-281 tubes in a full-wave rectifier configuration versus the single UX-280 used in the Radiola 66. Shunt resistors were added to the primary of the intermediate-frequency amplifier transformers to eliminate "double-peak" tuning, thereby preserving the advantage of critical coupling and making the tuning meter easier to use. The phonograph, tone arm, pick-up coil and volume control were similar to those found in the Radiola 47 phonograph.

The Radiola 67 was produced in two AC versions, one for 60 cycles and one for 25-40 cycles. It was on the market until the new Radiola line was introduced in September of 1930. According to Douglas a total of 4,939 were made, presumably including both AC versions.[14]

Radiola 21 (AR-1258)

The Radiola 21 was released on October 19, 1929, one of the last receivers to be released in the 1929-30 model year. The price for this table top radio was $85.25 with tubes and $69.50 without, one of the least expensive sets in the lineup. RCA, one of the few manufacturers to offer a battery-operated screen-grid TRF, continued their tradition of offering a full line of receivers for virtually all power line conditions—including rural areas without power lines. Although the set was officially released in October, it appeared in ads beginning in September 1929 where it was pictured with the rest of the 1929-30 lineup.[15] Beginning in November, the Radiola 21 and 22 were advertised in full page ads appearing in 17 farm papers throughout the country with a combined circulation of 32 million. The set was generally advertised in conjunction with the Model 103 table speaker.

The receiver was packaged in a relatively small attractive walnut-finished cabinet with a hinged top for access to the interior. Tuning and volume control knobs were located on the front panel, and an on-off toggle switch was located

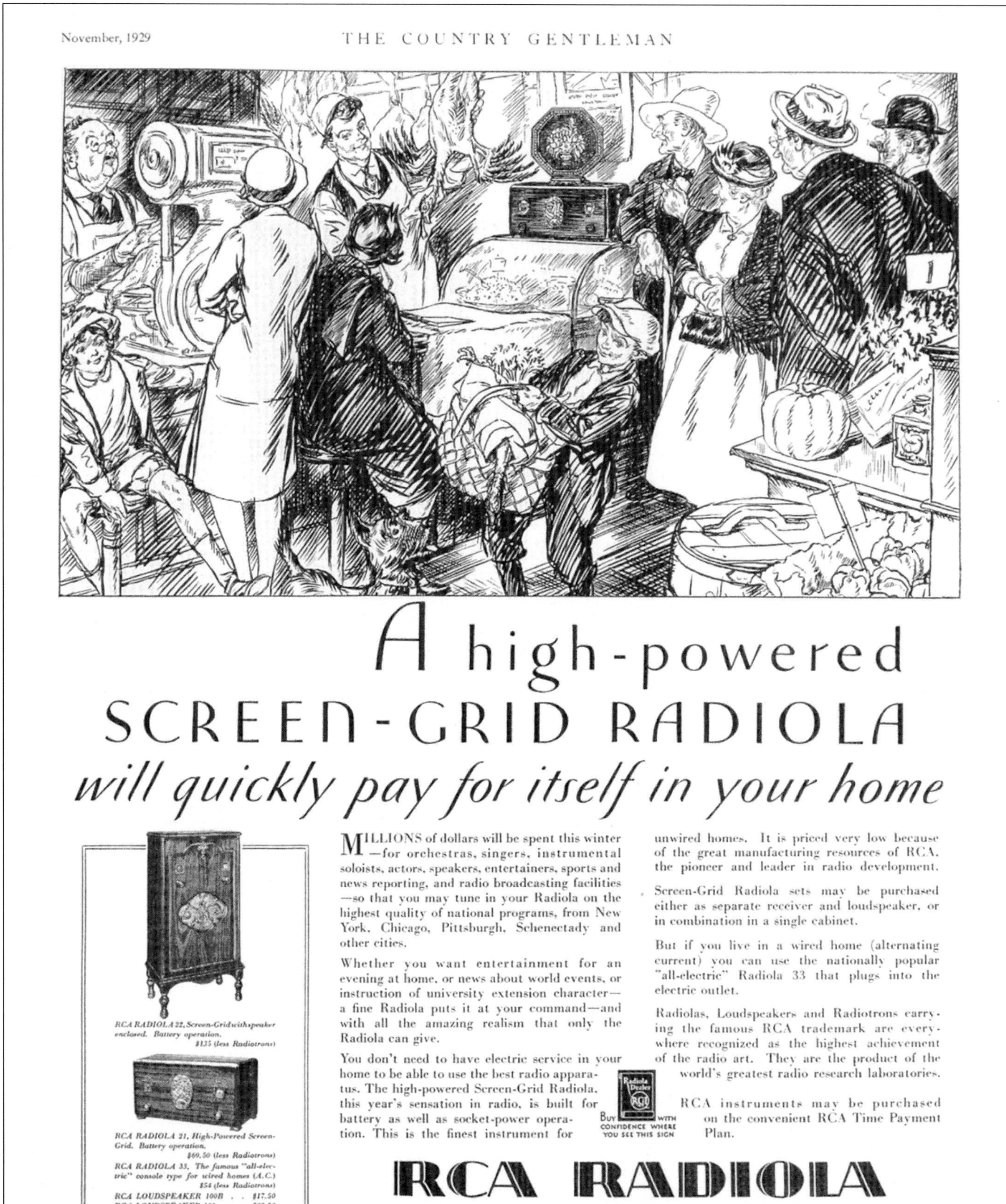

Ads for the Radiola 21 table model almost always appeared together with the Radiola 22 console. (*The Country Gentleman*, November 1929)

on the right side. The stencil-cut station selector dial was used to project illuminated dial numbers on a translucent window to identify stations as the station selector knob was rotated. A local-distant toggle switch was also located on the front panel to add a 230 pf condenser in the antenna circuit to make it resonant at 700 KHz, thereby making it more sensitive at lower frequencies to help tune in distant stations.

The circuit consisted of two screen-grid stages of amplification using two type UX-222 tubes, a grid-leak detector stage and audio amplifier stage using two type UX-112-A tubes, and an audio power output tube with the option of

The Radiola 21 was a compact screen-grid battery-operated radio with an attractive cabinet dominated by a flowery art-deco appliqué located in the center of the front panel (top). The Radiola 21 chassis used the two large electromagnetic shields shown in the back of this figure to prevent unwanted cross coupling in the RF amplifier stages (below).

either a UX-112-A or UX-171-A tube. The customer could choose the output tube depending on the type of speaker selected. The type UX-171-A had the advantage of greater audio power output, but it also drained the "B" battery at a rate of about 15% higher.

Two large electromagnetic shields were used to enclose and electrically isolate the RF amplifier stages to prevent feedback and unwanted regeneration. A fixed amount of regeneration was also added to the detector circuit to increase sensitivity, but no adjustment was necessary during operation.

In June 1930, RCA announced three new complementary tubes including the UX-230 general-purpose triode, the UX-231 power amplifier, and the UX-232 screen-grid tetrode. Shortly thereafter, RCA published rewiring instructions for adapting these tubes to the Radiola 21 and 22. The main advantages of the new tubes were the lower filament and plate currents, resulting in substantially longer battery life for both "A" and "B" batteries.[16]

The Radiola 21 was on the market into 1931, appearing in ads with the new 1930-31 lineup. The number of sets manufactured is unknown, but judging from the number on the auction market and the price, there must have been tens of thousands made.

Radiola 22 (AR-1265)

The Radiola 22 receiver was a Radiola 21 chassis with a Model 100-B speaker enclosed in a console, which was characterized as a "*beautiful Queen Anne cabinet, built of a variety of selected woods.*" While it was advertised in September, it was not officially released until November 7, 1929 at a list price of $150.75 including five Radiotrons and $135 without tubes. Like the Radiola 21, the Radiola 22 was on the market into 1931. It is not known how many of these sets were made, but they are rarely seen on the auction market.

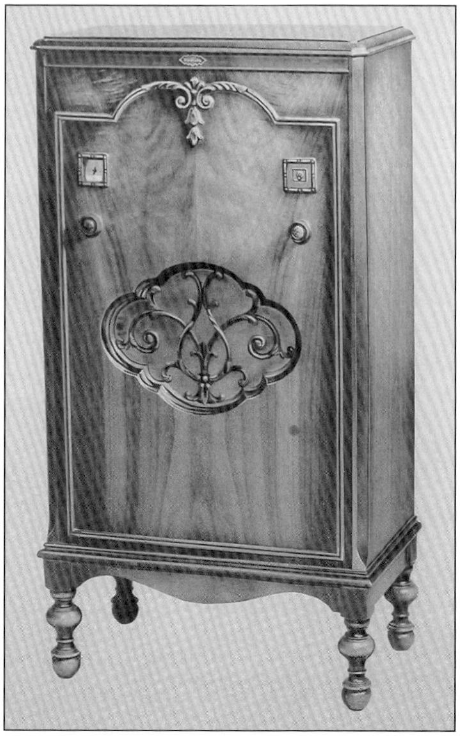

The Radiola 22 using the Radiola 21 chassis and an electrodynamic speaker was housed in an attractive Queen Anne cabinet built with a variety of hardwoods. (George H. Clark Collection, Smithsonian Institution)

Endnotes and References for Chapter 10

1 John W. Stokes, *Seventy Years of Radio Tubes and Valves*, (The Vestal Press Ltd., 1982) pp. 38-41.

2 William C. White, *The Early History of Electronics in the General Electric Company*, (William C. White Papers Collection, Schenectady Museum, Schenectady, New York, prepared 1953-1955) Chapter VI, p. VI-22.

3 Ibid., pp. VI-22-23.

4 "Strays from the Laboratory," *Radio Broadcast*, Vol. XIV, No. 5, March 1929, p. 300.

5 "New A. C. Tubes Announced, *Radio News*, Vol. 10, No. 10, April, 1929, p. 913-914.

6 "Questions the Trade Show Answered," *Radio Broadcast*, Vol. XV, No. 2, July 1929, p. 133.

7 *Radio News*, Vol. 10, No. 10, April, 1929, p. 971.

8 Written by Willis K. Wing, Editor of *Radio Broadcast* magazine in the introduction of the following article: Bert E. Smith, "Coupling Methods for the R. F. Amplifier," *Radio Broadcast*, Vol. XIII, No. 6, October 1928, p. 361.

9 "What the Screen-Grid Set Does," *Radio Broadcast*, Vol. XV, No. 4, August 1929, p. 212.

10 *Saturday Evening Post*, Vol. 202, No. 4, July 27, 1929, p. 42.

11 Alan Douglas, *Radio Manufacturers of the 1920's, Vol. 3 RCA to Zenith*, (The Vestal Press, Ltd. Vestal, New York, 1991) p. 52.

12 Ibid.

13 *RCA Victor Service Notes for 1929 – 1930,* (Service Division RCA Victor Company, Inc. Camden N.J., N.D.), Chapter on RCA Radiola 66, p. 7.

14 Alan Douglas, *Radio Manufacturers of the 1920's, Vol. 3 RCA to Zenith*, p. 51.

15 *Saturday Evening Post*, Vol. 202, No. 13, September 21, 1929, pp. 86-87.

16 *RCA Victor Service Notes for 1929-1930,* (Service Division, RCA Victor Company, Inc., Camden, New Jersey, N.D.), Chapter on RCA Radiolas 21 and 22.

14 *House & Garden*

CHOSEN
by
VICTOR
and
BRUNSWICK

Radiola
REG. U.S. PAT. OFF.
AN RCA PRODUCT

TWO great manufacturers—working independently and competitively—set out to find the radio receiver worthy to be combined with phonographs that have a world-wide fame to uphold.

They had every known kind and variety of radio set to choose from. They had every facility for research, test and exhaustive study. And they had a firm determination to combine with the Brunswick phonograph and the Victrola only that radio set which proved to be the leader today, and which carried full promise of holding that leadership through the years of tomorrow.

Both chose the Radiola.

Radiola Super-Heterodyne, the most famous of Radiolas. With the new power Radiotron as its sixth tube, it achieves new volume of tone and finer tone quality than ever.

RADIO CORPORATION OF AMERICA CHICAGO - NEW YORK - SAN FRANCISCO

Chapter 11

RCA is Chosen by Brunswick and Victor

1924-1929

As early as 1922, both the radio and phonograph industries contemplated combining radios with phonographs. The radio industry was interested in this concept to increase their market share, while the phonograph industry was interested just to maintain market share. In this regard, there was widespread concern reported in the press that a decrease in sales of phonographs during 1922 and 1923 was due to the increasing popularity of radio.[1] An excerpt from *The New York Times* quoted a *Standard Daily Trade Service* bulletin as follows:[2]

> *"The rapid growth in the popularity of radio in the United States during 1922 and 1923 had a decidedly unfavorable effect on the phonograph industry, taken as a whole."*
>
> *"Prior to the advent of radio, phonograph manufacturers had experienced a steady year-to-year-growth [sic] in their business. Not only was this expansion halted during the last two years, but most phonograph companies suffered a rather substantial loss in business in that period."*
>
> *"In an attempt to stem the tide of declining trade, the phonograph companies reduced prices sharply in 1923. Quotations on talking machines were cut on average of about 25 percent, while prices on records were reduced in proportion. But even these efforts to stimulate public interest in phonographs to its former pitch failed."*

Surely, both RCA and the Victor Talking Machine Company took note of this situation, because both began experimenting with combining radio receivers and phonographs into a single cabinet very early in the broadcast era. RCA wanted to combine their Radiolas with phonographs to expand its market share, while Victor wanted to combine radios—Radiolas or otherwise—with their Victrolas in an attempt to stem the decline in sales, which was popularly perceived to be caused in large part by the advent of radio.

Because of certain missteps on the part of Victor, Brunswick actually signed up with RCA first in 1924, and it was not until the following year that Victor finally signed up with RCA. A few months after the RCA-Victor agreement was signed in mid-1925, RCA took out large ads crowing about how both Brunswick and Victor had chosen RCA Radiolas for their respective phonographs. The agreements with Brunswick and Victor tuned out to be good for all three companies. RCA expanded its market share by selling equipment directly to the two phonograph companies, and the phonograph-Radiola combinations as well as the all-electric phonographs without radios were well received by the public. The arrangement between the three companies continued for the rest of the decade until RCA finally merged with Victor in 1929, and Brunswick went its own way.

Immediately following the agreement between Victor and RCA signed on May 16, 1925, RCA placed numerous large ads in national magazines proclaiming that both Victor and Brunswick had chosen RCA radio receivers for inclusion in their respective phonographs (*House and Garden*, October 1925).

The remainder of this chapter chronicles the interactions among RCA, Brunswick, and Victor leading up to the agreements in 1924 and 1925, identifies all known phonograph-Radiola combinations and stand-alone radios manufactured by Brunswick and Victor, and describes the Radiola models and configurations provided by RCA to the two companies. There has been no attempt to provide photographs of all the combinations offered by these two companies during this period because there are other publications, including contemporaneous advertisements, that cover most of these sets.

RCA Links Up With Brunswick and Victor

There are several versions of how RCA linked up with Brunswick and the Victor Talking Machine Company, but none better than the one by Alexander Magoun in his well-annotated dissertation entitled *Shaping the Sound of Music: The Evolution of the Phonograph Record, 1877-1950.*[3] While his version does not differ markedly from other versions,[4] it does contain a number of contemporaneous references—many from actual participants—which provide detail and insights not found in other versions.[5] The version presented here is consistent with the references cited, but with some embellishments based on material found in *The New York Times.*

Victor experimented with crystal sets but found that they did not have sufficient output to drive the acoustic horns used in the early victrolas.

Victor is Ambivalent about Radio

The Victor Company was initially ambivalent about combining radios with its Victrolas. In their view, the Victrola was a mature instrument with adequate reproduction capabilities, while radio had inferior sound with a number of technical problems—and perhaps was just a passing fad. Technical problems for radio reception in 1922 included inferior speakers, static, squeals from neighboring regenerative receivers, and the need for messy storage batteries and constant recharging. Nevertheless, to hedge their bets, Victor began to explore the feasibly of integrating radios into their Victrolas in 1922 by experimenting with both crystal sets and regenerative receivers.

A crystal set with a Victor label is pictured here and also in *Vintage Radio* by McMahon,[6] but there is little documentation on this or other Victor crystal sets, if any. While the crystal set was inexpensive and did not require batteries, it clearly could not suitably drive a horn with adequate volume. Victor developed several versions of a 4-tube regenerative receiver panel capable of driving a horn, one of which is shown in the *Victor Data Book.*[7] The same or similar radio panel integrated into a Victrola 80 cabinet is shown in *Look for the Dog.*[8] There were from 25 to 50 of these receivers placed in Victor Victrola 80 cabinets and sold as the VV-R-80 radio receiver in 1923. Apparently this set was introduced to register the Victor name for radios rather than as a serious attempt to introduce a new line of Victrola-radio combinations.

Victor must have recognized that the regenerative set they designed in 1922 would have infringed on RCA patents because they attempted to design another receiver in 1923 around RCA patents. According to RCA engineer Elmer Bucher, Victor even approached Wireless Specialty to see if they would design a receiver for them. Since Wireless Specialty was part of the RCA patent pool, they could not help Victor, and discussions were quickly terminated.

Victor then approached RCA directly to see about purchasing radios. RCA went off to develop a prototype set combining a Radiola with a Victor Victrola adapted to electric amplification using a magnetic pickup whose signal was sufficiently amplified by a Radiola to drive a special transducer attached to the Victrola's acoustic horn. However, before the meeting took place, word got back to RCA that Victor was not really all that interested in combining a "radio gadget" with a Victor Victrola. Sarnoff was said to be incensed, and directed the RCA staff to develop a prototype using a Brunswick phonograph, and to begin negotiations with Brunswick for supplying receivers to them. Brunswick had previously expressed interest in RCA Radiolas, and, in fact, had purchased

VICTROLA 80
Mahogany, Oak or American Walnut, $110.00

The first radio sold by Victor was a four-tube regenerative set integrated into the top panel of the standard Victrola No. 80 cabinet shown here, which was modified for the radio by removing the crank handle hole and the shelves for storing records. (Victor Catalog No. 7461—TQA—10-24), October 1924)

VICTROLA 400
Mahogany, $250.00
For Radio Installation, $265.00

VICTROLA 405
Walnut, $250.00
For Radio Installation, $265.00

VICTROLA 410
Mahogany, $300.00 For Radio Installation, $315.00

Victor decided against entering into an agreement with RCA to supply Radiolas for phonograph-radio combinations for the 1924-25 season, and instead offered four different models including the ones shown here with a universal blank so that the user could purchase a radio of his or her own choice. (Victor Catalog No. 7461—TQA—10-24), October 1924)

several thousand Radiola IX sets that were incorporated into their phonographs as an after-market addition.

According to Bucher, the meeting between Victor and RCA finally took place on February 19, 1924, but various accounts of events after the meeting do not agree. However, all accounts agree that when the dust settled, Victor decided not to pursue an arrangement with RCA—but Brunswick did! Brunswick was much more pragmatic and less concerned than Victor about who made their receivers. Brunswick signed an agreement with RCA dated March 1, 1924,[9] and so the 1924-25 line of Brunswick phonographs included a number of models with Radiolas, whereas Victor's did not.

Victor's somewhat feeble response to the RCA-Brunswick agreements was to offer the Victrola Models VV-S-215 and VV- and VE-S-400, -405, and -410 as part of the 1924-25 line with a universal blank panel advertised to accommodate "any radio receiving set."[10] Versions specially designed for radio installation were offered at $15 more than the standard versions. Models 215 and 400—both of which were designed to accommodate a radio—are shown in the *Victor Data Book*, one with a vertical panel in the front and one with a horizontal panel in the top left compartment.[11] However, the consumer had to buy the Victor Victrola, select another manufacturer's receiver separately, pay to have the receiver installed, and hope that the two were compatible—not unlike some computer-peripheral combinations today! Victor Models 215, 400, 405, and 410 were often equipped with Radiola IX radios as pictured in the last reference cited.

Sales of Victor phonographs during the 1924 Christmas season were far short of expectations and inventories at both Victor dealers and the factory piled up. Victor laid off a large number of employees, and almost half of the Victor dealers went out of business. The stock price decreased from a high 116½ early in 1925 to 65 just prior to the publication of the 1924 annual report during the second quarter of 1925. The Company was running short of cash, and so the quarterly dividend due on July 15, 1925 was suspended for the first time since 1901.[12]

The Victor Talking Machine Company was in deep trouble. The financial sector and the popular press clearly blamed Victor's troubles on radio, but whether this was actually the case is debatable. However, one thing was clear, Victor had been complacent in their position as the undisputed leader in the field of phonograph instruments—they had made no significant improvements in the basic sound quality of their Victrolas for many years. They had fumbled the opportunity to integrate RCA radios in their Victrolas, allowing Brunswick to steal that opportunity from right under their nose. The public had basically lost interest in buying their product.

Fortunately for Victor, the management by that time had transitioned from long-time Chairman and President Eldridge R. Johnson to a committee consisting of Director Edward E. Shumaker, Director and Treasurer Walter J. Staats, and Director and Vice-President E. R. Fenimore Johnson. These three individuals had the foresight to make two important decisions that ultimately saved Victor. The first was to introduce a new line of Victrolas for the 1925-26 season with new technologies including a new "Orthophonic" horn and sound box to dramatically improve fidelity, and Victrola-radio combinations using Radiolas. The second decision was to clear out existing stocks before they became obsolete by offering the existing inventories to their dealers with price cuts up to fifty percent, accompanied by an aggressive ad campaign.

Victor was in no position to develop all the new technologies on their own, so they approached RCA and Western Electric simultaneously for access to radio and horn technologies. The RCA-Victor agreement to supply Radiolas for Victor Victrolas was signed on May 16, 1925, and was widely reported in newspapers and trade magazines.[13] The Western Electric-Victor agreement for access to the reentrant exponential horn (later referred to by Victor as the Orthophonic horn or tone chamber) was signed a few days after the RCA-Victor agreement, but was not widely reported in the press.

Brunswick and Victor Compete

Brunswick, which had signed an agreement with RCA a year prior to the Victor agreement, had no serious competition for factory made radio-phonograph instruments during 1924-25 season. All that was about to change with Victor's new agreements with RCA and Western Electric in hand, and Victor's plans for a whole new line of Victrola-radio combinations for 1925-26.

Victor's strategy was to focus on their new Orthophonic horn and sound box by offering Orthophonic Victrolas with acoustical amplification for records, and Orthophonic Victrolas with Radiolas in which the Radiolas drove a transducer attached to the Orthophonic horn. Victor hedged its bets by offering a few all-electric Victrolas (which included cone speakers), but of the twelve new sets Victor offered in their December 1925 catalog for the 1925-1926 season, only three were all-electric—two were stand-alone Victrolas, and only one was a combination Victrola-Radiola (Hyperion). Brunswick's strategy—absent a new exponential horn to compete with Victor's Orthophonic horn—was to focus exclusively on RCA cone speakers, which required electrical amplification of signals produced by a new magnetic pickup device for the phonograph record. To emphasize their new all-electric technology, Brunswick dropped the term "phonograph" altogether in favor of the term "Panatrope."

Most of Brunswick phonographs and phonograph-Radiola combinations used a new magnetic pickup mechanism, which when amplified was able to drive the new RCA-supplied cone speakers.

The respective strategies for both companies required substantial development during 1925, and so the race was on to see who would come out with their new line for the 1925-26 season first. From an historical perspective, which invariably focuses on "firsts," the more interesting question was: who would come out with the first all-electrical amplification system for the phonograph—from magnetic pickup on the tone arm through a power amplifier to a cone speaker?

The first volley was fired by Brunswick with a startling announcement carried in the August 13, 1925 issue of *The New York Times* introducing the Panatrope, an all electric phonograph capable of reproducing sounds between 100 and 7000-8000 Hz—versus sounds between 1000 and 2000 Hz for existing phonographs.[14] The very next day—almost surely in response to the Brunswick announcement—Victor announced a new non-electrical system, later identified as the Orthophonic Victrola, that was claimed to give complete mechanical reproduction over the entire range of audible sound.[15] Within a week, Victor made a second announcement to the effect that they had invented an all-electric approach—and did not even bother to mention that their approach was actually "invented" by RCA or that it was essentially identical to the one Brunswick announced the week before.[16] The two Victor announcements when taken together give the clear impression that the Orthophonic principle was the main event for Victor, and the all-electric approach was just a side show.

For the next two months, both Brunswick and Victor made a number of public and private demonstrations of their respective technologies, Brunswick—their Panatrope, and Victor—their Orthophonic Victrola. The most widely advertised demonstration for Victor was a coordinated demonstration of the Orthophonic Victrola taking place at individual Victor dealers on November 2, 1925, a day characterized by Victor in newspapers across the country as Victor Day.[17] The most widely reported demonstration for the Brunswick Panatrope was at Aeolian Hall in New York City on November 11, 1925, where it was said that an audience filling the house repeatedly applauded musical performances of the Panatrope.[18]

Both companies continued to advertise and demonstrate their respective products, but there is no evidence that either was able to deliver any of the new products to the public until late November or early December when ads began to appear in *The New York Times*. Ads placed by both Victor and Brunswick dealers began to advertise limited store stocks of their respective products available for delivery before Christmas 1925. It would appear that both companies brought instruments to the market at about the same time.

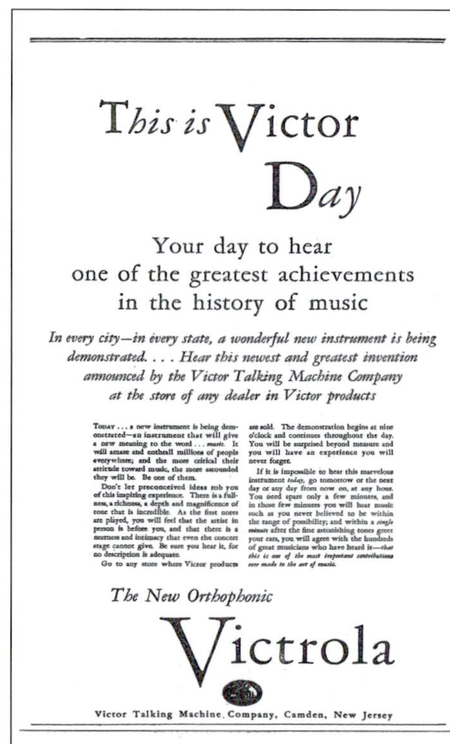

Victor placed ads in newspapers across the country proclaiming November 2, 1925 as Victor Day, the day on which they demonstrated their new Orthophonic Victrolas at Victor dealers *"In every city—in every state."* (*The New York Times*, Dec. 28, 1925, p.17)

Victor's strategy was to focus on their new Orthophonic tone chamber for both acoustical reproduction of records and as a reproducer for Radiola receivers used in Victrola-Radiola combinations. This ubiquitous figure of the Orthophonic horn appeared in many catalogs and advertisements of the day. (Victor Catalog No. 8976—*JQA*, November, 1926, p. 3)

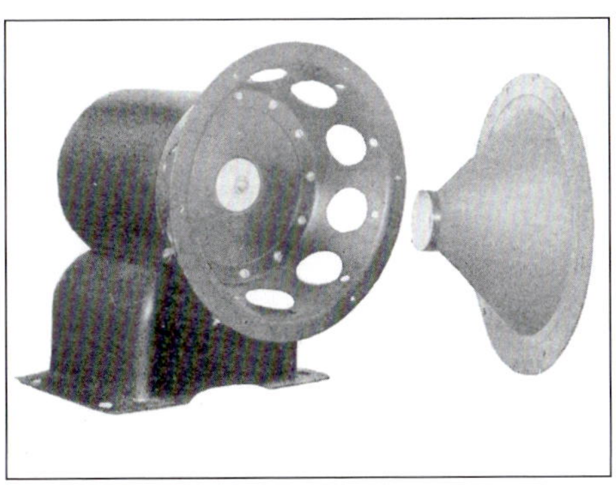

Brunswick's strategy—absent a new acoustic tone chamber—was an all-electric approach using both permanent-magnet and electrodynamic cone speakers supplied by RCA as a reproducer for both records and Radiola receivers. The 6" electrodynamic speaker shown here from the RCA Model 104 speaker was used in many Brunswick phonographs and phonograph-radio combinations. (*General Electric Review*, 1927, p. 71)

The public took to the new phonograph technologies, and the year 1926 was a banner year for both Victor and Brunswick—but particularly so for Victor, who earned back all of the money it lost in the catastrophic year of 1925. Victor's success was in no small part due to a new advertising agency and an extensive national ad campaign beginning circa October 1925—appearing most noticeably the *Saturday Evening Post*. Oddly, in stark contrast, Brunswick had no national advertising campaign for their new Panatropes for well over a year after their introduction. Although Brunswick had made extensive use of the *Saturday Evening Post* and other national magazines to advertise their 1924-1925 line of Brunswick Radiolas, their national ad campaign stopped abruptly in mid-1925 and did not begin again until September of 1926, amounting to a hiatus of more than a year!

Victor focused on the sets with the Orthophonic horn and sound box for the next several years supplemented with a few all-electric models, but by 1929, most of Victor's new offerings became all-electric. Brunswick, on the other hand, focused on the all-electric models, initially dropping their acoustic line all together. However, later in 1926 they introduced a new folded acoustic horn which they initially named a Prismatone to compete at the low-end with Victor's inexpensive Orthophonic Victrolas.

Who Was First With the All-Electric Phonograph?

While both companies brought their respective new lines to the market at about the same time, there has been some question about whether Brunswick or Victor was first with the *all-electric* phonograph. Robert Baumbach writes that some authors suggest Victor was the first to sell a "cone-speakered-equipped phonograph,"[19] while Gelatt, based on his chronology of events, seems to bestow that honor on Brunswick.[20] This debate has been fueled in no small part by RCA Victor's unofficial historian, Benjamin Aldridge, who makes a case in his book *The Victor Talking Machine Company* that Victor was first to market with the all-electric phonograph. However, it is difficult to gauge how strongly Aldridge believed his argument because of his concluding admonition: "*Under the circumstances, until better evidence is available, any claim for priority in electrical amplification should be made with discretion, if at all.*"[21] The "circumstances" that Aldridge used to support his contention are fanciful at best, and certainly contrary to the evidence presented here.

Aldridge starts out well enough by establishing a time line for shipment of Victor's first all-electric models:[22]

> *"On December 16th, Victor announced eight more models—two "Electrolas," four acoustical combinations, one electrical combination, and one electric-acoustical combination. Shipments were expected to start in a few days. However, the models with the magnetic speaker [i.e., all-electric models] didn't start until February [of 1926], and those with the dynamic speaker until the following September."* [23]

Aldridge went on to establish a time line for the shipments of the Brunswick Panatrope. He assumed that shipment of the Panatrope coincided with its earliest advertisement, which he believed did not occur until the fall of 1926:

> *"It has been known that the Panatrope was demonstrated to both the trade and the public before the Electrola. However the emphasis on "demonstrations" and the number of people who had "heard" the instrument, and the fact that it apparently was not advertised to the public until October 1926, leads to the conclusion that Victor had been first to make stock shipments."*

Aldridge's "fact" that the Panatrope was not advertised to the public until October 1926 was simply wishful thinking. While it is true that the Panatrope was not advertised in *national* magazines until September 1926, the Panatrope was, in fact, widely advertised in newspapers by retail stores as early as mid-December 1925 (see advertisement on page 364).[24] Not only that, these ads

One of the first ads offering immediate delivery of the Panatrope was placed by the Aeolian Company in *The New York Times* on December 21, 1925, almost nine months before the first ads for Panatrope appeared in national magazines.

with an artist's rendition of the P-3 Panatrope—an all-electric phonograph without a radio—clearly stated they were in stock and ready for delivery in December of 1925: *"The first models of the Panatrope are now on sale at Aeolian Hall. A very limited number for immediate delivery on orders as they are received."*[25] This ad was not an isolated ad. Scores of ads for the Panatrope and Panatrope-Radiola combinations appeared in *The New York Times*—and undoubtedly other major newspapers—over the next year and a half, well before the first ad for the Panatrope appeared in the *Saturday Evening Post* in September 1926.

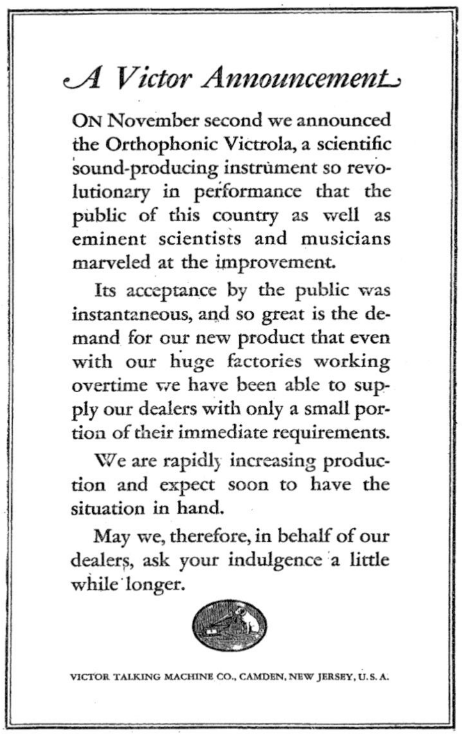

A Victor ad in the December 28, 1925 issue of *The New York Times* provides strong evidence that Victor offered only Orthophonic Victrolas in 1925—apparatus which were not of the all-electric type. (*The New York Times*, Dec. 28, 1925, p. 15)

More to the point, while ads for the all-electric Panatrope appeared in *The New York Times* in December of 1925, there were no such ads for the all-electric Victor models in *The New York Times*.[26] Ads for the new Victor line appearing in *The New York Times* placed by retail outlets in late 1925 stated that stocks were limited to two Orthophonic Victrola models (Credenza and Consolette).[27] Surely, if Victor had all-electric models in dealer showrooms before Christmas 1925, there would have been some dealer's ads for these models in the newspapers. Instead, Victor took the unusual step of publishing an ad in *The New York Times* shortly after Christmas apologizing to the public for the limited number of Orthophonic Victrolas for sale in 1925, promising there would be more in the following year.[28]

Assuming that Aldridge's statement that Victor's sets with magnetic speakers were first shipped in February 1926 is correct, the evidence clearly supports the claim that Brunswick was the first to have an all-electric phonograph on the retail market. While Brunswick apparently had the foresight to offer the first factory-made radio-phonograph combination and the first all-electric phonograph in the U.S., it must be remembered that RCA was the first to actually develop the all-electric phonograph technology and supply it to phonograph manufacturers—Brunswick and Victor alike.

Brunswick Radiola Apparatus

RCA provided Brunswick with a wide range of apparatus and associated licenses from mid-1924 to mid-1929, at which time the agreements were apparently terminated—almost certainly precipitated by the RCA-Victor merger in January 1929. While the agreements to provide apparatus may have been terminated, Brunswick continued to manufacture its own equipment under RCA licenses with an appearance similar to Radiolas.

RCA provided Brunswick with two distinctly different types of apparatus during the twenties: 1) Radiola receivers—first integrated with phonographs and later sold as stand-alone units, and 2) power amplifiers and cone speakers without radio receivers for use with phonographs. The power amplifiers—which were manufactured as a separate chassis number specifically for Brunswick phonographs—do not qualify as Radiolas and are not addressed here. These Brunswick models using power amplifiers without receivers were generally designated with a P followed by a one- or two-digit number such as P-3, P-11 (originally designated Model 10 without a P), P-13 and P-14. From the limited number of models designated with a P, it might appear that Brunswick offered a relatively small number of electrical phonographs without receivers, but that is not the case. For example, Brunswick actually offered seven sets in 1926 with the same P-3 designation: Gotham, Georgian, Oxford, Stratford, Beaux Arts, Chippendale, and Lombardi.[29] The model number P-3 apparently referred to the type of electrical amplifier, which was used in multiple cabinet styles with different features.

All known Brunswick phonographs which used Radiola receivers in the 1920's are listed in Table 11-1 and discussed briefly below. The entries in this table were found in one or more of the following four categories of contemporaneous sources: Brunswick catalogues and pamphlets, *Radio Retailing* magazine, advertisements from national magazines with a focus on those appearing in the *Saturday Evening Post*, and advertisements appearing in *The New York Times*. The primary sources used for each entry are given in the far right column of the table.

Brunswick used the three different names for their electric phonographs in as many years—changing the meaning of these names without any explanation, thereby confusing everyone in the process. In addition, Brunswick not only changed the model numbers occasionally, but they also assigned the same number to different sets, and different numbers to the same set. Had the Clio award for excellence in advertising existed in the twenties, Brunswick surely

Table 11-1. Brunswick phonographs with Radiola receivers

Year	Brunswick Model No.	Radiola Model No.	Receiver Type[1]	Power	Reproducer	Retail Price	Primary References[1]
1924-1926	Royal 30	Radiola III (AR-805)	Regenerative	Battery	Acoustic	$190	1924 Catalog
1924-1926	Westminster 35	Radiola III-A (AR-802)	Regenerative	Battery	Acoustic	$285	1924 Catalog SEP 1-24-25
1924-1926	Raleigh 100	Regenoflex (AR-817)	Regenerative	Battery	Acoustic	$450	1924 Catalog SEP 1-24-25
1924-1926	Tudor 160	Super-Het (AR-813)	Super-Het (6-tube cat.)	Battery	Acoustic	$550	1924 Catalog SEP 9-13-24
1924-1926	Hudson 260	Super-Het (AR-813)	Super-Het (6-tube cat.)	Battery	Acoustic	$600	1924 Catalog SEP 10-11-24
1924-1926	Empire 360	Super-Het (AR-813)	Super-Het (6-tube cat.)	Battery	Acoustic	$650	1924 Catalog SEP 9-13-24
1925-1926	60	Super-Het (AR-813)	Super-Het (6-tube cat.)	Battery	Acoustic	Unknown	SEP 2-21-25
1925-1926	460	Super-Het (AR-813)	Super-Het (6-tube cat.)	Battery	Acoustic	Unknown	SEP 4-25-25
1926-1927	PR-6	Radiola 25	Super-Het (6-tube cat.)	AC + A & C battery	Cone	$575	1926 Pamphlet
1926-1927	PR-16-C	Radiola 25	Super-Het (6-tube cat.)	AC	Cone	$775	1926 Pamphlet
1926-1927	PR-26-C	Radiola 25	Super-Het (6-tube cat.)	AC	Cone	$825	1926 Pamphlet
1926-1927	PR-128-C	Radiola 28	Super-Het (8-tube cat.)	AC	Cone	$1100	1926 Pamphlet
1926-1927	PR-36-C	Radiola 25	Super-Het (6-tube cat.)	AC	Cone	$875	1926 Pamphlet
1926-1927	PR-38-C	Radiola 28	Super-Het (8-tube cat.)	AC	Cone	$950	1926 Pamphlet
1926-1927	PR-46-C	Radiola 25	Super-Het (6-tube cat.)	AC	Cone	$925	1926 Pamphlet
1926-1928	PR-148-C (also 148)	Radiola 28	Super-Het (8-tube cat.)	AC	Cone	$925	1926 Pamphlet SEP 11-29-27 RR Aug. '26
1927	Cordova	Radiola 25	Super-Het (6-tube cat.)	Battery	Acoustic	$450[2]	1926 Pamphlet SEP 6-11-27
1927	Cordova	Radiola 28 (AR-998)	Super-Het (8-tube cat.)	Battery	Acoustic	$550[2]	1926 Pamphlet SEP 6-11-27
1927-1928	PR-138-C (also 138C)	Radiola 28	Super-Het (8-tube cat.)	AC	Cone	$945 AC $995 DC	1927 Catalog SEP 4-16-27
1928-1929	PR-17-8	Radiola 17 (AR-1084)	TRF	AC	Cone	$395	1928 Catalog RR Oct. '28
1928-1929	PR-17-8	Radiola 18	TRF	AC	Cone	$395	1928 Catalog RR Oct. '28
1928-1929	2KRO	Radiola 18	TRF	AC	Cone[3]	$250 $285	RR Nov. '28
1928-1929	3KRO	Radiola 18	TRF	AC	Cone	$395	1928 Catalog SEP 9-15-28 RR Nov. '28
1928-1929	3NC8	Radiola 62	Super-Het (8+1 tubes)	AC	Cone	$700	SEP 11-3-28 RR Nov. '28
1928-1929	3KR8	Radiola 18 (AR-746)	TRF	AC	Cone	$750	RR Nov. '28
1929	3KR6	Radiola 18	TRF	AC	Cone	$450	SEP 2-25-29
1929	3NW8	Radiola 64	Super-Het (8+3 tubes)	AC	Cone	$795	RR Jul. '29

1 Abbreviations: catacomb (cat.); *Saturday Evening Post* (SEP); *Radio Retailing* (RR)

2 The prices are from Brunswick pamphlet Form 1109-A FFHM dated 11-10-26; an optional electric motor was an additional $35

3 The Radiola used a cone speaker while the record player used the acoustic horn.

Brunswick sales pamphlets are the best source of information on the early Panatrope and Panatrope-Radiola combinations.

While the Brunswick phonograph models were never advertised with the BR prefix, a number of the nameplates affixed to Brunswick phonographs did contain the BR prefix.

would *never* have won. The paragraphs that follow are arranged with a view to sorting out the confusion.

Brunswick Radiola

Brunswick began advertising the "Brunswick Radiola" in national magazines circa September 1924, and also published a catalog entitled *Retail Price List* for dealers containing six models which were assigned the following numbers: 30, 35, 100, 160, 260, and 360. While neither the advertisements nor the catalog contained any reference to names for these sets, a photograph from the *Radioana Collection* shows all these six sets with the names listed on the back; these six names are reproduced in the top portion of Table 11-1. In early 1925, Brunswick advertised two additional sets under the Brunswick Radiola name—the Brunswick Radiola No. 60 and Radiola 460, but did not use a name or specify the price in their advertisements.

While the model numbers were described in ads by a two- or three-digit number without any preceding letters, the numbers appearing on the name tags attached to each set were preceded by the letters "BR"—for example, BR-30. This is just one example of the inconsistencies that appeared in the model designations used by Brunswick in the twenties.

Brunswick Panatrope and Panatrope with Radiola

From mid-1925 to the end of the year, RCA and Brunswick worked on developing a line of Brunswick phonographs with components that were totally electric from the phonograph pickup to the loudspeaker. Whereas the earlier models used a mechanical pickup arm connected directly to a horn for the phonograph and a separate electric transducer located at the end of the horn for the radio, the

Panatrope used a magnetic pickup coil on the phonograph arm which fed directly into the amplifier portion of a Radiola receiver, which in turn drove a cone-type speaker. Some Panatrope models were sold without a radio receiver, for which RCA supplied a power amplifier unit without a receiver to amplify the output of the magnetic pickup mechanism. The output of other radios could also be attached to this type of Panatrope to achieve better sound quality.

While Brunswick advertised the Panatrope as early as December 1925 in newspapers, it began advertising the Panatrope in earnest at the national level in September 1926, declaring it in the *Saturday Evening Post* and other national magazines as the "*Most Marvelous Musical Invention in [the] World's History.*" The advertisements specifically characterized Panatropes as operating from the AC lighting circuit with no batteries or outside wires required. The first two Panatropes appearing in national ads were the Model 10, which incorporated an RCA power amplifier without a Radiola receiver to amplify the phonograph, and the model PR-148-C, which incorporated a Radiola 28 receiver in conjunction with a Brunswick phonograph.

The Model 10 Panatrope appearing in the September 11, 1926 issue of the *Saturday Evening Post* apparently suffered an identity crisis because the model number changed twice in as many years. The set was first introduced in the April 1926 issue of *Radio Retailing* as the Model P-3. Brunswick apparently intended to designate all such Panatropes with P-3, because Brunswick prepared a pamphlet with seven different P-3 models, none of which was the same as the one illustrated in the *Saturday Evening Post*. The same model was designated P-11 in the May 14, 1927 issue of the *Post* and in a 1927 Brunswick catalog. Of course, it is possible that there were some minor changes in design along the way that gave rise to the different designations, but in general, each Brunswick model had its own unique cabinet, and this set had the same cabinet with three different designations. This is not the only set that Brunswick advertised with different numbers. For example, a Panatrope with Radiola initially advertised as the PR-148-C was later advertised as Model 148.

Prismatone

At the same time that Brunswick introduced the Panatrope in national magazines, it also introduced "*another musical instrument, as yet unnamed,* for bringing out the music of the new records. In tonal quality and its ability to*

The first nation-wide ads for the Panatrope—similar to this one declaring the Panatrope to be the "most marvelous musical invention in [the] world's history"—were placed in popular magazines beginning in September of 1926, long after the Panatrope was put on sale in December of 1925. (unknown magazine)

Brunswick displayed their 1924-1925 line of phonograph-Radiola combinations in their New York show room window, which included from left to right: No. 100 Raleigh, No. 160 Tudor, Beaux Arts, No. 260 Hudson, No. 360 Empire, No. 30 Royal, and No. 35 Westminster. (George H. Clark Collection, Smithsonian Institution)

MOST MARVELOUS MUSICAL INVENTION IN WORLD'S HISTORY

The BRUNSWICK PANATROPE

utilizes newly discovered electrical processes to bring you music and radio of a beauty hitherto unknown

The Brunswick Panatrope is the world's first purely electrical reproducing musical instrument. The new electrical method of recorded music reproduced on the Panatrope resembles the photographic process so closely that this amazing discovery is often called "music by photography."

The Brunswick Panatrope, Model 10. Finished in highly figured walnut. Price $650. (Slightly higher west of Rockies.)

GIORGIO POLACCO
Director Chicago Civic Opera Company, says:
"I am spellbound by this wonderful achievement of science, which, through electrical principles, has at last made it possible for music lovers to hear their favorite artists absolutely as perfectly as if they were in their presence."

THE PANATROPE is the joint achievement of four of the most widely-known institutions in the fields of electricity and acoustics: Radio Corporation of America, General Electric Company, Westinghouse Electric & Manufacturing Company, and The Brunswick-Balke-Collender Company.

Amazes critics

Although perfected less than a year ago, this new reproducing musical instrument has been heard by more than four million persons. When first demonstrated, it was given front-page space by the great metropolitan newspapers as the most important musical development of the age. Critics were no less enthusiastic. Nothing in the whole world of music ever created such profound interest and attention.

The Brunswick Panatrope may be had either alone or in combination with the Radiola Super-heterodyne. Thus it puts at your finger-tips all recorded music and the programs of radio as well. Operates entirely from the light socket; no batteries or outside wires needed.

Another triumph

Brunswick has also developed another musical instrument, as yet unnamed,* for bringing out the music of the new records. In tone quality and its ability to reproduce the entire musical scale, we believe this instrument represents a very great advance over anything in existence, save the Panatrope. It dispenses with the electrical mechanism of the Panatrope, and its prices are lower, ranging from only $115 to $300 (slightly higher west of the Rockies).

Certainly you must hear these wonder instruments. Your musical judgment will confirm the verdict of the critics that there is nothing in the field of musical reproduction equal to them. It is hardly necessary to suggest that before buying any musical instrument or radio, you hear the Brunswick Panatrope, the Brunswick Panatrope & Radiola and the new Brunswick musical instrument as yet unnamed.*

Brunswick dealers will gladly demonstrate these instruments. If there is no Brunswick store near you, write us.

**$5,000 for a name*

To find a suitable name for the new Brunswick instrument described above, we offer 3 prizes totaling $5,000 for the best names submitted with slogan not exceeding 10 words describing its music. Write for free booklet giving all details. Address Dept. P-101.

Brunswick

PANATROPES · RADIOLAS · RECORDS

THE BRUNSWICK-BALKE-COLLENDER CO., General Offices: 623 S. Wabash Ave., Chicago *reproduce the entire musical scale, this instrument represents a very great advance over anything in existence, save the Panatrope."* The asterisk in the advertisement referred to a contest with three prizes totaling $5,000 that Brunswick offered to the public to name the new musical instrument. The new musical instrument was nothing more than the standard Brunswick phonograph mechanism with a newly designed folded horn to give a better bass response. It was characterized in their later advertisements as a a Panatrope of the exponential type, but in fact, the horn was not exponential. The new instrument was actually a low-cost phonograph with an acoustically amplified horn designed to compete with the improved frequency response offered by Victor's new Orthophonic Victrola.

The winners were announced in the April 16th issue of the *Saturday Evening Post*, with the first prize going to one Mrs. Mildred A. Bux for the Brunswick Prismatone—which she described in her winning entry as *"the instrument of colorful music."* In this ad it was stated that this new musical instrument "reproduces recorded music by new and advanced principles of *acoustical* reproduction," and that it was an entirely different type of instrument than the Panatrope.

The first ad proclaiming the newly-named Prismatone appeared in the June 11, 1927 issue of the *Saturday Evening Post* along with three models—the *Madrid*, the *Cortez*, and the *Cordova*. The first two were standard wind-up phonographs with the new "exponential" horn, but the Cordova was described as a Prismatone with a 6- or 8-tube Radiola Superheterodyne, which would have contained the Radiola 25 and Radiola 28 respectively. This Prismatone was strikingly similar to the Panatrope, except that the radio was battery operated and it used the new exponential horn that was part of the phonograph rather than the new RCA electrodynamic speaker used with the Panatrope. In this regard, the Prismatone was very similar to the original Brunswick Radiolas of 1924-25, but with an improved horn.

Brunswick Panatrope Redefined

The Prismatone name was used once more in ads for the *Seville* Model, and then in September 1927, the *Hidalgo*, a low-priced phonograph without any electronics, was advertised as a Panatrope. The trade name Prismatone was never to be seen again, and without a word of explanation, all Brunswick phonographs suddenly became Panatropes! There were now three types of Panatropes—those of the "exponential type," those of the "electrical type," and those of the "Panatrope with Radiola" type.

The Panatropes with power amplifier amplifiers (no receivers) were given a designation beginning with the letter P, while the Panatropes with Radiolas were initially given a designation beginning with PR. There were at least four different "P" designations used for the Panatropes of the electrical type—P-3, P-11, P-13, and P-14—and eleven different "PR" designations used for Panatropes with Radiolas. This system was far too rational, so for the 1928-29 season Brunswick changed the designation system for Panatropes with Radiolas to a single number followed by either two letters and another number or three letters—and of course, the designation system for the 1929-30 season was changed again to a letter followed by two numbers—or in some cases, just two numbers without a letter.

Brunswick Radios

Beginning in 1928, Brunswick began to manufacture stand-alone radios without phonographs—offerings which appear to have been in direct competition with RCA. These radios had a distinct resemblance to RCA Radiola models, but unlike the chassis in the "Panatrope with Radiola" sets, these Brunswick radios did not have name tags with RCA or Radiola designations—only a reference to a licensing agreement with RCA in the fine print on the label. Despite the fact that there was no RCA or Radiola name,

Table 11-2. Brunswick Radios with Radiolas

Year	Brunswick Model Number	Brunswick Model Name	Receiver Type	Radiola Model Number	Retail Price	References
1928-1929	5KR	None	TRF	Radiola 18	$115 Less tubes	1928 Catalog RR Nov. '28 SEP 9-15-28
1928-1929	5NO	None	Super-Het 9-tube	Radiola 60	$175 Less tubes	1928 Catalog RR Nov. '28 SEP 10-20-28
1928-1929	5KRO	None	TRF	Radiola 18 (100A Spk'r)	$215 Less tubes	1928 Catalog RR Nov. '28 SEP 9-15-28
1928-1929	5KR6	None	TRF	Radiola 18 (106 Spk'r)	$250 Less tubes	NYT 4-3-29
1928-1929	5NC8	None	Super-Het 9-tube	Radiola 62	$375 Less tubes	1928 Catalog RR Nov. '28 SEP 10-20-28

Note: Abbreviations: *Saturday Evening Post* (SEP); *Radio Retailing* (RR), *The New York Times* (NYT)

these receivers were in fact manufactured by RCA for Brunswick. The radios Brunswick offered for the 1928-29 season are shown in Table 11-2 along with the corresponding Radiola model number used in each.

RCA was in serious negotiations to merge with Victor at this time, and so it would seem logical that RCA and Brunswick were negotiating to part ways even before the 1928-29 season. At the end of 1928 Brunswick negotiated the purchase of Bremer-Tully—a radio company holding licenses from RCA and Hazeltine among others—thus assuring Brunswick a source of competitive radio apparatus. Bremer-Tully was absorbed into Brunswick and supplied their radio apparatus beginning with the 1929-30 season. Beginning in the 1929-30 season, Brunswick Panatrope advertising no longer made any reference to Radiola or RCA—from that point on Brunswick phonographs with radios were advertised as "Brunswick Combination Panatrope with Radio" rather than "Panatrope with Radiola."

Reconciliation to other Brunswick Radiola References

The tables of Brunswick phonographs and radios equipped with RCA Radiolas presented in this chapter were developed from Brunswick catalogs and pamphlets, contemporaneous advertisements in national magazines such as the *Saturday Evening Post*, and major newspapers, most notably *The New York Times*. It is of interest to compare these compilations with several well-known publications listing Brunswick apparatus equipped with radios and electrical phonographs including the *Perpetual Trouble Shooter's Manual* by John F. Rider, *Radio Collectors Guide* by Morgan McMahon, and *The Radio Collector's Directory and Price Guide* by Robert E. Grinder. Based on a review of these documents, it appears that the compilation here is complete as it relates to *Radiola* equipment used in Brunswick radios and phonographs. However, the same cannot be said for the three references.

Rider's *Trouble Shooter's Manual* may be complete in so far as the chassis are concerned, but not the model numbers. So, for example, the manual has the entry "Superheterodyne" for the AR-813 chassis used in Models 60, 160, 260, 360, and 460—without identifying or listing these individual models. Oddly, this manual does not include any of the following models which appear in a Brunswick sales pamphlet dated 11-20-1926: PR-16-C, PR-26-C, PR-128-C, PR-36-C, PR-38-C, and PR-46-C.[30]

There are a number of entries in the Rider manual that do not appear here, but none of these use Radiolas. For example, the entries P-11, P-13 and P-14 correspond to electric amplifier and speaker systems without radio receivers that RCA made specifically for Brunswick phonographs, but which did not have a Radiola analog. In addition, there are entries such as R1, S-14, S-21, etc. for apparatus designed and manufactured by Brunswick after the agreements between RCA and Brunswick lapsed. These units can easily be identified in their advertising by the phrase "Panatropes with Radios" which replaced the phrase "Panatropes with Radiolas."

A Brunswick pamphlet dated Nov. 20, 1926 surfaced recently with six "Panatrope-with-Radiola" models which have not been previously chronicled (PR-16-C, PR-26-C, PR-36-C, PR-38-C, PR-46-C, and PR-128-C).

McMahon's *Radio Collector's Guide* is missing the same model numbers PR-16-C, PR-26-C, PR-128-C, PR-36-C, PR-38-C, and PR-46-C as those missing from the Rider's guide. In addition it is missing the PR-6, which does appear in the Rider's Manual. McMahon's guide contains only one model number with Radiola apparatus not found in the table here—an apparatus described as Model 148. A Model 148 was indeed advertised in the October 6, 1928 issue of the *Saturday Evening Post* on page 148, but it is actually identical to the PR148-C—a fact which can be verified by examining the PR148-C from the advertisement on page 99 of the November 26, 1927 issue of the same magazine. This was not the first time Brunswick changed the model number in their advertising. In this regard, the PR-138-C was also advertised as Model 138C. It should also be noted that the apparatus listed in this guide dating from 1929 and 1930 after the termination of agreements with RCA do not appear in Tables 11-2 or 11-3 because they do not contain Radiola receivers.

Grinder's *Radio Collectors Directory and Price Guide* is advertised to contain apparatus from 1921 to 1965, but as far as the Brunswick apparatus are concerned, the book contains only two Brunswick models dating earlier than 1928—Model PR-138 and PR-148. Actually, the book contains less than half the models with Radiola receivers actually produced by Brunswick in the period between 1924 and 1929. There are no models with Radiola receivers in this book that are not contained in the above two tables.

With regard to the over-all completeness of these tables, it is worthwhile to point out that in addition to the production models listed, Brunswick and their dealers made a limited number of specialty items which are not listed here. They incorporated standard Radiola chassis in a variety of cabinets for demonstration purposes, special sales, furniture upgrades, and high-end sets made to order. For example, the Brunswick Salon in New York City advertised Panatropes in conventional or period cabinets from $90 to $5000. In another ad, Gimbels was authorized to sell the 5KR in an exclusive cabinet model with legs containing a Radiola 100-A speaker.

The Beaux Arts model appearing in the photograph on page 368 with the rest of the 1924 Brunswick line appears to be a demonstration model using a Radiola Super-Heterodyne AR-812 chassis in a Beaux Arts cabinet—a cabinet that was first manufactured and marketed in 1920, and later used for one of the P-3 Panatropes offered in the 1925-26 season. Perhaps the AR-812 Superheterodyne panel was integrated into the Beaux Arts cabinet for purposes of demonstrating a Brunswick-Superheterodyne combination before the smaller AR-813 Superheterodyne panel was available from RCA.

The Missing 1926-27 Season Models

There are six Brunswick and Radiola combinations—PR-16-C, PR-26-C, PR-128-C, PR-36-C, PR-38-C, and PR-46-C—that have gone virtually unnoticed because they do not appear where one would expect to find them: national magazine advertisements, *Radio Retailing* Magazine and other trade magazines, Rider's *Trouble Shooter's Manual* and other books cataloging radios and radio-phonograph combinations, books on Brunswick, and Brunswick catalogs. This author has also not seen them either for sale or at auction. These Radiola-phonograph combinations came to the attention of this author through a single Brunswick sales pamphlet that surfaced recently.

Why Brunswick kept these sets such a secret is a mystery. These sets were released just at the time Brunswick initiated a large advertising campaign in

Brunswick
Panatrope and Radiola

Brunswick
Panatrope and Radiola
Model PR-6 [Panatrope with six-tube Radiola Super-Heterodyne] $575
Finished in Adam Brown Mahogany or Walnut. Width 39 ins. Height 37 ins. Depth 22 ins. Operates directly from the electric light socket, in conjunction with six "A" and one "4½C" dry batteries, doing away entirely with expensive "B" batteries.
(NOTE: If desired, a Gould Uni-Power or other good "A" battery eliminator may be used, thus making this instrument operate entirely from the light socket.)

Brunswick
Panatrope and Radiola
Model PR-16-C [Panatrope with six-tube Radiola Super-Heterodyne] $775
Finished in Two-tone Adam Brown Mahogany or Walnut. Width 41 ins. Depth 22 ins. Height 37¼ ins. Operates on Alternating Current, 110 volts, 50 to 60 cycles.

Brunswick
Panatrope and Radiola
Model PR-36-C [Panatrope with six-tube Radiola Super-Heterodyne] $875
Model PR-38-C [Panatrope with eight-tube Radiola Super-Heterodyne] 950
Finished in Adam Brown Mahogany or Walnut. Width 32½ ins. Height 50½ ins. Depth 22¼ ins. Operates on Alternating Current, 110 volts, 50 to 60 cycles.

national magazines beginning in September of 1926, most noticeably in the *Saturday Evening Post*, and yet there is no mention of any of these sets in that magazine. The cynic might say that the Brunswick brochure with these sets was just wishful thinking, and that the sets were not actually manufactured or distributed. However, such is not the case. Both the PR-6C and PR-16C, for example, were pictured in an ad placed by the Aeolian Company in the December 15, 1926 edition of *The New York Times*.[31] Undoubtedly, a further search of the newspapers would turn up the other models.

The Brunswick Radiola Hardware

The Radiolas supplied to Brunswick by RCA had the same electrical circuits as the respective counterparts marketed by RCA, but in many cases had different mechanical layouts for the chassis and/or panels. While the panels and numbers on the earliest Radiolas supplied to Brunswick—Radiola III, IIIA, and Regenoflex were essentially the same as those sold to the public by RCA, the subsequent Radiolas made for Brunswick were distinctly different from their respective Radiola counterparts sold by RCA and were assigned different AR numbers.

Perhaps the most different and interesting panels made specifically for Brunswick were those using the standard 6- and 8-tube superheterodyne circuits sealed in the infamous catacombs. The first superheterodyne panel made specifically for Brunswick was the AR-813 panel based on the AR-812 Super-Heterodyne. The new panel was roughly square in order to fit into one half of the phonograph width—as compared to the AR-812 with a length-to-width ratio of almost two to one. The catacomb was behind the panel in such a fashion that the panel had to be lifted to access the vacuum tubes. GE also made a phonograph panel based on the AR-812 Super-Heterodyne in which the

The AR-812 panel (above) used in the Radiola Super-Heterodyne at a width of 19-1/4 inches was too wide to fit into most phonographs, and so RCA had GE repackage this receiver into a panel designated AR-813 (below) having a width of 13 inches specifically to fit into Brunswick phonographs.

tubes protruded through the top of the panel, but it is not known if this panel was used in a commercial phonograph-Radiola combination.

The next four panels made for Brunswick were based on the 6-tube superheterodyne catacomb made for the Radiola 25 and the 8-tube superheterodyne catacomb made for the Radiola 28. These panels can be easily differentiated from the AR-813 panels by the two thumbwheel tuning dials and the vacuum tubes which protrude through the panel to facilitate replacement. There were two versions of both 6- and 8-tube sets, one for battery operation and one for AC operation. The four types can be distinguished from each other by the number of tubes protruding through the panel. The battery versions of the panels have the full count of 6 and 8 tubes, while the AC versions have one less tube because the output tube is mounted on the power amplifier chassis, so that the last tube socket on the catacomb is blank. Unfortunately, the tubes were concealed by a metal covering, so the number of tubes cannot be readily determined without removing the cover.

The remaining Radiola models used in the Brunswick line including Radiolas 17, 18, 62, and 64 were mounted on metal chassis, and were physically similar to the respective Radiolas sold to the public by RCA. However, the controls were often mounted differently on some models such as the Brunswick PR-17-8, where the controls are mounted on the turntable panel.

Brunswick panels with tubes protruding through the front panel such as the PR-138-C panel shown here were invariably covered with a trim plate to conceal the tubes.

This unidentified phonograph has a Super-heterodyne panel similar to the AR-813, with the notable exception that the six UV-199 tubes protrude through the face of the panel. (George H. Clark Collection, Smithsonian Institution)

Victor Radiola Apparatus

RCA provided Victor with a wide range of apparatus and associated licenses from mid-1925 through 1929 when Victor merged with RCA and the two entities jointly manufactured phonographs sold under the RCA Victor name. RCA provided the Victor Company with the same two distinctly different types of apparatus during the twenties that they provided to Brunswick: 1) Radiola receivers which were either integrated with Victor Victrolas, or in one case, for use in a stand-alone Victor radio receiver (designated Model R-20), and 2) power amplifiers used to drive either the Victor Orthophonic horn using a special electric driver or RCA-supplied cone speakers. The power amplifiers designed specifically for Victor Electrolas do not qualify as Radiolas and are not addressed here.

All Victor apparatus which used Radiola receivers in the 1920's are listed in Table 11-3 and discussed briefly below. The Victor Victrolas are well documented as compared to Brunswick phonographs, and so the entries in this table were taken from the five Victor catalogs dated between December of 1925 and November of 1928,[32] and three books written by Robert W. Baumbach—two editions of *Look for the Dog*, and *Victor Data Book*.[33] Unfortunately, the data from the three sources do not always agree. The following paragraphs explain the entries in the table and identify some of the discrepancies.

While the model designations for the Victor Victrolas are somewhat more straightforward than for the Brunswick phonographs, there are certain subtleties that must be understood. Victor used a variety of terms to describe their Victrolas depending on the particular technology employed in each set. The three main categories of Victor instruments identified in the Victor catalogs are described below.

Table 11-3. Victor Victrolas and radios from the 1920's with factory-equipped Radiola receivers

Year	Model	Radiola Model	Radio Power	Reproducer	Motor	No. Mfg'd.	Price	N
1925-1926	Alhambra I (7-1)	20	Battery	Orthophonic	Spring	5,047	$350	1
1925-1926	Alhambra II (7-2)	25	Battery	Orthophonic	Spring	993	$425	1
1925-1926	Borgia I (9-3)	28	Battery	Orthophonic	Spring	1,595	$675	1
1925-1926	Borgia II (9-2)	28	AC	Orthophonic	Induction Electric	2,505	$1000	2
1925-1926	Florenza (9-1)	25	Battery	Orthophonic	Spring	3,984	$550	1
1925-1926	Florenza X (9-1)	25	Battery	Orthophonic	Induction Electric	20	$585	1
1925-1926	Florenza ? (9-1)	25	Battery	Orthophonic	Universal Electric		$605	1
1925-1928	Hyperion (15-1)	28	AC	Radiola 100 or 6" Dynamic Cone	Induction Electric	2,092	$900	3
1926-1928	VV 7-3	20	Battery	Orthophonic	Spring	9,187	$375 ($325)	1
1926-1928	VE 7-3 X	20	Battery	Orthophonic	Induction Electric	4,021	$410 ($360)	1
1926-1928	VE 7-3	20	Battery	Orthophonic	Universal Electric		$430 ($380)	1
1926-1928	VV 9-15	28	Battery	Orthophonic	Spring	833	$600	1
1926-1928	VE 9-15 X	28	Battery	Orthophonic	Induction Electric	1,747	$635	1
1926-1928	VE 9-15	28	Battery	Orthophonic	Universal Electric		$655	1
1926-1928	VE 9-40 E	28	AC	Orthophonic	Induction Electric	5,136	$1000	2
1926-1927	R-20 Radio only	20	Battery	Orthophonic	None	6,382	$135	4
1927-1928	VV 7-10	16	Battery	Orthophonic	Spring	5,829	$275	1
1927-1928	VE 7-10 X	16	Battery	Orthophonic	Induction Electric	725	$310	1
1927-1928	VE 7-10	16	Battery	Orthophonic	Universal Electric		$330	1
1927-1928	VV 7-25	17	AC	Orthophonic	Spring	4,072	$385	1
1927-1928	VE 7-25 X	17	AC	Orthophonic	Induction Electric	4,053	$420	1
1927-1928	VV 7-30	20	Battery	Orthophonic	Spring	3,881	$325	1
1927-1928	VE 7-30 X	20	Battery	Orthophonic	Induction Electric	181	$360	1
1927-1928	VE 7-30	20	Battery	Orthophonic	Universal Electric		$380	1
1927-1928	VE 9-25 E	28	AC	6" Dynamic Cone	Induction Electric	1,037	$1150	3
1927-1928	VE 9-55 E	28	AC	6" Dynamic Cone	Induction Electric	2,039	$1550	5
1928-1929	VV 7-11	18	AC	Radiola 100-A & Horn	Spring	12,109	$250	6
1928-1929	VE 7-11 X	18	AC	Radiola 100-A & Horn	Induction Electric	14,350	$285	6
1928-1929	VE 7-26 X	18	AC	Radiola 100-A	Induction Electric	7,622	$425	3
1928-1929	VE 7-26 A	18	AC	8" Dynamic Cone	Induction Electric	8,823	$475	3
1928-1929	VE 9-16 X	18	AC	6" Dynamic Cone	Induction Electric	2,109	$750	3
1928-1929	VE 9-18 X	64	AC	8 "Dynamic Cone	Induction Electric	3,126	$925	3
1928-1929	VE 9-54 E	64	AC	8 "Dynamic Cone	Induction Electric	2,941	$1350	5
1928-1929	VE 9-56 E	64	AC	8 "Dynamic Cone	Induction Electric	258	$1750	5

1. Orthophonic Victrola-Radiola
2. Orthophonic Victrola-Orthophonic Electrola-Radiola
3. Electrola-Radiola
4. Radiola
5. Automatic Electrola-Radiola
6. Victrola-Radiola

Orthophonic Victrola

Victor introduced their Orthophonic Victrola in 1925, and so all the Victrolas using the Orthophonic horn and sound box with acoustical amplification (as opposed to electric amplification) from 1925 to 1929 were described as *Orthophonic Victrolas*. These sets did not contain Radiolas, and so are not of interest to the subject of this book.

Electrola

Electrola was the term used by Victor for Victrolas with electrical amplification (i.e., power amplifiers)—whether the reproducer was the Victor Orthophonic horn activated by a magnetic transducer or a cone speaker supplied by RCA. *Orthophonic Electrolas* were Victrolas that had electrical amplifiers to drive the Orthophonic horn. Electrolas with radio receivers always had the Radiola name appended, such as *Electrola-Radiola*. Electrolas without Radiolas are not of interest to the subject of this book and are not treated further.

Radiola

Radiola by itself and in combinations with Victrola, Orthophonic, and/or Electrola were used to describe units equipped with RCA Radiola receivers—each combination describing a specific combination of reproduction technologies. *Radiola* by itself was used only for the R-20, the only stand-alone radio offered by Victrola prior to the merger with RCA in 1929. The *Orthophonic Victrola-Radiola* was a radio-Victrola combination that provided acoustical reproduction for the records directly through the Orthophonic horn, and electrical amplification for the Radiola receiver in conjunction with a magnetic transducer to drive the Orthophonic horn. The *Electrola-Radiola* provided electrical amplification for both the records and the radio in conjunction with a cone speaker supplied by RCA. The *Orthophonic Victrola-Orthophonic Electrola-Radiola* provided the option of acoustic reproduction or

The R-20, the only stand-alone Radiola offered by Victor, used the Radiola 20 circuit with a panel that had five tubes, one separated from the other four closely-spaced tubes. (George H. Clark Collection, Smithsonian Institution)

Much of the data regarding Victor Victrola-Radiola combinations was taken from the five annual Victor catalogs from 1924 through 1928.

electrical amplification of the records through the Orthophonic horn, and power amplification of the Radiola receiver to drive the Orthophonic horn. The *Automatic Electrola Radiola* was an Electrola Radiola that changed the records automatically. The *Victrola Radiola* provided acoustic reproduction of the records through an Orthophonic horn, and electrical amplification of the Radiola receiver with a power amplifier which drove with a cone speaker. The 7-11 was the only Victrola-Radiola combination in this category. This set is simply an aggregation of two independent units—a Radiola 18 with a Model 100 speaker and a Victrola with non-Orthophonic horn resembling apparatus from a Victor Victrola portable. A phonograph record and the radio could actually be played simultaneously.

The specific term applied to each instrument in the Victor catalogs from 1925 to 1929 is indicated by the number in the column to the far right of the table. The majority of sets by far were Orthophonic Victrola Radiolas until the 1928-29 season when all-electric Electrola Radiolas became the majority. The move away from the Orthophonic horn reflected the maturing of the dynamic cone speaker and the increased public interest in electrical amplification.

Model Year

The schema used for describing the model year used here is somewhat different than those used by Baumbach in his books, which do not always agree with each other, and on occasion do not agree with Victor catalogs and advertisements. Rather than attempting to identify specific years that the Victor sets were actually sold or shipped, the entries in Table 11-3 correspond to the Victor model year (or years) as determined by the year or years that each set appeared in the applicable annual catalogs published by Victor.

Victor published a catalog towards the end of each of the four calendar years which described the new sets for the four model years of interest. For example, the catalog dated December 1925 contained sets for the model year 1925-26, the catalog dated November 1926 contained sets for the model year 1926-27, etc. All sets listed in Table 11-3 appeared in catalogs for one model year only with two notable exceptions: 1) the Hyperion (Model 15-1) appeared for a record three models years (1925-26 through 1927-28), and 2) all Victrola combinations with Radiolas appearing in the catalog for the 1926-27 season were held over and appeared again in the catalog for the 1927-28 season.

Model Designations

Models in the new 1925 Orthophonic Victrola line were designated by names having a decidedly Spanish flavor such as Alhambra and Florenza. Subsequent models were assigned a one or two digit number followed by a dash and two additional numbers in lieu of a name, and indeed, even the named sets from 1925 were later given model numbers, presumably to give some sense of order to the entire Orthophonic line. It is highly unlikely that any of these numbers assigned retroactively ever appeared on Victrola name plates. Except for the notable exception of the Hyperion (later designated 15-1), the first number for all Victor Victrolas with Radiolas began with either a "7" or a "9." The price range for models beginning with a "7" ranged from $250 to $475, while prices for models beginning with a "9" ranged from $550 to $1750. Thus, models beginning with "7" were the "Fords" of Victrolas with Radiolas, while those beginning with "9" were the "Cadillacs."

While only the names or numbers were used in Victor catalogs and advertising for the Radiola Victrola combinations, the dealers and technical manuals used two prefix letters in front of the numbers, either VV or VE. The letters VV originally stood for Victor Victrola, but in the Radiola Victrola line the second letter also designated the type of motor—the V was used to indicate a wind-up spring motor while an E was used to indicate an electric motor. However, there were two types of electric motors used in many of these sets, a more expensive universal motor operating on both AC and DC current, and a less-expensive induction disk motor operating on AC current only. To distinguish between the two types of electric motors, an X or E suffix was added at the end of the model designation for units that contained the less expensive induction motor.

The prefixes and suffixes appearing in Table 11-3 are taken from Baumbach's *Victor Data Book*, with several notable exceptions. First, the suffix for both the VE 9-16 and VE 9-18 was X rather than E, and second, the suffix for the VE 9-56 was E rather than X which appears in the book. That the VE 9-56 had the E designation should be no surprise because it was identical to the VE 9-54 E except for the elaborate Chinese Chippendale cabinet used for the 9-56 E versus the walnut cabinet with a Spanish influence used for the VE 9-54 E. It should be noted that Baumbach did not mention the scheme—if any—that Victor used for distinguishing between the universal motor and the induction motor for the Florenza instrument, and so a question mark is entered in the table for the suffix representing the universal motor. Indeed, the Victor catalogs referenced here do not even mention the electric motors for this set. Since such a small number were shipped with an electric motor (approximately 20 versus 3,984 with a spring motor), perhaps there was no distinguishing suffix.

While there is no doubt that both the E and X suffixes were used to designate the induction motor, it is not clear why two different letters would be needed to designate the same induction motor. Robert Baumbach suggests that the "E" suffix appearing on sets between 1926 and 1928 was actually used to indicate electrical amplification, thereby distinguishing these sets from earlier sets with acoustical amplification. He goes on to explain that all sets with electrical amplification also had an electric motor, so an "X" suffix in addition to the "E" suffix already on these sets was not required. However this explanation is not entirely satisfactory, because while it is true that all sets with an "X" suffix had acoustical amplification, there were also sets with *electrical* amplification that had the "X" suffix as well (e.g.VE-26 X , VE 9-16 X).

Prices

The prices listed in Table 11-3 are the introductory prices as they appeared in the four Victor catalogs previously identified. The list prices for these sets did not change from catalog to catalog—with the notable exception of Model 7-3, which was reduced by $50 sometime during 1927. This change is indicated in Table 11-3 by the prices in parenthesis below the introductory prices for three Model 7-3 sets which have different prices due to the different types of motors. Other Victor Victrolas had price increases, but they were not equipped with Radiolas. For example, It should also be noted that the price of the Credenza model was increased by $25 from the time it was first advertised in the newspapers in late 1925 to the time it first appeared in a Victor catalog.

Many of the sets were offered with more than one type of motor—wind-up spring, Universal Electric (AC/DC), and/or Induction Disk Electric (AC only). Models with Induction Disk motors were uniformly priced $35 above models with the spring motor, while sets with the Universal Electric motor were priced $55 above models with the spring motor.

As a general rule, the higher the number, the higher the price and presumably the quality, as well. As with any rule, there are several minor exceptions such as the 7-30 priced at $325 while the 7-25 was priced at $385.

Number Manufactured

The number of sets manufactured is taken from the latest issue of *Look for the Dog*, which differs slightly from information in the *Victor Data Book*, but which presumably has the latest data. The data in this book are generally divided into sets with spring motors and electrical motors, but not subdivided further into sets with Universal Electric motors versus Induction Disk Electric motors. Consequently, production quantities for sets with the two different types of electric motors are lumped together as one entry in Table 11-3.

The Victor Radiola Hardware

The Radiolas supplied to Victor were similar to those supplied to Brunswick, although the sets made for Victor were given different AR numbers than the respective sets made for Brunswick, and had noticeably different markings and layout details. Excluding the 1924-25 season, the Radiola models supplied to Victor were the same as those supplied to Brunswick with several notable exceptions. First, since Victor focused on the Orthophonic sets while Brunswick focused on all-electric sets, Victor used more types of battery sets than Brunswick. In particular, Victor used battery-operated Radiolas 16 and 20 in four different Victrolas and the R-20 radio, whereas Brunswick did not use these two Radiola models in any of their sets. Second, since Brunswick made a number of stand-alone radios in the late 1920's whereas Victor did not, Brunswick used the Radiola 60 and 62 and Victor did not. Brunswick also used the Radiola 62 in one Victrola-radio combination, whereas Victor did not. Both used the Radiola 64 in Victrola-Radiola combinations.

Radiolas supplied to Victor were electrically similar but physically different than both the respective Radiola models RCA provided to Brunswick and the models RCA sold to the public. The differences were most obvious on the Victor panels utilizing the Radiola 20, 25, and 28 circuits. These panels were square in shape and designed such that the tubes extended through the face of the panel to facilitate replacement, but unlike the Brunswick panels, there were no tube covers. As a result, one can easily distinguish among the various panels at a glance based on tube count. The general rule of thumb is as follows: panels with 5 and 6 tubes spaced close together are based on the Radiola 25, while panels with 7 and 8 tubes spaced close together are based on the Radiola 28. Panels with an even number of tubes are battery operated, while panels with an odd number of tubes are AC operated. As explained previously, the missing tube on each of the two AC catacombs is the output tube, which was located on the power amplifier chassis.

The panel made for Victor based on the Radiola 20 is notable because the use of a Radiola 20 model was unique to Victor. It was used on three different Victrola-Radiola combinations—Victor Models 7-1, 7-3, and 7-30. The Victor Radiola 20 panel had an appearance similar to the Radiola 25 with the large thumb wheels and five tubes protruding through the face of the panel. The

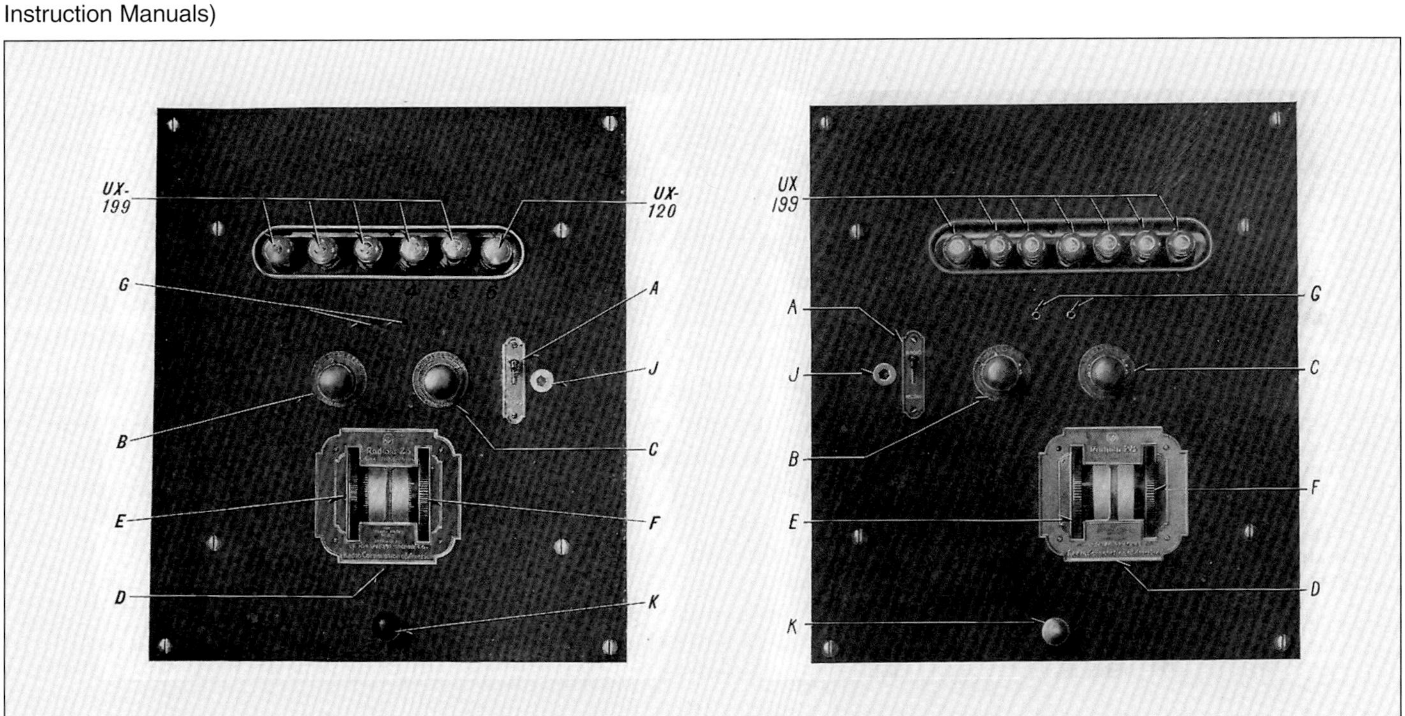

The Victor panel from the Florenza Victrola-Radiola (left) combination has six closely-spaced tubes visible from the top of the panel, readily identifying it as a Radiola 25 (which had 5 or 6 tubes) and battery operated (which had an even number of tubes). The Victor panel from the Borgia II Victrola-Radiola (right) combination has seven closely-spaced tubes visible from the top of the panel, readily identifying it as a Radiola 28 (which had 7 or 8 tubes) and AC operated (which had an odd number of tubes). (Victor Borgia II and Victor Florenza Instruction Manuals)

The Victor Hyperion Victrola-Radiola combination used the same massive Radiola 28 panel that RCA offered in their Radiola 28, rather than the new smaller panel offered by RCA specifically for Victor Victrolas such as Borgia I and Borgia II. (George Clark Collection, Smithsonian Institution)

Radiola 20 panel can be most easily distinguished from Victor's Radiola 25 panel by the large separation of one tube from the other four (see panel in the photo of the R-20 radio). In contrast, the five tubes appearing on the Radiola 25 are all spaced close together. The spacing of the fifth tube on this panel mimicked the spacing on the chassis of the Radiola 20 sold by RCA.

The Radiola 16, unique to Victor, was used in only one Victrola-Radiola model, the battery-operated Victor Model 7-10. The radio panel in this set was similar to the panel in the Radiola 16 marketed by RCA, differing only in the width of the panel and the presence of Victor Talking Machine on the name plate. The Radiola 17 and 18 used by Victor in four different Victrola-Radiola sets were unremarkable in that the panels used in the Victrola were similar to the panels on the respective sets marketed by RCA.

No discussion of the Victor Victrola-Radiola combinations would be complete without commenting on the Hyperion, the first all-electric Victrola-Radiola combination introduced by Victor circa February 1926. The main production used an RCA Model 100 permanent magnetic speaker, but late in the production the 6" electrodynamic speaker used in the RCA Model 104 speaker was used in the Hyperion. The massive Hyperion was based on RCA's Radiola 28, but it did not use the square Radiola 28 panel used on subsequent Victor models such as the Borgia II, a set introduced in the same year as the Hyperion. Instead, the Hyperion incorporated the Radiola 28 chassis mounted on the slopping wood panel found in the Radiola 28 that RCA distributed.

Endnotes and References for Chapter 11

1 "Phonograph Sales Drop," *The New York Times*, June 18, 1922, p. 40; "Radio Now Linked with Phonograph," *The New York Times*, May 3, 1924, p. 24.

2 *The New York Times*, May 3, 1924, p. 24.

3 Alexander B. Magoun, *Shaping the Sound of Music: The Evolution of the Phonograph Record, 1877-1950*, (PhD Dissertation, University of Maryland, 2000); available from UMI Dissertation Services, Ann Arbor Michigan.

4 Roland Gelatt, *The Fabulous Phonograph*, (Appleton-Century, New York, 1965); another version that should not be missed can be found in Chapter 8 of *The Victor Talking Machine Company* written by Benjamin L. Aldridge, who worked at Victor from 1918 until he retired at age 65. An on-line version of this rare book published in 1964 can be found on the David Sarnoff Library web site at www.davidsarnoff.org.

5 In addition to writings of well-known participants such as Eldridge Johnson, Raymond Sooy, Benjamin Aldridge, and David Sarnoff, Magoun references extensive writings of RCA Engineer Elmer Bucher which bear directly various facets of RCA's attempts to combine Radiolas with phonographs. Bucher's writings are contained in a number of volumes deposited at the David Sarnoff Library, Princeton, New Jersey—the first four volumes written in the early 1940's, which draw from the RCA archives that no longer exist, being the most useful.

6 Morgan E. McMahon, Vintage Radio, (Vintage Radio, Rolling Hills, Estates, CA, 3rd Edition, 1981), p. 67.

7 Robert W. Baumbach, *Victor Data Book*, (Mulholland Press, Inc., Los Angeles, CA, 2003), p. 157.

8 Robert W. Baumbach, *Look for the Dog*, (Mulholland Press, Inc., Los Angeles, CA, Revised Limited Edition, 2005), p. 108.

9 "Offer 130,000 Shares of Brunswick Balke," *The New York Times*, November 27, 1924, p. 33; "New Contract for Radio," *The New York Times*, March 30, 1928, p. 34.

10 Victor Advertisement, *The New York Times*, November 28, 1924, p. 11; *Victrola*, Victor Catalog Form 7461—TQA—10-24, October 1924; Victrola No. 210 was not specifically designed for a radio, but a number of companies built radio panels for this model.

11 Robert W. Baumbach, *Victor Data Book*, pp. 179, 208-209.

12 "Victor Co. Defers Common Dividend," *The New York Times*, June 19, 1925, p. 28.

13 "Victor Co. to Enter the Radio Business," *The New York Times*, May 20, 1925, p. 27.

14 "Invent Disk to Play Whole Symphonies," *The New York Times*, August 13, 1925, p. 22.

15 "Victor Co. Produces A New Record Also," *The New York Times*, August 14, 1925, p.14.

16 "Tells Details of Aid Radio Gave Victor," *The New York Times*, August 30, 1925, p. 14.

17 "This Victor Day," *The New York Times*, November 2, 1925, p.17.

18 "Music Reproducer Meets Severe Test," *The New York Times*, November 12, 1925, p. 15.

19 Robert W. Baumbach, *Look for the Dog*, (Mulholland Press, Inc., Los Angeles, CA, 6th Printing 1999), p. 156.

20 Roland Gelatt, *The Fabulous Phonograph*, p. 325.

21 Benjamin L. Aldridge, *The Victor Talking Machine Company*, Edited by Frederic Bayh (RCA Sales Corporation, Camden, New Jersey, 1964), p. 95.

22 Benjamin L. Aldridge, *The Victor Talking Machine Company*, pp. 94-95.

23 Aldridge used the term "magnetic speaker" for one with a permanent magnet, and "dynamic speaker" for one with an electromagnet. Victor's earliest all-electric models used the magnetic speaker. However, contrary to Aldridge's assertion about the dynamic speaker first appearing in September, it actually first appeared circa April 1926 in the Tuscany model.

24 For example, see the following ads placed by the Aeolian company and the Brunswick Salon for Panatropes: *The New York Times*, December 16, 1925, p. 17; *The New York Times*, December 20, 1925, p. 26; *The New York Times*, December 21, 1925, p. 15; *The New York Times*, December 22, 1925, p. 12.

25 "The Panatrope," *The New York Times*, December 21, 1925, p. 15.

26 Victor placed an ad for the all-electric Hyperion Victrola (with a permanent magnetic speaker) in the November 21, 1925 edition of the *Saturday Evening Post*, which, because of Aldridge's comment that shipments of sets with the magnetic speaker were delayed until February, does not imply that Hyperions were actually delivered at the time of the ad.

27 Victrola Ad, *The New York Times*, November 22, 1925, p. 23.

28 "A Victor Announcement," *The New York Times*, December 28, 1925, p. 15.

29 *Brunswick Panatrope DeLuxe Styles*, Brunswick Pamphlet, Form 794 FPM, March 8, 1926.

30 *Brunswick Panatrope and Radiola*, Brunswick Pamphlet, Form 1030-A2 ADM , November 20, 1926.

31 Aeolian Company Ad, *The New York Times*, December 15, 1926, p. 17. Note that one set in this ad was identified as PR-6C while the same set was identified as PR-6 without the C in the Brunswick pamphlet.

32 *The New Victor Line*, Catalog Form No. 8400, December 1925; *The New Orthophonic Victrolas Electrolas and Radio Combinations*, Catalog Form No. 88976-JQA, November 1926; *The New Orthophonic Victrolas Electrolas and Radio Combinations*, Catalog Form No. 99579-TQA, August 1927; *The New Orthophonic Victrolas Electrolas and Radio Combinations*, Catalog Form No. 9719-JQA, November 1927; *Victor Presents*, Catalog Form No. 762-IQA, November 1928.

33 Robert W. Baumbach, *Victor Data Book*; Robert W. Baumbach, *Look for the Dog*, (6th Printing, 1999); Robert W. Baumbach, *Look for the Dog*, (Revised Limited Edition, 2005).

In 1930, RCA consolidated its engineering, manufacturing and marketing activities at the Victor Talking Machine Co. plant with 34 buildings in Camden NJ. Note the woman modeling for this picture in the phonograph cabinet on the roof. (George H. Clark Collection, Smithsonian Institution)

Afterwords

1930-1935

In the first decade after incorporation, RCA transformed itself from a relatively insignificant company with annual revenues of less than $2 million derived from commercial transoceanic and marine communication services in 1920 to an industrial giant in 1929 with annual revenues of $182 million—the majority of which were derived from sales of home entertainment and sound motion picture equipment rather than from its original business base of worldwide commercial communications. The largest segment of this business by far was the sale of broadcast radio apparatus consisting of receivers and vacuum tubes (see Table AF-1).

Table AF-1. RCA revenue from sale of radio apparatus[1]

Year	Revenue from Gross Sales	Total Revenues
1920	$520,000*	$1,920,000*
1921	$1,468,919	$4,160,844
1922	$11,286,489	$14,830,856
1923	$22,465,090	$26,394,789
1924	$50,747,202	$54,848,131
1925	$46,251,785	$50,405,144
1926	$56,093,504	$61,157,286
1927	$56,651,658	$65,418,620
1928	$86,900,152	$101,851,603
1929	Not reported	$182,137,738

*Calculated from data appearing in the 1921 Annual Report

Despite the incredible success RCA experienced during this period, it was plagued by inefficiencies that were caused first by having three separate organizations (RCA, GE, and Westinghouse) responsible for the design and selection of the apparatus, and then by having both GE and Westinghouse responsible for manufacturing the apparatus in the ratio of 60%-40% respectively. RCA had solved the problem of how to meet the required ratio by having both companies manufacture the same equipment to the same drawings in the specified ratio, but this approach had the effect of increasing the inefficiencies even further.

By 1927, RCA had begun to explore the concept of consolidating engineering and manufacturing in one location. RCA, GE and Westinghouse were all in agreement for the need to consolidate operations, and formed a committee on October 27, 1927 headed by David Sarnoff to study the matter.[2] On April 14, 1928, this committee recommended that a new company be formed—suggesting the name Radio Manufacturing Company—in which GE and Westinghouse would own all of the common stock in a proportion of 60% to 40% respectively, with RCA having the right to purchase up to 20% of the stock.

The idea of consolidating manufacturing at the Victor Talking Machine Company in Camden, New Jersey was an obvious one based on RCA's ongoing relationship with Victor in which RCA had been supplying radio equipment for

Victor Victrolas since 1925. The fit was good, not only because RCA needed a location with existing manufacturing facilities such as those at Camden to consolidate manufacturing, but also because Victor needed RCA's radio technology and patent pool for its Victrola products, which were becoming increasingly more dependent on radio receivers, amplifiers, and electrical pickup mechanisms for their Victrola and record business.

RCA and Victor had preliminary discussions about combining forces in late 1927 and early 1928 which led to a meeting between David Sarnoff and Victor Company President E. E. Shumaker in July of 1928. At that meeting, Shumaker proposed to David Sarnoff that RCA buy the entire Victor Company—including all of the production facilities and test equipment, a complete cabinet factory, and extensive marketing channels in the phonograph business throughout the world. RCA would also have access to an extensive pool of talent under contract to Victor for making records that RCA could use in their broadcast and sound picture activities.

RCA Acquires Victor Talking Machine Company

After a period of negotiations between RCA and Victor, an agreement was eventually struck and RCA shareholders approved the acquisition of Victor at a shareholder meeting on February 27, 1929. RCA gained controlling interest of Victor on March 15, 1929 at a price of $154 million with a combination of RCA stock and $32 million in cash. The $32 million was loaned to RCA by GE and Westinghouse, of which approximately $19.5 million was used to retire the preferred stock of Victor and the remainder was an advance to Victor for working capital, presumably to expand and improve the facilities. The price was high considering that the public had paid only $53 million for Victor two years before, but GE provided the cash and RCA contributed only stock. Fortunately, RCA's stock—often cited as one of the hottest stocks of this period—had also been advancing rather smartly, offsetting the impact of the increase in Victor stock.

Two separate companies were formed in April 1929 to conduct the business of RCA's radio merchandise Sales Department and the former Victor Talking Machine Company: the Radio-victor Corporation of America (victor with a small "v") and the Audio Vision Appliance Company (AVA).[3] Radio-victor became a wholly-owned subsidiary of RCA organized as a Maryland corporation on April 25, 1929, while AVA became a wholly owned subsidiary of GE and Westinghouse organized as a New Jersey corporation on April 29, 1929. AVA was owned in the ratio of 60%-40% by GE and Westinghouse respectively, with a yet-undetermined fraction of ownership to accrue to RCA. The assets and operations of the Victor Company were transferred into AVA with the exception of the Victor marketing organization, which was integrated into Radio-victor along with the commercial marketing organization of RCA.

While RCA had acquired Victor, it still had not achieved its goal of unifying engineering, manufacturing, and marketing. Where there had been three separate organizations, there were now five, causing an increase in chaos.[4] Even the marketing organizations of Victor and RCA in Radio-victor acted independently, each selling their respective products through their respective sales channels. The pressure increased to unify radio research, engineering, manufacturing and marketing.

David Sarnoff Unifies Radio R&D, Marketing and Manufacturing

During the summer of 1929, David Sarnoff pressed for full unification of the radio business within the highest management circles of RCA. On October 4, 1929, the RCA Board of Directors approved in principle the complete unification of radio research, engineering, manufacturing, and marketing activities at Camden. Before the end of the year, both GE and Westinghouse also concurred with the unification plan, and to this end the entire company was restructured. On December 26, 1929, the RCA Victor Company, Inc. was organized as a wholly-owned subsidiary of RCA, and began doing business on January 1, 1930. RCA Victor took over the assets and responsibilities of both AVA and Radio-victor, and the remaining assets of the Victor Talking Machine Company. AVA was formally dissolved on November 7, 1930, and Radio-victor was dissolved on July 14, 1931. All research, manufacturing and marketing operations were moved to Camden, requiring the transfer of a number of key RCA, GE and Westinghouse engineers—the GE and Westinghouse engineers becoming RCA employees.

The reorganization of RCA did not stop with RCA Victor. RCA also incorporated the RCA Radiotron Company, Inc. on December 26, 1929 as a wholly owned subsidiary to research, design, manufacture and market Radiotron vacuum tubes. The vacuum tube plants of Westinghouse at Indianapolis, IN and GE at Harrison, NJ became part of Radiotron effective January 1, 1930. The research activities of Westinghouse at East Pittsburgh, PA and GE at Schenectady, NY were transferred to the Radiotron tube plant at Harrison, which became the headquarters of Radiotron.

As a result of the unification and reorganization, there was a wholesale transfer of assets and personnel from GE and Westinghouse to RCA, including partial ownership in existing subsidiaries. In addition, income from licenses and the 20% profits that accrued to GE and Westinghouse as a result of the original agreements with RCA to manufacture receivers and tubes now accrued directly to RCA. Finally, Westinghouse and GE agreed to cancel the $32 million advanced to RCA to purchase Victor, plus accrued interest. In order to balance accounts, there was a wholesale change in the ownership of RCA in favor of GE and Westinghouse. According to the RCA Annual Report for 1930, Westinghouse and GE received a total of 6,580,375 shares of Radio Corporation of America Common Stock—amounting to the total of all the Common Stock issued and outstanding at the time! The shareholders approved the authorization of 15 million common shares on May 6, 1930, the final step necessary to consummate the deal. This transaction constituted one of the largest transfers of assets and ownership in corporate America.

RCA Becomes a Holding Company

As a result of restructuring and the creation of new subsidiaries, RCA announced to the shareholders in the 1930 Annual Report that it had "*become largely a holding company...engaged in coordinating its diversified interests in every field of radio development, in exercising general supervision over financial matters and research activities, in administering patent licenses and the collection of royalties and in protecting the broad legal rights of its various subsidiaries.*" In this report RCA also published an RCA Organization Chart identifying nine subsidiaries with a listing of the primary activities and responsibilities of each.

Such a dramatic change in the organization and operations of the Company called for an equally dramatic shakeup in personnel assignments. At the January 3, 1930 Board Meeting of RCA Directors, long-time Chairman of the Board Owen D. Young resigned and was elected as Chairman of the newly created Executive Committee. General James Harbord then resigned as President of RCA and was elected the new Chairman of the Board, whereupon Vice-President David Sarnoff was elected President of RCA. David Sarnoff was also Chairman of the Board of the three subsidiaries involved in broadcast radio products—RCA Victor Company, Inc., RCA Radiotron Company, Inc., and E. T. Cunningham, Inc. E. E. Shumaker, the previous President of Victor Corporation became the President of RCA Victor. Thus, David Sarnoff at age 38 became one of the most powerful executives at RCA.

RCA ORGANIZATION CHART

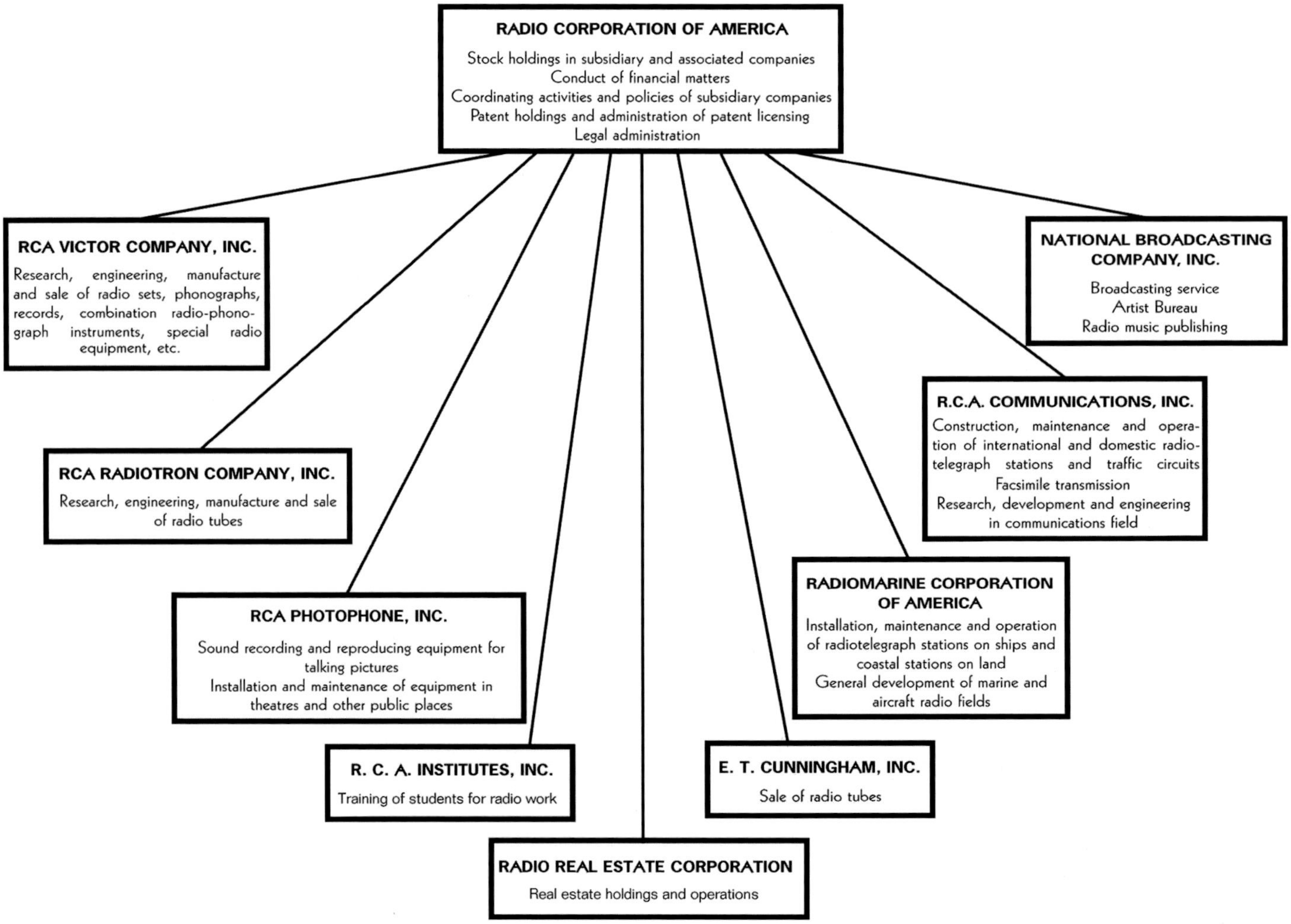

As a result of the restructuring and creation of new subsidiaries, RCA became a holding company by the end of 1930 with a total of nine subsidiaries. (RCA Annual Report for 1930)

RCA Becomes Independent as a Result of Antitrust Litigation

The acquisition of the Victor Talking Machine Company and the unification under RCA did not go unnoticed by the Department of Justice. Even before the acquisition of Victor and unification at Camden, there were a number of influential people in both government and industry that believed the cross-licensing agreements among RCA, GE Westinghouse, and AT&T and others constituted an illegal monopoly. There had been a long history of allegations and several investigations into the monopoly issue. As early as 1922, the Federal Trade Commission had been investigating a complaint charging that GE had set up RCA in an attempt to establish a monopoly for the sale and manufacture of radios. Just before this investigation was completed, the House of Representatives of the Sixty-seventh Congress passed House Resolution 548 on March 3, 1923 directing the Federal Trade Commission (FTC) to obtain information required to help Congress enact legislation regulating the radio industry.[5] The FTC was directed to gather and report on facts with regard to ownership of patents, contracts and leases designed to restrict the manufacture and sale of apparatus or give exclusive rights in reception and transmission of interstate communications, and to supply information regarding potential violations of antitrust legislation. This legislation caused the FTC to extend and widen its investigation and to focus more on RCA.

In the end, the FTC spent almost two years investigating RCA and associated companies, during which time RCA opened its records—correspondence, files, meeting minutes, contracts, and license agreements to the FTC.

The FTC issued its long awaited report with their findings on December 1, 1923,[6] and while RCA characterized the report as being highly complimentary to RCA in many respects, still the FTC filed a complaint in January of 1924 alleging RCA and its associated companies used unfair methods of competition against their rivals. The hearings associated with this complaint dragged on for another four years, with extensive testimony taken all over the United States which ultimately filled over 4000 pages.[7] Finally, on December 19, 1928, the FTC dismissed its complaint against RCA and associated companies on charges of a monopoly and unfair competition.

However, by 1930 the mood of the country had changed. The country was in a depression, there was a new administration in the White House,[8] and the Justice Department had just won an antitrust suit against Standard Oil of Indiana and other gasoline companies for patent pooling—circumstances not unlike those involving RCA and the associated radio companies. The ink had hardly dried on the final agreements consummating the RCA reorganization when the Justice Department filed an antitrust suit against RCA on May 13, 1930 naming RCA, GE, Westinghouse, AT&T and others as co-defendants.[9] It was ironic that the government was instrumental in setting up RCA as a monopoly in 1919—a fact the government conceded—and now here they were prosecuting RCA for the very cross-licensing agreements they encouraged a short ten years earlier.

The legal proceedings progressed very slowly, and while both sides were preparing for trial, the Supreme Court reversed the decision against Standard Oil on April 13, 1931.[10] Suddenly, the Justice Department became as interested in settling as the defendants. In any settlement, the government insisted that GE and Westinghouse divest themselves of their stock holdings in RCA, and that the patent agreements be rearranged so that Westinghouse and GE could compete with RCA—but all agreed it had to be done in such a way that RCA remained viable in order to protect its 85,000 shareholders.

Negotiations dragged on well into March of 1932, at which time the Justice Department increased the scope of the action by including international activities and naming subsidiaries of RCA, GE and Westinghouse involved in foreign activities as additional defendants.[11] Negotiations among RCA, Westinghouse, GE and the Justice Department intensified but still dragged on for much of 1932. Trial dates were set and subsequently postponed as negotiations continued. A trial date of October 8, 1932 was set and then postponed "*to afford representatives of the government and the radio group to consider new proposals designed to meet some of the principal points at issue.*"[12] Another trial date was set for November 15, but once again postponed to November 21—this time with a stipulation that if a decree was not submitted, the case would go to trial on November 21.[13]

The issues under negotiation were complex, revolving around how Westinghouse and GE would divest their respective interests in RCA without depressing RCA stock, how Westinghouse and GE would be adequately compensated, how RCA could be relieved of the substantial debt it had incurred, and how the patent agreements would be modified so all could compete fairly in the future. The story of the intense negotiations that went right down to the last minute before the scheduled trial is given in a riveting account by Gleason Archer in his book *Big Business and Radio*.[14]

A final settlement was reached just in time, and a consent decree dated November 21, 1932 was signed by all parties.[15] Details of the final settlement appeared in both the RCA and GE Annual Reports for 1932, and also in *The New York Times* and Archer's book.[16] The key provisions can be summarized as follows. RCA gave up its exclusivity such that Westinghouse, GE and AT&T could compete with RCA but only after a two and one half year period following the decree to give RCA time to adjust to the new competition. The licenses for manufacturing radios were still available to all, but each had to pay appropriate royalties to the owners for their use. RCA obtained license agreements to manufacture transmitters and transmitter tubes—rights they did not have before. RCA transferred ownership of the building at 570 Lexington in New York City, thus relieving them of $200,000 in annual operating losses. RCA was released from approximately $18 million in debt through a combination of cancellation of debt, issuance of debenture bonds, and transfer of real estate. Westinghouse and GE were required to liquidate their holdings in RCA stock in such a manner that it did not depress the stock price. RCA was able to keep all of the gains it had made during unification in 1930—including cancellation of the $32 million debt incurred in the Victor acquisition. Westinghouse and GE retained the stock they had received during unification in 1930, but the price in 1932 at the time of the consent decree was only $6—versus the $40 share price at the time of unification.

At the meeting of the Board of Directors on November 22, 1932, all Directors associated with outside radio companies resigned—with the exceptions of Owen D. Young, Chairman of the Board of General Electric, and Andrew W. Robertson, Chairman of the Board of Westinghouse, who were allowed by the terms of the consent decree to remain. As a result, the overwhelming majority of the Board consisted of outside independent directors, and at this point RCA finally became a truly independent company.

Wall Street analysts believed that RCA was stronger as a result of the split-up, even though it no longer had the financial backing of GE or Westinghouse.[17] RCA retained patent rights to radio equipment of the two companies albeit on a non-exclusive basis, but it was still the only company that could offer complete licenses on all required radio patents. RCA's debt and building lease was cancelled, leaving it in a solid financial position. Neither GE nor Westinghouse could manufacture radio equipment until May of 1935, leaving RCA free from additional competition for several years. It was clear that RCA got the better end of the bargain in the process, largely due to David Sarnoff's vision and ability to persuade. David Sarnoff had guided RCA through a very difficult period in RCA's history, and he was now the man of the hour. The history of RCA for the next forty years would continue to be closely associated with his name.

Radio Receiver Marketing and Manufacturing Undergo Major Changes

On April 26, 1929—shortly after the formation of Radio-victor Corporation of America—RCA began to use Radio-victor Corporation of America in lieu of the more familiar Radio Corporation of America on advertising, sales brochures, and other documents associated with Radiolas. This name change first appeared in ads beginning in June 1929,[18] but by the end of the year it found its way on all documents and even onto radio receiver instructions and name tags affixed to the apparatus. Despite the name change, RCA retained the ubiquitous RCA monogram with RCA letters and the zigzag lightning flash within the circle—which in later years was often referred to as the "meatball" logo. Radio manufacturing was not affected by this name change—both GE and Westinghouse continued to manufacture sets for RCA in the usual ratio of 60%-40%.

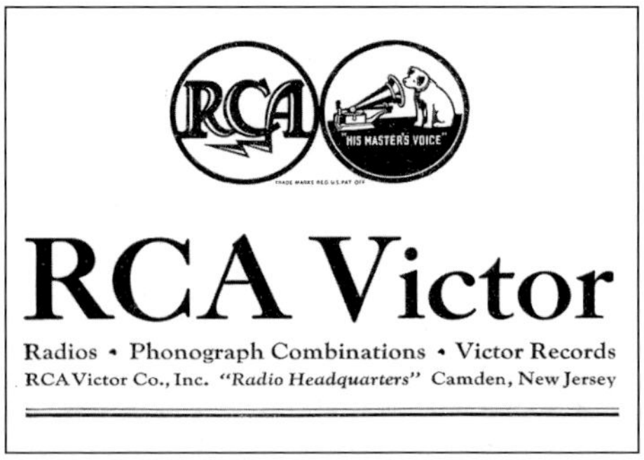

RCA Victor created its new logo by combing the RCA "meatball" logo with the Victor Nipper dog logo. (*Time*, July 3, 1933, p. 1)

The Radio-victor name was short-lived because on December 26, 1929, the RCA Victor Company, Inc. was created as a wholly owned subsidiary of RCA, and began to do business under that name on January 1, 1930. Almost immediately, the name was changed once again on documents and name tags associated with home entertainment equipment including Radiola receivers—this time from Radio-victor to RCA Victor. The ubiquitous RCA meatball monogram was retained, joined by the equally ubiquitous Victor logo consisting of the Gramophone and Nipper dog shortly after consolidation was completed—which involved a wholesale transfer of personnel from GE and Westinghouse to Camden. One of the first, if not the first, use of the combined RCA-Nipper monogram was on RCA's consolidation announcement which

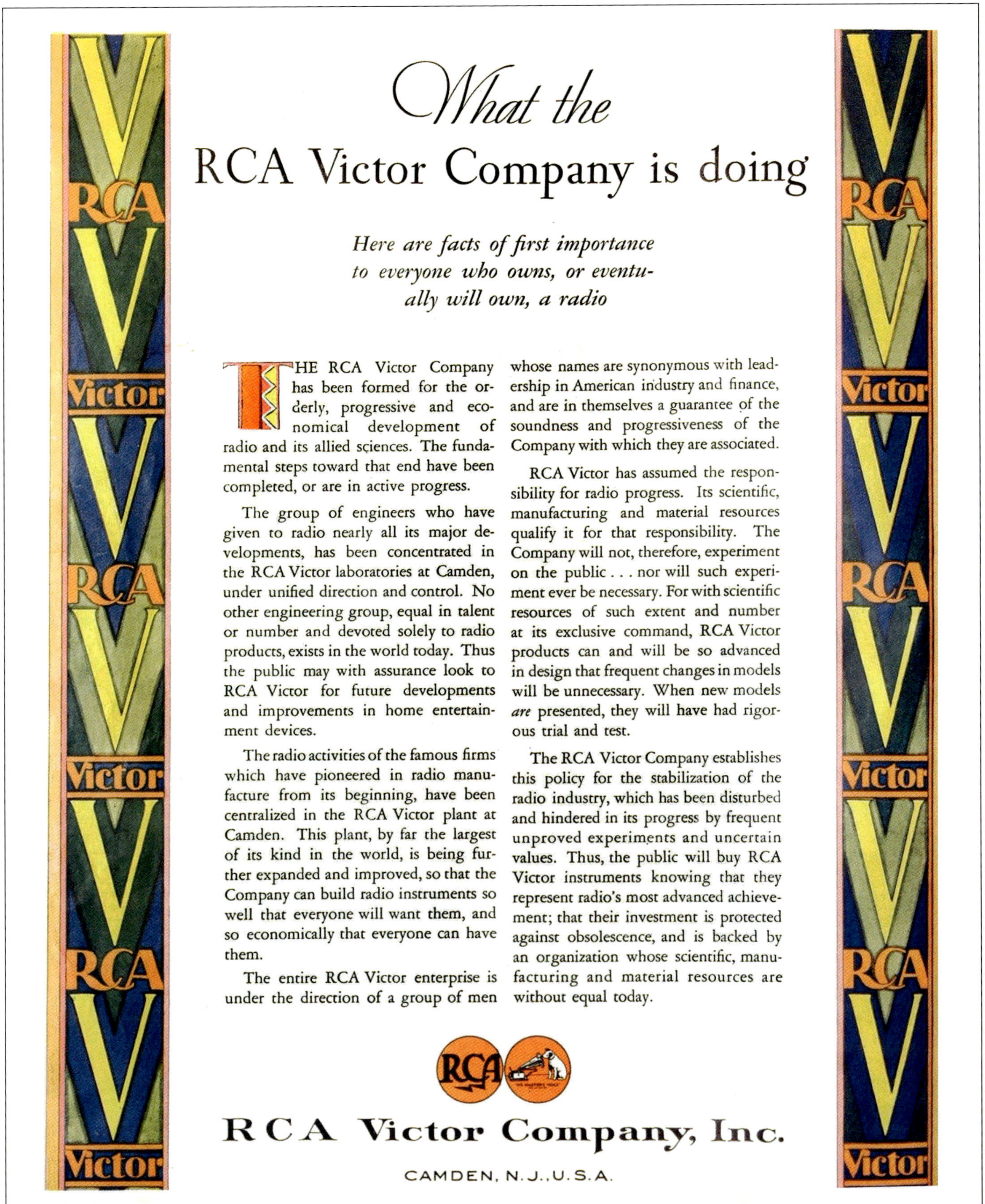

The first public announcement in June 1930 upon completion of RCA's consolidation in Camden is interesting not only because of the early appearance of the combined RCA-Victor logo, but because it indicated that the consolidation process—which began in January 1930—took approximately six months to complete. (*Saturday Evening Post*, June 21, 1930, back cover)

This image was part of a four-page ad placed by GE in the *Saturday Evening Post* introducing their first broadcast radio receiver line with the GE name to the public. (*Saturday Evening Post*, September 6, 1930, pp. 62-65)

appeared in national magazines in June of 1930.[19] While the combined logo appeared as early as mid-1930, it did not appear regularly on RCA Victor ads until 1931, and was not registered as a trademark until April 25, 1933.[20]

By the terms of the consolidation agreements among RCA, GE and Westinghouse, RCA Victor began to make receivers for all three entities beginning with the 1930-31 radio season. RCA Victor continued to market its receivers under the Radiola name, while GE and Westinghouse began to market their receivers under their respective company names but with different model numbers and cabinet styles. Both Westinghouse and GE took out large ads in the September 1930 *Radio Retailing* magazine to introduce their new lines to the trade. GE promised "*the largest radio announcement that has ever appeared in the Saturday Evening Post and Collier's*. Its ad strategy was to emphasize quality associated with the GE brand name and ubiquitous monogram within a circle. Westinghouse announced "*the biggest publicity-getting and sales-building idea of the season*" with a $10,000 Radio Idea Contest. Its ad strategy was a "radio-cabinet-design" contest which could only be entered at Westinghouse radio dealers—thereby bringing potential customers into dealer

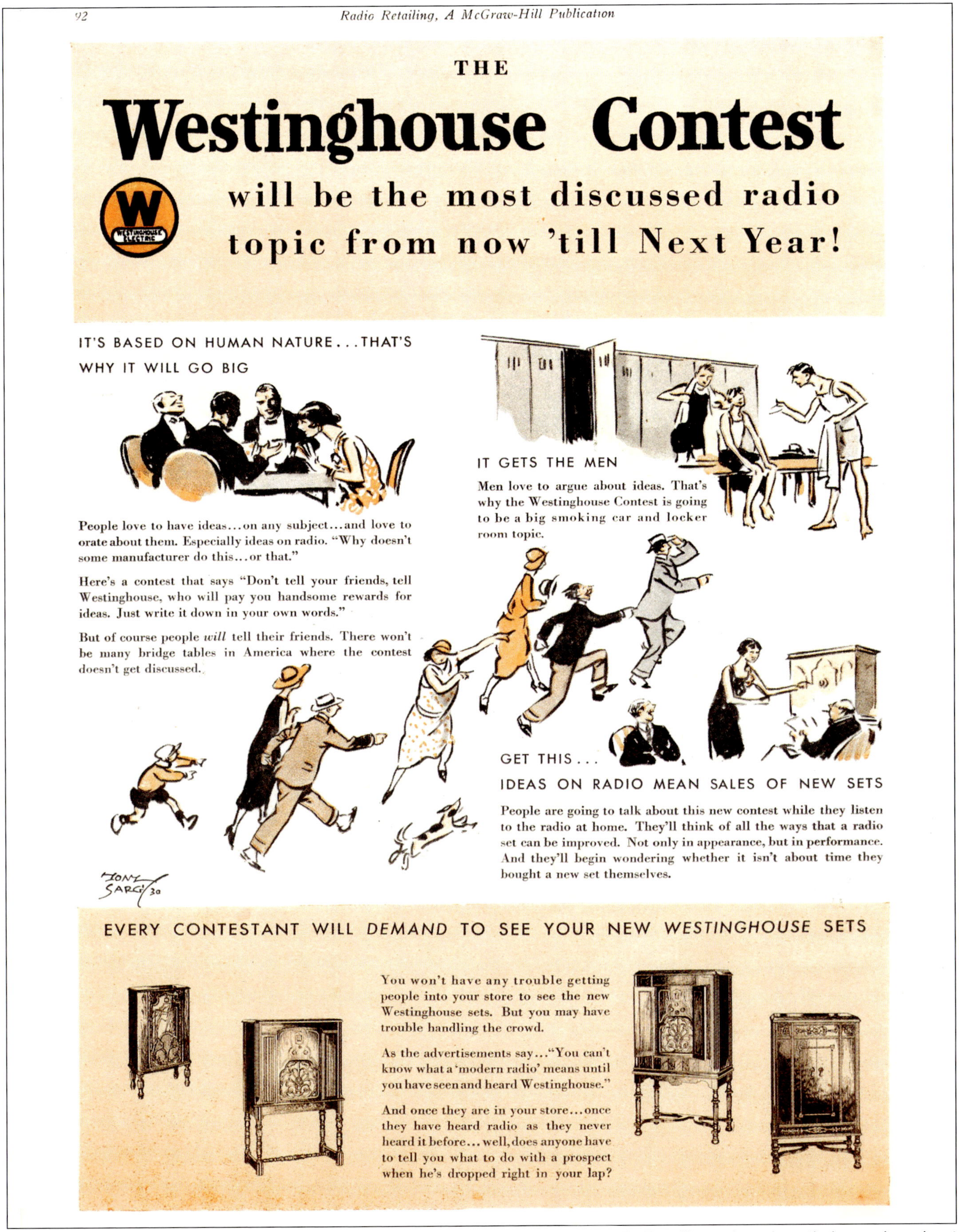

This image was the first page of a four-page ad placed by Westinghouse in *Radio Retailing* to introduce their new broadcast receiver line with the Westinghouse name to the trade. (*Radio Retailing*, September 1930, pp. 90-93)

Table AF-2. RCA Victor Radiolas manufactured in Camden and equivalent GE and Westinghouse models

RCA Victor Radiola Model	GE Model	Westinghouse Model	Receiver Type	Receiver Power
21	B-1	-	Screen Grid TRF	Battery
22	B-2	-	Screen Grid TRF	Battery
42	-	-	Screen Grid TRF	Socket power
48	T-41	WR-4	Screen Grid TRF	Socket power
80	H-31	WR-5	Superheterodyne	Socket power
82	H-51	WR-6	Superheterodyne	Socket power
86	H-71	WR-7	Superheterodyne	Socket power

stores. Notably, neither company's ads made any reference to the fact that RCA would be manufacturing their radios.

Models based on Radiola designs marketed by the three companies in the 1930-31 radio season are shown in Table AF-2. The RCA Victor Radiola models are shown in the first column while the equivalent GE and Westinghouse models are shown in the second and third columns. This season would be the last that RCA Victor would use the Radiola name exclusively for all its receivers. Receivers for the 1931-32 season would be advertised as RCA Victor sets with model names beginning with an R, RE, or RAE followed by a number—for example RCA Victor RAE-26.

RCA Victor continued to design and manufacture sets for GE and Westinghouse, but as a part of the consent decree dated November 21, 1932, GE and Westinghouse agreed to divest themselves of all ownership interests in RCA, and further, agreed not to manufacture radio hardware for a period of two and a half years. Thus, RCA had several years to develop a first class industrial complex in Camden to research, design, and manufacture radios without additional competition from GE or Westinghouse. The RCA models and equivalent GE and Westinghouse models using radio equipment manufactured by RCA during this period are shown in a cross-index on Page XI in the front of the RCA Victor Service Data Book, Volume 1.

GE began to manufacture its own receivers for sale in the fall of 1935.[21] One of the first ads for the new line manufactured by GE appeared in the *Saturday Evening Post* in August of 1935 featuring Models A-63 and A-125.[22] Likewise, Westinghouse also began to manufacture their own receivers in the same time frame, with the first national ads appearing circa September 1935.[23] RCA did not appear to suffer from the new competition, and all three competed successfully in the broadcast receiver market for many decades, first in radio and later in television.

Thus ended the Golden Age of Radiola—the decade of the twenties in which RCA had focused almost exclusively on Radiola radio receivers. A new age began with the decade of the thirties, one in which RCA through the RCA Victor organization expanded from radio receivers into phonographs, records, combination radio-phonograph instruments, special radio equipment and eventually television. While the Radiola name would occasionally appear on radios for decades to come, the RCA Victor name and logo would take center stage.

OCTOBER 17, 1931 THE LITERARY DIGEST 1

Announcing sensational new
RCA Victor Radio-Phonograph

Featuring new record that plays 30 minutes (about 4 times as long as present type) and FIVE other revolutionary developments . . . all at the former price of a radio alone!

1

New 9-tube Super-Heterodyne Radio with Pentode Tubes and Automatic Volume Control.

2

New electrically recorded and amplified phonograph music.

3

New automatic record changer—that will play ten records as long as you like.

4

New long-playing records—15 minutes of music on each side, or half hour per record.

5

New chromium needles—good for 100 playings—also improve tone.

6

New acoustically balanced cabinet—carefully designed for fine tone effect.

The New RCA Victor Radio-Phonograph, Model RAE-26, Price $247.50

TODAY . . . a glorious new type of home entertainment is here! Today with the new RCA Victor radio-phonograph, you have at your command not only America's $70,000,000 worth of broadcasting programs—not only a $500,000 library of the world's finest music—but also—you can now enjoy *almost two hours* of record music without touching the instrument.

You can now enjoy a marvelous new type of record that plays 15 minutes on each side—30 minutes in all. A record far more true in reproducing tone.

You can now enjoy a new RADIO—a 9-tube Super-Heterodyne—finer than any other previous RCA Victor radio. It employs the new RCA Victor Synchronized Tone System, affording radio entertainment heretofore considered impossible.

A new type of record changer is here, too! . . . capable of playing any 10 regular-size records (old type or new) as long as you desire—with no effort on your part.

A new type of phonograph is here with a new method of electrical amplification—almost uncanny in its realistic effect. And all these features come to you in this new instrument for only $247.50—less than *half* the former price of such an instrument. By all means, hear the new radio-phonograph at your RCA Victor dealer's today . . . hear the new long-playing records . . . and hear the many other new RCA Victor instruments . . . radios from $37.50 to $179 complete . . . combinations from $129.50 to $995.

RCA Victor Co., Inc. "RADIO HEADQUARTERS"
Camden, N.J. *A Radio Corporation of America Subsidiary*

All you do is select the records . . . the mechanism does the rest . . . 35 minutes of music from present type records . . . nearly two hours—almost four times as much—from new long-playing type. Machine also plays individually any record you now own.

RCA Victor

RADIO PHONOGRAPH-COMBINATIONS

VICTOR RECORDS

Beginning with the 1931-32 season, RCA no longer used the Radiola name exclusively, most often referring to their models by the RCA Victor name followed by several letters and numbers such as "RCA Victor Model RAE-26." (*The Literary Digest,* October 17, 1931, p. 1)

Endnotes and References for Afterwords

1 Data taken from RCA Annual Reports for the years 1921 to 1929.

2 Gleason L. Archer, *Big Business and Radio*, (The American Historical Company, Inc., New York, 1939) p. 340.

3 Frederick O. Barnum III, *"His Masters Voice" in America*, (General Electric Company, Camden, New Jersey, September 1991) p. 155.

4 Gleason L. Archer, *Big Business and Radio*, pp. 345-6.

5 *Report of the Federal Trade Commission on the Radio Industry in Response to House Resolution 548, Sixth-seventh Congress, Fourth Session*, (Government Printing Office, Washington DC, 1924) p. 10.

6 *Report of the Federal Trade Commission on the Radio Industry in Response to House Resolution 548, Sixth-seventh Congress, Fourth Session*, (Government Printing Office, Washington DC, 1924).

7 *Federal Trade Commission vs. General Electric Company, et. al., Radio Case, Brief of Facts*, (Sidney C. Ormaby Company, 217 Broadway, New York, 1928) Vol. 1-8.

8 Herbert Hoover, elected 31st President of the U.S. in 1928, established the antitrust division of the Justice Department to prosecute unfair competition and restraint of trade cases.

9 "Trust Suit Filed on Radio Compacts," *The New York Times*, May 14, 1930, p. 1.

10 "Government Loses Oil Crack Case," *The New York Times*, April 14, 1932, p. 13.

11 "Widens Radio 'Trust" Suit'," *The New York Times*, March 9, 1932, p. 16.

12 "Radio Suit Deferred for New Parleys," *The New York Times*, October 9, 1932, p. 1.

13 "Radio Suit Waits Until Next Monday," *The New York Times*, November, 16, 1932, p. 25.

14 Gleason L. Archer, *Big Business and Radio*, pp. 352-386.

15 "Anti-Trust Decree to End Radio Group," *The New York Times*, November 22, 1932, p. 1; Decree Dissolving Radio Group Under Anti-Trust Law, *The New York Times*, November 22, 1932, p. 17.

16 "General Electric Details R.C.A. Deal," *The New York Times*, March 28, 1933, p. 27; Gleason L. Archer, *Big Business and Radio*, pp. 364-386.

17 "R.C.A. Seen stronger Following Split-Up," *The New York Times*, November 27, 1932, p. 1.

18 *Saturday Evening Post*, Vol. 201, No. 52, p. 78.

19 *Saturday Evening Post*, Vol. 202, No. 51, June 21, 1930, rear cover.

20 The pair of RCA and Victor monograms appearing together was registered as Trade-Mark No. 302,712 on April 25, 1933 by the RCA Victor Company, Inc.

21 One of the last sets manufactured by RCA for GE was the D-50 auto radio advertised in the Saturday Evening Post as late as May 25, 1935.

22 *Saturday Evening Post*, Vol. 208, No. 5, August 3, 1935, p. 36-37.

23 For example, see *Saturday Evening Post*, Vol. 208, No. 11, September 14, 1935, p. 38.

Appendix A

RCA Receiver Apparatus Characteristics

The basic characteristics of both broadcast and marine radio receiver apparatus manufactured by General Electric, Westinghouse, Wireless Specialty, and Marconi for sale or lease by RCA are summarized in this appendix. Broadcast radio apparatus listed here are limited to those which appeared in one or more of the following during the period of 1920 to 1930: published RCA price sheets, RCA or RCA distributor catalogs, and ads offering apparatus for sale in magazines or newspapers. Excluded are prototypes or preproduction models for which there is no evidence that they were offered for sale through RCA distribution channels. Marine radio receiver apparatus listed include receivers known to have been manufactured for and used by RCA for marine applications in the early 1920's. The IP-500 and IP-501 are included with marine receivers although they were also sold to the public for broadcast reception. Notes explaining table entries and abbreviations follow the tables.

Broadcast Radio Apparatus

RCA broadcast receiver apparatus are listed in three tables: A-1) broadcast radio receiver apparatus, A-2) loudspeakers, and A-3) other accessories. Receivers listed in Table A-1 are organized as follows: 1) early Westinghouse sets with Type Numbers only, 2) sets with the Aeriola name, 3) sets with the Radiola name but without numbers or numerals, 4) sets which were given AA, AR, or ER designations in lieu of names, 5) Radiolas with Roman numerals, and 6) Radiolas with Arabic numbers. Apparatus within the first four groups are listed in approximate chronological order, while the remaining two groups consisting of Radiolas with Roman numerals or numbers are listed in ascending numerical order.

Like any good schema, there are exceptions: the Balanced Amplifier follows the companion Radiola III, the Super-Heterodyne is placed in proximity to the nearly identical Super-VIII, and the Regenoflex is placed in proximity to the nearly identical Radiola X. This particular organization has a pleasing symmetry, allows for rapid location of each model, and yet retains an approximate chronological order. In Table A-2, early loudspeaker apparatus are listed chronologically, while later apparatus are listed in ascending numerical order. In Table A-3, other accessories for receivers are grouped according to antennas, detectors, and power/amplifier units.

Marine Radio Apparatus

Marine receivers manufactured for sale or lease by RCA are listed in table A-4. The Marconi 106 was discontinued before RCA was incorporated, replaced by the Marconi 106B. After the RCA-Marconi merger in late 1919, GE continued to manufacture the 106B for RCA, and subsequently designed and manufactured 106C and 106D versions as well. The duplex radio-telephone receiver, while not manufactured in quantity, was included in the table because it is particularly relevant to the subject of RCA broadcast receivers and because it was the first superheterodyne designed by GE—forshadowing the development of the Radiola Super-Heterodyne in 1923-24. Specifically excluded from this list are the barrage receivers used at Riverhead, L. I. (Radio Central) for early transoceanic and/or transoceanic marine use because they were distinctly different from broadcast receivers.

Table A-1. Broadcast radio receiver apparatus manufactured by GE, Westinghouse (W), and Wireless Specialty Apparatus (WSA) for sale by RCA between 1920 and 1929 appearing in RCA price lists, RCA/Distributor catalogs, and/or published advertisements

Model	Model No.	Type[a]	No. of Tubes[b]	Mfgr.	Date of Release	Discont. By	No. Mfg.[c]	Price ($)	Comments
DA	DA	Det/Amp	3	W	Dec. 1920	Jul. 1923	10,800	**$65**	2-stage amplifier + detector
RA	RA	Tuner	0	W	Dec. 1920	Jul. 1923	10,800	$65	Regen. recv'r using DA or DB Xtal
RC	RC	Regen.	3	W	Dec. 1920	Jul. 1923	65,600	**$125**	RA + DA in one cabinet
Aeriola Jr.	RE	Crystal	0	W	Jul. 1921	Dec. 1922	**24,249+**	$25	1st receiver for broadcast radio
Aeriola Sr.	RF	Regen.	1	W	Dec. 1921	Mar. 1924	**55,102+**	$75	1st regen. recv'r for broadcast radio
Aeriola Sr. Amplifier	AC	AF Amp	2	W	Sep. 1922	Mar. 1924	~4000	$68	2-stage amplifier for Aeriola Sr.
Aeriola Grand	RG	Regen	4 + 4	W	Mar. 1922	Dec. 1922	5,200	$325	Resistance-coupled amplifier
Radiola Grand	RG	Regen.	4	W	Dec. 1922	Mar. 1924	11,100	$350	Transformer-coupled amplifier
Radiola Senior	RF	Regen.	1	W	Feb. 1923	Mar. 1924	3,300	$65	Aeriola Sr. renamed Radiola Sr.
Radiola AC	AC	AF Amp	2	W	Feb. 1923	Mar. 1924	5,800	$68	Aeriola Sr. Amp renamed Radiola AC
Radiola AR	AR	RF Amp	3	W	Jan. 1923	Mar. 1924	2,500	**$80**	Used in combinations with RA/DA/RT
Radiola RT	RT	Coupler	0	W	Jan. 1923	Mar. 1924	~2,500	$45	Used in combinations with RA/DA/AR
Radiola RS	RS	Regen.	2	W	Jan. 1923	Mar. 1924	**4613**	$85	Radiola Senior + 1-stage amplifier
Radiola Special	None	Regen.	1	WSA	Feb. 1923	Mar. 1924	10,000	$62.50	Converted from AR-1375
AA-484	AA-484	AF Amp	2	WSA	Apr. 1922	Aug. 1922	~100's	**$45**	2-stage breadboard amplifier
AA-485	AA-485	Det/Amp	6	WSA	Aug. 1922	Mar. 1924	~100's	**$160**	RF amp + detector + audio amp
AR-1300	AR-1300	Crystal	0	GE	Apr. 1922	Feb. 1923	Unknown	$50	Regen. receiver with AA-1400 amp
AA-1400	AA-1400	Det/Amp	3	GE	Apr. 1922	Feb. 1923	Unknown	**$75**	Detector + 2-stage amplifier
AA-1520	AA-1520	RF Amp	3	GE	Aug. 1922	Mar. 1924	Unknown	**$80**	Untuned 3-stage RF amplifier
AR-1375	AR-1375	Crystal	0	WSA	Apr. 1922	May 1923	**635**	$40	1st set bearing Radiola name
AR-1382	AR-1382	Crystal	0	WSA	Aug. 1922	Feb. 1923	Few	$70	Used as tuner with AA-485 amplifier
ER-753	ER-753	Crystal	0	GE	Apr. 1922	Aug. 1922	~1000's	$18	Vertical and Horizontal Wedge
ER-753A	ER-753A	Crystal	0	GE	Aug. 1922	Mar. 1924	Unknown	$25	Redesign of ER-753 in wood case

[a]Abbreviations: **Crystal** - Crystal receiver; **Det**. - Detector; **Regen**. - Regenerative receiver; **SG** - Screen Grid tubes; **S. Het.** - Superheterodyne; **TRF** - Tuned-Radio-Frequency receiver; **URF** - Untuned-Radio-Frequency receiver

[b]Number of tubes used in receiver stages shown first followed by number of rectifier and/or ballast tubes; bold font indicate sets operated from lighting circuit

[c]Numbers in bold font are from *Radio Manufacturers of the 1920's* by Douglas; other numbers are the author's best estimate from serial numbers and auction results (See Appendices D and E)

Table A-1 (continued)

Radiola I	ER-753A	Crystal	0	GE	May 1923	Mar. 1924	**7,900**	$25	ER-753A renamed Radiola I
Radiola II	AR-800	Regen.	2	GE	Dec. 1922	Mar. 1924	**9,594**	$97.50	Portable set in a wood box
Radiola III	AR-805/RI	Regen.	2	W	2/2/24	Sep. 1926	293,000	$35	Similar to Radiola RS
Balanced Amp	AR-806/AP	AF Amp	2	W	3/4/24	Sep.1926	55,000	$30	Push-pull amp for Radiola III
Radiola III-A	AR-802/RL	Regen.	4	W	3/27/24	Feb. 1927	**164,624**	$65	Radiola III + Balanced Amp in 1 box
Radiola IV	AR-880	Regen.	3	GE	Dec. 1922	Mar. 1924	10,400	$275	1st self-contained set incl. batteries
Radiola V	AR-885	Regen.	3	GE	Dec. 1922	Mar. 1924	**21,138**	**$132.50**	AR-1300 + AA-1400
Radiola V-A	AR-885A	Regen.	3	GE	Mar. 1923	Mar. 1924	See V	$142.50[d]	AR-1300 + AA-1400A
Radiola VI	AR-895	URF	6	GE	Dec. 1922	Mar. 1924	**2,774**	**$162.50**	AA-1520 + AA-1400
Radiola VII	AR-905	TRF	5	WSA	Sep. 1923	Jan. 1924	**1,692**	$290	Step-tuned RF transformers
Radiola VII-A	AR-905	TRF	5	WSA	Sep. 1923	Jan. 1924	Few	$???	Console version of Radiola VII
Radiola VII-B	AR-907	TRF	5	WSA	Jan. 1924	Feb. 1925	**2,569**	$275	Replaced Radiola VII; same as R. IX
R. Super- Heterodyne	AR-812	S. Het.	6	GE	3/4/24	Jun. 1926	**148,300**	**$220**	1st RCA superhet; sealed catacomb
Radiola Super-VIII	AR-810	S. Het.	6	GE	3/18/24	Feb. 1927	20,000	$425	Console version of Super-Het
Radiola IX	AR-1404	TRF	5	WSA	Nov. 1923	Feb. 1925	Unknown	$215	Same circuit as Radiola VII-B
Radiola Regenoflex	AR-817/RJ	Regen.	4	W	4/25/24	Sep. 1925	≥14,000	**$150**	Regen. receiver with reflex stage
Radiola X	AR-816/RK	Regen.	4	W	3/21/24	Sep. 1925	**14,000**	$245	Regenoflex set with speaker
Radiola 16	AR-924	TRF	6	W/GE	8/8/27	Apr.1929	**85,320**	$82.75	New uni-control battery TRF
Radiola 17	AR-927	TRF	**6 + 1**	W/GE	9/15/27	Apr. 1928	**179,917**	$157.50	New uni-control AC TRF
Radiola 18	AR-936	TRF	**6 + 1**	W/GE	4/19/28	Jan. 1930	**250,000**	$115	Replacement for Radiola 17
Radiola 18 DC	AR-891	TRF	**6**	W/GE	12/18/28	Jan. 1930	Incl.	$125	DC version of Radiola 18
Radiola 20	AR-918	TRF	5	W/GE	11/6/25	Sep. 1928	**135,121**	$115	New 5-tube TRF design
Radiola 21	AR-1258	TRF-SG	5	W/GE	10/19/29	1930+	Unknown	$85.25	1st RCA battery screen-grid
Radiola 22	AR-1265	TRF-SG	5	W/GE	11/7/29	1930+	Unknown	$150.75	Console version of Radiola 21
Radiola 24	AR-804	S. Het.	6	GE	5/12/25	Dec. 1925	**7,079**	$195	6-tube catacomb portable
Radiola 25 Option 1 Radiola 25 Option 2	AR-919 AR-919	S. Het. S. Het.	6 +1 6	W/GE W/GE	9/10/25 9/10/25	Jul. 1927 Jul. 1927	**94,815**	$165 $165	6-tube battery catacomb[e] 6-tube battery catacomb[e]

[d] Advertised price included tubes, batteries and headphones

[e] When Operated on AC power with Model 104 Radiola loudspeaker using AC Package UP-971 @$35, output tube is moved to Model 104 power/amplifier chassis

Table A-1 (continued)

Model	Model No.	Type[a]	No. of Tubes[b]	Mfgr.	Date of Release	Discont. By	No. Mfg.[c]	Price ($)	Comments
Radiola 26	None	S. Het.	6	W	6/23/25	Jul. 1927	**19,507**	$225	6-tube catacomb portable
Radiola 28 Option 1 Radiola 28 Option 2	AR-920 AR-920	S. Het. S. Het.	8 + 1 8	W/GE W/GE	Aug 1925 Aug 1925	Sep. 1928 Sep. 1928	**84,978**	$260 $260	New 8-tube battery superhet[f] New 8-tube battery superhet[f]
Radiola 28 A-C Oper.	AR-969	S. Het.	**7**	W/GE	9/29/25	Sep. 1928	See 28	Unknown	AC-only version of Radiola 28[g]
Radiola 30	AR-921	S. Het.	**8 + 3**	W/GE	Nov 1925	Feb. 1927	**10,835**	$575	1st AC-only set; used Radiola 28
Radiola 30A	AR-926	S. Het.	**8 + 3**	W/GE	8/7/27	Apr. 1929	**17,363**	$495	Similar to Radiola 30
Radiola 30A DC	AR 912	S. Het.	**11**	W/GE	9/14/27	Sep. 1928	See 30A	$495	DC version of Radiola 30A
Radiola 32	AR-925	S. Het.	**8 + 3**	W/GE	7/27/27	Sep. 1928	Unknown	$895	Radiola 28 in a console
Radiola 32 DC	AR-928	S. Het	**11**	W/GE	9/15/27	Sep. 1928	Unknown	$895	DC version of Radiola 32
Radiola 33	AR-784	TRF	**6 + 1**	W/GE	3/26/29	Sep. 1930	**236,961**	$86.50	Metal cabinet with optional legs
Radiola 33 DC	Unknown	TRF	**7**	W/GE	7/29/29	Sep. 1930	See 33	$79.75	DC version of Radiola 33
Radiola 41	AR-782	TRF	**6 + 1**	W/GE	12/15/28	Sep. 1929	Unknown	$238.50	Console; similar to Radiola 17
Radiola 41 DC	AR-871	TRF	**10**	W/GE	3/1/29	Sep. 1929	Unknown	$255	DC version of Radiola 41
Radiola 44	AR-594	TRF-SG	**4 + 1**	W/GE	5/29/29	Sep. 1930	Unknown	$128.50	1st RCA screen-grid receiver
Radiola 46	AR-596	TRF-SG	**4 + 1**	W/GE	5/29/29	Sep. 1930	**97,633**	$197.50	DC version of Radiola 44
Radiola 46 DC	AR-597	TRF-SG	**6**	W/GE	11/6/29	Sep. 1930	Incl.	$212.25	Console version of Radiola 44
Radiola 47	AR-1147	TRF-SG	**4 + 1**	W/GE	10/7/29	Sep. 1930	**15,120**	$293.50	Radio-phono combo; R. 46 chassis
Radiola 48	Unknown	TRF-SG	**6 + 1**	RCA Vtr	Nov 1930	Mid-1931	Unknown	**112.50**	AC screen-grid TRF
Radiola 50	AR-910	TRF	**6 + 1**	W/GE	3/14/28	Jun.1928	**3,220**	$285	Console version of Radiola 17
Radiola 51	AR-904	TRF	**6 + 1**	W/GE	6/13/28	Sep. 1929	**6,188**	$195	Console version of Radiola 18
Radiola 51 DC	AR-903	TRF	**6**	W/GE	12/18/28	Sep. 1929	See 51	$205	Console version of Radiola 18 DC
Radiola 60	AR-954	S. Het.	**8 + 1**	W/GE	8/13/28	Sep. 1930	**135,336**	$210	New 8-tube superhet design
Radiola 62	AR-982	S. Het.	**8 + 1**	W/GE	9/14/28	Jan. 1930	**19,972**	$410	Radiola 60 in a console
Radiola 64	AR-894	S. Het.	**9 + 2**	W/GE	10/23/28	Sep. 1930	**7,406**	$608	Radiola 60 in a console with AVC
Radiola 66	AR-598	S. Het.	**7 + 1**	W/GE	7/18/29	Sep. 1930	Unknown	$246.50	New 7-tube superhet in a console
Radiola 67	AR-1168	S. Het.	**8 + 2**	W/GE	9/28/29	Sep. 1930	**4,939**	$733.00	1st RCA phono; Radiola 66 chassis
Radiola 80	Unknown	S. Het.	**8 + 1**	RCA Vtr	Nov 1930	Mid-1931	Unknown	**$142.50**	AC screen-grid superheterodyne
Radiola 82	Unknown	S. Het.	**8 + 1**	RCA Vtr	Nov 1930	Mid-1931	Unknown	**$179.50**	Radiola 80 with tone control
Radiola 86	Unknown	S. Het.	**8 + 1**	RCA Vtr	Nov 1930	Mid-1931	Unknown	**$285**	Radiola 80 plus phonograph

[f]When operated on AC power with Model 104 Radiola loudspeaker using AC Package UP-972 @$35, output tube is moved to Model 104 power/amplifier chassis

[g]Radiola 28 (A-C Operated) was a factory-made AC-only set designed for use only with loudspeaker Model 104 which contained the power amplifier tube

Table A-2. Radiola loudspeakers appearing in RCA price lists, RCA/Distributor catalogs, and/or advertisements which were marketed by RCA between 1920 and 1929

Model	Model No. or Type	Type	No. of Tubes	Mfgr.	Date of Release	Discon'd By	No. Mfg.	Price ($)	Comments
Loud Speaker Attachment	LS	Tone Arm Transducer	0	W	Mid-1921	Mar. 1924	Unknown	$18	Made for Victrola and Grafonola phonograph arms
Vocarola	Type I, LV	Horn	0	W	Jan. 1922	Sep. 1923	~10,000	$30	Based on automobile horn; wall mounted or table top
FH	Type FH	Horn	0	W	Sep.1923	Feb. 1925	Unknown	$36.50	1st appeared with the Radiola VII; no diaphragm adjustment
UZ-1320	UZ-1320	Horn	0	GE	Nov. 1923	Oct. 1924	Unknown	$36.50	1st GE horn; knurled-nut adjustment for diaphragm
UZ-1325	UZ-1325	Horn	0	GE	Oct. 1924	Feb. 1927	Unknown	$25	Improved UZ-1320; upright lever to adjust diaphragm
100	UZ-915	Cone Type, Perm. Magnet	0	GE	Sep. 1925	Jul. 1927	~200,000+	$35	1st permanent loudspeaker; table model
100-A	UZ-1076	Cone Type, Perm. Magnet	0	GE/W	5/25/27	1930 +	~300,000+	$35	Improved Model 100; Desk-top model
100-B	UZ-783	Cone Type, Perm. Magnet	0	GE/W	3/26/29	Sep. 1930	~10,000s	$22	Designed for Radiola 33; desk-top model
102	UZ-913	Cone Type, Perm. Magnet	0	GE	Sep. 1925	Jul. 1927	~1000s	$140	Model 100 used with Uni-Rectron for AC power
103	UZ-749	Cone Type, Perm. Magnet	0	GE/W	8/17/28	Sep. 1930	~100,000s	$37.50	Tapestry loudspeaker; table-top model
104	UZ-914	Cone Type, Dynamic	**1 + 4**	GE	Sep. 1925	Sep. 1928	Unknown	$245	1st dynamic loudspeaker; floor model
104 DC	Unknown	Cone Type, Dynamic	**4**	GE/W	June 1927	Sep. 1928	Unknown	$310	DC version of Model 104
105	UZ-1082	Cone Type, Dynamic	**1 + 2**	GE/W	3/8/28	Nov. 28	~1000s	$350	Contained power supply for generic receiver
106	UZ-642	Cone Type, Dynamic	0	GE/W	11/26/28	Sep. 1930	~10,000s	$88	Similar to 105 but without the power supply for a receiver
106 DC	Unknown	Cone Type, Dynamic	0	GE/W	4/12/29	Sep. 1930	Unknown	$65	DC version of Model 106

Table A-3. Pertinent broadcast radio accessories appearing in RCA price lists, RCA/Distributor catalogs, and/or advertisements which were marketed by RCA between 1920 and 1929

Model	Model No. or Type	Type	No. of Tubes	Mfgr.	Date of Release	Discon'd By	No. Mfg.	Price ($)	Comments
Antenna Outfit For Receiving	AD	Long wire (outside)	0	W	Oct. 1921	Sep. 1923	Unknown	$7.50	Included 150' wire, ground lightning arrestor, hardware
Radio Receiving Antenna	AG-788	Long wire (outside)	0	GE	Apr. 1922	Mar. 1924	Unknown	$7.50	Essentially a copy of the Westinghouse AD kit
Radiola Loop Antenna	AG-814	Loop (rotating)	0	GE	Apr. 1924	Sep. 25	Unknown	$12	Designed for the AR-812 Super-Heterodyne
Collapsible Loop Antenna	AG-1380	Loop (fixed)	0	WSA	Jun. 1922	Feb. 25	Unknown	$18	Used with UC-1820 variable condenser
Collapsible Loop Antenna	AG-1385	Loop (rotating)	0	WSA	Sep. 1923	Feb. 25	Unknown	$18	AG-1380 antenna with a rotating base
Crystal Detector Antenna	DB	Perikon & Crystal	0	W	Aug. 1921	Mar. 1924	Unknown	$6.50	Dual detectors with a condenser under the base
Cleartone Crystal Detector	UD-1430	Crystal	0	WSA	May 1923	Mar. 1924	Unknown	$1.60	Inexpensive crystal detector of the "open type"
Cleartone Crystal Detector	UD-1432	Crystal	0	WSA	May 1923	Mar. 1924	Unknown	$2.30	Similar to UD-1430 but of the "glass enclosed" type
Loading Coil	CB	N/A	0	W	c. 1921	Jul. 1923	Unknown	$6.00	Used with RC
Variable Condensor	UC-1820	N/A	0	WSA	Jun. 1922	Mar. 1924	Unknown	$7.70	Used with AG-1380 and AG-1385 loop antennas
Rectigon Battery Charger	2.5 Ampere 6 Ampere	Battery charger	0	W	c. 1923	c. 1921	Unknown	$18 $28	First sold by Westinghouse in 1921, later by RCA
Tungar Battery Charger	2-Ampere 3/5 Ampere	Battery Charger	0	GE	c. 1923	c. 1922	Unknown	$18 $28	Priced to match Westinghouse offering
Uni-Rectron	AP-935	Power/ Amplifier	**3**	GE	Sep. 1925	Dec. 1927	Unknown	$105	Power amplifier for 102 and other loudspeakers
Duo-Rectron	AP-937	"B"-Battery Eliminator	3	GE	Sep. 1925	Dec. 1927	Unknown	$65	"B"-battery eliminator operated from the AC line
"B" Eliminator	AP-1080	"B"-Battery Eliminator	0	GE/W	3/5/28	Apr. 1929	Unknown	$48	Similar to AP-937, but with disc rectifiers
Adapter Socket	UR-556	Tube	0	GE	Sept. 1925	Unknown	Unknown	$1.50	Adapts UX-120 to UV-199 socket
Adapter Socket	UR-589	Tube	0	GE	May 1923	Unknown	Unknown	$1.25	Adapts UV-199 to standard Navy base

Table A-4. Marine/wireless receivers manufactured for and sold or leased by RCA in the early 1920's

Model	Type	Intro Date	No. Tubes	Mfgr.	Price	Comments
106	Crystal	Early 1915	0	Marconi	Unknown	Manufacturing discontinued before GE acquired Marconi
106B	Crystal	Late 1919	0	Marconi & GE	Unknown	106 with two crystal detectors and a new "squared" cabinet style
106C	Crystal or Regen.	Early 1922	0[a]	GE	Unknown	Additional connectors and controls added for an external regenerative detector
106D	Crystal or Regen.	Late 1922	0[a]	GE	Unknown	Modification kit for retrofitting 106, 106B and 106C; used with external regenerative detector and amplifier (e.g. AA-1400)
IP-500	Crystal or Regen.	1919	0[a]	WSA	$595	Commercial equivalent of Navy SE-1220
IP-501	Crystal or Regen.	1919	1	WSA	$550	Commercial equivalent of Navy SE-1420
IP-501A	Crystal or Regen.	1922	3	WSA	Unknown	Commercial equivalent of Navy SE-1440; Combined IP-501 + Triode B in one cabinet; made specifically for RCA Marine
Triode A One-step amp	Detector + amp	1919	2	WSA	$190	Detector plus one-step amplifier controller for IP-500
Triode B Two-step amp	2-stage amp	1919	2	WSA	$95	Two-step amplifier for IP-501 (detector tube is contained within the IP-501)
Triode C Eaton Oscillator	Oscillator	c. 1919	0	WSA	$13.50	Used as part of a regenerative receiver, or as an rf generator
ER-1620	Crystal or Regen.	Early 1922	3	GE	Leased	First new marine set designed by GE for RCA: AR-1529 tuner, AD-1527 detector, two AA-1528 amplifiers, AD-692 crystal detector, and CW-834 headphones
AA-1400	Det. + 2-step amp	1922	3	GE	$75	Used with the Marconi 106D to form a regenerative receiver; consists of one AD-1527 plus two AA-1582 amps in one box
2RF/D/21F/3-8/3RDA/A	Superhet.	1922	8	GE	Prototype	Prototype duplex radio-telephone receiver; first superheterodyne made by GE

[a] Required a separate unit containing at least one tube for regenerative reception, and may have also contained additional tubes for audio amplification.

Notes on Broadcast Receiver Apparatus Entries

The basic data summarized in tabular form include the following: name, model number, type of unit or circuit, number of tubes, manufacturer, release date, date discontinued, number manufactured, and suggested retail price at the date of release. Separate entries are made for AC versus DC lighting circuits, but not for different types of sets within the AC category (e.g., 60 cycles versus 25 cycles) or within the DC category (e.g., 110 versus 220 volts). Also included is a short comment about each entry to provide some descriptive fact or feature.

Model

The receiver models listed in Column 1 appear exactly as they do on the receiver name plates and/or instruction sheets. In some cases, there are minor differences in the model designation appearing on the name plate versus the instruction sheet. For example, Radiola VII-B appears with a "dash" on the instruction sheet and in RCA correspondence, while there is no dash on the receiver name plate. In these cases, the model names found on instructions and other documents were used in favor of those on the name plate.

In a few cases where the model name was changed by RCA on both the receiver and the instruction sheet—even without any corresponding change in design—there is one entry for each name. This occurred three times circa mid-1922 when RCA decided to use the Radiola trade name for all their apparatus including the Aeriola Sr., Aeriola Sr. Amplifier, and the ER-753A, whose names were changed to Radiola Senior, Radiola AC and Radiola I respectively. Despite the name change, all three keep their original designations—Westinghouse "Type RF," Westinghouse "Type AC," and General Electric Model "ER-753A" respectively.

RCA also changed the name of the Aeriola Jr. and RC to Radiola Junior and Radiola RC, but these sets were discontinued before the Radiola name found its way onto the receivers themselves or their instruction sheets. These two sets did appear in advertisements and catalogs with the new Radiola name, but because the new name did not appear on either the sets or the instructions, these two sets do not appear in the table with the Radiola name. Note that while Westinghouse used the abbreviated form Jr. and Sr. with their Aeriola sets, RCA often spelled out Junior and Senior when used with the Radiola name.

Loudspeaker models began with phrase "Radiola Loudspeaker Model," Radiola Loudspeaker," "Loudspeaker Model," or "RCA Loudspeaker" followed by a Westinghouse "Type" designation, an RCA UZ designation, or a model number consisting of three digits. The particular form used depended on whether the name appeared in an ad, on the instructions sheets, in RCA catalogs, or on product name plates. A similar diversity of names was used on other accessories including antennas, detectors and amplifier/power units.

Model Number

Virtually all broadcast radio receiver apparatus sold by RCA in the twenties were assigned either a Westinghouse two-letter "Type" designation or a GE/RCA model number beginning with an AA, AR, or ER followed by three or four digits—and in some cases an additional "alteration" letter. In at least five instances, both the Westinghouse and GE/RCA designations were assigned, and in two cases—the Radiola Special and Radiola 26—no designation was assigned. The type and/or model numbers are shown in Column 2; in five entries the RCA model numbers are separated from the two-letter Westinghouse Type designation by a slash.

Receiver accessories including loudspeakers were also assigned either a Westinghouse Type designation or an RCA designation. RCA designations were as follows: UZ for loudspeakers, AG for antennas, UD for detectors, and AP for amplifier/power units.

Type

The apparatus type is identified in Column 3. The abbreviations used in the receiver table are as follows: **Crystal**–crystal set; **Det.**–detector; **Regen.**–regenerative receiver; **SG**–screen grid tubes; **S. Het**. or **S. H.**–superheterodyne receiver; **TRF**–tuned-frequency receiver; **URF**–untuned-radio-frequency receiver.

Number of Tubes

The number of tubes used in each apparatus is indicated in Column 4. For battery sets, each tube represented a separate stage, but in sets designed for the AC power line, there were always one or two additional tubes used for rectification and/or voltage regulation. The general custom at the time was to characterize a set by the number of tubes used in the receiver function–excluding the rectifier tubes. Following this custom, the first entry refers to the number of tubes used in the receiver stages, while the second entry following the plus sign refers to the number of tubes used as a ballast or rectifier. Radiolas 25 and 28 were first offered with ballast tubes which were removed in later sets–hence the two entries Option 1 and Option 2 with a different total number of tubes. All sets that operate from the power line–both AC and DC–are indicated by the bold font on the tube entries.

Manufacturer

All broadcast receiver apparatus marketed by RCA in the twenties were manufactured by General Electric (GE), Westinghouse (W), and/or Wireless Specialty Apparatus (WSA). The manufacturer of each set is indicated in Column 5. In order to meet the 60%-40% manufacturing agreement between GE and Westinghouse, RCA arranged for both companies to manufacture sets produced circa late 1925 and thereafter. Sets manufactured by both companies are indicated by the double entry separated by a slash. WSA was involved in manufacturing sets for RCA after 1926 in a relatively minor way by helping with sub-assemblies, sets with special requirements, etc. Beginning in 1930, RCA Victor (RCA Vtr) began to manufacture radio and phonograph apparatus for RCA.

Dates of Release and Discontinuation

The details of how these dates were selected are addressed in Appendix C. Suffice it to say that the entries in the "release date" column with specific dates refer to official RCA release dates as identified by documents found in the George H. Clark *Radioana Collection*, while all other dates–which include only month and year–are this authors best guess based on criteria presented in Appendix C.

Number Manufactured

Most of the information on the number of sets manufactured and/or sold by RCA comes from Alan Douglas' *Radio Manufacturer's of the 1920's, Vol. 3*, which in turn was obtained from the RCA Sales Corporation in Indianapolis, Indiana. Unfortunately, data are not available on all apparatus–particularly loudspeakers and other accessories–and so the author has made educated guesses in a number of cases based on serial number data, number of sets available on the auction market and price. The Douglas data appears in bold font so that they can be readily distinguished from the author's entries based on either serial number data or "best-guess estimates" denoted by the approximate symbol " ~ " preceding the number.

Price

The prices in Column 9 represent the advertised retail prices at the time sets were first introduced. Prices include tubes wherever that data was available. In a number of cases RCA sold the equipment without tubes and did not advertise a price with tubes; in these cases the prices are shown in a bold font. In other cases, particularly for the high-end models, the only prices advertised were those which included additional items such as batteries and battery chargers. At the low end, the price of crystal sets most often included headsets.

Notes on Marine Receiver Entries

There were far fewer types of marine receiver apparatus manufactured for RCA than broadcast radio apparatus in the 1920's—probably less than a dozen distinctly different receiver types. The marine receivers appearing in Table A-4 were selected because they were either offered for sale by RCA, or in the case prototypes, because they have relevance to the design of broadcast radios.

Models

There were basically three categories of sets—those that were derived from the Marconi 106 designed earlier by American Marconi for commercial marine service well before the creation of RCA in late 1919, those that were derived from the receivers manufactured by Wireless Specialty Apparatus (WSA) for the Navy during WWI, and those that were designed by GE specifically for RCA.

Type

The early marine receivers were all crystal sets. The later sets were regenerative receivers to increase sensitivity, but all used crystal detectors as a backup for emergencies.

Dates

The dates these sets were introduced into service are not as well known as the corresponding dates for broadcast revivers. In many cases, the dates are only known to within about six months.

Tubes

In most of the early regenerative receivers, the tuner was housed in a cabinet separate from the detector and/or audio amplifier cabinets; consequently the sets themselves (e.g., 106C, 106D, IP-500) did not have tubes.

Manufacturer

All early marine sets used by RCA were manufactured by American Marconi, General Electric (GE) or Wireless Specialty Apparatus (WSA). The 106B was first manufactured by the Marconi Aldene operations for RCA, and then later by GE after Marconi was moved and integrated into GE in Schenectady.

Price

The price of many of the marine receivers is not known—in fact, some of these sets were for lease only and were not offered for sale. Prices that do appear in the table are known because they were offered for sale to the public by RCA in their 1922 catalog.

Appendix B

Vacuum Tubes Used in Radiola Receivers

This appendix provides a list of the vacuum tubes used in the Radiola receivers covered in this book. The Radiolas are conveniently divided in four groups, each with identifiably different tube technologies:

- Early Battery Sets (audio output power ≤ 100 mw)
- Later Battery and DC Lighting Circuit Sets (audio output power > 100 mw)
- Early AC Lighting-Circuit Sets (before AC-filament tubes)
- Later AC Lighting-Circuit Sets (with AC-filament tubes)

Pertinent comments on receivers in each category appear in the following paragraphs preceding each table.

Early Battery Sets (without power amplifier tubes)

Battery sets are divided into two categories—early sets using low-power amplifier tubes with output powers of less than 100 mw, and later sets using tubes with output powers of 100 mw or more that were specifically designated as power amplifiers. The early sets using the low-power tubes in audio output stages were designed for use with either earphones or horn speakers, while the later sets with higher-power output tubes were designed primarily for use with cone speakers.

Tubes for the early battery sets can be divided into two basic categories including 1) ballast tubes, and 2) amplifier and/or detector tubes. The Aeriola Grand was the only early battery set that used ballast tubes. Most tubes in the "detector and/or amplifier" category were used for both applications, but some tubes performed better—and therefore were used more—in one application than the other. For example the UV-200 was described by RCA as a "detector only" tube.

The earliest tube sets addressed here—the RC receiver and DA amplifier—were first offered by Westinghouse without tubes, but the set was designed for and often sold with Moorhead tubes. These sets and others that were sold with different tube options have multiple entries in the tables with the abbreviation "Opt." followed by a number for each set of different tube types used. The DA (either sold by itself or as an RC), Radiola V, and Radiola VI were sold with the most number of different types of tubes at three each (the later Radiola V was sometimes referred to as the V-A). It should be noted that a number of sets were also sold with Cunningham tubes which had different numbers than the RCA equivalents indicated in the table, but these tubes did not represent different tube types.

There were other sets offered by RCA without tubes such as the AA-484 amplifier, in which case the dealer—and not RCA—often selected the tube type to be sold with the set. A large number of dealer catalogs were reviewed to identify the various options appearing in the tables, but with the large number of dealers and potential combinations, there is no way to be certain that the options presented in the tables are complete.

Table B-1. Early battery sets (before power amplifier tubes)

Receiver Type	Total No.	Amplifier and/or Detector								Bal.[1]
		Moor-head	WD-11	WD-12	WD-21	UV-199	UV-200	UV-201	UV-201-A	WD-800
DA/RC Opt. 1	3	3[2]								
DA/RC Opt. 2	3						1	2		
Radiola RC Opt. 3	3			3						
Aeriola Sr.	1		1							
Aeriola Sr. Amp	2		2							
Aeriola Grand	4 + 4				4					4
Radiola Grand	4		4							
Radiola Senior	1		1							
Radiola AC	2		2							
Radiola AR Opt. 1	3							3		
Radiola AR Opt. 2	3								3	
Radiola RS	2		2							
Radiola Special	1					1				
AA-484	2							2[3]		
AA-485	6						1	5		
AA-1400	3						1	2		
AA-1520	3							3		
Radiola II	2					2				
Radiola III	2		2							
Balanced Amp	2		2							
Radiola III-A	4		4							
Radiola IV	3					3				
Radiola V Opt. 1	3						1	2		
Radiola V Opt. 2	3			3						
Radiola V/V-A	3					3				
Radiola VI Opt. 1	6						1		5	
Radiola VI Opt. 2	6			6						
Radiola VI Opt. 3	6					6				
Radiola VII	5					5				
Radiola VII-A	5					5				
Radiola VII-B	5					5				
R. Super-Het.	6					6				
R. Super-VIII	6					6				
Radiola IX	5					5				
R. Regenoflex	4		4							
Radiola X	4		4							
Radiola 24	6					6				
Radiola 26	6					6				

[1]Ballast tubes
[2]Initially offered by Westinghouse without tubes but designed for and often sold with Moorhead tubes
[3]Initially offered by RCA without tubes but designed for use with the UV-201

Later Battery and DC Lighting Circuit Sets (using power amplifier tubes)

Later battery sets and sets operated from the DC power line are grouped together because of commonality in tubes used for both classes. There were basically three types of tubes used for these sets: amplifiers and/or detectors, power amplifiers, and ballasts or regulators. Tubes designated as power amplifiers were used only in the final audio output stage—with the notable exception of the UX-112-A, which was used not only as a power amplifier in the audio output stage, but also as an amplifier and/or detector in other stages.

The three power amplifier tubes listed in the table (UX-112-A, UX-120 and UX-171-A) were rated by RCA with maximum undistorted power outputs of 120 mw, 110 mw and 790 mw respectively. Of the four tubes listed here as amplifier and/or detectors, three were triodes while the fourth, a UX-222, was a screen-grid tetrode used primarily as an RF amplifier. Ballast tube UV-877 was initially used in early Radiolas 25 and 28 but eliminated from later sets.

Table B-2. Later battery and DC lighting-circuit sets (with power amplifier tubes)

Receiver Type	Total No.	Amplifier and/or Detector				Power Amplifier			Bal.[1]
		UX-199	UX-201-A	UX-112-A	UX-222	UX-112-A	UX-120	UX-171-A	UV-877
Radiola 16	6		5			1			
Radiola 20	5	4					1		
Radiola 21 Opt. 1	5			2	2	1			
Radiola 21 Opt. 2	5			2	2			1	
Radiola 22 Opt. 1	5			2	2	1			
Radiola 22 Opt. 2	5			2	2			1	
Radiola 25 Opt. 1	6+1	5					1		1
Radiola 25 Opt. 2	6	5					1		0
Radiola 28 Opt. 1	8+1	7					1		1
Radiola 28 Opt. 2	8	7					1		0
Radiola 18 110[2] V DC	6			5				1	
Radiola 30A 104 V DC	11	7						4	
Radiola 32 104 V DC	11	7						4	
Radiola 33 110 V DC	7			5				2	
Radiola 33 220 V DC	6			5				1	
Radiola 41 110[2] V DC	10			6				4	
Radiola 46 110 V DC	6			2	2			2	
Radiola 51 110[2] V DC	6			5				1	

[1]Ballast tube UV-877 was supplied with early Radiolas 25 and 28, but eliminated from later sets
[2]Nominal voltage of 110 V; specified voltage range was 107.5-127.5 V DC

Early AC Lighting-Circuit Sets (before AC-filament tubes)

Radiolas operating from the AC lighting circuit are divided into two categories—early sets without AC tubes requiring relatively large and expensive power supplies, and later sets with AC-filament tubes which dramatically reduced the cost and size of the power supplies. The early AC lighting circuit sets without AC-filament tubes are listed in table B-3 where the tubes have been divided into four categories: Detector/Amplifier (D/A), Power Amplifier, Rectifier, and Regulator.

Radiola 28 was manufactured as a battery set which could be converted to an AC set (addressed in the previous section), but the Radiola 28 (A-C Operated) set was manufactured for use only with the AC lighting circuit. The Radiola 28 (A-C Operated) set was factory wired and sold with the Model 104 loudspeaker which contained the power amplifier tube. This set used eight tubes in the receiving circuit with seven located in the Radiola 28 cabinet and a single power amplifier tube in the Model 104 loudspeaker cabinet.

It was a common practice at the time to characterize the number of tubes in a set by the number used in the receiving circuits—exclusive of rectifier or regulator tubes. Following this convention, the total number in the receiving circuit is indicated by the first number in the "Total No." column.

Table B-3. Early AC lighting-circuit sets (before AC-filament tubes)

Receiver Type	Total No.	D/A	Power Amplifier			Rectifier		Ballast Tube	
		UX-199	UX-120	UX-171	UX-210	UX-216-B	UX-281	UV-876	UV-886
R. 28 A-C Operated[1]	7	7							
Loudspeaker 104	1+3				1	2		1	
Radiola 30 50-75 Hz	8+3	7			1	2		1	
Radiola 30 40-45 Hz	8+3	7			1	2			1
Radiola 30A	8+3	7		1			2	1	
Radiola 32	8+3	7			1		2		1

[1]Radiola 28 AC-Operated was factory wired for the Model 104 loudspeaker, which contained the power amplifier

Later AC Lighting-Circuit Sets (with AC-filament tubes)

The later sets operated from the AC lighting circuit used AC-filament tubes that reduced the size and cost of the power supply. Three types of amplifier/detector tubes were used in these sets including the UX-226 with an AC-filament electron emitter, the UY-227 with a separate cathode emitter, and the UY-224 screen-grid tube, also with a cathode emitter.

New power amplifier tubes were developed in rapid succession, each with greater output power than the last to drive larger speakers and produce greater undistorted volumes. The maximum power ratings of the four tubes shown in the table were 700 mw, 1.5 w, 1.6 w, and 4.6 w respectively. The actual power delivered to speakers in the sets listed in the table depended upon plate voltage.

Two types of rectifier tubes were used in these sets, a full-wave rectifier designated UX-280, and a half-wave rectifier tube designated UX-281. Both had higher power ratings than their predecessors in order to provide the additional power required by the larger power amplifier tubes.

Table B-4. Later AC lighting-circuit sets (with AC-filament tubes)

Receiver Type	Total No.	Amp and/or Detector			Power Amplifier			Rectifier		
		UY-224	UX-226	UY-227	UX-171-A	UX-210	UX-245	UX-250	UX-280	UX-281
Radiola 17	6+1		4	1	1				1	
Radiola 18	6+1		4	1	1				1	
Radiola 33	6+1		4	1	1				1	
Radiola 41	6+1		4	1		1			1	
Radiola 44	4+1	3					1		1	
Radiola 46	4+1	3					1		1	
Radiola 47	4+1	3					1		1	
Radiola 50	6+1		4	1	1				1	
Radiola 51	6+1		4	1	1				1	
Radiola 60	8+1			7	1				1	
Radiola 62	8+1			7	1				1	
Radiola 64	9+2			8				1		2
Radiola 66	7+1			6			1		1	
Radiola 67	8+2			7				1		2

Appendix C

Release and End Dates For Radiola Receivers

The dates that Radiola receivers were released and discontinued are summarized in this appendix. These dates can be used to put the receivers in their proper historical context and to estimate the time they were on the market—a good indicator of both commercial success and relative rarity. In general, apparatus on the market for a shorter period of time were rarer than those on the market for a longer time. The data are presented in tabular form in Table C-1 where the release dates, the end dates, and the number of months on the market are given. The entries in this table and the methodology used to determine these entries are discussed in the paragraphs below.

Release Dates

Both historians and collectors alike relish dating antique receivers to the year, if not the month or day, if possible. While it is not always possible—nor is it generally necessary—to date apparatus to the day or even to the month, still there certainly is a need to date equipment, if for no other reason than to classify, group and compare them. The problem with radios is that, unlike people, they are not born on a specific day. Unless the manufacturer specifics a "release date," there are a number of dates in the life cycle of a radio that could be chosen as the release date, such as the date on the name plate of the first production run, the date on the earliest distributor price list, the date of a trade show where the apparatus is first introduced, the date it is shipped to the distributor or dealer, the date the first advertisement appears, etc.

Fortunately, RCA was in the practice of naming specific release dates, at least for the later receivers they marketed from mid-1923 onward. RCA historian George Clark has assembled a list of many of these dates, which have been entered into Column 2 of the table. Clark made two types of entries in his list, one with a specific date, and the other with the month and year only. The entries with a specific date are assumed to be the official release dates designated by RCA management, while the entries with month and year are assumed to be Clark's best guess of the "release" date of those sets for which management did not designate an official release date.

The "official" release dates appear to be the dates on memos from the RCA Sales Department to RCA Wholesale Distributors announcing the new apparatus. These memos were typically one or two pages containing a description of the set, the price structure including retail and dealer discount prices, delivery dates, and other information deemed pertinent by the RCA Sales Department. A small number of these memos have been found in the *Radioana Collection*. For example, a memo entitled "Subject: Radiola VII" from E. E. Bucher to RCA Wholesale Distributors dated September 17, 1923 stated in part:

> *"In announcing the new broadcast receiver — Radiola VII — delivery of which is expected to begin September 25th, we wish to offer a word of explanation to distributors."*

From this memo, one can infer a release date of September 17, 1923 for the Radiola VII.

Table C-1. Radiola receiver apparatus release dates, dates by which receivers were discontinued, and the estimated number of months on the market

Model	Release Dates[a] (G. H. Clark)	Release Dates (Author)	Discontinued By	Months For Sale
DA	Not listed	Dec. 1920	Jul. 1923	31
RA	Not listed	Dec. 1920	Jul. 1923	31
RC	**Jan. 1922**	Dec. 1920	Jul. 1923	31
Aeriola Jr.	No date	Jul. 1921	Dec. 1922	18
Aeriola Sr.	**Feb. 1922**	Dec. 1921	Mar. 1924	39
Aeriola Sr. Amplifier	Sep. 1922	Sep. 1922	Mar. 1924	18
Aeriola Grand	Mar. 1922	Mar. 1922	Dec. 1922	9
Radiola Grand	Not listed	Dec. 1922	Mar. 1924	14
Radiola Senior	Not listed	Feb. 1923	Mar. 1924	13
Radiola AC	Not listed	Feb. 1923	Mar. 1924	14
Radiola AR	Jan 1923	Jan. 1923	Mar. 1924	14
Radiola RT	Not listed	Jan. 1923	Mar. 1924	14
Radiola RS	Jan. 1923	Jan. 1923	Mar. 1924	14
Radiola Special	Feb. 1923	Feb. 1923	Mar. 1924	13
AA-484	Not listed	Apr. 1922	Aug. 1922	4
AA-485	Not listed	Aug. 1922	Mar. 1924	19
AR-1300	Not listed	Apr. 1922	Feb. 1923	10
AA-1400	Not listed	Apr. 1922	Feb. 1923	10
AA-1520	Not listed	Aug. 1922	Mar. 1924	19
AR-1375	Not listed	Apr. 1922	May 1923	13
AR-1382	Not listed	Aug. 1922	Feb. 1923	6
ER-753	Not listed	Apr. 1922	Aug. 1922	4
ER-753A	Not listed	Aug. 1922	Mar. 1924	9
Radiola I	Not listed	May, 1923	Mar. 1924	10
Radiola II	Dec. 1922	Dec. 1922	Mar. 1924	15
Radiola III	Feb. 2, 1924	Feb. 2, 1924	Sep. 1926	29
Balanced Amp	Mar. 4, 1924	Mar. 4, 1924	Sep. 1926	28
Radiola III-A	Mar. 27, 1924	Mar. 27, 1924	Feb. 1927	35
Radiola IV	Dec. 1922	Dec. 1922	Mar. 1924	15
Radiola V	Dec. 1922	Dec. 1922	Mar. 1924	15
Radiola V-A	March 1923	Mar. 1923	Mar. 1924	12
Radiola VI	Dec. 1922	Dec. 1922	Mar. 1924	15
Radiola VII	**Feb. 1923**	Sept. 1923	Jan. 1924	4
Radiola VII-A	Not listed	Sept. 1923	Jan. 1924	4
Radiola VII-B	**Jan. 1923**	Jan. 1924	Feb. 1925	13
R. Super-Heterodyne	Mar. 4, 1924	Mar. 4, 1924	Jun. 1926	27
Radiola Super-VIII	Mar. 18, 1924	Mar. 18, 1924	Feb. 1927	35
Radiola IX	Nov. 1923	Nov. 1923	Feb. 1925	15
Radiola Regenoflex	Apr. 25, 1924	Apr. 25, 1924	Sep. 1925	17
Radiola X	Mar. 21, 1924	Mar. 21, 1924	Sep. 1925	18
Radiola 16	Aug. 8, 1927	Aug. 8, 1927	Apr. 1929	20
Radiola 17	Sep. 15, 1927	Sep. 15, 1927	Apr. 1928	7

Table C-1 (continued)

Radiola 18	Apr. 19, 1928	Apr. 19, 1928	Sep. 1929	17
Radiola 18 DC	Dec. 18, 1928	Dec. 18, 1928	Sep. 1929	9
Radiola 20	Nov. 6, 1925	Nov. 6, 1925	Sep. 1928	31
Radiola 21	Oct. 19, 1929	Oct. 19, 1929	After 1930	14+
Radiola 22	Nov. 7, 1929	Nov. 7, 1929	After 1930	14+
Radiola 24	May 12, 1925	May 12, 1925	Dec. 1925	7
Radiola 25	Sep. 10, 1925	Sep. 10, 1925	Jul. 1927	22
Radiola 26	Jun. 23, 1925	Jun. 23, 1925	Jul. 1927	25
Radiola 28	Aug. 1925	Aug. 1925	Sep. 1928	35
Radiola 28 AC	Sep. 29, 1925	Sep. 29, 1925	Sep. 1928	34
Radiola 30	**Feb. 15, 1925**	Nov. 1925	Feb. 1927	20
Radiola 30A	Aug. 7, 1927	Aug. 7, 1927	Apr. 1929	20
Radiola 30A DC	Sep. 14, 1927	Sep. 14, 1927	Sep. 1928	9
Radiola 32	Jul. 27, 1927	Jul. 27, 1927	Sep. 1928	14
Radiola 32 DC	Sep. 15, 1927	Sep. 15, 1927	Sep. 1928	12
Radiola 33	Mar. 26, 1929	Mar. 26, 1929	Sep. 1930	18
Radiola 33 DC	Jul. 29, 1929	Jul. 29, 1929	Sep. 1930	14
Radiola 41	Dec. 15, 1928	Dec. 15, 1928	Sep. 1929	9
Radiola 41 DC	Mar. 1, 1929	Mar. 1, 1929	Sep. 1929	6
Radiola 44	May 29, 1929	May 29, 1929	Sep. 1930	16
Radiola 46	May 29, 1929	May 29, 1929	Sep. 1930	16
Radiola 46 DC	Nov. 6, 1929	Nov. 6, 1929	Sep. 1930	10
Radiola 47	Oct. 7, 1929	Oct. 7, 1929	Sep. 1930	11
Radiola 50	Mar. 14, 1928	Mar. 14, 1928	Jun. 1928	3
Radiola 51	Jun. 13, 1928	Jun. 13, 1928	Sep. 1929	15
Radiola 51 DC	Dec. 18, 1928	Dec. 18, 1928	Sep. 1929	9
Radiola 60	Aug. 13, 1928	Aug. 13, 1928	Sep. 1930	13
Radiola 62	Sep. 14, 1928	Sep. 14, 1928	Jan. 1930	16
Radiola 64	Oct. 23, 1928	Oct. 23, 1928	Sept. 1930	23
Radiola 66	Jul. 18, 1929	Jul. 18, 1929	Sept. 1930	14
Radiola 67	Sep. 28, 1929	Sep. 28, 1929	Sept. 1930	12

[a]Bold fonts indicate errors in dates identified by Clark which have been corrected in Column 3

Unfortunately, Clark's list of release dates is not complete and it contains a few significant errors, which are identified by the bold fonts in Column 2. The following scheme was used to verify Clark's release dates and identify release dates for receivers not on Clark's list. First, all available RCA correspondence and documents in the *Radioana Collection* were reviewed to identify specific release dates designated by RCA. These dates took preference over all other dates. Next, the following dates were identified:

- Month and year of RCA distributor price list in which apparatus first appeared.
- Month and year of trade magazine (*Radio Retailing, Radio Merchandising,* or *Town Crier*) where apparatus was first introduced.
- Month and year of national magazine (*The Wireless Age*, *Radio*, *Radio News*, *Popular Radio*, *Saturday Evening Post,* or *National Geographic*) in which advertisement first appeared.

Clark's dates were checked by comparing his entries with the earliest of the above dates—which for the most part agreed with Clark's dates—in which case it was assumed that Clark's dates were accurate. There were five cases where Clark's dates were clearly incorrect, as indicated by the bold font in Column 2 and as summarized here:

RC and Aeriola Jr.
Clark used the dates that RCA began to market these receivers, but they were actually released and marketed earlier by Westinghouse before RCA assumed that responsibility.

Radiola VII
Clark's date may have been the original release date, but the Radiola VII was seriously delayed as indicated by a memo found in the *Radioana Collection* from RCA Sales Manager E. E. Bucher to distributors which clearly identified the release date as September 25, 1923.

Radiola VII-B
Clark's date of January 1923 is far too early—perhaps the year 1923 may have been a typographical error. A memo from RCA Sales Manager E. E. Bucher in the *Radioana Collection* clearly identified the release date as January 10, 1924.

Radiola 30
Clark's date of February 15, 1925 is far too early—the first ads for this set appeared in November 1925.

The same methodology was used to identify release dates for those apparatus not on Clark's list. The author's best estimate of all release dates appear in Column 3.

While the methodology relied on the aforementioned dates, several other dates were considered: date of first trade show where apparatus was displayed, the date on the apparatus name plate, and the date of the first version of the instruction manual or card. In the end, these dates were generally regarded as being unreliable indicators of release dates, but, they were useful in many cases. For example, the above methodology when applied to the Radiola II and RS would indicate that these sets were not released until May 1923, the date of their earliest entry in the RCA price list—well after Clark's release dates of December 1922 and January 1923 respectively. However, the dates on the instruction cards for both sets—and the manufacturing date on the RS name plate (10-16-22)—indicated a much earlier release date. In this case, an RCA memo dated January 1922 clearly showed that both sets were on sale to the public in Bamberger's Department Store in New York during January 1923, thus confirming the accuracy of Clark's release dates. It appears that RCA released these two sets early in the model year but did not advertise them widely until later in the year, hoping to clear out inventories of obsolete sets first.

End Dates

RCA discontinued their sets for one of three basic reasons: to replace a defective set with one having similar capabilities and appearance, to make way for new sets introduced at the beginning of a new model year, and to eliminate excessive inventories that were obsolete or were no longer selling. Except for the replacement of a defective set, RCA rarely if ever announced the discontinuance of a particular set—consequently, there is generally no official discontinuance date. Often, both the dealers and the public first discovered that a particular receiver was being discontinued when there was a clearance sale at Gimbels or other similar retailer—much to the consternation of the dealers who often were caught with a substantial inventory of discontinued sets that often had to be sold below their cost.

Other than the highly visible clearance sale, the discontinuation of RCA apparatus was signaled by the disappearance of the set from national advertising and/or the deletion from RCA distributor and dealer price lists. The dates associated with these events give an approximate indication of when the apparatus was discontinued. However, these dates are not highly accurate because ads for each set did not always appear monthly and because the RCA price list was not published each and every month. Also, the discontinuation of a set was an event that actually took place over several months as the distributors and dealers continued to sell their stock, even if at a reduced price.

The RCA price lists are the best indication that the apparatus was being discontinued. Approximately 28 price sheets covering a 10 year period from 1921 to 1930 were available for this dating. It is not known how many additional price sheets were published or whether a new price sheet was introduced after each set was discontinued. As a result, there is some granularity in the results–the average elapse time between each price sheet used in this dating is 4.3 months (ten years divided by 28 price lists). Fortunately, a number of these price sheets were published at the time the new receivers were introduced–which often corresponded to the time frame that the old-model receivers were discontinued (see Table C-2). The author's best estimate of the date receivers were discontinued is entered in Column 4 of the table. Strictly speaking, these data should be interpreted as *the month by which* the apparatus was discontinued–not necessarily *the month of* discontinuation.

Table C-2. Dates in which the new RCA line for each radio season first appeared in RCA price lists, advertising, and at radio shows used by RCA to introduce the new lines

RCA Season	1st Time New Line Appeared in RCA Price Lists	Primary Radio Show Used by RCA	1st Advertising of New RCA Line	Month/Year Used for New RCA Season
1922	April 1, 1922	Mar. 7-11, 1922 Amateur Radio Show	June 1922 RCA Radio Catalog	June 1922
1923	May 1, 1923	Dec, 21-30, 1922 American Radio Exposition	February 1923 *Saturday Evening Post*	February 1923
1924-25	April 15, 1924	New RCA line missed the fall radio shows	March 1924 Many magazines	Late February or early March 1924
1925-27	Sept. 1, 1925	2nd Annual Radio World's Fair @ NYC Armory Week of Sept. 14	November 1925 *Radio Merchandising*	September 1925 delivery of sets (delayed)
1927-28	July 15, 1927	4th Annual Radio World's Fair @ Madison Square Garden – Sep. 19-23, 1927	June 1927 *Town Crier*, September 1927 *Saturday Evening Post*	July-September 1927
1928-29	Oct. 3, 1928	5th Annual Radio World's Fair @ Madison Square Garden – Sep. 17-21, 1928	September 1928 Trade Magazines *Radio Retailing* & *Town Crier*	September 1928
1929-30	Oct. 1, 1929	June 1929 Chicago Show	July 1929 (two screen-grid receivers only)	June 1929 (two screen-grid receivers only)
		6th Annual Radio World's Fair @ Madison Square Garden – Sep. 23-27, 1929	September 1929 (whole line)	September 1929 (whole line)

Time on the Market

The last entry in the table is an estimate of the time each apparatus was on the market. This time was calculated by taking the difference between the dates in the third and fourth column. The period of time has been rounded to a whole month by assuming the 15th of the month for each entry–so for example, the period from a specific date in March of 1925 to March of 1926 is assumed to be 12 months). Not surprisingly, the sets with the longest times on the market are also the sets with the largest number manufactured. Conversely, the sets with the shortest time on the market tend to be the ones with the fewest manufactured, and are certainly among the rarest.

For example, sets on the market for more than 30 months include the RA/DA/RC, Aeriola Sr., Radiola III-A, Radiola Super-VIII, Radiola 20, and Radiola 28. These were among the most popular sets of the time, and a very large number were manufactured—accounting for their general availability on the auction and flea markets today. At the other end of the spectrum, sets on the market for less than ten months include the Aeriola Grand, AA-484, AR-1382, ER-753, Radiolas VII and VII-A, Radiola 17, Radiola 18 DC, Radiola 24, Radiola 30A DC, Radiolas 41 and 41 DC, and Radiola 50. Of these sets, four were discontinued early because of defects (Aeriola Grand, Radiolas VII and VII-A, Radiola 17, and Radiola 50). These four sets and the ER-753 were removed from the market earlier than planned and replaced by another set. The DC sets were generally on the market for a shorter time because in many cases they were developed later than their AC counterparts. With the notable exception of the Radiola 17, these sets are among the rarerest, and do not appear often on the markets of today.

Appendix D

Serial Number Data

The serial numbers appearing on early broadcast radio receivers are of interest to the historian and collector, not only to help quantify the number of sets manufactured but also to follow the progression of design changes and place them in proper chronological order. Collectors often use the serial numbers to identify the early production units, which for some reason they generally find more desirable. Serial numbers seem to be of more interest for the early receivers before RCA standardized manufacturing at GE and Westinghouse—perhaps because there is less known about them due to lack of documentation, or perhaps there were significantly more design changes that occurred before standardization.

Westinghouse began the practice of assigning serial numbers sequentially to their receivers beginning with the RA and DA units in late 1920, while GE did not begin to assign serial numbers to broadcast receivers until the Radiola IV was released in December of 1922. Oddly, both companies were selective about the apparatus to which they assigned serial numbers. Westinghouse did not assign serial numbers to Aeriola Jr., Aeriola Sr., and Aeriola AC Amplifier, while GE did not assign serial numbers to ER-753, ER-753A, AR-1300, AA-1400, AA-1520, or Radiolas I and II. It is not clear whether serial numbers were assigned to Radiolas V and VI—numbers have been observed on some but not all of these sets, but it is not clear that these numbers are serial numbers. In sharp contrast, Wireless Specialty Apparatus assigned serial numbers to all the sets they manufactured for RCA with the possible exception of the AR-1375 crystal set.

Serial numbers were affixed to RCA radio apparatus in a number of different ways: stamped into metal name plates which in turn were riveted or screwed to panels or cabinets, stamped directly into the bottom of the wood cabinets (inside and outside) or into the underside of interior wood panels, stamped directly into the metal chassis or struts, stamped into the edge of the bakelite front panel, stamped on serial number strips which were riveted to the chassis, and printed on paper decals which were affixed to the back or bottom of the cabinet or housing. The serial numbers are easy to spot on most sets, but the locations are not obvious on some sets—most notably those manufactured by Wireless Specialty.

The serial numbers presented below have been collected over a four-year period from auctions, radio swap meets, and fellow collectors. Obviously, the sample size is directly related to the availability of sets which in turn is related to the number of sets manufactured. In many cases the sample size is large enough to make an accurate estimate of the number manufactured, while in other cases the data is useful only to infer a lower bound. Collecting of serial numbers by the author has focused on radios manufactured between 1920 and 1925 prior to standardization of manufacturing by Westinghouse and GE.

Westinghouse Serial Numbers

Serial numbers for Westinghouse sets manufactured before the 1925-27 season—the first season when manufacturing at Westinghouse and GE was standardized—can be most conveniently divided into three groups: 1) sets manufactured between December 1920 and December 1923 when serial numbers were assigned sequentially, 2) Radiola III, III-A, and Balanced Amplifier sets manufactured during the 1924-25 season with overlapping serial numbers manufactured in four distinct production runs, and 3) the remaining sets manufactured during the 1924-25 season, including Radiolas Regenoflex, X and 26.

Westinghouse Serial Numbers Dating From 1920–1923

Westinghouse attached metal name plates with production dates and serial numbers on ten different broadcast receiver models released between December 1920 and December 1923—the RA, DA (either alone or paired together in a single cabinet as the RC), DR, Aeriola Grand, Radiola Grand, Radiola Sr., Radiola AC Amplifier, Radiola RS, Radiola AR and Radiola RT. For some reason Westinghouse did not put name plates or serial numbers on the Aeriola Jr., Aeriola Sr., or Aeriola Sr. AC Amplifier.

RA and DA sets dominated the serial numbers in this time frame, which were assigned from a low of 100 or less—probably beginning with "1"—to a high of about 143,000. There are two small gaps (9,858–11,794 and 45,447–50,188) and one large gap (118,302–140,239) in the RA/DA serial number data where about 28,000 serial numbers for sets other than the RA and DA were assigned. Another 7000 or so of these other sets were assigned serial numbers between 143,000 and 150,000 (approximately) which fall above the highest numbers observed for the RA/DA sets. Very few other sets were actually interleaved with the RA and DA sets—it is estimated that only 1,200 or about 25 % of the Radiola AC serial numbers were actually interleaved with those of the RA/DA sets.

The serial number samples collected for Westinghouse sets in the 1920-23 period are shown in Table D-1—the top half consisting of RA/DA sets are arranged in order of increasing serial numbers, and the bottom half consisting of the DR, Aeriola Grand, and Radiolas AC, Senior, Grand, AR, RS, and RT sets (referred to collectively as the "other eight sets") are also arranged in order of increasing serial numbers using the lowest serial number sample found for each model in the ordering process. There were six different manufacturing dates for the RA/DA sets which appear in the fourth column, the last one having a significant gap in serial numbers between 118,302 and 140,239. The number of RA and DA sets inferred from the serial number data is shown in the fifth column of the table. The total number of RA/DA sets with serial numbers is estimated to be 113,400 (the sum of all RA/DA sets listed in the fifth column) but because virtually all the 31,400 DA's with a date of 5-29-22 were paired in RC sets with RA's that were not assigned serial numbers, the total number is approximately 144,800 (113,400 RA/DA sets with serial numbers plus 31,400 RA sets without serial numbers).

While serial numbers for all eight of the other sets conveniently appear in three definite gaps (see shaded rows in Table D-1) in the RA/DA serial number data, they are for the most part interleaved with each other. Consequently, it is more difficult to estimate the number of these eight individual models manufactured from the existing data base. What can be determined with some certainty from the existing data base is the total number of non-RA/DA sets. Based on serial number data, it is estimated that there are a total of about 28,000 sets in the three RA/DA gaps (1,500 + 4,500 + 22,000), 7000 sets above the highest DA serial number (143,000 to 150,000), and 1,500 AC sets interleaved with RA/DA sets—for a grand total of total of 36,500 for the eight other sets. Implicit in this estimate is the assumption that serial numbers were assigned without substantial gaps, and that there were no sets other than those identified here that were assigned serial numbers. Based on the relatively uniform density of data points across the entire range of samples, and the absence of substantial gaps in the samples for the ten sets as a group, this assumption seems to be reasonable.

As evident from the table, while numbers for the eight other sets are interleaved with each other, still there are large blocks of serial numbers grouped together for each set, so it is possible to make an educated guess as to the number of each manufactured. The author's estimates of the number of sets manufactured for each of the eight models appear in the fifth column of the table. The number of sets in this column is equal to the total estimate of 36,500, but there is some uncertainty as to the specific numbers as noted in the following paragraphs.

The three Aeriola Grand data points are by themselves insufficient to make any guess about the number, but the very large number of RA/DA data points indicates there are two significant gaps in a range of serial numbers shared only by the Aeriola Grand and DR sets. The first gap of about 4,500 brackets one of the Aeriola Grand serial numbers and no others, and so it is assumed that the gap of 4,500 is filled with Aeriola Grands. The other gap of about 1,500

Table D-1. Serial number samples collected for Westinghouse sets in the period 1920-1923

Range of Serial Number Samples	Sample Size	Model Type of Sets[a]	Mfg. Date	Est. No.	Comments
134 – 1,616	26	RA/DA	11-30-20	1,700	Pittsburgh sets
1,846 – 3,394	9	DA	5-18-21	1,700	
3,781– 5,128	8	RA	9-28-21	1,700	
6,102 – 9,857	19	RA/DA	8-2-21	4,900	
9,858 – 11,794	N/A	Gap of ~1,500 in RA/DA data[b]	N/A		~700 Aeriola Grand's ~800 DR's
11,795 – 45,446	75	RA/DA	12-14-21	34,000	
45,447 – 50,188	N/A	Gap of ~ 4,500 in RA/DA data[c]	N/A		~4,500 Aeriola Grand's
50,189 – 87,719	75	RA/DA	12-14-21	38,000	
88,309 – 118,302	54	DA	5-29-22	28,800	1,500 AC's interleaved with DA's; RA's with no serial numbers paired with DA's as RC's
118,303 – 140,238	N/A	Gap of ~22,000 in RA/DA data	N/A		Gap filled with R. Senior Grand, RS, AC, AR, RT
140,239 – 142,835	13	DA	5-29-22	2,600	No RA's with SN
10,518 – 10,836[d]	7	DR	?-26-21	800	All in one small DA gap
11,162; 46,580; 46,738	3	Aeriola Grand	10-27-21	5,200[b,c]	In two small DA gaps
112,443 – 118,875 127,648 – 142,012	8 12	Radiola AC Amplifier	10-16-22	1,200 4,600	25% interleaved in DA 75% in large DA gap
119,962 – 122,222 137,962 & 148,847	10	Radiola Senior	10-16-22	3,300	Almost all in large DA gap
121,700 – 122,092 132,699 – 135,279 144,783 – 148,421	3 7 8	Radiola Grand	10-16-22	11,100	Half in large DA gap, half above highest DA
122,317; 122,406; 123,055; 123,490	4	Radiola AR	10-4-22	2,500	All in large DA gap
127,433 – 127,525 131,116 – 132,551 149,949 – 150,572	4 7 3	Radiola RS	10-16-22	5,300	75% in large DA gap
132,718; 132,869; 132,890; 149,184; 149,374	5	Radiola RT	10-16-22 12-11-22	~2,500	Not interleaved in DA

[a]Best estimate based on serial number data summarized in the first column
[b]Gap of ~1,500 in RA/DA data believed to be filled by 700 Aeriola Grand's and 800 DR's
[c]Gap of ~4,500 in RA/DA data believed to be filled by 4,500 Aeriola Grand's
[d]Two DR's with serial numbers 48,786 and 48,772 have been observed

brackets serial numbers observed on one Aeriola Grand and six DR sets but no others. Based on the close grouping of the six DR serial numbers in a range of about 700 numbers it is estimated that there were 800 DR sets manufactured with serial numbers in this gap, leaving a space of about 700 serial numbers for Aeriola Grand's. The total of 5,200 Aeriola Grand's may seem to be on the high side, but there is no other information indicating the number of these sets made.

The estimate of 11,100 Radiola Grand sets deduced from this data may also seem to be on the high side, but just two groupings of Radiola Grand serial numbers alone with seven and eight samples respectively—and without any interleaving of other sets—account for 6,200 sets alone. The remaining sets are highly interleaved with other sets, and so the estimated number is more uncertain, but another 4,900 sets is not unreasonable. The total number of 11,100 fits nicely with the projected number of 5,200 Aeriola Grand's manufactured. About 5,000 or so Radiola Grand's would have been made to replace a large fraction of the 5,200 ill-fated Aeriola Grand's—mostly for existing consumers who turned their sets back in for Radiola Grands—and the additional 6,100 or so Radiola Grand's would have been manufactured for new customers who bought the set while it was on the market during the following year.

The grouping of three data points for the AR cannot be directly extrapolated to the 2,500 sets shown, but once again it occupies a gap of approximately 2,500 serial numbers where there are no other sets. Thus, it has been assumed that the gap is filled with other AR sets. The RT is problematic because some sets have been observed with serial numbers while others have a name plate without a serial number stamped in the blank space reserved for one. These sets also have at least two different manufacturing dates. Since there are no large gaps at the locations of the RT serial numbers, the number cannot be inferred from gap sizes. Consequently, it has been assumed that the number of RT sets manufactured was equal to the number of AR sets. There is really no justification for this assumption other than the fact that the two sets were to be used together and were manufactured at the same time.

Several other comments are in order. First, the number of Radiola AC Amplifier sets (5,800) seems high in relation to the number of Radiola Senior sets (3,300), but then the use of the AC Amplifier was not confined to just the Radiola Senior. It would have been purchased for use with other receivers made by RCA—and other manufacturers as well. Second, the number of RS receivers predicted with this methodology (5,300) agrees reasonably well with the number of 4,613 reported independently by Douglas. With the high degree of interleaving and the relatively low sample sizes available for the sets other than the RA/DA, error margins of at least 15 % using the serial number data would be expected.

Radiola III, III-A, and Balanced Amplifier (BA) from 1924-25

Serial numbers for the Radiola III, III-A and BA were stamped into the bottom of the wood cabinets corresponding to what is believed to be four distinct production runs: the first with five digits beginning with the digit 4 or 5, and the remaining three with six digits beginning with the digit 2, 4, or 6. As indicated in Table D-2, only Radiola III was manufactured in the first run, all three in the second run, Radiolas III and III-A in the third run and only Radiola III in the fourth run. The groupings of serial numbers and the number of samples in each group are also shown in the table. In addition there are a few "outliers" to this scheme—for example one Radiola III was observed with the serial number 79,143, two BA's were observed with serial numbers 300,049 and 663,069, and one Radiola III-A with a serial number 490,485, well outside the range indicated in the table. The outliers appear to be so few in number that they do not affect the analysis described in the following paragraphs.

Table D-2. Serial number data for Radiola III, III-A, and Balanced Amplifier (BA)

Mfg. Run	Range of Serial Numbers	Sample Size	Radiola Model	Estimated No. (With Interleaving)	Estimated No. (No Interleaving)
1	40,196 – 51,176	13	III	11,000	11,000
2	200,686 – 297,790	123	III	55,800	100,000
	200,738 – 295,518	64	III-A	28,600	100,000
	201,456 – 254,306	37	BA	15,600	55,000
3	400,048 – 498,592 402,657 – 465,441	109 32	III III-A	77,300 22,700	100,000 66,000
4	600,053 – 681,489	50	III	82,000 (Gaps ignored)	82,000 (Gaps ignored)
Alt. 4[a]	600,053 – 608,790 625,389 – 626,181 633,252 – 681,489	10 3 37	III	58,000 (Total reduced by gaps sizes)	58,000 (Total reduced by gaps sizes)
All	Total	428	All	293,000	514,000

[a]Alternative Run 4 represents the number of sets assuming that no sets were manufactured with serial numbers in the large gaps observed in the data samples, an assumption which does not seem to be realistic.

There are virtually no gaps in the serial number data for any of the production runs except for the last run in which there were two significant gaps of 16,599 and 7,071 in the Radiola III samples (see shaded line in Table D-2). There were no other known sets with serial numbers in this range, nor was there any obvious reason for Westinghouse to interrupt the serial number sequence, and so the reason for these gaps remains somewhat of a mystery—although there are several plausible explanations. For example, a number of sets originally manufactured as Radiola III's were put in other cabinets such as Brunswick Radiola-phonograph combinations and larger after-market cabinets with loudspeakers. This occurred after the last manufacturing run, and so it is logical that these sets would have been taken preferentially from the last run. Radiola III's were also exported to South America, Europe and the Far East. For example, RCA provided components of the Radiola III to SATT, a Swedish company who manufactured sets with an appearance identical to the Radiola III but operating at different wavelengths (see Appendix H). One of the components provided by RCA was the Radiola III box, and one has been observed by this author with a serial number from the last production run stamped on the bottom along with the RCA initials.

In the end, one must either assume that the sets corresponding to the gaps were never manufactured in the first place or that they were produced but for some reason were removed from the market from which samples were collected. Since there are reasonable explanations as to why a number of sets could have been removed from the American Radiola III market (from which samples were gathered), and there is no plausible explanation as to why numbers would have been arbitrarily skipped, it seems reasonable to conclude that there were in fact 82,000 Radiola III's manufactured in the 4th production run—the number which is included in the totals shown in the bottom line of the table.

The total number of each model deduced from the serial number data is strongly dependent on whether or not the Radiola III, III-A, and/or BA serial numbers are interleaved. If the numbers are interleaved then there is no duplication of serial numbers and the number of sets in each production run is equal to the range of the production run—for example, production Run 2 which includes all three sets would have a total 100,000 sets. If, on the other hand, the numbers for each model are sequential (not interleaved), then the serial numbers would be duplicated, and the total number of sets in a given production run would be much greater—for example, production Run 2 without interleaving would have 100,000 Radiola III's, 100,000 Radiola III-A's, and 55,000 Balanced Amplifiers.

The most direct evidence to demonstrate the absence of interleaving would be the existence of two different model types with the same serial number. However, with the existing data base of just 428 samples and a total potential population of 500,000 serial numbers assigned to the three different models, the probability of finding two identical serial numbers is small—and indeed no duplication has been found.

Fortunately, another approach to determine the presence or absence of interleaving exists. It is based on the fact that if the serial numbers for a given set are assigned sequentially and the samples of serial numbers from the total population of sets are collected without bias, then the serial number samples will be distributed uniformly—that is to say, the density of sample points will not vary significantly over the entire range. This is the situation that would exist if there were no interleaving. However, if two or more sets are interleaved and the interleaving is not uniform across the full range of serial numbers, then the density of serial number samples for the two sets will not be uniform over that same range. For example, in the third production run, the Radiola III serial number samples exist over the entire range of ~100,000 sets (400,048–498,592), while the Radiola III-A samples exist only over two thirds of the range (402,657–465,441). Thus, if the serial numbers were interleaved, there should be substantially more Radiola III's with serial numbers in the last one third (above 465,441) where there are no Radiola III-A's than those in either the first third (400,000–433,333) or the second third (33,334–465,441). It turns out that the Radiola III density of serial number samples is virtually constant over the range, strongly suggesting that the serial numbers were not interleaved.

Not only is the density of Radiola III samples constant across the range of manufacturing Run 4, it is constant across all manufacturing runs! Despite the fact that there are different sets and different numbers of sets occupying the four manufacturing runs shared by the Radiola III, the density of Radiola III sets in this data base is virtually constant at about 1 sample per every 1,000 serial numbers. For example, in Run 1 there are 13 samples covering a range of approximately 11,000 serial numbers (1.1 samples per thousand), in Run 2 there are 123 samples covering a range of 100,000 (1.2 samples per thousand), in Run 3 there are 109 samples covering a range of approximately 100,000 (1.1 per thousand), and in Run 4 there are 50 samples covering several ranges of about 58,000 serial numbers (.86 samples per 1000). If the serial numbers were interleaved, one would expect a higher density of Radiola III sets in Run 1 and 4 which is not shared by any other model, and lowest in Run 2 which is shared by both the Radiola III-A and BA. This is clearly not the case, so one can only conclude that the serial numbers for the Radiola III, III-A and BA were not interleaved.

It is of interest to compare the number of sets manufactured as predicted from the serial number data of Table D-2 with those reported by Douglas from RCA records. According to Douglas in *Radio Manufacturers of the 1920s, "RCA figures state that either 164,624 or 'nearly 500,000' Radiola IIIA's were produced."* According to Table D-2, there were 166,000 Radiola III-A's produced assuming no interleaving (100,000 + 66,000), but only 51,300 with interleaving (28,600 + 22,700). Thus, RCA records support the no-interleave option. With regard to the 500,000 figure, there is no way that there could have been 500,000 Radiola III-A's because the maximum possible number of all three sets from the serial number data is 514,000. Perhaps the figure of 500,000 cited by Douglas actually refers to the combined total number of Radiola III, III-A, and BA sets manufactured—not just the total number of Radiola III-A's.

The figure of 103,701 Radiola III's cited by Douglas is far too low under any circumstances. Estimates of the number of Radiola III's without interleaving is 293,000, and even with interleaving the number would be at least 226,100.

Further, it is not credible that the number of Radiola III-A's was greater than the number of Radiola III's. The number of Radiola III's must be greater than the number of Radiola III-A's for several reasons: 1) the Radiola III was produced in all four production runs while the Radiola III-A was produced in only one run and part of a second, 2) the number of Radiola III serial number samples collected far exceed those of the Radiola III-A, and 3) the rate at which Radiola III's appear on the auction and flea markets far exceed the rate for Radiola III-A's.

Radiolas Regenoflex, X and 26 from 1924-1925

In contrast to the Radiola III/III-A/BA, the numbers of samples collected for the Radiola Regenoflex and X were insufficient to make accurate projections of the number of sets manufactured. The reasons for the small sample sizes are due not only to the relatively small number of sets manufactured, but also to the obscure location of the serial number such that the number was not generally included with set descriptions. Only six samples of the Regenoflex were collected with serial numbers ranging between 200,586 and 213,469. One could infer from this limited data that at least 14,000 sets were manufactured, but it could have been more. The situation for the Radiola X is more complicated—the four serial number samples that were collected have six digits with the first digit being either a "1," a "2" or a "4." The number of sets manufactured cannot be inferred from this data, but according to Douglas, there were 14,000 Radiola X's produced.

In contrast to the Regenoflex and Radiola X, the number of Radiola 26 sets can be determined with some accuracy because a relatively large number of samples were collected—48 serial numbers ranging between a low of 201,169 and a high of 219,936. There were no other sets that could have been interleaved with the Radiola 26, and the data have a relatively constant density of points across the entire range with no obvious gaps. One can confidently conclude from the data that approximately 20,000 sets were manufactured, and indeed, Douglas reports that RCA records show 19,507 manufactured.

General Electric Serial Numbers

Sets for which GE assigned serial numbers before standardization of manufacturing in the 1925-27 season can be conveniently divided into two groups: 1) sets manufactured for the 1923 season which include only Radiola IV's, and 2) superheterodyne sets manufactured during the 1924-25 season which include Radiola Super-Heterodynes, Super VIII's, AR-813 panels, and Radiola 24 portables.

General Electric Serial Numbers Dating from the 1923 Season

The first set to which GE assigned serial numbers was the Radiola IV released in December 1922 for the 1923 season. Individual serial numbers were engraved into a bakelite panel on the top left of the set accessible by lifting the top lid. Serial numbers from 48 sets have been collected, with a low of 293 and a high of 10,693. The serial number samples are uniformly distributed across the range without any substantial gaps, and so it can be safely assumed that approximately 10,700 to 11,000 sets were manufactured.

The two remaining sets manufactured by GE for the 1923 season were the Radiola V and VI. The Radiola V used the AR-1300 and AA-1400 while the Radiola VI used the AA-1400 and AA-1520, all of which had been manufactured the year before without serial numbers. There were no serial numbers on these individual units used in the Radiola V and VI, but numbers with one and two digits have been found stamped into the top of the wood base of some—but not all—Radiola V's and VI's. The numbers are identical in size and style to the serial numbers used on the Westinghouse Radiola III/III-A/BA sets, and later on the GE AR-813 panels, as discussed in the next section, but they are not believed to represent serial numbers.

General Electric Serial Numbers Dating from 1924-25

During the 1924-25 season, GE made four sets and one panel with serial numbers: Radiola Super-Heterodyne (AR-812), Super-VIII, Radiola 24, and the AR-813 superheterodyne panel used in Brunswick sets and the unadvertised AR-813 Super-Heterodyne console radio described in Chapter 6. These sets were manufactured in five production runs with six different ranges of serial numbers as shown in Table D-3. What is most striking about this table is that GE production runs 2–6 were assigned six-digit numbers beginning with an odd digit (1, 3, and 5), while the Westinghouse sets listed in Table D-2 were assigned six-digit numbers beginning with an even digit (2, 4, and 6). Collectively, the serial numbers assigned by the two companies fit neatly and tightly together without overlap, and so one must conclude that this did not happen by chance but rather by direction from RCA. It would also appear that the limited number of sets produced with five-digit numbers (Westinghouse's Radiola III and GE's Super-VIII) constituted trial runs as a precursor to the main production runs that were to follow. Serial numbers for these limited production runs did not overlap.

Table D-3. Serial numbers for set manufactured by GE dating from 1924-25

Mfg. Run	Range of Serial Numbers	Sample Size	Radiola Model	Estimated No.
1	10,310 – 10,788	3	Super-VIII	1,000
2	101,037 – 172,943	77	R. Super-Heterodyne (AR-812)	73,000
	174,263 – 187,113	8	Radiola Super-VIII	16,000
	191,925 – 196,803	3		3,000
	188,948 – 198,545	4	R. Super-Heterodyne (AR-812)	8,000
3	301,310 – 325,488	8	AR-813 Panel*	26,000+
4	350,294 – 395,601	22	R. Super-Heterodyne (AR-812)	50,000
5	501,019 – 532,324	15	R. Super-Heterodyne (AR-812)	33,000
6	540,269 – 548,068	16	Radiola 24	8,500

*Used in both Brunswick phonographs and Radiola Super-Heterodyne AR-813 radio consoles

With the notable exception of the Super-Heterodyne and the Super-VIII, the serial numbers among the various sets are clearly not interleaved. Most of the six digit serial numbers beginning with "1" were assigned to the AR-812 Super-Heterodyne with two notable exceptions. First, there is a definite gap of approximately 16,000 serial numbers in the Super-Heterodyne samples from 172,943 to 188,948 which is filled with Super-VIII serial number samples. Second, the Super-Heterodyne and Super-VIII serial number samples are clearly interleaved in the range of about 189,000 to 199,999, and one can infer from the ratio of the number of serial numbers collected for each set that there were about 3000 Super-VIII sets and 8000 Super-Heterodynes manufactured with serial numbers interleaved in this range. Therefore, the total number of Super-VIII sets as indicated in Table D-3 is 1,000 from the first production run and 19,000 from the second production run for a grand total of 20,000.

If one assumes the number of Super-Heterodynes manufactured in the 4th and 5th production runs is based on the full range of serial number samples, the number of Super-Heterodynes manufactured during the 2nd, 4th and 5th production runs respectively would be 81,000 + 50,000 + 33,000 for a grand total of 164,000. However, there are relatively large gaps in the serial number samples in first half of both the 4th and 5th production runs of the Super-Heterodyne. There is no good reason to believe that serial numbers were skipped in the manufacturing process, but if they were, the number manufactured in these two runs would be about 25,000 and 16,000 respectively, and the total number would only be 122,000—a number that is far too small when compared to the 148,300 appearing in RCA records as reported by Douglas.

There are several plausible explanations of what might have happened to the "missing sets" in the data base of the 4th and 5th runs if indeed they had been manufactured. First, a number of these could have been shipped out to foreign markets such as South America. Another possibility is that a number of unsold sets at the end of the production cycle were disassembled for parts in lieu of clearing inventory by selling them at a huge discount to Gimbels, for example, in the way the RC and Radiola III sets were cleared. In contrast to the Radiola III and RC sets whose parts were obsolete towards the end their life, the Super-Heterodyne sets still had valuable parts which could have been used in the manufacturing of the Radiola 25 for the following season. Alternatively, the catacombs—which were retailing for about $72 each as replacements—could have been removed and sold to RCA dealers. The 164,000 figure inferred from the full range of serial number samples is about 10% higher than the RCA figure—a respectably small margin of error (assuming the RCA number is correct).

The AR-813 panel was originally designed for the Brunswick phonograph, but a number in excess of those needed for Brunswick phonographs were manufactured, and the excess was used in the AR-813 Super-Heterodyne console. Those used in the phonographs had serial numbers below 325,000 (approximately), while ones used in the radio had numbers above 325,000. The small sample size collected for the AR-813 Super-Heterodyne is insufficient to make an estimate of the number of these AR-813 console radios.

The estimated of number of Radiola 24's, the last set covered by this table, is relatively straightforward. The number of sets estimated from the 16 serial number samples is the difference between the highest and lowest serial number rounded up to 8,500—a reasonable number considering that 19,500 of the competing but more popular Radiola 26 sets were manufactured.

Wireless Specialty Apparatus Serial Numbers

All the sets manufactured by Wireless Specialty for RCA were assigned serial numbers, and further, it is believed that they were all assigned sequentially beginning with 1—with the possible exception of the Radiola VII-B. Because the locations of the serial numbers on some sets are well hidden, the locations are given in Table D-4. Most of these sets are sufficiently rare that an insufficient number of serial number samples have been collected to make estimates of the number manufactured. The limited serial numbers collected to date are shown individually in the fourth column.

Table D-4. Serial numbers and estimated number of sets inferred from serial number data observed on Wireless Specialty Apparatus broadcast radio apparatus manufactured for RCA

Model	Serial Number Location	No. of Samples	Serial Numbers	Est. No. of Sets
AA-484	Name plate on bakelite board	1	281	≥281
AA-485	Name plate on front wood panel	1	566	≥566
AR-1375	Stamped into top edge of front panel	1	2,217	See text
Radiola Special	Stamped into top edge of front panel	11	3,293; 4,267; 4,558; 5,246; 6,235; 6,391; 6,814; 7,017; 7,997; 8,476; 9,896	10,000
Radiola VII	Stamped into top of interior wood base	3	191; 594; 1,286	≥1,286
Radiola VII-B	Stamped into under side of wood panel	2	12,506; 12,554	See text
Radiola IX	Stamped into under side of wood panel	3	167; 2123; 2327	≥2327

Based on the limited serial number data availabe, all one can say about the AA-484 and AA-485 is that there must have been at least 281 and 556 sets manufactured respectively. Neither of these models has been reported for sale in auctions or flea markets for a number of years.

With regard to the AR-1375 and the Radiola Special, there are serial numbers stamped into the top edge of the front panel on all of the eleven Radiola Specials and the one AR-1375 set observed by this author. One AR-1375 has been reported without a serial number, as well. Since the highest serial number observed to date is 9,896, it is logical to assume that there were at least 10,000 AR-1375 and Radiola Special sets manufactured. However, according to RCA records reported by Douglas, there were only 635 AR-1375 sets manufactured, a number far too small to account for the observed serial number of 2,217 on the one AR-1375—not to mention the very large number of Radiola Specials converted from AR-1375's that are known to exist both from serial number evidence and appearances in auctions and flea markets over the years.

From the two serial number samples for the Radiola VII, one can conclude that at least 1,286 sets were made. This number is not inconsistent with RCA data reported by Douglas indicating that 1,692 were manufactured—a number which is supported by other RCA documents. Serial numbers for the Radiola VII-B would indicate that at least 12,554 were made, but this number is at odds with the 2,569 reported by Douglas—and indeed the small numbers of sets appearing on the auction and flea markets today do not support such a large number manufactured. Recognizing that the Radiola VII-B and Radiola IX used the identical receiver panel, it is possible that the serial numbers for the two sets were assigned in one sequence such that the 12,554 sets implied by the serial numbers included both the Radiola IX and the Radiola VII-B. However, there is a problem with this postulate because there were not a large number of Radiola IX sets manufactured either, so the total of both Radiola VII-B and IX sets was almost certainly less than 12,554. Yet another possible explanation is that WSA began the serial number sequence for the Radiola VII-B with 10,000 rather than 1, in which case the serial number data would suggest that only 2,554 sets were manufactured—consistent with the number appearing in RCA records reported by Douglas.

Appendix E

Relative Availability of Radiola Apparatus

The availability of apparatus is always of interest to both the collector and historian, but it is not always easy to quantify. Two obvious metrics to gauge availability are the number manufactured and the number appearing on popular auction markets. Both metrics have their advantages and disadvantages. The number manufactured is a good metric because it is does not change with time, but it does not necessarily relate to the number available today. Further, the number manufactured is not known for all sets. On the other hand, the number available at auction is a direct metric of availability, but only for that year and for the particular auction site. This metric may change over time and may vary from one auction site to another.

These two metrics are often—but not always—correlated. Some sets are so popular that they are actually harder to obtain and more expensive than other sets with smaller production runs. Further, some sets such as the larger consoles were disproportionately discarded as compared to the smaller sets, which were easier to store in basements or attics. Also because of their size they do not appear as often in on-line auctions. Nevertheless, the two metrics will be shown to be reasonably well correlated for most apparatus.

The number of sets originally manufactured and the approximate number appearing at auction on a popular auction web site in 2004 are shown in the 2nd and 3rd columns of Table E-1. The measure of availability used here is based on the number sold at auction in 2004, which is shown in Column 3. Four terms have been selected to describe relative availability, which appears in the last column: 1) **plentiful**—more than ten sold at auction in a year, 2) **available**—between one and ten sold at auction in a year, 3) **scarce**—none sold at auction in a year, but can be purchased by networking with other collectors, and 4) **very rare**—only a few are know to exist, and it is likely that only the most avid and well-heeled collector will be able to acquire one. The distinction between "plentiful" and "available" is admittedly somewhat arbitrary—apparatus which are "plentiful" can be found and acquired almost immediately, while apparatus which are "available" may take more time to acquire—perhaps as long as a few months or even a year. The distinction between "scarce" and "very rare" is based on the author's personal experience in attempting to acquire the apparatus falling in these two categories.

The availability of Radiola loudspeakers is shown in Table E-2. While there is insufficient information on the number manufactured, the number appearing at auction can be used to characterize availability.

Correlation of the Metrics

It is obvious that the two metrics used here—the number manufactured and the number appearing at auction in a given year—are correlated at some level. The apparatus appearing most often at auction were also manufactured in the greatest quantities, and the ones appearing least often or not at all were manufactured in the smallest quantities. There are not sufficient data to perform a rigorous correlation analysis, but a simple measure of correlation can be made by comparing the ratios of the two metrics. If the ratios for all apparatus were identical, the correlation would be "perfect." The degree to which the ratios deviate indicates the degree of noncorrelation. The ratios of the two

Table E-1. The relative availability of Radiola receiver apparatus

Model	No. Mfg.[a]	No. Sold at Auction in 2004	Correlation of Metrics	Relative Availability
DA	10,828	4	0.7	Available
RA	10,828	4	0.7	Available
RC	65,587	38	1.2	Plentiful
Aeriola Jr.	**24,249+**	25	≤2.0	Plentiful
Aeriola Sr.	**55,102+**	91	≤3.3	Plentiful
Aeriola Sr. Amplifier	Unknown	10	—	Available
Aeriola Grand	5,200	0	—	Very Rare
Radiola Grand	11,100	7	1.3	Available
Radiola Senior	3,300	6	3.6	Available
Radiola AC	5,800	5	1.72	Available
Radiola AR	2,500	0	—	Scarce
Radiola RT	~2,500	0	—	Scarce
Radiola RS	**4613**	8	3.5	Available
Radiola Special	10,000	5	1.0	Available
AA-484	~100's	0	—	Very Rare
AA-485	~100's	0	—	Very Rare
AR-1300	Unknown	2	—	Available
AA-1400	Unknown	4	—	Available
AA-1520	Unknown	0	—	Scarce
AR-1375	**635**	0	—	Very Rare
AR-1382	Few	0	—	Very Rare
ER-753	Unknown	2	—	Available
ER-753A	Unknown	Incl. with Radiola I	—	Available
Radiola I	**7,900**	5	1.3	Available
Radiola II	**9,594**	14	2.9	Plentiful
Radiola III	293,000	158	1.1	Plentiful
Balanced Amp	55,000	14	0.5	Plentiful
Radiola III-A	**164,624**	69	0.8	Plentiful
Radiola IV	10,400	10	1.9	Available
Radiola V	**21,138**	14	1.3	Plentiful
Radiola V-A	Incl. with V	4	Incl. with V	Available
Radiola VI	**2,774**	2	1.4	Scarce
Radiola VII	**1,692**	1	—	Scarce
Radiola VII-A	Few if any	0	—	Very Rare
Radiola VII-B	**2,569**	1	0.8	Scarce
R. Super-Heterodyne	**148,300**	32	0.4	Plentiful
Radiola Super-VIII	20,000	3	0.3	Available
Radiola IX	Unknown	0	—	Scarce

Table E-1 (continued)

Model	No. Mfg.[a]	No. Sold at Auction in 2004	Correlation of Metrics	Relative Availability
Radiola Regenoflex	≥14,000	6	≤0.9	Available
Radiola X	**14,000**	2	0.3	Available
Radiola 16	**85,320**	2	1 .5	Plentiful
Radiola 17	**179,917**	70	0.8	Plentiful
Radiola 18	**250,000**	134	1.1	Plentiful
Radiola 18 DC	Incl. with 18	0	—	Scarce
Radiola 20	**135,121**	41	0.6	Plentiful
Radiola 21	Unknown	6	—	Available
Radiola 22	Unknown	1	—	Available
Radiola 24	**7,079**	4	1.1	Available
Radiola 25	**94,815**	14	.3	Plentiful
Radiola 26	**19,507**	21	2.2	Plentiful
Radiola 28	**84,978**	6	0.1	Available
Radiola 28 AC-Only	Unknown	0	—	Scarce
Radiola 30	**10,835**	0	—	Scarce
Radiola 30A	Incl. with 18	0	—	Scarce
Radiola 30A DC	See 30A	0	—	Scarce
Radiola 32	Unknown	0	—	Scarce
Radiola 32 DC	Unknown	0	—	Scarce
Radiola 33	Incl. with 18	25	0.2	Plentiful
Radiola 33 DC	See 33	0	—	Scarce
Radiola 41	Unknown	0	—	Scarce
Radiola 41 DC	Unknown	0	—	Scarce
Radiola 44	Unknown	9	—	Available
Radiola 46	**97,633**	4	0.08	Available
Radiola 46 DC	See 46	0	—	Scarce
Radiola 47	**15,120**	1	—	Available
Radiola 50	**3,220**	0	—	Scarce
Radiola 51	Incl. with 18	1	—	Available
Radiola 51 DC	See 51	0	—	Scarce
Radiola 60	**135,336**	77	1.1	Plentiful
Radiola 62	**19,972**	3	0.3	Available
Radiola 64	**7,406**	0	—	Scarce
Radiola 66	Unknown	7	—	Available
Radiola 67	**4,939**	2	0.8	Available

[a]Figures in bold font are taken from Douglas' *Radio Manufacturers of the 1920's*; other entries are the authors estimates based on serial number data from Appendix D; figures preceded by ~ are simply the authors best guess

metrics for each set are shown in the 4th column, adjusted so that half are greater than one and half are less than one. Thus, on average, the closer the ratio is to one, the better the correlation between the two metrics. No ratio is entered in the 4th column when there were none sold or the number manufactured is unknown. The adjustment is obtained by multiplying the ratio of "number at auction" to "number manufactured" for each apparatus by 2000—implying that *on average*, for every 2000 of each apparatus manufactured, one appeared at auction in 2004. For example, there were 250,000 Radiola 18 sets manufactured, and so one would expect that 250,000/2000 = 125 sets would have appeared at the selected auction site in 2004—very close to the actual number of 134.

It is interesting to note that most of the ratios fall between 0.5 and 2.0, indicating the two metrics are reasonably well correlated (to within plus or minus a factor of two), although there are not enough data to place confidence limits on the correlation. There were substantially more Aeriola Jr. and Aeriola Sr. sets sold at auction than would be expected from the number manufactured as indicated in RCA records—but that number most likely reflects those sold by RCA and not those manufactured and sold by Westinghouse before RCA assumed responsibility for marketing Westinghouse sets. There were also more RS sets appearing at auction than expected, and indeed the serial number evidence suggests that about 5,300 were made. There are several significant outliers including Radiolas 28, 33, 46—for which far too few sets appeared at auction in relation to the number reportedly manufactured. One might argue that this result would be expected for consoles and floor models which may have been preferentially discarded over the years relative to table models (due to size, weight and storage constraints), but it does not explain the relatively low number of table model Radiola 33 sets.

Table E-2. The relative availability of Radiola loudspeaker apparatus

Model	No. Mfg.	No. Sold at Auction in 2004	Correlation of Metrics	Relative Availability
Loudspeaker Attachment	Unknown	0	--	Scarce
Vocarola	Unknown	1	--	Scarce
FH	Unknown	3	--	Available
UZ-1320	Unknown	17	--	Plentiful
UZ-1325	Unknown	76	--	Plentiful
100	Unknown	102	--	Plentiful
100-A	Unknown	161	--	Plentiful
100-B	Unknown	14	--	Plentiful
102	Unknown	0	--	Scarce
103	Unknown	67	--	Plentiful
104	Unknown	3	--	Available
104 DC	Unknown	0	--	Scarce
105	Unknown	0	--	Scarce
106	Unknown	10	--	Available
106 DC	Unknown	0	--	Scarce

Appendix F

RCA Model Designation Schema

RCA recognized the need for a schema to designate radio apparatus from the smallest component to the largest radio receiver or transmitter system. The numbering scheme needed to be simple, yet contain sufficient information to identify the function and assembly level. RCA adopted the scheme that GE had been using for their apparatus at the time RCA was created, almost certainly with some modification. George Clark captured the essence of this system and documented it in a memo which was found in the *Radioana Collection.* His document is reproduced on the next three pages without change or edit.

The model designation system was quite simple in principle: a) two letters, known as "Designation Letters" followed by b) a number, known as the "Serial Number," and occasionally followed by c) a letter, known as the "Alteration Letter," used when the apparatus differed slightly from the preceding designs. Clark's memo is divided into three sections corresponding to a), b), and c), with a detailed explanation of each.

The number and letters appearing on RCA apparatus often did not correspond to this system for two reasons: 1) the RCA designation system was created before the RCA-Westinghouse agreements, and Westinghouse continued to use their quite different designation system for several years until manufacturing at GE and Westinghouse was unified, and 2) design practices changed, occasionally requiring changes in the system. Several examples of these changes and the resulting impact on the designation system follow.

In the early days, a complete transmitter or receiver generally consisted of several boxes, and so the system consisting of the entire collection of boxes was assigned the letter E for "equipment" followed by an R for a receiver or T for a transmitter (e.g., ER or ET). Each of the boxes forming the equipment was considered to be an "assembly" so the boxes were assigned an A followed by an R for a tuner (receiver), a second A for an amplifier, or a D for a detector. However, as time passed the receivers were housed in a single unit such that the letters E (for equipment) and A (for assembly) became redundant. A choice had to be made, and so the letter E was dropped in favor of the letter A. There were only two or three receivers assigned the E designation, while most of the rest were assigned the A designation.

The second example is the serial number designation. In the early days, the three digit number following the first two letters was referred to as the "serial number," a number which was the same for all sets of a given model. RCA did not assign different serial numbers to each set—although both Wireless Specialty and Westinghouse did. Beginning with the Radiola IV in 1923, GE also begin to assign individual serial numbers to each set, and so the individual number assigned to each set of a given model became the serial number, while the three or four digit number following the first two letters became part of the model number.

The Radio Corporation's Model Designation consists of:-

a. Two letters, known as "Designating Letters".
b. A numeral, known as the "Serial number".
c. A letter, known as the Alteration Letter, where the apparatus differs slightly from preceding designs.

An example of a Model Designation for a piece of apparatus whose design is not dependent on the supply circuit used, would be:-

UV 200 A

UV are the Designating Letters;
200 is the Serial Number
A is the Alteration Letter.

a. Designating Letters.

The first letter of the two Designating Letters indicates the mechanical form of the apparatus. This letter is one of the following five:-

P. meaning PIECE. This is used for apparatus which is not complete in itself, and which, in itself, serves no complete major function. Thus, a quenched gap gasket, a key contact, a motor field coil, are <u>pieces</u>.

A <u>piece</u> may not contain as a component part another device to which a model designation has been assigned. Thus, a so-called "spark gap unit", formed by two sparking surfaces and a gasket, to all of which "piece" model designations have been assigned, is really a unit. The complete quenched gap, which comprises a number of spark gap units, a spark gap mounting, etc., is therefore an <u>Assembly</u>.

U. meaning UNIT. This refers to a device which is complete in itself, and which performs a definite major function without the addition of other apparatus. This applies especially to apparatus designed for use on the Unit Assembly principle. A variable condenser in a case, a hand key complete on base, are <u>units</u>.

A unit may have as a component part a device to which a model designation has been assigned, provided this device is a piece. A Unit cannot have another unit as a component part. Such a combination of two units would be an <u>assembly</u>.

A. meaning ASSEMBLY. This consists of more than one unit, or an aggregation of units and parts, mounted on a common panel or within a common case. Thus, a panel type spark transmitter is an <u>assembly</u>, a tuner is an <u>assembly</u>.

S. meaning SYSTEM. This is an aggregation of two or more units, or assemblies, or units and assemblies, which are grouped together because, in the case in question, they all work together to perform a definite function. Thus an aggregation of tuners, couplers, condensers, etc. all in separate boxes, would form a Tuning <u>System</u>.

Mar. 22-6A

E. meaning EQUIPMENT. This refers to a complete working aggregation which in itself is sufficient for radio transmitting, radio receiving, or any major sub-operation thereof. Transmitting equipment means, in general, the combination of the power supply and the purely radio apparatus used in transmission, not including the antenna-ground system; receiving equipment means the combination of the purely radio apparatus used in receiving, the detector, and the listening or recording devices, but not including the antenna-ground system.

The <u>second letter</u> of the two designating letters indicates the electrical classification under which the apparatus is classed. This second letter may be any one of the following:-

A. Amplifier apparatus. All types of amplifiers, Includes resonance relays and mechanical amplifiers, Does not include the vacuum tubes proper used in tube types of amplifiers.

C. Condensers, transmitting and receiving. Includes both complete units and condensers assembled as part of other apparatus. Protectors made up for such use exclusively and furnished with terminals will be listed under protector, even tho they consist of a condenser only.

D. Detector, (except detector vacuum tube). Includes all forms of crystal detectors, magnetic detectors, coherers, etc.; test buzzer; Tone tester. Includes heterodyne, synchronous detector, RAF detector, AF detector, etc.

F. Arc transmitters and component parts thereof not listed under separate heading. Thus, wave changers for arcs are listed under L; anodes for arcs under F.

G. Antenna; ground; counterpoise; towers; masts. Does not include insulators.

H. Radio alternators. Includes magnetic modulator. Includes tuner, if designed <u>first</u> for use with alternator.

I. Insulators.

J. Teleautomatic systems.

K. Radio compass. Radio navigational aids. Pilot cable.

L. Inductances, transmitting and receiving, Couplers, loading inductors; wave-change couplers; wave-change switches.

M. Meters. Power meters. Radio frequency meters.

P. Power equipment, Batteries; motor-generators; switchboards; gasolene engine equipment; oil-engine equipment; switches; fuse protectors; transformers; motor starters; resistors; rheostats.

Q. Station equipment. Such apparatus as may be mounted at will about the station, and not necessarily attached to the main radio equipment. Send-receive switches; antenna selector switches; antenna grounding switches; keys; jacks for telephones.

R. Tuners (receiving); receivers as a whole.

S. Spark transmitters and parts not classified under other heads. Includes transmitting tuners if <u>first</u> designed for use with spark equipment.

T. Tube transmitters and parts not classified under other heads. Includes transmitting tuners, if <u>first</u> designed for use with tube equipment.

V. Vacuum tubes, transmitting and receiving.

W. Wave meters (radio frequency meters); measuring devices; test sets; audibility meters; meggers.

X. Miscellaneous apparatus not clearly coming under any of the above groups.

Z. Aural or visual indicating means to convey intelligible signals to the receiving operator. Includes telephone head sets; telephone cords; phonographic recorders; undulators; photographic recorders; ink writers, etc..

b. Serial Number.

The serial number assigned to each piece of apparatus is the main designating feature, enabling the apparatus to be looked up serially in the Model Designation Catalogue.

c. Alteration Letter.

When a device is redesigned, the new design will in general fall into two distinct classes:-

a. Where the new design interchanges electrically and mechanically with its predecessor.
b. Where the new design differs electrically, or mechanically, or both, from its predecessor.

In the latter case, the new design will be assigned a totally new Model Designation, reference being made in the catalogue to the previous design. This is necessary in order that there shall be no confusion as to whether the new device has the same wave length range as the preceding one, or can be bolted down on the same bed plate, or can be inserted in a receiving panel frame to replace an old unit.

If the new design interchanges both electrically and mechanically with the preceding issue, where such electrical or mechanical features are ones that could possibly affect identity of operation or of location, the first re-design will have the Letter A following the serial number; the second re-design the Letter B, and so on.

Appendix G

The Art of Radiola Advertising

From the outset, RCA depended heavily on advertising to promote its products and services. During the first decade RCA used a variety of advertising venues including newspapers, magazines, sales brochures and catalogs, and other advertising display material such as signs, tapestries, blotters, and the Radiotron Boy. All of these items, particularly the printed material, provide a permanent record of equipment descriptions and the time frames in which they were actively marketed. The focus of this appendix is on the commercial artists and the creative and colorful art they created to advertise Radiolas in the 1920's.

In the early years, RCA ads focused on transoceanic communications—the original motivation for creating RCA. It was not until the end of 1921 when RCA assumed responsibility for marketing Westinghouse receivers that RCA began to advertise broadcast radios in earnest. Even then, the advertising for transoceanic communications was clearly more elaborate than ads for radio receivers. For example, Radiola ads in 1922 were in black and white and were confined to one page or less, while ads for "World Wide Wireless" radiograms—such as the ad in the November 18, 1922 edition of the *Saturday Evening Post*—not only covered two full pages but were also in color. It would not be until 1923 when Radiola ads for receivers were on a par with ads for transoceanic communications, and not until Christmas of 1923 when Radiola ads would begin to appear in color.

RCA placed their receiver ads in three types of magazines—those directed at the general public as exemplified by the *Saturday Evening Post*, *Literary Digest*, *Collier's* and *Liberty*; those directed at rural areas without electrification as exemplified by *Country Gentlemen*, *Capper's*, and *Successful Farming*; and those directed at radio amateurs and other radio enthusiasts as exemplified by *Radio*, *Radio News*, *The Wireless Age* and *Popular Radio*. Almost all of the Radiola ads created in the 1920's can be found in one or more of these magazines.

The progression of ads in terms of sophistication, complexity and color can be seen in the succession of chapters of this book covering the time span from 1922 when broadcast receiver ads first appeared to 1929 when these ads reached a zenith. The earliest ads are in black and white, and focus on the radio receiver equipment, with little artistic content. The black and white ads became more sophisticated, often with a stylized genre evocative of impressionism and signed by the artist. Color ads began to appear, first around Christmas of 1923, and then on a regular basis thereafter. Often a color ad would appear in one magazine while the same ad would appear in another magazine in black and white.

The early ads had a limited range of color, often appearing to have a distinct sepia flavor. By the mid-twenties, the ads began to appear in full color, often with a distinct impressionistic flavor. A few years later the ads took on an art-deco style reflecting the roaring twenties life style and the flapper era. These later ads were the most colorful and most numerous, and therefore probably the most memorable.

The Artists

Almost from the outset, RCA used graphic artists to create drawings for their ads, and indeed many of the magazines had a group of such illustrators who could be called upon to support their clients as needed. Undoubtedly RCA also contracted with illustrators directly to create their ads, because a number of these illustrations appeared in multiple magazines. While many other companies commissioned ads with the artist's signatures in the margins or the corners of the drawings well before 1920, it was not until sometime in mid-1923 when Radiola ads began to appear with the artist's signature. From that point on, the artist's signature more often than not appeared in the illustrations used in Radiola ads.

While there is no known documentation focusing on the illustrators RCA used in the twenties, many of them can be identified from their signatures in the ads. The artists which have been identified in this way are listed in Table G-1 along with the Radiola models appearing in the ads and the approximate year in which they appeared. Many of these illustrators were artists in their own right, and went on to display in art galleries and museums.

Perhaps the earliest Radiola ad with the artist's signature in the margin is one for the Radiola V by Roy H. Still which appeared in the May 12, 1923 issue of the *Saturday Evening Post.* Little is known about Roy Still except that he was a relatively prolific illustrator for a number of companies in the early 1920's. His illustrations appeared in several early Radiola ads—all in black and white. The first colored illustration in an RCA ad for broadcast receivers appears to be one by Elbert M. Jackson placed in the *Saturday Evening Post* on December 15, 1923 for Christmas.

Artists whose works appeared repeatedly in Radiola ads include David Hendrickson, William Rienecke, Everett Shinn, and Katharine Sturges in particular. While the works of these artists appeared in advertisements for only three or four different Radiola models, they created several different illustrations for each of those models. Most if not all appeared in both color and black and white. Saul Tepper, Henry Albert Botkin, and Karl Godwin created illustrations for only one or two Radiola ads, but they appeared often and were particularly striking.

Many artists had interesting and varied careers quite apart from the world of commercial art. Everett Shinn was one of the founders of the Ashcan School, a pioneering group of artists that portrayed New York in a fresh and stark uncompromising style. Shinn was a master of pastels and used that medium in portraying New York. Henry Botkin was a cousin of George Gershwin and the two often worked together—Gershwin composing opera and Botkin painting life of the American Negro in the South. Saul Tepper produced illustrations for most of the major magazines of the time, as well as advertisements for major corporations including General Motors, Coca-Cola, Mobil, Texaco, GE and RCA. He was elected into the Society of Illustrators Hall of Fame in 1980.

The list of illustrators used by RCA in the accompanying table is extensive but not complete. Obviously, this author has not seen all the ads published by RCA, and of those that the author has seen, there are a few in which the illustrator's signatures are undecipherable. The artists that do appear in Table G-1 represent a good cross section of those used by RCA, and their illustrations provide a full range of the style and content of the graphics used to illustrate both RCA ads in particular and magazine articles of the day as well.

About half of the artists listed in the table appeared in one or more editions of *The Illustrator in America* by Walt Reed, a book chronicling the art of illustrating magazines, books and newspapers of the twentieth century. Reed selected only the best artists for his book, and so many of RCA's illustrators share pages in this book with the icons of illustration including the likes of Norman Rockwell, Charles Russell, Howard Pyle, and Frederic Remington.

This advertisement for the Radiola V by Roy Still was one of the first–if not the first–Radiola ad created for RCA with the artists signature in the margin (upper left under the bench). (*Saturday Evening Post*, May 12, 1923, p. 146)

Table G-1. A sampling of the graphic artists used by RCA for Radiola ads in the 1920's

Artist	Life Span	State	Radiola Ads	Style	Ref.
Beckhoff, Harry	1901 - 1979	NJ	Radiola 20 (1927)	Cartoons	1,2,4
Booth, Hanson	1885 - 1944	IN	Regenoflex	Line Drawings	1,3,4
Botkin, Henry Albert	1896 - 1983	NY	Radiola III (1925)	Impression-ism	1,4
Frame, Walter K.	1885 - 1957		Radiola 44	Art Deco	1,3
Godwin, Karl	1893 - 1962	NY	Radiola 28	Modern	1,4
Guild, Lurelle V.A.	1898 - 1985	NY	Radiola 66 (1930)	Realism	1,4
Hendrickson, David	1896 - 1973	CA	Radiola X (1924) Radiola 20 (1926) Radiola 16 (1928) Radiola 60 (1929) Radiola 22 (1930)	Line Drawings	1,2,3,4
Jackson, Elbert M.	1896 - 1962	NY	Radiola II (1923)	Americana	1,2,4
Kernan, Joseph F.	1878 - 1958	NY	Regenoflex (1924) Regenoflex (1924)	Realism	1,2,4
Patterson, Russell	1896 - 1977	NY	Loudspeaker 104	Modern	1,2,3,4
Perley, Myron	1883 - 1939	USA	Radiola 28	Art Deco	1,3,4
Rienecke, William	1892 - 1937	NY	Radiola 62 (1928) Radiola 64 (1928) Radiola 33 (1929)	Art Deco	1,3,4
Seaford, John Albert	1858 - 1936	IN/ MA	Radiola Super-Het. (1925)	Americana	1,4
Seaton, Walter W.	1895 - 1957	CA	Radiola 28 (1927) Radiola 46 (1929)	Modern	1,4
Shinn, Everett	1876 - 1953	NY	Radiola 28 (1926) Radiola 26 (1927) Radiola 30 (1926)	Modern	1,2,4
Still, Roy H.	1904 - ?	NY	Radiola II (1923) Radiola V (1923) Radiola RC (1923)	Sketch	1,3
Stoops, Herbert Morton	1887 - 1948	CT/ NY	Radiola 28 (1927) Radiola 25 (1927)	Americana	1,2,4
Sturges, Katharine	1890 - 1979	Chi, IL	Radiola 47 (1929) Radiola 67 (1929)	Art Deco	1,2,3,4
Tepper, Saul	1899 - 1987	NY	Radiola 20 Radiola 60 (1929) Radiola 33 (1929)	Impressionism	1,2,3,4

[1]Lisa Reinhardt, ed., *Davenport's Art Reference & Price Guide, Gold Edition*, (Gordon's Art Reference, Inc., Phoenix, AZ, 2003)
[2]Walt Reed, *The Illustrator in America 1860-2000*, (The Society of Illustrators, Inc., New York, N.Y., 2001, 3rd edition)
[3]John Okolowicz, "Radio Art," *Radio Age*, Vol. 22, No. 2, February 1997, pp. 1-8
[4]*Ask Art*, Internet web site at http://www.askart.com/

Appendix H

Other Radiolas

The Radiola trade name was used by a number of companies other than RCA both in the United States and abroad in the 1920's. While RCA made Radiolas for export to Latin America, Japan and other regions, companies other than RCA also manufactured and marketed receivers using the Radiola trade name—primarily in the U.S., Canada, Australia, England, France, Sweden, and Austria. Many of the companies outside the U.S. using the Radiola name were sanctioned or licensed to do so by RCA, but none was in the U.S.—except for Brunswick-Balke-Collender and Victor Talking Machine Companies, who sold Radiola apparatus supplied by RCA—most often combined with their phonographs.

The objective of this appendix is to chronicle the wide spread use of the Radiola trade name by a number of companies other than RCA in the twenties. No attempt has been made to identify every company that used the Radiola name—or even every country in which they were manufactured or marketed. Examples of Radiola apparatus and/or ads from selected companies are shown in this appendix.

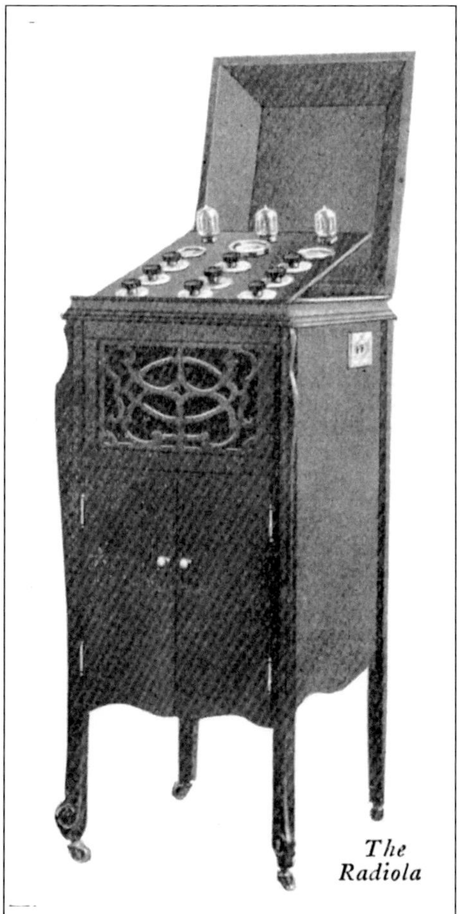

A Radiola manufactured by a relatively unknown company, Wilson-McGuire of San Francisco, was a rather impressive-looking receiver in a mahogany cabinet with a storage battery case below, evocative of the Aeriola Grand. (*Radio*, January 1922, p. 24)

United States

The Radiola trade name began to appear on radios in the U.S. circa January 1922, shortly before what is believed to be the first documented use of the name on an RCA receiver in June of 1922.[1] There were at least three documented instances of use by others. Not surprisingly, the use of this name by others stopped a few months later—almost certainly as a result of the registration of Radiola as a trademark by the U.S. Patent Office on March 27, 1922.

Wilson & McGuire Co.

Radiola as a trade name was used in three distinctly different ways by companies other than RCA in the United States. The first example is what one might expect—a relatively unknown company with the name of Wilson & McGuire Co. Inc. manufactured a set with the Radiola name. In a January 1922 advertisement, Wilson & McGuire was described as "*the new shop of Wilson-McGuire Co., Inc., San Francisco, formerly the McGuire Radio Laboratory.*"[2] It appears that the company was nothing more than a small owner-operated company with a new partner.

The Wilson-McGuire Radiola was a rather impressive-looking receiver in a mahogany cabinet with a storage battery case below, evocative of the Aeriola Grand. It was described as a non-regenerative receiver with three stages of power amplification. Tuning was accomplished by "*a new and original method,*" and the sound was produced by a Magnavox reproducer. It was advertised in two consecutive issues of *Radio* magazine and then abruptly disappeared, never to be seen again.[3] It is not known whether the radio did not sell, whether the company had financial troubles, or whether RCA made some threat or other accommodation with the company which caused them to take it off the market.

Radiola Wireless Corporation

The second example of a Radiola is more interesting than the first. A set with the Radiola trade name was manufactured by a company with the same name—

The Radiola Wireless Corporation incorporated on March 14, 1922 in New York manufactured a crystal set with the trade name Radiola. (Robb Leamy collection)

The Radiola Wireless Corporation name is evident on the name plate attached to the top of its Radiola crystal set. (Robb Leamy collection)

none other than the Radiola Wireless Corporation! The Radiola Wireless Corporation was incorporated on March 14, 1922 in New York, and later changed its name to Aerovox Corporation effective March 2, 1932. It dissolved voluntarily on October 4, 1945. It is not known what action, if any, RCA may have taken against Radiola Wireless for using Radiola in the corporation name, but they surely would have taken action against the company for using the Radiola trade name on their radio receivers. Since the company kept the Radiola name until 1932, it is likely that RCA did not or could not force them to change the company name.

The receiver shown here is a small unassuming crystal set with a name plate sporting the Radiola trade name and the Radiola Wireless Corporation name. It is not known how many other radios, if any, were manufactured by the Radiola Wireless Corporation.

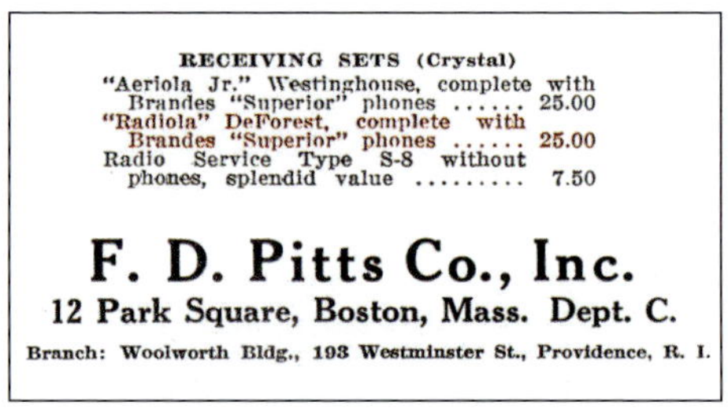

RECEIVING SETS (Crystal)
"Aeriola Jr." Westinghouse, complete with Brandes "Superior" phones 25.00
"Radiola" DeForest, complete with Brandes "Superior" phones 25.00
Radio Service Type S-8 without phones, splendid value 7.50

F. D. Pitts Co., Inc.
12 Park Square, Boston, Mass. Dept. C.
Branch: Woolworth Bldg., 193 Westminster St., Providence, R. I.

Beginning in February 1922, F. D. Pitts began to advertise a De Forest crystal set described as a "Radiola" De Forest, complete with Brandes "Superior" phones for $25.00 (color added for emphasis). (*The Wireless Age*, February 1922, p. 6)

De Forest Radio Telephone and Telegraph Company

The third example, a De Forest crystal set distributed by the F. D. Pitts Co. under the Radiola name, is markedly different in nature than the first two examples. The F. D. Pitts Company distributed radio equipment for a number of companies including Westinghouse, De Forest, Brandes, RCA, Grebe, and

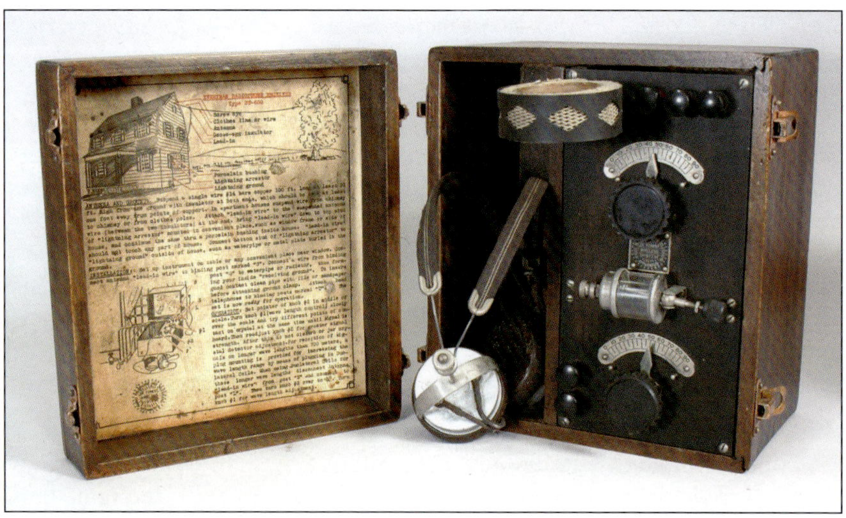

The De Forest Everyman crystal set, which sold for $25 with a Brandes Superior phones, matches the description of the "Radiola" De Forest appearing in Pitts Co. ads in early 1922.

Murdock, to name a few. Beginning in the February 1922 issue of *The Wireless Age*—RCA's own house organ—F. D. Pitts began to advertise a De Forest crystal set described as a "Radiola" De Forest, complete with Brandes "Superior" phones for $25.00. The same ad also appeared in the March and April issues of *The Wireless Age*, as well as in other magazines directed at the radio experimentalist and broadcast radio listener.

While there is no known set appearing in a De Forest catalog described as a Radiola, the only De Forest crystal set sold at that time matching the description in these ads, compete with a Brandes Superior phones for $25, was the De Forest Everyman set. Convincing evidence that the set in question was the Everyman appeared in the May issue of *The Wireless Age* where, the "Radiola" crystal set for $25 in the F. D. Pitts ads had been replaced by one with the Everyman name at $25. It was almost certainly not a coincidence that the Radiola name was discontinued in advertising by F. D. Pitts shortly after March 7, 1922, the date that the Radiola trademark was officially registered in the United States.

Canada

At the time RCA was incorporated in the United States in late 1919, both General Electric and Westinghouse had been operating Canadian subsidiaries for a number of years—Canadian General Electric (CGE) Company Limited and Canadian Westinghouse (CW) Company Limited respectively. In the years following the introduction of broadcast radio in Canada in early 1922, both CW and CGE began to manufacture Radiola sets that were virtually identical to those manufactured by their parent companies for RCA in the U.S. In addition, they also designed and manufactured a few receivers for which there were no U.S. counterparts—although these sets were not given the Radiola name.

Why RCA initially ceded the Canadian radio receiver market to the Canadian subsidiaries of the two companies that manufactured its sets is not known by this author. Perhaps there were export issues, perhaps the economics resulting from licensing arrangements were more favorable than setting up a distribution system in Canada, or perhaps the American parent companies simply did not want to compete with their Canadian subsidiaries, albeit indirectly. RCA did not begin to operate in Canada until 1929 when it took control of the Canadian subsidiary of the Victor Talking Machine Company as part of the acquisition of Victor's parent company in the U.S. and renamed it RCA Victor of Canada, Ltd.

Canadian Westinghouse (CW)

It appears that Canadian Westinghouse got the jump on Canadian GE because an early catalog prepared in late 1922 or early 1923 advertised the complete line of Westinghouse sets from the Aeriola Jr. and Sr. through the Aeriola Grand and Aeriola RS receivers. It is notable that most of the early equipment sold by Canadian Westinghouse used the Aeriola name, and none used the Radiola name. It was not until 1924 when CW produced models with the Radiola name in quantity, although it is possible that some of the early sets manufactured under the Aeriola name were changed to Radiola sometime during the 1923 season. A 1924 CW catalog (HC-12)—virtually identical to the RCA catalog in the U.S. for the same year—lists the following Radiolas: III, III-A, Balanced Amplifier, X, Super-Heterodyne, Super-VIII, Regenoflex, and loudspeakers UZ-1325 and FH.

The Radiola III-A shown on the next page is one example of a Radiola set made by Canadian Westinghouse. At first glance it appears to be identical to the U.S. counterpart, but in fact, it is different in minor details—both externally and internally. The four most noticeable differences on the exterior are 1) the Canadian Westinghouse monogram in the circle on the top of the set versus the RCA monogram, 2) details of the print style, 3) the decal or name plate on the side identifying the respective manufacturers, and 4) the absence of a serial number stamped on the bottom of the box on Canadian sets. The interior of the

The panel of the Canadian Westinghouse Radiola III-A was virtually identical to the RCA version except for the Westinghouse monogram versus the RCA monogram, and details of print style.

two sets at first appears to be identical, but there are clearly differences. The most obvious difference is the 180-degree rotation of two of the audio transformers. It appears that the Canadian Westinghouse designers wanted to put their own imprimatur—however small—on the Radiola sets they manufactured for the Canadian market.

The interior components of the Canadian Westinghouse Radiola III-A were virtually identical to the RCA version, but the orientations of two of the three audio transformers were rotated by 180 degrees.

The name plate of the Canadian Westinghouse Radiola III-A differed significantly from the RCA version—most notably making no reference to an RCA license or heritage.

In addition to the Radiola models, which had Radiola model numbers, designs, and external appearances virtually identical to their U.S. counterparts, CW also marketed several sets of their own design. While these sets had an appearance similar to the U.S. Radiola counterparts, the designs were different and most importantly, the Radiola name was not used. Three such receivers appearing in a Canadian distributor catalog published by A. Carey & Sons dated November 1925 were designated Westinghouse 53, Westinghouse 55, and Westinghouse Eight Super-Heterodyne. A Westinghouse-manufactured loudspeaker designated Westinghouse-Brandes Type H also appeared in this catalog.

The CW Model 53 was similar in appearance in the Radiola III and in function to the Radiola III-A except that it had three tubes, the third tube—a power amplifier UX-120—replaced the two WD-11 tubes which were used in a push-pull circuit in the Radiola III-A. Model 55 was similar in appearance to the Radiola 20 but with certain differences in the regenerative circuit, which apparently was the reason the Radiola name and number was not used. The Westinghouse Eight Super-Heterodyne was similar in appearance to the Radiola 28, but with a slightly different circuit described as a "*coupled radio frequency application to the loop*" to increase sensitivity.

Canadian General Electric (CGE)

It was not until late 1924 that CGE entered the radio market beginning with the Radiola III, III-A and Balanced Amplifier receivers.[4] In the 1925-26 time frame, CGE manufactured the Radiola Super-VIII and Radiola Super-Heterodyne; Radiolas 20, 24, 25, 28, and 30; CGE Models 93, I93, and 205; Radiola loudspeakers 100, 102, 104, and UZ-1325; and the Radiola loop CGE 228. They also manufactured a version of the Uni-Rectron Model AP-935 and Duo-Rectron Model AP-937. The external appearance of the CGE Model 93 was virtually identical to the Westinghouse Model 53, and the CGE model I93 was the CGE Model 93 in a larger cabinet with space for the batteries. The CGE 205 was not similar to any Radiola set in either appearance or circuitry. It had a vague resemblance to the Radiola 20, but did not use thumb wheel tuning, and it had only one stage of TRF amplification, a detector, and three stages of audio amplification. All models listed here except the CGE 205 and Radiola 24 appeared in a CGE catalog entitled *Radiolas Radiotrons and Accessories* published in late 1925 or early 1926.

The Canadian GE Radiola III shown here is one example of a Radiola set made by Canadian GE. The external appearance of the set was virtually identical to the RCA Radiola III except for the CGE monogram on the panel,

The panel of the Canadian GE Radiola III was similar to the RCA version except for the GE monogram versus the RCA monogram, UV tube sockets vs. WD-11 tube sockets, and metallic surfaces plated with a gold metal versus nickel.

The interior components of the Canadian GE Radiola III differed from those of the RCA version, but none more than the capacitors in the antenna circuit which were formed by conducting layers on a fiber board in the GE version, versus discrete tubular capacitors in the RCA version.

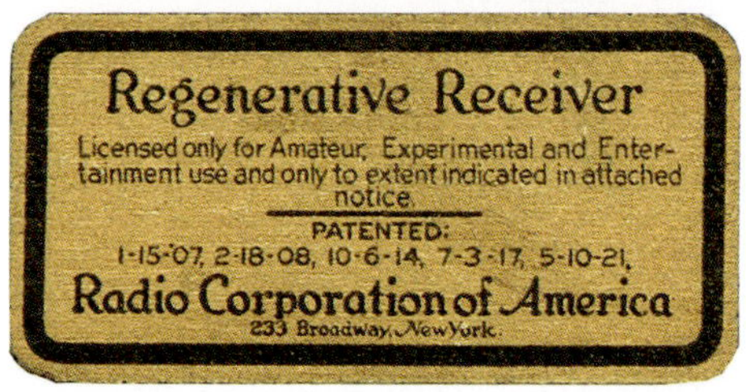

The decal of the Canadian GE Radiola III differed from the RCA version—most notably making no reference to an RCA license or heritage.

the gold-toned finish of the metal parts, the CGE decal on the box, and the UV-199 tubes. However, the interior components were quite different from those in the RCA Radiola III counterpart—in contrast to components in the Canadian Westinghouse Radiola III-A which were identical. For example, the CGE Radiola III with UV-199 tubes used UV tube sockets rather than the WD tube sockets of the RCA Radiola III, and the audio transformer, while similar in appearance to that of the RCA Radiola III, had different terminal locations.

The most radically different component was the resistor/capacitor network used in the antenna circuit—which in the CGE Radiola III consisted of conducting strips laid out on a flat fiber board evocative of the modern PC board—in stark contrast to the discrete tubular components used in the RCA Radiola III and III-A shown in photographs of Chapter 6. The capacitors were formed by sandwiching a conducting layer between fiber boards covered with smaller individual conducting layers on the outer surface facing the interior of the box (see accompanying figure). The difference in components used by CGE—coupled with the similarity of components used by CW—is further evidence that suggests CW's parent, American Westinghouse, made the Radiola III and III-A for RCA in the U.S..

The CW Model 53 and CGE Model 93 Conundrum

Canadian Westinghouse Model 53 and Canadian GE Model 93 are virtually identical in appearance, circuit design, and layout of components—even though there was no three-tube Radiola sold in the U.S. which could have been used as a pattern for either set. The exterior appearance of these sets is similar to the Radiola III except for one additional tube, and the circuits of both are similar to the Radiola III-A—but with two UX-199 tubes used as the regenerative detector and first audio amplifier, and one UX-120 power amplifier tube used in place

The Canadian Westinghouse Model 53 and Canadian GE Model 93 are strikingly similar despite the fact they were made by two competing companies and there was no equivalent three-tube set marketed by RCA in the U.S. after which these sets could have been patterned.

of the two-tube push-pull arrangement in the audio output stage of the Radiola III-A. The similarity to Radiola III or III-A ends there, however, because while the interior components of the two sets are very similar to each other, they are both quite different from the RCA Radiola III or III-A. Totally unlike the Radiola III, the interior of both sets are dominated by a metal can—reminiscent of the Radiola Super-Heterodyne catacomb—in which the two audio transformers are located. While the metal can was not filled with wax, it was sealed with rivets that had to be drilled out to access the audio transformers, which do not have individual metal shields like the ones used on the Radiolas. Other components were similar to each other but were distinctly different from those used in the Radiolas, as well.

The striking similarly in sets made by the two competing companies in the absence of a U.S. Radiola counterpart presents a conundrum that begs for an

The interior of the Westinghouse Model 53 and GE Model 93 are striking similar to each other, but distinctly different from the Radiola III or III-A.

explanation. This conundrum has been noted by others, and it has been suggested that the similarity was a result of collaboration between the two competing companies.[5] This explanation is not entirely satisfactory because it was well known that the American parent companies were fierce competitors, and therefore it is unlikely that the two Canadian subsidiaries with close ties to their respective parents would have cooperated to the extent that would have been necessary to produce these receivers—both with the same unique components and strange catacomb-like layout.

An alternative explanation is suggested by the discovery of a photograph of a three-tube prototype in the George Clark *Radioana Collection* manufactured for RCA by Westinghouse. A photograph of this prototype with an external appearance identical to Models 53 and 93 is reproduced on page 233 of this book. It appears that RCA may have contemplated marketing such a three tube set—but obviously did not. It is logical to assume that RCA licensed this design to both Westinghouse and GE in Canada—as it did for other of its Radiola designs—and that both Canadian subsidiaries opted to manufacture and market a set based on the American Westinghouse prototype developed for RCA.

France

Société Française Radio-Électrique (SFR)

SFR was founded in 1910 by the French engineer Emile Girardeau for the purpose of making receivers and transmitters for the French army. In 1922, SFR founded the first private French radio broadcast company with the name "Radiola", and also started to manufacture radios for civilian use with the Radiola and Radiolina brand names. SFR did not have any known connection with RCA or their manufacturers at the time. In 1932, the civilian radio branch of SFR was sold to Philips, and the Radiola name is still present today on some Philips television sets sold in France. It became one of the largest French radio manufacturers, and the company supplies electronics to the French army, the aerospace industry, and was the first on the market with cordless telephones.

During the 1920's, SFR specialized in tube radios, although they made a few crystal sets as well. All the early SFR Radiolas are now very prized on the French

This rare crystal set manufactured by SFR in France bears the Radiola name.

The Société Française Radio-Électrique (SFR) sold a number of radios under the trade name "Le Radiola," as indicated in this ad from the early 1920's.

vintage radio market, but the crystal sets such as the one shown here are much harder to find than the tube models. SFR Radiola sets can also be found in early catalogs and advertisements such as the one shown here. Note that one of the sets in this ad is designated a Radiola Concert receiver, the same designation appearing on early sets manufactured by Wireless Specialty for RCA.

Australia

Amalgamated Wireless Australasia Ltd. (AWA)

In 1913, Amalgamated Wireless Australasia Ltd. (AWA) was formed by an amalgamation between Marconi's Wireless Telegraph Company, Ltd. and the Australian Wireless Company. Ernest Fisk, one of the founders of the Company, was originally a wireless operator for British Marconi. The Company's logo was almost identical to the RCA logo with the World Wide Wireless globe and lightning flashes, except that the words "Worldwide Wireless" appeared at the top of the globe rather than the bottom to allow for a depiction of Australia in the southern hemisphere.

AWA watched the broadcast radio craze develop in the U.S. in 1922, and urged the government to license broadcast stations in Australia. AWA succeeded in this endeavor by early 1923, and the popularity of broadcast radio in Australia was immediate. The system Australia first adopted was significantly different from that in the U.S. in that the Australian sets were fixed to receive only one station and were then sealed. The potential listener could subscribe to a specific station with a specific frequency for a fee and then buy a set fixed to that frequency. As a result, the sets manufactured by AWA were, of necessity, different in design than U.S. Radiolas. The sealed set approach was not successful, and it was not long before the system was converted to open sets with two tiers of broadcast stations—sometime in mid-1924.

Amalgamated Wireless (Australasia) Ltd.

The Amalgamated Wireless Australasia Ltd. (AWA) formed by an amalgamation between Marconi's Wireless Telegraph Company and the Australian Wireless Company created a logo which is very similar to an early RCA logo.

The AWA Radiola Senior 6, a six-tube superheterodyne battery set with a sloping front panel and protruding antenna designed for use with UX-199 tubes was similar to the RCA Radiola 25. (undated AWA catalog entitled *The New Radiolas for Better Broadcast Reception*)

From the very outset, AWA manufactured and marketed a line of Radiola receivers that often had an appearance and/or a model name evocative of the Radiolas sold by RCA in the U.S. Over the succeeding decades, AWA manufactured a very large number of Radiola models, certainly many more different models than RCA marketed in the U.S. An AWA catalog from the last half of the 1920's contains a number of Radiola receivers—some of which are evocative of Radiola models in the U.S.—such as the AWA Radiola Senior 6, a superheterodyne battery set similar to the RCA Radiola 25 with a sloping front panel, a rotating antenna protruding from the top of the cabinet, and six UX-199 tubes.

England

British Thomson-Houston

British Thomson-Houston, popularly referred to as B.T.H., made a line of receivers with the Radiola name. B.T.H. was formed in 1896 with the capital held mainly by American Thomson-Houston (part of GE of America) and partly by British and French bankers. As a result, B.T.H. had access to patents and licenses from GE in America.

Two examples of B.T.H Radiolas are shown here, the first a Type C, Form A Radiola crystal set that does not look like any RCA crystal set. The second set pictured in a B.T.H. ad is a three-tube set designated as a Radiola III, which actually looks very much like the RCA Radiola IV. B.T.H. made other sets under the Radiola name, but not a large number.

The British Thomson-Huston Company produced a Type C, Form A Radiola crystal set that does not look like any RCA crystal set.

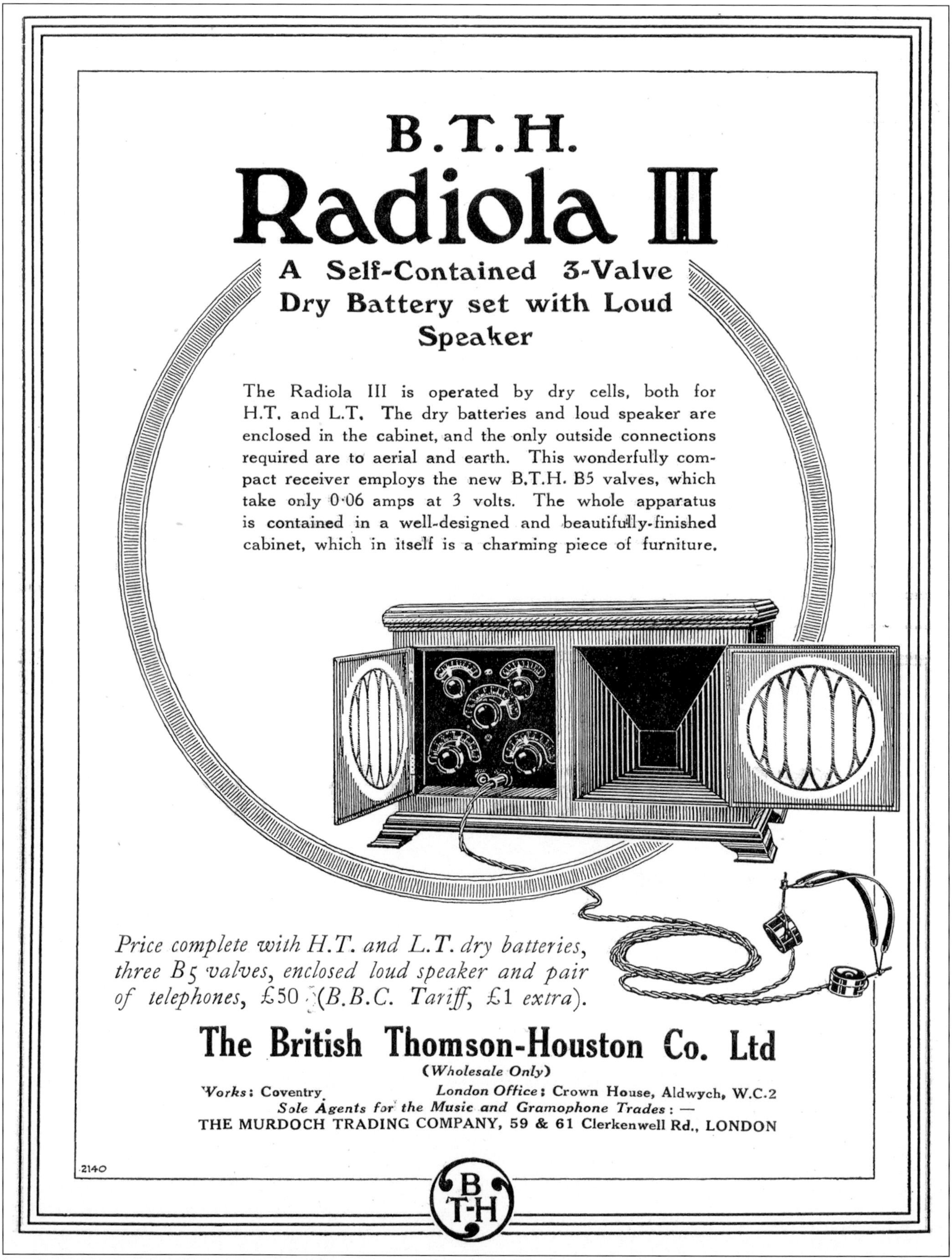

The British Thomson-Huston Company produced Radiola receivers such as the three-tube Radiola III appearing in this ad which is strikingly similar to the RCA Radiola IV.

The Swedish company Svenska Radio Aktiebolaget (SRA) manufactured a line of Radiola radio receivers beginning in the early 1920's, including this set sold under the Radiola Kristall trade name.

Sweden

Svenska Radio Aktiebolaget

The Swedish company Svenska Radio Aktiebolaget (Aktiebolaget is Swedish for Corporation), popularly known as SRA, manufactured a line of radio receivers beginning in the early 1920's with the trade name Radiola. SRA was formed in 1919 by heavy equipment maker ASEA, boiler and gas equipment maker AGS, and telephone manufacturer LM Erickson to focus on research and development in the radio field.

SRA began making crystal sets in 1924 when broadcast transmissions began in Sweden. The Company became interested in valve receivers and soon began to manufacture a line a receivers with vacuum tubes and later with loudspeakers. SRA continued to make radios under the Radiola name for a number of decades thereafter. At some point, British Marconi acquired an ownership position in the company, and it may have been that connection that caused or permitted SRA to use the Radiola name.[6] It has also been rumored that British Marconi warned SRA of RCA's intention to register the Radiola trade name so that SRA could register the Radiola trade name in Sweden first.

An example of an SRA crystal set with the Radiola Kristall trade name is shown here. The set itself is stamped with the stylized monogram containing the letters SRA, and while the body of the set is not stamped with the Radiola name, the detector is. Advertisements for this crystal set appeared circa 1925 for 20 Kr. including the detector (not all SRA crystal sets were sold with the detector). Numerous examples of Radiola sets by SRA—most with tubes—can be found on the Internet, most notably on the Radiomuseum web site.[7]

Svenska Aktiebolaget Trådlös Telegrafi (SATT)

While SRA manufactured Radiolas that were not RCA receivers, Svenska Aktiebolaget Trådlös Telegrafi (SATT) manufactured RCA receivers that were not Radiolas! SATT was a Swedish company owned by Swedish Telefunken that was licensed by both RCA and Telefunken to manufacture several sets very

A SATT brochure of the resistance coupled E 305 has a schematic diagram for the audio amplifier which is most likely identical to the Westinghouse DR configuration. (Bengt Svensson collection)

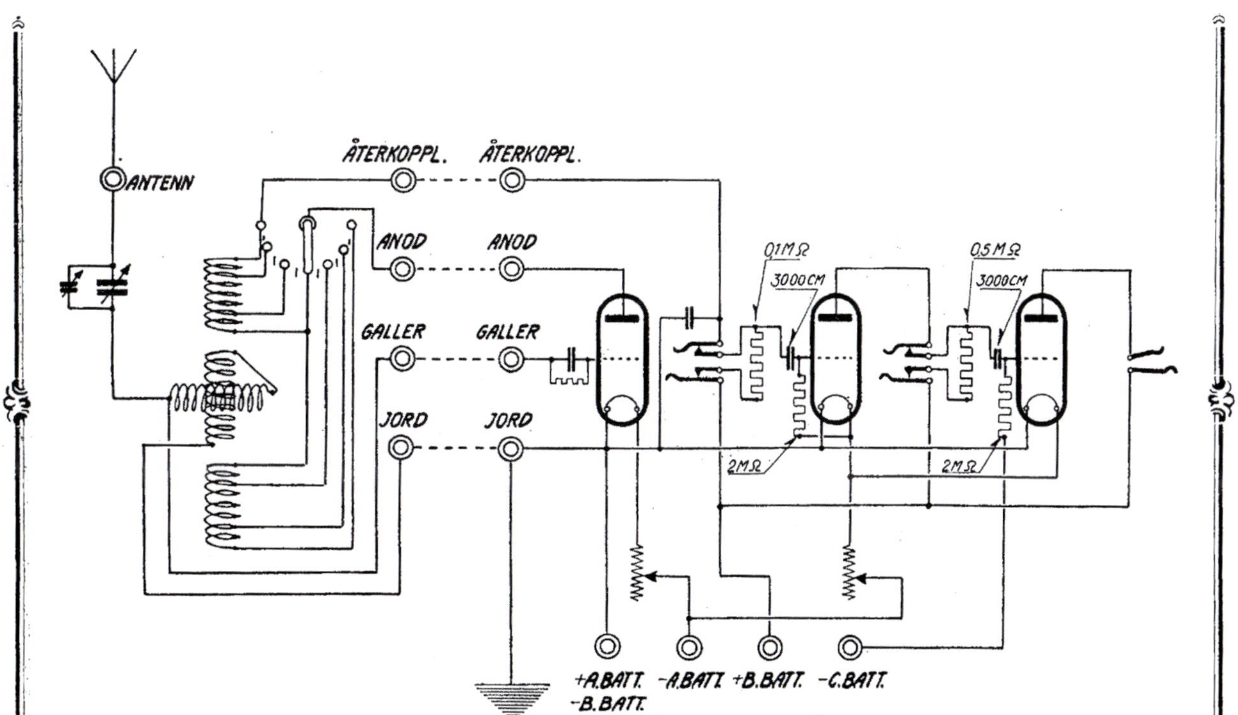

Principschema, rundradiomottagare typ E 305,

This SATT brochure for the E 304 makes it clear that it was licensed or sanctioned by both Telefunken-koncernen of Tyskland (Germany) and Radio Corporation of America. (Bengt Svensson collection)

similar in appearance and design to those marketed by RCA in America—most notably two versions of the RC and two versions of the Radiola III. German Telefunken was licensed to manufacture the Telefunkon D (Model 304) in Germany for export only, and Telefunken and RCA in turn licensed SATT to manufacture the E 304 and E 305 (a resistance-couple version) for domestic consumption in Sweden. Both sets pictured on the SATT brochure shown here are identical in appearance to the German Telefunken D (Model E 304) shown in Chapter 3 of this book. Note the prominent reference to Telefunken-koncernen of Tyskland (Germany) and Radio Corporation of America on the brochure—a sure sign of the licensing agreement.

The SATT Model E 305 was identical in appearance to the E 304, a major difference being the amplifier which was resistance coupled rather than transformer coupled—just like the Westinghouse/RCA DR. The E 305 also had a different range of operating wavelengths than the E 304. A schematic diagram from the SATT Type E 305 manual reproduced here on page 451 shows the resistance coupling circuit, and more importantly gives the values of the coupling resistors and capacitors—undoubtedly the same values as those used

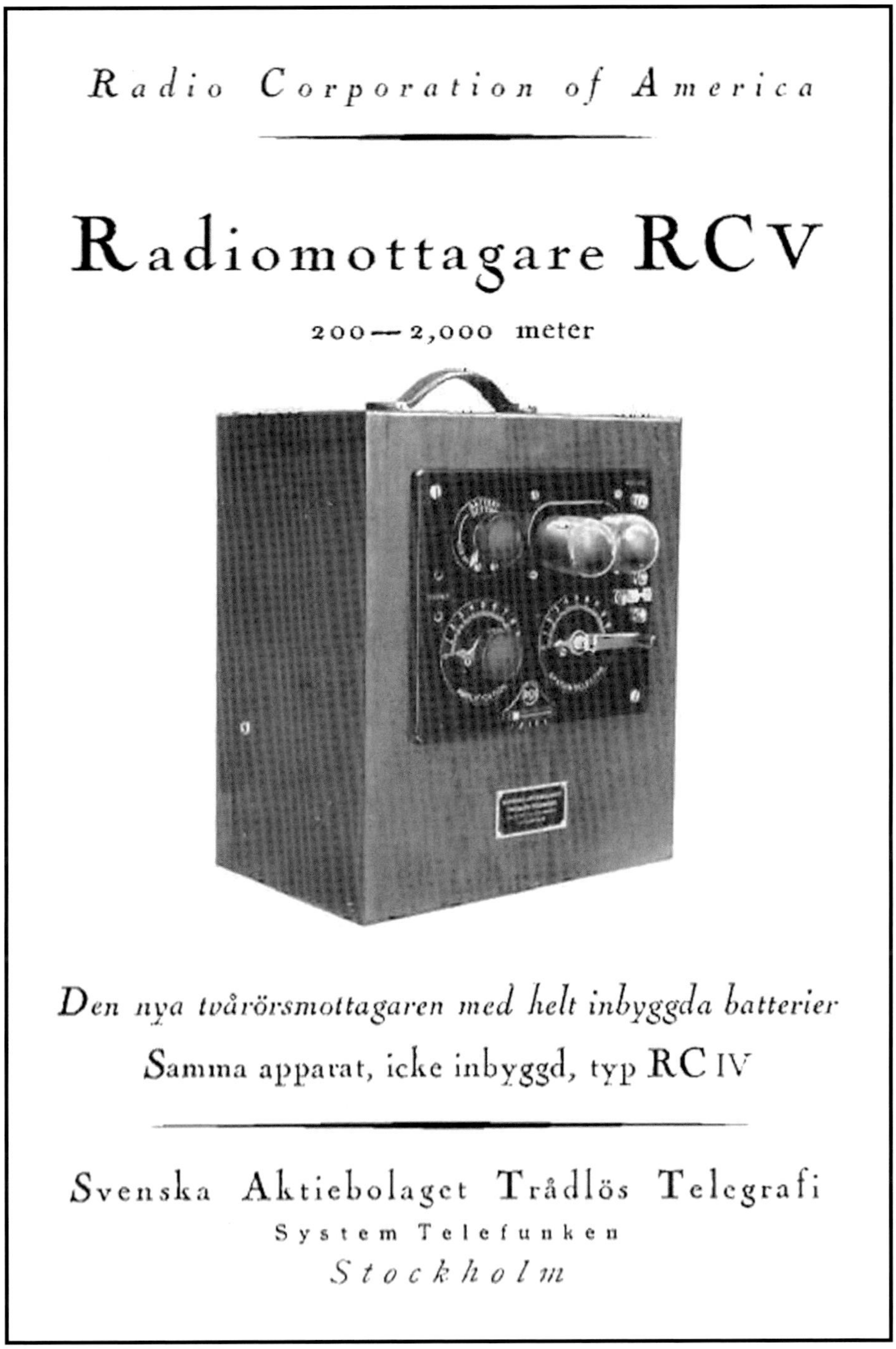

The SATT RCv was essentially the same as the RCiv except that it was placed in a larger box sufficient in size to fit the batteries. (Bengt Svensson collection)

in the RCA DR amplifier for which no factory schematic is known to exist (0.1M and 0.5M for the 1st and 2nd plate resistors respectively, 2M for both grid resistors, and 3000 pf for both coupling capacitors).

The second pair of sets manufactured by SATT had the appearance of the Radiola III, the Model RCiv being identical in appearance to the Radiola III except that there was a small lever jutting from the panel where the Radiola III name would have been on the RCA model to adjust the wavelength to four ranges: 200—350 m, 300—1,300 m, 1,200—1,700 m, and 1,400—2,000 m. While the RCA name still appeared on the panel, a SATT nameplate was placed over the Radiola decal to cover the Radiola name—almost certainly because SRA had reserved the Radiola trade name in Sweden. It would appear that RCA supplied Radiola III hardware to SATT, who in turn modified them by adding switches and coils for the necessary wavelengths. Model RCv was identical to the RCiv

SATT received Radiola III hardware from RCA, removed the Radiola name from the panel of the Radiola III and replaced it with a four position wavelength switch, covered over the Radiola decal with the SATT name, and renamed it the Model RCIV for sale in Sweden. (Anders Widell collection)

RCA Radiola decal on the RCIV with a SATT name plate. (Anders Widell collection)

except that it was placed in a larger cabinet with a handle which also contained the batteries. It is somewhat ironic that this Radiola III look-alike was assigned the model name RC, while the RC look-alikes were assigned the model designations E 304 and E 305.

Austria

Radiola

In 1924, William Wohleber from Heidelberg, Germany formed a small company based in Austria named "Radiola" with the objective of manufacturing radio apparatus to satisfy the demand for broadcast receivers as the broadcast radio craze hit Europe.[8] The Company manufactured sets under different names such as Aeriola, Aero Phon, Radioglobe, Radiola and Radiopa. The company changed its name to Minerva in 1927 because the name Radiola was protected in the U.S. and France. Several early Minerva Radiola sets appear on the Radiomuseum website.

Endnotes and References for Appendix H

1 The first documented used of the Radiola trademark by RCA is believed to be on the AR-1375 crystal set, which appeared in photographs on pages 16-17 of the RCA catalog *Radio Enters the Home*, (Radio Corporation of America, Marketing Department, New York City) June 1, 1922.

2 *Radio*, Vol. IV, No. I, January 1922, p. 24.

3 The last issue in which the ad appeared was *Radio*, Vol. IV, No. II, February 1922, p. 47.

4 D. A. Challoner, "Major C.L. Richardson, Canadian Radio Engineer," *The Cat's Whisker*, (Bulletin of the Canadian Vintage Wireless Association) Vol. 8, No. 3, 1978, pp. 6-17; this article was reprinted in Robert P. Murray, Editor, *Early Development of Radio in Canada 1901-1930*, (Sonoran Publishing, LLC, Chandler, AZ, June 2005), pp. 136-142.

5 Robert P. Murray, ed., *The early Development of Radio in Canada*, (Sonoran Publishing, LLC, Chandler, AZ, June 2005) p.134.

6 The author does not know if Marconi obtained a minority interest in SRA before SRA produced their first Radiola receiver, and if so, whether this interest brought about the use of the Radiola name.

7 http://www.radiomuseum.org/.

8 See Minerva Radio Austria web site: www.minervaradio.com.

Graybar ads such as this one did not contain the Radiola name nor did they reference RCA in any way. (*Saturday Evening Post*, December 15, 1928)

Appendix I

Graybar Electric Receivers

On December 22, 1925 the Western Electric Company, the manufacturing arm of AT&T since 1881, announced that its electrical supply business had been segregated and incorporated as a wholly owned subsidiary named Graybar Electric Company, Inc.[1] The name of the new company was derived from Gray and Barton, the original partnership formed by Elisha Gray and Enos M. Barton in 1869, which was subsequently reorganized as the Western Electric Manufacturing Company in 1872 and renamed the Western Electric Company in 1881. The new company was to be responsible for the distribution of non-telephone electrical equipment manufactured by Western Electric and others.

It was also announced that in addition to motors, generators and other items manufactured by Western Electric, Graybar would sell any radio equipment which the Western Electric Company might manufacture. This portion of the announcement was significant because AT&T and its subsidiary Western Electric were involved in protracted negotiations with RCA regarding the relative rights of the organizations to manufacture broadcast radio receiver apparatus. Approximately six months after the formation of Graybar, AT&T and RCA settled their differences, with AT&T agreeing to leave the broadcast radio business. Not only did AT&T agree not to manufacture broadcast radio equipment, they also announced their intent to sell broadcast station WEAF to RCA as well.[2]

While the agreements between RCA and AT&T—which also included subsidiaries Western Electric and Graybar—precluded Western Electric and Graybar from manufacturing radio equipment, there was apparently nothing in these agreements to preclude Graybar from distributing radio equipment manufactured by a third party. Graybar was eager to enter the lucrative radio field using its existing wholesale outlets in 70 principal cities, and so Graybar negotiated with RCA—and presumably with other manufacturers—to provide radio apparatus that could be distributed under the Graybar Electric name.

On August 16, 1928, Graybar announced its intent to enter the radio receiving set field by September 1928 with five new receivers, four of which were to operate from the light socket.[3] The four AC-operated receivers—Models 300, 310, 330, and 340—were introduced to the radio trade in the September, 1928 issue of *Radio Retailing*.[4] Ads for the public soon began to appear in national magazines such as the *Saturday Evening Post*. It is interesting to note that Graybar did not identify RCA as the supplier of the radio apparatus in these announcements or advertisements, but it became evident to all that RCA had been selected as the supplier when the new Graybar sets appeared on the market with decals clearly stating that the radio apparatus was provided by the Radio Corporation of America.

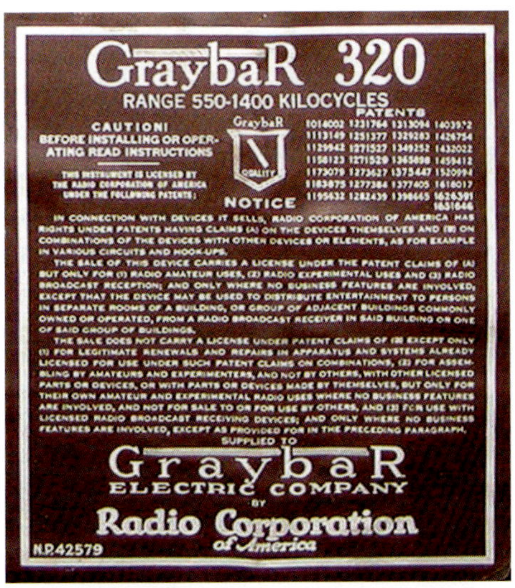

Graybar decals such as this one did not carry the Radiola name, but did state that the apparatus was supplied to the Graybar Electric Company by the Radio Corporation of America. (Courtesy of Robert Fathauer)

RCA's arrangement with Graybar was similar to those with Brunswick and Victor in that RCA supplied the radio apparatus while Graybar provided their own cabinets—which in many cases were significantly different in appearance than those of the equivalent Radiola sets distributed by RCA. One significant difference between the Graybar arrangement and the Brunswick/Victor agreements was that the sets sold by Graybar had no reference to the Radiola name, even though the equipment supplied to Graybar was essentially identical to that sold by RCA in their Radiolas. Thus, these Graybar radios do not

The cabinets for the Graybar sets such as this Graybar Model 320 were significantly different in appearance than the equivalent Radiola sets distributed by RCA—in this case the Radiola 51. (Courtesy of Robert Fathauer)

qualify as Radiola receivers.

Graybar offered thirteen models with Radiola apparatus supplied by RCA during the three model years as identified in Table I-1. This list is thought to be complete—at least for the time frame indicated—because it contains all the Radiolas distributed by RCA in this time frame except for battery-operated Radiolas 21 and 22, and Radiolas 47 and 67 Radiola-phonograph combinations.

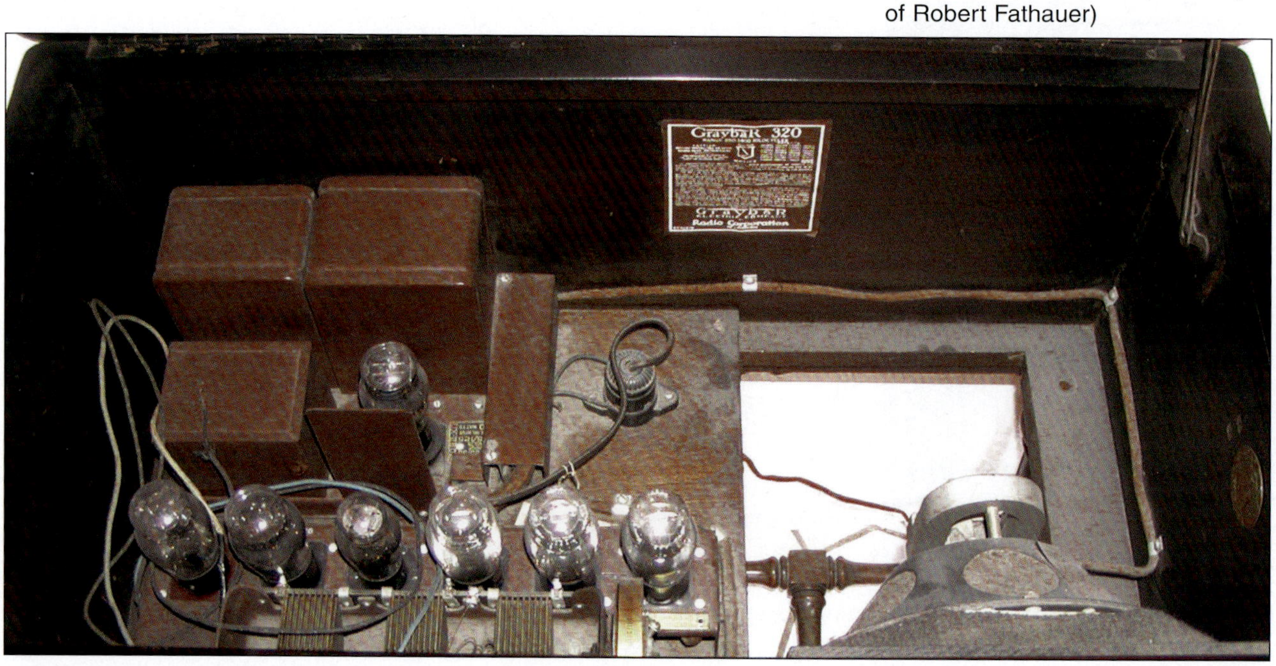

The apparatus used in the Graybar sets such as this model 320 were virtually identical to the apparatus in the equivalent RCA set—in this case the Radiola 51 which used the Radiola 18 chassis and RCA Model 100-A loudspeaker. (Courtesy of Robert Fathauer)

The escutcheon plates for the Graybar Model 320 were similar but not identical to the plates on the Radiola 51. (Courtesy of Robert Fathauer)

According to the RCA Victor Service Data Book, there were no Graybar sets using Radiola 21 or 22 apparatus, and there is no indication that Graybar marketed radio-phonograph combinations at that time. The prices of the Graybar sets were advertised without tubes, and the initial prices were virtually identical to those of the electrically equivalent Radiola sets sold by RCA.

Table I-1. Receivers distributed by Graybar Electric using Radiola apparatus as indicated supplied by RCA

Graybar Model Year	Graybar Electric Model	Radiola Model Apparatus Used
1928-29	300	16
1928-29	310	18
1928-29	311	33
1928-29	320	51
1928-29	330	60
1928-29	340	62
1929-30	500	44
1929-30	550	46
1929-30	600	66
1930-31	678	48
1930-31	700	80
1930-31	770	82
1930-31	900	86

Endnotes and References for Appendix I

1 "Western Electric Splits its Business," *The New York Times*, Dec 23, 1925, p. 25.

2 "WEAF to Merge With Radio Station WJZ; A.T. & T. Will Quit the Broadcasting Field" *The New York Times*, July 22, 1926, p. 1.

3 "Graybar to Sell Radio Receivers," *The New York Times*, August 17, 1928, p. 25.

4 "Graybar Receiving Sets," *Radio Retailing*, Vol. 8, No. 3, September 1928, p. 86.

Selected Bibliography

BOOKS AND ARTICLES

Aldridge, Benjamin L., *The Victor Talking Machine Company*, Edited by Frederic Bayh (RCA Sales Corporation, Camden, New Jersey, 1964).

Alexanderson, E. F. W. "Transoceanic Radio Communication, General Electric, Vol. XXIII, No. 10, October 1920, pp.794-803.

"An Almost Unlimited Field for Radio Telephony," *Radio News*, February, 1920, p. 434.

"America's Busiest Radio Station," *The Wireless Age*, Vol. 9, No. 12, September 1922, p. 1955.

Anderson, John M., "A Sketch of Early Radio Vacuum-Tube Research and Development at the General Electric Company," *The AWA Review*, Vol. 4, 1989, pp. 6-31.

Anderson, John M., "Supplying Tubes, Sets, and People to RCA: The General Electric Connection," *The AWA Review*, Vol. 5, No. 1, 1990, pp. 1-20.

Annual Report of the Directors of Radio Corporation of America, (Radio Corporation of America, New York, 1920 through 1933).

"Anti-Trust Decree to End Radio Group," *The New York Times*, November 22, 1932, p. 1.

Archer, Gleason L., *Big Business and Radio*, (The American Historical Company, Inc., New York, 1939).

Archer, Gleason L., *History of Radio to 1926*, (The American Historical Society, Inc. New York, 1938).

Armstrong, Edwin "The Story of the Super-Heterodyne," *Radio Broadcast*, Vol. 5, No. 3, July 1924, p. 199.

Armstrong, Edwin H., "The Super-Heterodyne—Its Origin, Development, and Some Recent Improvements," *Proc. IRE*, Vol. 12, No. 5, October 1924, pp. 539-552.

Bacon, Christopher, "Commemorating the 75th Anniversary of Radio Central," *The AWA Review*, Vol. 10, 1996, pp. 150-187.

Baker, W. R. G. and I. F. Byrnes, "Duplex Radio Telephone Transmitter with Special Reference to Tests on the S. S. America," *General Electric Review*, Vol. XXV, No. 8, August 1922, pp. 477-483.

Banning, William Peck, *Commercial Broadcasting Pioneer, The WEAF Experiment 1922 - 1926*, (Harvard University Press, Cambridge, Mass., 1946).

Barnouw, Erik, "A Tower in Babel," *A History of Broadcasting in the United States, Volume 1—to 1933*, (Oxford University Press, New York, 1966).

Barnum III, Frederick O., *"His Masters Voice" in America*, (General Electric Company, Camden, New Jersey, September 1991).

Bartrum, Graeme, "Marconi v. British Radio Telegraph and Telephone Company: The Patent Case that Changed the World," *The AWA Review*, Vol. 13, 2000, pp. 39-80.

"Battleship Type Receiver," *Radio Amateur News*, Vol. 1, No. 2, August 1919, pp. 56-57.

Baumbach, Robert W., *Look for the Dog*, (Mulholland Press, Inc., Los Angeles, CA, 6th Printing 1999).

Baumbach, Robert W., *Look for the Dog*, (Mulholland Press, Inc., Los Angeles, CA, Revised Limited Edition, 2005).

Baumbach, Robert W., *Victor Data Book*, (Mulholland Press, Inc., Los Angeles, CA, 2003).

Beverage, Harold H., "Duplex Radiophone receiver on U.S.S. George Washington," *General Electric Review*, Vol. XXIII, No. 10, October 1920, pp. 807-812.

Boyd, William M., "The RCA Radiola Grand," *Antique Radio Classified*, Vol. 7, No. 3, March 1990, pp. 6-11.

Brown, Stephen C. "A Comparison-Contrast of the Aeriola Grand and Radiola Grand," *The Old Timer's Bulletin*," Vol. 35, No. 1, February 1994, pp. 6-10, 58.

Bucher, Elmer E., "The Alexanderson System for Radio Communication," *General Electric Review*, Vol. XXIII, No. 10, October 1920, pp. 813-839.

Bucher, Elmer E., *Practical Wireless Telegraphy*, (Wireless Press, Inc., New York, 1921).

Bucher, Elmer E., *Vacuum Tubes in Wireless Communications*, (Wireless Press, Inc., New York, 1921).

Bucher, Elmer E., *The Wireless Experimenters Manual*, (Wireless Press, Inc., New York, 1918).

"The Censorship of Messages," *The Wireless Age*, Vol. 1, No.12, September 11, 1914, pp. 964-966.

"The Censorship Situation," *The Wireless Age*, Vol. 2, No. 1, October, 1914, pp. 8-10.

Clark, George H., *Radioana*—a compendium of historical documents collected and prepared by RCA historian George Clark entitled the *George H. Clark Radioana Collection*, which is deposited in the Archives Center, National Museum of American History, Behring Center, Smithsonian Institution, Washington, D. C.

Commercial and Government Radio Stations of the United States, Department of Commerce, Bureau of Navigation, (Washington D.C Government Printing office, June 30, 1922 Edition).

"Commercial Radio Across Atlantic and Pacific Opens," *The Wireless Age*, Vol. 7, No. 6, March 1920, p. 10.

"A Concert by Wireless," *QST*, January, 1917, p. 26.

Condon, Bill, "The Moorhead Story," *Tube Collector*, Vol. 5, No. 2, April, 2003, pp. 3-26.

Condon, Bill, "Westinghouse Tube Development Leading to the WD-11," *Tube Collector*, Vol. 6, No. 2, April, 2004, pp. 2-10.

Conrad, Frank, "Radio Receiving Equipment," *IRE Proceedings*, Vol. 10, No. 6, Dec. 1922, pp. 426-439.

"The Control of Wireless," *The Wireless Age*, June 1917, p. 651.

Corkutt, Bill, "The Mighty Superhet," *Antique Radio Classified*, Vol. 17, No. 7, July 2000, pp. 4-7.

"Dangers of the Proposed New Radio Bill," *The Wireless Age*, March 17, 1917, pp. 374-387.

Davis, H. P., "The Early History of Broadcasting in the United States," in *The Radio Industry: The Story of Its Development* (Chicago/New York: A. W. Shaw Company, 1928; reprinted by Arno Press in 1974) p. 192.

"Decree Dissolving Radio Group Under Anti-Trust Law," *The New York Times*, November 22, 1932, p. 17.

Denby, Edwin, Sec. Navy, Navy Department (under direction of), *History of the Bureau of Engineering Navy Department During the World War*, Office of Naval Records and Library, (Government Printing Office, Washington, D.C. 1922).

Douglas, Alan, "General Electric Radio," *Radio Age*, Vol. 5, No. 8, October 1979, pp. 1-5.

Douglas, Alan, "Radio Corporation of America," *Radio Age*, Part I: Vol. 8, No. 8, October 1982, pp. 1-3; Part II & III: Vol. 8, No. 9, November 1982, pp. 3-5; Part IV: Vol. 8, No. 10, December 1982, pp. 2-3; Part V: Vol. 9, No. 1, January 1983, pp. 4-6; Part VI: Vol. 9, No. 2, February 1983, pp. 4-6; Part VII: Vol. 9, No. 5, May 1983, pp. 1-3.

Douglas, Alan, *Radio Manufacturers of the 20's, Vol. 3 – RCA to Zenith*, (The Vestal Press, Ltd, Vestal, New York 13851, 1991).

Douglas, Alan, "Technology Crosses the Atlantic," *Radio Age*, Vol. 16, No. 3, January 1990, pp. 4-5.

Douglas, Alan, "The Western Electric 3B, and Other Mysteries," *Radio Age*, Vol. 15, No. 10, October, 1990, pp. 8-10.

Douglas, Alan, "Westinghouse Radio Patents," *Radio Age*, Vol. 15, No. 12, December 1990, pp. 10-12.

Douglas, Alan, "The Westinghouse Type RC and the RA – DA – DR – TR – AR – Part I," *Antique Radio Classified*, Vol. 8, No. 6, June 1991, pp. 4-6 (see Peckham, Lauren for Part II).

Douglas, Alan, "Who Invented the Superheterodyne?," *Proceedings of the Radio Club of America*, Vol. 64, No. 3, Nov. 1990, (on-line version: http://antiqueradios.com/superhet/).

Douglas, George H., *The Early Days of Radio Broadcasting*, (McFarland & Co., Inc. Jefferson, North Carolina, and London, 1934).

Douglas, Susan, *Inventing American Broadcasting 1899-1922*, (The John Hopkins University Press, Baltimore and London, 1987).

Eichberg, Robert, "Progress of Radio Receiving Equipment," *Radio-Craft*, Vol. IX, No. 9, March 1938, pp. 546-550.

"8XK," *QST*, November 1920, p.8.

"8KX, Pittsburgh, PA.," *QST*, September 1920, Vol. IV, No. 2, pp. 32-34.

"Election Returns Flashed by Radio to 7,000 Amateurs," *The Electrical Experimenter*, January, 1917, p. 650.

Elskamp, Richard and Alan Douglas, "Westinghouse Short-Wave Broadcasting," *The Old Timers Bulletin*, Vol. 18, No. 4, March 1978, pp. 18-19.

Espenschied, Lloyd, "Discussion of 'A History of Some Foundations of Modern Radio-Electronic Technology,'" *Proc. IRE*, Vol. 47, No. 47, July 1959, pp. 1253-1258.

Federal Trade Commission vs. General Electric Company, et. al., Radio Case, Brief of Facts, (Sidney C. Ormaby Company, 217 Broadway, New York, 1928).

"Filing Radiograms with the Postal," *The Wireless Age*, Vol. 10, No. 1, October 1922, p. 55.

Frank Frimmerman, "Frank Frimmerman's Station," *Radio Amateur News*, February 1920, pp. 420-421.

Gelatt, Roland, *The Fabulous Phonograph*, (Appleton-Century, New York, 1965).

"General Electric Details R.C.A. Deal," *The New York Times*, March 28, 1933, p. 27.

Gernsback, Hugo, "Amateur Radio Restored," *The Electrical Experimenter*, June, 1919, page 131.

Goldsmith, Alfred N., "Progress in Radio During 1924," *General Electric Review*, Vol. XXVIII, No. 1, January 1925, pp. 37-62.

Goldsmith, Alfred N., "Progress in Radio During 1925," *General Electric Review*, Vol. XXIX, No. 1, January 1926, pp. 70-77.

Goldsmith, Alfred N., "Progress in Radio During 1926," *General Electric Review*, Vol. 30, No. 1, January 1927, pp. 67-72.

Goldsmith, Alfred N., "Progress in Radio During 1927," *General Electric Review*, Vol. 31, No. 1, January 1928, pp. 64-69.

Goldsmith, Alfred N., "Progress in Radio During 1928," *General Electric Review*, Vol. 32, No. 1, January 1929, pp. 74-81.

Goldsmith, Alfred N., "Progress in Radio During 1929," *General Electric Review*, Vol. 33, No. 1, January 1930, pp. 69-79.

"Government Ownership of Wireless," *The Wireless Age*, February 1917, pp. 300-308.

"Government Seizes Whole Radio System [-] Navy Takes Over All Wireless Plants It Needs and Closes All Other," *The New York Times*, April 8, 1917, p. 2.

"Graybar to Sell Radio Receivers," *The New York Times*, August 17, 1928, p. 25.

"Graybar Receiving Sets," *Radio Retailing*, Vol. 8, No. 3, September 1928, p. 86.

Grinder, Robert E., Radio *Collectors Directory and Price Guide, 1921 – 1965*, (Sonoran Publishing, LLC, Chandler, AZ, June 1995).

Henry, Walter J., "Latest Developments in Audio Frequency Amplifiers," *Radio Amateur News*, Vol. 1, No. 4, October 1919, p. 161.

History of the Bureau of Engineering Navy Department During the World War, under direction of Edwin Denby, Sec. Navy, Navy Department, Office of Naval Records and Library, (Government Printing Office, Washington, D.C. 1922) pp. 110-115.

Horle, L. C., "Navy Receiving Equipment," *QST*, Part I: Vol. III, No. 2, July 1920, pp. 12-14, 17; Part II: Vol. IV, No. 1, August 1920, pp. 8-10.

Howeth, Captain Linwood S., USN (Retired), *History of Communications-Electronics in the United States Navy*, (United States, Government Printing Office, Washington, 1963).

"Invent Disk to Play Whole Symphonies," *The New York Times*, August 13, 1925, p. 22.

Jensen, Pete R., *In Marconi's footsteps – Early Radio*, (Kangaroo Press Pty Ltd. Kenthurst, Australia, 1994).

Kintner, S. M., "Pittsburgh's Contribution to Radio," *Proc. IRE*, Vol. 20, No. 12, December 1932, pp. 1849-1862.

Kreuzer, James H, and Felicia A., "Marconi – The Man and His Apparatus," *The AWA Review*, Vol. 9, 1995, pp. 6-96.

"Letter from RCA production designer W. L. Carlson to one E. A. Laport with subject: FIRST SUPER HETERODYNE RECEIVERS," *Old Timers Bulletin*, Vol. 12 No. 1, June 1971, p. 21.

List of Wireless Telegraph Stations of the World, Department of the Navy, Bureau of Equipment, Corrected to September 1, 1909, (Government Printing Office, Washington, D.C., 1909).

Little, Donald G., "Dr. Conrad founds KDKA," *American Heritage*, (Vol. VI, No. 5 August, 1955) pp. 71-73.

Luetz, Charles R., *Modern Radio Reception*, (Experimenters Information Service, Inc., New York City, 1924).

Lyon, Ed, ed., "KDKA's First Broadcast – How We Did It," (from Frank Hill's interview with Donald G. Little), *Radio Age*, Vol. 21, No. 11, November 1996, p. 4-6.

MacKnight, Worth, "The Marconi Type 106 Tuner," *The Electrical Experimenter*, June 1919, pp. 112-113.

Magoun, Alexander B., *Shaping the Sound of Music: The Evolution of the Phonograph Record, 1877-1950*, (PhD Dissertation, University of Maryland, 2000–available from UMI Dissertation Services, Ann Arbor Michigan).

"Marconi Absorbs United Wireless; Inventor's Suit for Infringement Comes to an End and Merger Is Announced," *The New York Times*, March 26, 1912, p. 15.

"The Marconi Quit Against Simon," *The Wireless Age*, Vol. 3, No. 3, December 1915, p. 209.

"Marconi Station Seized by Daniels," *The New York Times*, September 26, 1914, p. 4.

"Marconi to Sell and Lease ship Equipment," *The Wireless Age*, Vol. 6, No. 4, January 1919, p. 8.

"Marconi Sells Apparatus," *The New York Times*, November 28, 1918, p. 11.

"The Marconi Valve Receiver," *Modern Electrics*, February 1912, Vol. 4, No. 11, pp. 779-780.

"Marconi Wireless Telegraph Co. of America v. De Forest Radio Telephone & Telegraph Co. et al.," *The Federal Reporter*, Vol. 236, (West Publishing Co., St. Paul, 1917) pp. 942-955.

"Marconi Wireless Telegraph Co. of America v. De Forest Radio Telephone & Telegraph Co. et al.," *The Federal Reporter*, Vol. 243, (West Publishing Co., St. Paul, 1917) pp. 560-567.

"Marconi Wireless Telegraph Co. of America v. De Forest Radio Telephone & Telegraph Co. et al.," *The Federal Reporter*, Vol. 261, (West Publishing Co., St. Paul, January-March, 1920) pp. 393-395.

Marconi Wireless Telegraph Company of America, Report of the Directors and Statement of Accounts, 1902-1918, (Annual reports).

Mayes, Thorn, "History of American Marconi Company," *Old Timer's Bulletin*, Vol. 13, No. 1, June 1972, pp. 11-18.

Mayes, Thorn L., *Wireless Communication in the United States, The Early Development of American Radio Operating Companies*, (The New England Wireless and Steam Museum, Inc., East Greenwich, R. I. 02818, 1989).

McMahon, Morgan E., ed., *Radio Collector's Guide 1921-1932*, (Antique Electronic Supply, Tempe, AZ, 1981).

McMahon, Morgan E. ed., *Vintage Radio*, (Vintage Radio, Rolling Hills Estates, CA, 3rd Edition, 1981).

"Melba Entertains Europe by Wireless Telephone," *Telephony*, July 10, 1920, p. 19.

Minton, John P., "Recent Advancement in Loud Speaker Reproduction," *The Wireless Age*, Vol. XII, No. 10, July 1925, pp. 17-18, 51-52.

Morley, Vincent, "Adapting Dry Cell Tubes to the Radiola RC," *The Wireless Age*, Vol. 4, No. 1, October 1923, pp. 46-47.

Murray, Robert P., Editor, *Early Development of Radio in Canada 1901-1930*, (Sonoran Publishing, LLC, Chandler, AZ, June 2005).

"Music Reproducer Meets Severe Test," *The New York Times*, November 12, 1925, p. 15.

"Name Receivers for Wireless Company," *The New York Times*, August 6, 1912, p. 12.

"New A. C. Tubes Announced," *Radio News*, Vol. 10, No. 10, April, 1929, pp. 913-914.

"New Contract for Radio," *The New York Times*, March 30, 1928, p. 34.

"The New Marconi Standard Equipments," *The Wireless Age*, Vol. 3, No.1, October 1915, pp. 28-34.

The New Orthophonic Victrolas Electrolas and Radio Combinations, Catalog Form No. 88976-JQA, November 1926.

The New Orthophonic Victrolas Electrolas and Radio Combinations, Catalog Form No. 99579-TQA, August 1927.

The New Orthophonic Victrolas Electrolas and Radio Combinations, Catalog Form No. 9719-JQA, November 1927.

"A New and Powerful Wireless Company," *The Wireless Age*, Vol. 7, No. 2, November, 1919, p. 10.

"The New Radio Legislation," *The Wireless Age*, January 1917, pp. 226-240.

The New Victor Line, Catalog Form No. 8400, December 1925.

"New York's Amateur Show," *The Wireless Age*, Vol. 9, No. 7, April, 1922, p. 21.

"New York Radio Central Station," *The Wireless Age*, Vol. 7, No. 10, August 1920. pp. 10-12.

"News and Music to be Transmitted by Wireless Telephone," *Electrical Review and Western Electrician*, November, 18, 1926, p. 880.

"Offer 130,000 Shares of Brunswick Balke," *The New York Times*, November 27, 1924, p. 33.

Okolowicz, John, "Radio Art," *Radio Age*, Vol. 22, No. 2, February 1997, pp. 1-8.

"The Opening of Radio Central," *The Wireless Age*, Vol. 9, No. 3, December 1921, p. 21.

"Our German Cables Cut at the Azores," *The New York Times*, August 6, 1914, p. 6.

Patterson, Donald O., "Harry William Houck 1897," *Radio Age*, Vol. 15, No. 2, February 1990, pp. 1-4.

Patterson, Don, "History of Golden Leutz," *Radio Age*, Part I: Vol. 12, No. 6, June 1986, pp. 1-4; Part II: Vol. 12, No. 7, July 1986, pp. 4-5; Part III; Vol. 12, No. 8, August 1986, pp. 3-5.

Paul, Floyd A., *Horn Speaker Encyclopedia*, (Floyd A. Paul, Glendale, CA, 1989).

Paul, Floyd A., "Radio Horn Speakers that RCA Didn't Make," *The Old Timer's Bulletin*, Vol. 28, No. 3, Nov. 1987, pp. 24-25.

Paul, Floyd A., and Ray Thompson, "RCA Radio Horn Speakers," *The Old Timer's Bulletin*, Vol. 30, No. 3, November 1989, p. 37.

Peckham, Lauren A., "The Westinghouse Type RC and the RA – DA – DR – TR – AR – Part II," *Antique Radio Classified*, Vol. 8, No. 6, June 1991, pp. 7-9 (see Douglas, Alan for Part II).

"Phonograph Sales Drop," *The New York Times*, June 18, 1922, p. 40; "Radio Now Linked with Phonograph," *The New York Times*, May 3, 1924, p. 24.

"Postal Raids Show Vast Stock Frauds," *The New York Times*, November 22, 1910, pp. 1-2.

Priess, William H., "The Eaton Oscillator and Eaton Circuit Driver," *Radio Amateur News*, Vol. 1, No. 5, Nov. 1919, p. 216.

Radio Apparatus for Amateur and Experimental Use with Instructions for Continuous Wave Operation, (Radio Corporation of America catalog, New York, September 1, 1921).

"Radio Concerts," *Radio News*, September, 1920, p. 133.

"Radio Connects Ship and Shore Telephone," *The Wireless Age*, Vol. 10, No. 3, December 1922, pp. 41-43.

"Radio Corporation of America Buys International Company," *The Wireless Age*, Vol. 9, No. 1, October 1921, p. 15.

"Radio Corporation of America Plans Commercial Wireless," *The Wireless Age*, Vol. 7, No. 4, January 1920, p. 7.

"Radio Corporation Holds First Meeting," *The Wireless Age*, Vol. 7, No. 9, June 1920, p. 7.

Radio Enters the Home, (Radio Corporation of America catalog, Marketing Department, New York City) June 1, 1922.

The Radio Industry: The Story of Its Development (Chicago/New York: A. W. Shaw Company, 1928; reprinted by Arno Press in 1974).

Radio Laws and Regulations of the United States, (Department of Commerce, Bureau of Navigation; July 27, 1914 Edition, Government Printing Office, Washington, D.C., 1914).

"Radio Music," *Scientific American Monthly*, October, 1920, p. 178.

"Radio Now Linked with Phonograph," *The New York Times*, May 3, 1924, p. 24.

Radio Receiving Apparatus for Short Wave Telegraphy and Telephony, Westinghouse Circular 1641, N.D. (circa 1921).

Radio Stations of the United States, Edition July 1, 1914 (Department of commerce Bureau of Navigation Radio Service, Government Printing Office, 1914).

Radio Stations of the World, (Department of Commerce, Bureau of Navigation, Government Printing Office, Washington, many editions).

"Radio Suit Deferred for New Parleys," *The New York Times*, October 9, 1932, p. 1.

"Radio Suit Waits Until Next Monday," *The New York Times*, November, 16, 1932, p. 25.

Radio Supplies and Equipment, Robertson-Cataract Electric Company Catalog No 22, 1922-1923, (Reprinted by the Vestal Press Ltd., Vestal, New York 13850, copyright 1922).

Radio Telegraph and Telephone Equipment, (Wireless Specially Apparatus Company catalog, Boston, Mass, 1919).

"Radio Vacuum-Tube Litigation is Settled," *Radio Amateur News*, Vol. 1, No. 2, August 1919, p. 77.

"R.C.A. Makes Tubes Available for Ship Work," *The Wireless Age*, Vol. 9, No.11, August 1922, p. 53.

"R.C.A. Seen Stronger Following Split-Up," *The New York Times*, November 27, 1932, p. 1.

RCA Victor Service Data Volume I 1923-1937, (RCA Victor Division of Radio Corporation of America, Camden, NJ, 1st Edition, 1944).

RCA Victor Service Notes for 1923 – 1928, (Service Division, RCA Victor Company, Inc., Camden NJ., N.D.).

RCA Victor Service Notes for 1929-1930, (Service Division, RCA Victor Company, Inc., Camden, NJ, N.D.).

"Re-Broadcasting, a New Era in Radio," *Radio News*, Vol. 5, No. 9, March 1924, pp. 1242.

"Recently Discovered," *Radio Age*, Vol. 18, No. 7, July 1993, p. 7.

Reed, Walt, *The Illustrator in America 1860-2000*, (The Society of Illustrators, Inc., New York, N.Y., 2001).

Reinhardt, Lisa, ed., *Davenport's Art Reference & Price Guide, Gold Edition*, (Gordon's Art Reference, Inc., Phoenix, AZ, 2003).

Report of the Federal Trade Commission on the Radio Industry in Response to House Resolution 548, Sixty-seventh Congress Fourth Session, December 1, 1923, (Government Printing Office, Washington, DC, 1924).

Rider, John F., *Perpetual Trouble Shooter's Manual* Volume I, (John Rider, New York, 1933).

Rogers, W. W., "Will Short Waves Revolutionize Broadcasting?" *The Wireless Age*, Vol. XI, No. 6, p. 39.

Schiffer, Michael Brian, *The Portable Radio in American Life* (The University of Arizona Press, Tucson & London, 1991).

Secor, Winfield H., "Radio Detector Development," *Electrical Experimenter*, Vol. IV, No. 9, January, 1917.

Seeley, Ward, "What 'via RCA' Means to a Message," *The Wireless Age*, November 1922, Vol. 10 No. 2, pp. 57-60.

Seitz, Frederick, *The Cosmic Inventor Reginald Aubrey Fessenden (1866-1932)*, (American Philosophical Society, Independence Square, Philadelphia, 1999).

"Shipowners Fight Marconi Co. Rates," *The New York Times*, October 14, 1914, p. 17.

"Send Election Returns by Wireless Telephone," *Electrical Review*, November 6, 1920, p. 733.

"Siasconset Station Reopened," *The Wireless Age*, Vol. 2, No. 5, February, 1915, p. 392.

Sleeper, M. B., *How to Make Commercial Type Radio Apparatus*, (The Norman H. Henley Publishing Co. New York, 1922).

Smith, Bert E., "Coupling Methods for the R. F. Amplifier," *Radio Broadcast*, Vol. XIII, No. 6, October 1928, p. 361.

Sobel, Robert, *RCA*, (Stein & Day, New York, 1986).

"Some Developments in the Electrical Industry During 1921," *General Electric Review*, Vol. XXV, No. 1, January 1922, pp. 29-32.

"Specifications for Broadcast Receivers for 1924," RCA Research Department Report, January 16, 1923.

Sterling, George *The Radio Manual*, (New York: D. van Nostrand Company, Inc., 1928).

Stokes, John W., *Seventy Years of Radio Tubes and Valves*, (The Vestal Press Ltd., Vestal, New York 13851, 1982).

"Strays," *QST*, December, 1920, p.48.

"The Story of KDKA (Once Upon a Time)" *Radio Age*, Vol. 2, No. 5, May 1976, pp. 1-4.

"The Story of KDKA – Part II," *Radio Age*, Vol. 2, No. 6, June 1976, pp. 3-4.

"Tells Details of Aid Radio Gave Victor," *The New York Times*, August 30, 1925, p. 14.

"$10,000,000 Marconi Company," *The New York Times*, November 23, 1899, p. 1.

"This Victor Day," *The New York Times*, November 2, 1925, p. 17.

Thorn, Ralph G., MD, "A Contribution to Aeriola/Radiola Grand History," *Radio Age*, Part I: Vol. 14, No. 3, March 1988, pp. 3-4; Part II: Vol. 14, No. 4, April 1988, pp. 2-3.

"Thousand Word-a-Minute World-Wide Radio Station," *The Wireless Age*, Vol. 7, No. 10, July 1920, p. 7.

"Trust Suit Filed on Radio Compacts," *The New York Times*, May 14, 1930, p. 1.

Tyne, Gerald F. J., *Saga of the Vacuum Tube*, (Prompt Publications, Berkeley Heights, N.J., 1977).

van der Lande, Tor, "Radioer Jeg Har Mott [Radios I have Met]," *Hallo Hallo*, NR. 53 (1/96), Mars 1996, pp. 36-38 (Norwegian periodical).

"A Victor Announcement," *The New York Times*, December 28, 1925, p. 15.

"Victor Co. Defers Common Dividend," *The New York Times*, June 19, 1925, p. 28.

"Victor Co. to Enter the Radio Business," *The New York Times*, May 20, 1925, p. 27.

"Victor Co. Produces A New Record Also," *The New York Times*, August 14, 1925, p.14.

Victor Presents, Catalog Form No. 762-IQA, November 1928.

"The Vocarola for Radiophone Reception," *The Wireless Age*, Vol. 9, No. 9, June 1922, p. 61.

"Voices on Shore Find Ships at Sea," *The New York Times*, December 22, 1929, p. XX11.

"Washington Radio Amateurs Hear Radio Concert," *Radio Amateur News*, May, 1920, p. 646.

"WEAF to Merge With Radio Station WJZ; A.T. & T. Will Quit the Broadcasting Field" *The New York Times*, July 22, 1926, p. 1.

Weagant, Roy A. "Some Recent Marconi Sets," *The Wireless Age*, Vol. 1, No. 7, April 1914, pp. 546-552.

Wenaas, Eric P., "Wireless Specialty Apparatus Nomenclature Gone Awry," *The AWA Review*, Vol. 18, 2005, pp. 25-48.

"Western Electric Splits Its Business," *The New York Times*, Dec 23, 1925, p. 25.

"Westinghouse Company Enters Wireless Field," *Electrical Review*, October 16, 1920, p. 615.

"Westinghouse to Cover Country with Radio Entertainment," *Electrical Review*, December 10, 1921, p. 887.

Westinghouse 1894-1969 East Pittsburgh Divisions' 75th Anniversary (Westinghouse, 1969).

"What the Screen-Grid Set Does," *Radio Broadcast*, Vol. XV, No. 4, August 1929, p. 212.

White, Thomas H., *United States Early Radio History*, (On-line articles and extracts about early radio and related technologies at http://earlyradiohistory.us/, July 2006).

White, William C., *"The Early History of Electronics in the General Electric Company*," (William C. White Papers Collection, Schenectady Museum, Schenectady, New York, prepared 1953-1955).

White, W. C. "Early Tube Development at GE," *Tube Collectors Association Special Publication No. 10*, December 2005.

"Who's Who in Amateur Wireless," *QST*, February, 1921, pp. 39, 45, 48.

"Wide Expansion of Trans-Ocean Wireless Assured," *The Wireless Age*, Vol. 8, No. 1 October, 1920, pp. 7-8.

"Widens Radio 'Trust' Suit," *The New York Times*, March 9, 1932, p. 16.

"Wireless Companies Consolidate," *Electrical Review and Western Electrician,* May 21, 1910, p. 1044.

"A 'Wireless' Concert," *The Electrician*, June, 18, 1920, p. 658.

"Wireless Men Guilty," *The New York Times*, January 10, 1913, p. 8.

"Wireless Supplants Cable [-] Government Gives Contract to Massie Company for Block Island Service," *The New York Times*, January 13, 1912, p. 1.

Wolkonowicz, John P., "The Radiola Superheterodyne Story," *Radio Age*, Part 1: Vol. 3, No. 9, November 1977, pp. 1-2; Part 2: Vol. 3, No. 10, December 1977, pp. 1-3; Part 3: Vol. 4, No. 1, January 1978, pp. 2-4; Part 4: Vol. 4, No. 3, March 1978, pp. 2-3; Part 5: Vol. 4, No. 4, April 1978, pp. 7-9.

Wolkonowicz, John P. with Alan Douglas, John Terrey and Richard G. Wolven, "RCA's Intended Models for 1923-1924 Season 'What Might Have Been,'" *Antique Radio Classified*, Vol. 16, No. 6, June 1999, pp. 4-11.

PERIODICALS

Antique Radio Classified, Vol. 1, May 1984 – present.

The AWA Journal. See The Old Timer's Bulletin.

The AWA Review, Vol. 1, 1986 – present, (published annually by the Antique Wireless Association).

The Electrical Experimenter, Vol. 1, May 1913 – Vol. 8, July 1920 (name changed to *Science and Invention* beginning with the August 1920 issue).

Electrical Review (also *Illustrated Electrical Review*), Vol. 1, February 1882 – Vol. 53, 1908, (absorbed *Western Electrician* circa November 7, 1908 and changed name to *Electrical Review and Western Electrician*).

Electrical Review and Western Electrician, Vol. 53, November 1908 – Vol. 79, December 1921, (*Western Electrician*, Vol. 1, 1887 – Vol. 43, 1908 and *Electrical Review* combined to form *Electrical Review and Western Electrician* on November 7, 1908; became *Electrical Review and Industrial Engineer* beginning with Vol. 80 No. 1 in January of 1922).

The Electrician and Mechanic, Vol. 17, July 1906 – Vol. 25, December 1913, (was originally *Amateur Work* beginning with Vol. 1, November 1901; later merged with *Modern Electrics* in January 1914 to form *Modern Electrics and Mechanics*).

General Electric News, Vol. 4, 1920 – Vol. 13, 1929 (covers only the period most relevant to this book).

General Electric Review, Vol. XXIII, 1920 – Vol. XXXII, 1929 (covers only the period most relevant to this book).

MAARC Newsletters, Vol. 1, August, 1984 – Vol. 11, May 1994, (MAARC absorbed *Radio Age* and began to publish Vol. 11, No. 5 in June 1994 under the title *Radio Age – The Newsletter of the Mid-Atlantic Antique Radio Club*).

Modern Electrics, Vol. 1, April 1908 – Vol. 6 December 1913, (merged with *Electrician and Mechanic* in January 1914 to form *Modern Electrics and Mechanics*).

Modern Electrics and Mechanics, Vol. 28 (January 1914 – June 1914) (six issues only).

The New York Times , 1899 – 1934, (covers only the period for contemporaneous references relevant to this book).

The Old Timer's Bulletin, Vol. 1, January 1960 – Vol. 45, October 2004; *The AWA Journal*, Vol. 46, January 2005 – present, (name changed to *The AWA Journal* in 2005; published quarterly by the Antique Wireless Association).

Pacific Radio News, Vol. 1, January 1917 – Vol. 3, October 1921, (Publication suspended June 1917 – December 1919 due to WWI; name changed to *Radio* beginning with November 1921 issue).

Popular Radio, Vol. 1, May 1922 – Vol. 13, March 1928.

QST, Vol. 1, December 1915 – present, (publication suspended October 1917 – May 1919 during WWI).

Radio, Vol. 3, November 1921 – Vol. 31, February/March 1947 (formerly *Pacific Radio News*).

Radio Age, Vol. 1 October 1975 – Vol. 18, December 1993; *Radio Age – The Newsletter of the Mid-Atlantic Antique Radio Club*, Vol. 19, June 1994 – present, (the Mid-Atlantic Antique Radio Club (MAARC) purchased rights to Radiola Age from Don Patterson, Publisher of *Radiola Age*).

Radio Amateur News (see *Radio News*).

Radio Broadcast, Vol. 1, May 1922 – Vol. 16, April 1930, (absorbed by *Radio Digest* in Sept 1930).

Radio Digest Illustrated, Vol. 1, April 1922 – Vol. 30, March 1933, (weekly publication; absorbed *Radio Digest* in Sept 1930).

Radio Merchandising, Vol. 1, June 1922 – Vol. 24, May 1928.

Radio News, Vol. 1, July 1919 – Vol. 12, Vol. 40, July 1948, (Vol. 1 titled *Radio Amateur News*; subtitle included "*and Short Wave Radio*" in the 1930's; name was changed to *Radio and Television News* in 1948).

Radio Service Bulletins, Vol. 1, January 1915 – Vol. 83, June 1932, (website address: (http://www.fcc.gov/mb/audio/decdoc/radio_service_bulletins.html).

Radio Retailing, Vol. 1, January 1925 – Vol. 24, March 1939, (name changed to *Radio and Television Retailing* in 1939).

Saturday Evening Post, Vol. 194, July 1921 – Vol. 205 June 1933, (covers only the period for contemporaneous advertisements relevant to this book).

The Town Crier, Vol. I, June 1923 – Vol. XIII, December 1929, (RCA publication for its distributors and retailers; covers only the period most relevant to this book).

Tube Collector, Vol. 1, February 1999 – present.

The Wireless Age, Vol.1, October 1913 – Vol. 12, August 1925 (merged into *Popular Radio* in 1925).

General Index

Index to Radios, Phonographs and Accessories by Model

Bold pages correspond to titled subsections